Real Estate Concepts

The essential reference tool for all real estate, property, planning and construction students.

Edited by Professor Ernie Jowsey of Northumbria University, *Real Estate Concepts* provides built environment students with an easy-to-use guide to the essential concepts they need to understand in order to succeed in their university courses and future professional careers.

Key concepts are arranged, defined and explained by experts in the field to provide the student with a quick and reliable reference throughout their university studies. The subjects are conveniently divided to reflect the key modules studied in most property, real estate, planning and construction courses.

Subject areas covered include:

- Planning
- Building surveying
- Valuation
- Law
- Economics, investment and finance
- Quantity surveying
- Construction and regeneration
- Sustainability
- Property management

Over the 18 alphabetically arranged subject specific chapters, the expert contributors explain and illustrate more than 250 fully cross-referenced concepts. The book is packed full of relevant examples and illustrations and after each concept further reading is suggested to encourage a deeper understanding. This book is an ideal reference when writing essays and assignments, and revising for exams.

Ernie Jowsey is Professor of Property and Real Estate at Northumbria University. He is the author of a number of books including *Real Estate Economics* and *Modern Economics* with Jack Harvey.

Real Estate Concepts

Real Estate Concepts

A handbook

Edited by Ernie Jowsey
with contributions from staff at
Northumbria University

LONDON AND NEW YORK

First published 2015
by Routledge
2 Park Square, Milton Park, Abingdon, Oxon OX14 4RN

Simultaneously published in the USA and Canada
by Routledge
711 Third Avenue, New York, NY 10017

Routledge is an imprint of the Taylor & Francis Group, an informa business

British Library Cataloguing in Publication Data
A catalogue record for this book is available from the British Library

Library of Congress Cataloging-in-Publication Data
Real estate concepts : a handbook / edited by Ernie Jowsey with contributions from staff at Northumbria University.
pages cm
Includes bibliographical references and index.
1. Real estate business--Great Britain . 2. Real property--Great Britain. 3. Real estate development--Great Britain. 4. Commercial real estate--Great Britain. I. Jowsey, Ernie.
HD596.R43 2014
333.330941--dc23
2013050184

ISBN: 978-0-415-85741-3 (hbk)
ISBN: 978-0-415-85742-0 (pbk)
ISBN: 978-0-203-79764-8 (ebk)

Typeset in Bembo
by GreenGate Publishing Services, Tonbridge, Kent

Printed and bound in Great Britain by
TJ International Ltd, Padstow, Cornwall

Contents

Figures

Tables

Contributors

All contributors are from the Department of Architecture and Built Environment at Northumbria University, Newcastle upon Tyne, UK.

Graham Capper BSc MSc MRICS CEnv

Julie Clarke BA (Hons) PGCED

Andy Dunhill BSc Hons MRICS

Stuart Eve BSc (Hons) MSc PCAPL MRICS

Dom Fearon MRICS

Minnie Fraser BSc MRICS AHEA

Hannah Furness MRICS

Barry J. Gledson BSc(Hons) PGCert FHEA

Paul Greenhalgh BSc (Hons) PhD MRICS FHEA PGCED

Cara Hatcher BSc (Hons) MRICS LLM

John Holmes BA MA MBEng CEnv

Richard Humphrey FCIOB CRSA FCMI FIoD MAPM PGCertAPL FHEA

Eric Johansen MCIOB MAPM

Lynn E. Johnson BSc (Hons) MRICS

Ernie Jowsey PhD MA BA (Econ) PGCE FHEA

Kenneth Kelly BSc (Hons) MRICS

Mark Kirk BSc (Hons) PGCE

Rachel Kirk DPhil FHEA

Sara Lilley BSc MSc PGHEP

Carol Ludwig PhD MRTPI MSc BSc (Hons) PGCert

David McGuinness BA (Hons) MSc PGDip FHEA MeRSA

Simon Robson DBA MBA PCAP BSc (Hons) FHEA FRICS

Glenn Steel BSc MRICS MBA

Jane Stonehouse BSc (Hons) MRICS

Becky Thomson BA (Hons) DipSurv MRICS

John Weirs MCIOB MA PGCE

Rachel Williams LLB (Hons) PGCert PGDip

Cheryl H. Williamson MBA PCAP MRICS

Preface

This book is intended to be of use to undergraduate and professional students of real estate, surveying, land management, estate management, housing, planning and all property-related courses. It may also be of interest to anyone working in the broad fields of real estate, property management, building surveying, construction management, property investment and property development.

One of the questions that university lecturers in these subjects are often asked is: 'Is there one book we can buy that covers everything?' Of course there is no such book, because there are far too many component subjects and vast quantities of further reading in the form of books, academic journal articles, web pages, law reports and professional journal articles. *Real Estate Concepts* should be invaluable, however, in providing an introduction to the most important subjects in all land and property fields.

The structure of the book is very straightforward: there are eighteen chapters, covering all of the major subjects studied on real estate surveying courses by focusing on the most relevant concepts to understand in each subject. Each concept identifies key terms and further reading so that more in-depth understanding can be achieved. The concepts have been chosen and written by staff from the Department of Architecture and Built Environment at Northumbria University, all of whom are experts in their field and most of whom have considerable professional experience.

My thanks are due to all twenty-eight contributors, who delivered the concepts with admirable good humour while under considerable pressure from their day jobs and their students! My thanks must also go to Ed Needle from Routledge for his patience throughout this project. While there are many contributors to *Real Estate Concepts*, responsibility for any errors remains with myself.

Ernie Jowsey
Professor of Property and Real Estate
University of Northumbria

Abbreviations

AAP	Area Action Plan
ABC	Acceptable Behaviour Contract
ABI	area-based initiative
ABS	asset-backed security
AC	average cost
ACCC	Australian Competition and Consumer Commission
ACM	asbestos-containing materials
ADR	alternative dispute resolution
AGA	authorised guarantee agreement
AIB	asbestos insulation board
ALMO	arm's-length management organisation
AONB	Area of Outstanding Natural Beauty
AR	average revenue
ARPT	annual residential property tax
ARY	all-risks yield
ASB	anti-social behaviour
ASBO	Anti-Social Behaviour Order
ASR	alkali–silica reaction
B&ES	Building and Engineering Services Association
BCIS	Building Cost Information Service
BCO	British Council of Offices
BGS	British Geological Survey
BIM	building information modelling
BIS	Department for Business, Innovation and Skills
BOOT	build-own-operate-transfer
BoQ	Bill of Quantities
BOT	build-operate-transfer
BPF	British Property Federation
BPR	Business Protection Regulations
BRE	Building Research Establishment
BREEAM	Building Research Establishment Environmental Assessment Method
BRI	building-related illness
BSRIA	Building Services Research and Information Association
CABE	Commission for Architecture and the Built Environment
CAD	computer-aided design
CAPM	capital asset pricing model
CASBEE	Comprehensive Assessment System for Building Environmental Efficiency

CBA	cost–benefit analysis
CBD	central business district
CBL	choice-based lettings
CCA	Climate Change Agreement
CCS	Considerate Constructors Scheme
CDM	Construction Design and Management Regulations
CDO	collateralised debt obligation
CGT	capital gains tax
CHP	combined heat and power
CIBSE	Chartered Institution of Building Services Engineers
CIC	community interest company
CIL	Community Infrastructure Levy
CIOB	Chartered Institute of Building
CIRIA	Construction Industry Research and Information Association
CIS	Cooperative Insurance Services
CLG	company limited by guarantee
CLT	community land trust
CM	construction manager
CMBS	commercial mortgage-backed security
CoP	coefficient of performance
CP	contractor's proposal
CPD	continuing professional development
CPO	compulsory purchase order
CPR	Consumer Protection Regulations
CPRE	Campaign for the Protection of Rural England
CPSE	Commercial Property Standard Enquiries
CRAR	Commercial Rent Arrears Recovery
CRC	Carbon Reduction Scheme
CSR	corporate social responsibility
CVA	company voluntary arrangement
CVR	cost value reconciliation
CWP	combined heat and power
DAB	dispute adjudication board
D&B	design and build
DCF	discounted cash flow
DCLG	Department for Communities and Local Government
DCMS	Department for Culture, Media and Sport
DEC	Display of Energy Certificates
DECC	Department of Energy and Climate Change
Defra	Department for Environment, Food and Rural Affairs
DES	Department for Education and Skills
DETR	Department for Environment Transport and Regions
DFMA	Design for Manufacture and Assembly
Dft	Department for Transport
DPC	damp-proof course
DPD	development plan document
DPM	damp-proof membrane
DRB	dispute review board
DRC	depreciated replacement cost

DSS	Department of Social Security
DTI	Department of Trade and Industry
DV	development value
EA	Environment Agency
ECO	Energy Company Obligation
EIA	environmental impact assessment
EiP	examination in public
EH	English Heritage
EMS	environmental management system
EPBD	Energy Performance of Buildings Directive
EPC	Energy Performance Certificate
ER	employer's requirement
ERM	enterprise risk management
E(Rm)	expected return on the market portfolio
E(Rp)	expected risk premium
ESCO	energy supply company
EST	Energy Saving Trust
EUETS	EU Emissions Trading System
EUQ	element unit quantity
EUR	element unit rate
EUV	existing use value
EV	electric vehicle
EW	expert witness
EZ	enterprise zone
FGR	freehold ground rent
FiT	Feed-in Tariff
FM	facilities management
FMOP	fair maintainable operating profit
FRI	full repairing and insuring
FRS	Financial Reporting Standard
FSA	Financial Services Authority
GDP	gross domestic product
GGBS	ground granulated blast furnace slag
GIA	gross internal area
GNP	gross national product
GPDO	General Permitted Development Order
GPF	Growing Places Fund
GSHP	ground source heat pump
HAC	high alumina cement
HAWT	horizontal-axis wind turbine
HCA	Homes and Communities Agency
HER	Historic Environment Records
HMRC	Her Majesty's Revenue & Customs
HPA	Health Protection Agency
HPR	Heritage Protection Reform
HSE	Health and Safety Executive
HV	hope value
HVAC	heating, ventilation and air conditioning
IAQ	indoor air quality

IFC	Industry Foundation Class
IHD	in-home display
IPD	Investment Property Databank
IPO	Interim Possession Order
IPS	Industrial Provident Society
IRI	internal repairing and insuring
IRR	internal rate of return
IUCN	International Union for Conservation of Nature
IVA	individual voluntary arrangement
JCT	Joint ContractsTribunal
JESSICA	Joint European Support for Sustainable Investment in City Areas
JLL	Jones Lang LaSalle
JVC	joint venture company
KPI	key performance indicator
LABV	local asset-backed vehicles
LCA	life cycle assessment
LCC	life cycle costing
LCI	life cycle inventory
LDF	Local Development Framework
LEP	local enterprise partnership
LIBOR	London Inter-Bank Offered Rate
LPA	local planning authority
LTV	loan to value
LVT	land value tax
MBS	mortgage-backed security
MC	marginal cost
MCS	multiple chemical sensitivity
MES	minimum efficient scale
MIRAS	mortgage interest relief at source
MLR	Money Laundering Regulations 2007
MLRO	money laundering reporting officer
MMC	modern methods of construction
MMO	Marine Management Organisation
MPA	mineral planning authority
MPC	marginal private cost
MPP	marginal physical product
MPS	Mineral Policy Statement
MPT	modern portfolio theory
MR	market rent
MRP	marginal revenue product
MSA	Mineral Safeguarding Area
MSB	marginal social benefit
MSC	marginal social cost
MSW	manufactured solid waste
MV	market value
NAEA	National Association of Estate Agents
NAR	net annual return
NAV	net annual value
NCIS	National Criminal Intelligence Service

NDC	New Deal for Communities
NEC	New Engineering Contract
NEPA	National Environmental Policy Act
NIA	net internal area
NLUD	National Land Use Database
NPPF	National Planning Policy Framework
NPV	net present value
NRM	New Rules of Measurement
NSIP	nationally significant infrastructure projects
OFT	Office of Fair Trading
OLEV	Office of Low Emission Vehicles
OMV	open market valuation
OPC	Ordinary Portland Cement
OPEC	Organization of Petroleum Exporting Countries
OSM	off-site manufacture
PBP	period by period
PDL	previously developed land
PDR	permitted development rights
PFI	private finance initiative
PHEV	plug-in hybrid electric vehicle
PINS	Planning Inspectorate
PIP	Partership Investment Programme
PMA	Property Misdescriptions Act 1991
POCA	Proceeds of Crime Act 2002
PPC	production possibilities curve
PPE	personal protective equipment
PPG	Planning Policy Guidance
PPP	public–private partnership
PPS	Planning Policy Statements
PSBR	public sector borrowing requirement
PUT	property unit trust
PV	present value
QUANGO	quasi-autonomous non-governmental organisation
QS	quantity surveyor
R&D	research and development
RAR	risk-adjusted return
RDA	regional development agency
REIT	real estate investment trust
RFR	risk-free rate of return
RHI	Renewable Heat Incentive
RIBA	Royal Institute of British Architects
RICS	Royal Institution of Chartered Surveyors
RO	Renewable Obligation
ROC	Renewable Obligation Certificate
RPI	Retail Price Index
RSL	registered social landlord
RSS	Regional Spatial Strategies
RV	rateable value
SA	sustainability appraisal

SAR	suspicious active report
SBS	sick building syndrome
SCG	Statement of Common Ground
SCI	Statement of Community Involvement
SDLT	Stamp Duty Land Tax
SDRT	Stamp Duty Reserve Tax
SEA	strategic environmental assessment
SME	small and medium enterprise
SMM	standard method of measurement
SMR	Sites and Monuments Record
SOCA	Serious Organised Crime Agency
SPD	Supplementary Planning Document
SRB	single regeneration budget
SSAP	Statement of Standard Accountancy Practice
SUDS	sustainable urban drainage system
SVR	standard variable rate
TAGC	transfer as a going concern
TCC	Technology and Construction Court
TDC	total development cost
TDS	Tenancy Deposit Scheme
TIF	tax incremental financing
TLATA	Trusts of Land and Appointment of Trustees Act 1996
TPI	tender price index
TPO	Tree Preservation Order
TVG	town and village green
UBR	uniform business rate
UDC	Urban Development Corporation
UK GAAP	UK General Accepted Accounting Principles
UKAS	United Kingdom Accreditation Service
UKHCA	UK Homecare Association
UKVS	UK Valuation Statement
ULEV	ultra-low emission vehicles
URC	Urban Regeneration Company
UTF	Urban Task Force
UWP	Urban White Paper
VAT	Value Added Tax
VAWT	vertical-axis wind turbine
VM	value management
VOA	Valuation Office Agency
VR	virtual reality
VRS	Valuer's Registration Scheme
YP	year's purchase

1 Agency

Andy Dunhill, Jane Stonehouse and Rachel Williams

1.1 The inspection

Key terms: inspection; valuing; marketing; measure; confidentiality

It is crucial that agents inspect a property before valuing it or advising on marketing and they must understand what they have looked at. Make sure you take everything with you to carry out the inspection (e.g. a measuring device, camera, location plan, floor plans if available, appropriate clothes, and keys). When inspecting property always take care (see Concept 1.22).

The following is a suggested general checklist for inspecting property for marketing purposes:

- Address – be exact
- Date and time of inspection for future reference
- Weather
- Location
- Surrounding area
- Access by vehicles/pedestrians/public transport etc.
- Check architects floor plans if available
- Draw sketch plan if you do not have one
- Measure to RICS Code of Measuring Practice or appropriate local standard
- Exterior of the building – form of construction etc.
- Interior of the building – entrance, common areas, specification etc.
- Repair – general state
- Contamination – be aware of this and know when to involve an expert
- Services – what is available (gas, electricity etc.)
- User – what is it or could it be used for
- Town and Country Planning – what use is permitted or what could be obtained

All of this information will help you to advise your client on the best marketing strategy, enable you to prepare accurate letting or sales details and all other marketing information. If you make an error in your inspection this will flow through the whole marketing process and result in an error in the final transaction. Most organisations have a standard format for taking inspections notes depending on the type of property. The main thing is to adopt a logical and consistent approach to property inspections.

You will need to consider different factors depending on the type of building being inspected. The main categories are office, retail, industrial and residential; however, you may have cause to be involved with others, such as leisure (pubs, bars, restaurants), medical (hospitals, doctors surgeries), etc.

Before the visit make sure you know where you are going – which sounds simple and obvious but many surveyors have set off ill prepared. Establish whether the property is vacant or occupied. If it is vacant take the usual safety precautions and lock up afterwards. If it is occupied be clear as to why a report on marketing is required. There is nothing worse than going to a property, introducing yourself and explaining the reason for the visit only to find that the people on the ground know nothing about the plans of the owners of the business. It is OK if a relocation is planned but if that part of the business is to close the reception may be hostile. Confidentiality is crucial.

If acting for a receiver/liquidator, special care must be taken, especially where the business has not publicly gone into administration, or whatever is proposed, at the time of inspection. An agent may have to pretend to be there for a different purpose such as fire insurance valuation. There is little more sobering than to carry out an inspection of a property with a full workforce when you know they will most likely be redundant in the very near future, but they do not know.

Research the local market before the visit to see what deals have been done recently and what is currently on the market. An agent should have a reasonable idea of the market for that kind of property in that location because the occupying client will have – or at least in some cases think they have.

Be certain of exactly what is to be included in the marketing report (boundaries, parking etc.) and the tenure of the property. Drive around the area to get an up-to-date feel for it.

Even if architects' floor plans have been provided check them carefully. Buildings are often altered during their life. Plans are often for the original design, which may have changed during construction but 'as built plans' were never done or not made available.

It is not always possible to inspect all parts or aspects of a building. Some internal areas may be locked or otherwise inaccessible; make a note of this and refer to it in the report. If a building has had the water drained down for winter to avoid burst pipes that is sensible, but means it is not possible to check whether the supply is working. A property inspection for marketing purposes is not a structural building survey; however, it is crucial that an agent understands when to recommend that a detailed survey is needed.

An agent needs to take note of what repairs or other action should be done in order to best prepare the building for marketing. Sometimes simple and inexpensive redecoration can make a significant improvement and on the basis that first impressions count; a small level of investment can make a big difference.

Be very much aware of potential for alternative use to increase value; however, consider whether planning consent, if required, can realistically be obtained. Sometimes the interest of the owner is best served by splitting the property into two or more separate packages to be sold or let separately, and this should be spotted during the inspection.

Further reading

RICS (2011) *UK Commercial Real Estate Agency Standards*, RICS, Coventry, section 3.3.
www.isurv.com (accessed 02/12/2013).
www.rics.org/uk (accessed 02/12/2013).

1.2 Reporting to the client

Key terms: report; client; market; valuation; marketing plan; Energy Performance Certificate

The following is a suggested format of the main areas to be covered when reporting to a client. Most firms will have their own standard format to follow.

Introduction

Set out clearly what you understand your instructions are and the client's objectives in the marketing of the property concerned. This is to avoid future confusion or dispute. There will be occasions where your client needs to sell or let quickly. A common example is where a company goes bankrupt and you are instructed by the administrators/receivers to dispose of the surplus property assets/liabilities of the company.

Description

Describe the building and detail the areas. The amount of detail required will depend to some extent on the nature of the client you are reporting to. Some will want full details, others less so. You may be reporting to a client that either has never seen the property or has not done so for some time and is located a long way from the property.

If what you are looking at is part of a multi-occupied building you should make it clear what is and is not included in your client's interest, and hence your report. You should identify, for example, location, the form of construction, specification, areas in need of repair etc. This should be supplemented by a suitable location plan and photos where appropriate.

Strengths and weaknesses

Identify the strengths of the building that you will be able to focus on in your marketing. You should also identify the weaknesses but try to suggest appropriate solutions. For example, a lack of car parking in a city centre office building can be helped by acquiring some long term contract spaces in a nearby public car park, but you need to do some research. A client does not simply want to be told that they have a poor building. You need to be proactive, focus on the strengths and provide constructive solutions for the weaknesses. You will certainly need this when responding to enquiries about the property during marketing.

Recommendations

Following your appraisal of the building you may consider it appropriate to make some recommendations before marketing commences. This may include undertaking repairs to make it ready for occupation. Some simple but effective cleaning and decoration can have a significant effect especially with residential and office property. The first impressions of tenants/buyers can be crucial.

In some cases it may be worth considering obtaining planning consent for a different use or the carrying out of alterations in order to maximise value. If you are dealing with secondary property you may wish to consider recommending that your client uses one of the standard short form leases. The Law Society has one, as does the Royal Institution of Chartered Surveyors (RICS). Check their respective websites for the latest version. These should keep costs down for all concerned.

The market generally

You should be able to give a good indication of the strength, or otherwise, of the market for that type of property in that location at that point in time and where you see it heading during the marketing period. This will influence the potential time needed to secure a transaction, the strategy to be adopted, the likely terms of any deal etc. If the market is improving time is on your side and you should be able to generate a high level of interest and negotiate a good deal for your client. If, however, the market is in decline you need to be realistic and ensure that your client appreciates this. If you ask too high a price/rent you may miss the market and end up following it down, but never quite reaching a low enough level to do a deal.

Valuation – rent or capital

You must give a very clear recommendation of an appropriate asking price/rent/premium in the case of a leasehold interest allowing for a negotiating margin. In most cases buyers/tenants like to feel they got a reduction or negotiated you down. There are times when quoting an asking price/rent at a relatively low level (but not in a misleading way) can be a very effective tactic. The aim is to generate a high level of interest and then seek best offers. This is likely to work with a property where you anticipate a high level of interest, and sometimes where the market is uncertain and it is difficult to value. It can also work in a slow market where the low starting price suggests a good deal can be had. This tactic can generate the interest, which is key. Any valuation recommendations must be realistic and not inflated as a tactic to secure an instruction.

Marketing plan

It is good practice to provide full details and costs of your marketing plan at the reporting stage. If, however, you are not certain to receive the instruction then it may be more appropriate to be brief and provide full details later. It is not always sensible to give away all your ideas until you have got the job. This will include your strategy, advertising proposals, draft marketing particulars etc. All of this must be agreed with your client before marketing commences.

Energy Performance Certificate

You must explain the legal requirement to obtain one of these before an agent is permitted to market a property. It is the responsibility of the landlord/client to obtain the Energy Performance Certificate (EPC), but most agents will have contacts with suitable assessors to do the work.

Fees

You should set these out clearly. Fees may be:

- a fixed sum;
- a percentage of the rent/ price;
- a minimum fee;
- incentive-based to encourage the agent to secure a higher rent/price, i.e. the percentage may increase above a certain level.

Other costs

Depending on the marketing plan there are normally other costs that an agent will wish to incur and which they will intend to charge to the client on top of their fee for negotiating the transaction. Such costs may be advertising in magazines, boards etc. If any such costs are to be incurred by the agent and then recharged to the client, it is a legal requirement in the UK for the agent to obtain approval first, and it is good practice to set out these costs fully in advance and obtain client approval for costs, dates and all detail proposed.

Reservations

It is advisable to make it clear to your client what you have not done. For example you are unlikely to have carried out a full building or structural survey. You may well not have tested any of the services or used the lift if the power was off.

Conclusion

It is good practice to end the report advising your client that no further action will be taken until you receive their written instructions to proceed, and again this is a legal requirement in the UK.

Terms of agency

In the UK there are different bases of agency agreement (joint, sole etc.) and these must be set out clearly to the client. Many firms will detail this in a separate letter from the marketing report relating to the property concerned (see Concept 1.4).

Other

It is helpful to include an Ordnance Survey-based location plan and photographs where appropriate.

Further reading

RICS (2011) *UK Commercial Real Estate Agency Standards*, RICS, Coventry, sections 2.3, 2.4 and 3.2–3.4.
www.isurv.com (accessed 02/12/2013).
www.rics.org/uk (accessed 02/12/2013).

1.3 Terms of engagement

Key terms: instructions; fees; calculation; disbursements; Consumer Protection Regulations; Business Protection Regulations

The basis of the instruction should be clearly set out with your client before any action is taken. This is beneficial to both parties. If it is agreed in writing there is little room for dispute. The issues to be agreed are clear confirmation of your instruction, i.e. what you should and perhaps should not do, together with fees. It is normal for this information to be provided in or supplementary to the initial marketing report to the client.

An agent has a statutory responsibility to set out and agree the terms of agency in writing with a client before proceeding with any work. This was introduced by the Estate Agents Act 1979 and supplemented by subsequent Estate Agents Regulations in 1991 (Provision of Information; Undesirable Practices and Specified Offences). In 2008 the Consumer and Business Protection Regulations were introduced (CPR/BPR). These are general regulations and were only effectively applied to the property sector in 2012 with publication of the Office of Fair Trading (OFT) guidance on property sales, OFT 1364 September 2012.

It is suggested you should agree:

1 The amount of the fee or a method by which it is to be calculated. This should be set out clearly. Fees may be:

- A fixed sum.
- A percentage of the rent. In the UK it is normally based on a full year's rent or the average over a period of years if, for example, a rising rent is agreed. In the USA it is common to relate the fee to the length of lease agreed.
- A percentage of the price.
- A minimum fee to cover the costs of dealing with lower value properties where it may not otherwise be cost effective to accept the instruction. In the UK this is a variable figure depending on location and the size of firm. Few firms will take an instruction at less than £1,000 and in many cases this will be substantially higher.
- Incentive based to encourage the agent to secure a higher rent/price, i.e. the percentage may increase above a certain level.
- Subject to the market. There used to be fixed percentage fees in the UK in the 1970s, but this is now illegal.

2 When the fee is to be paid, which would normally be at the point where an unconditional legal contract is signed. It may also be at completion.

3 Any additional costs over and above the fee and which are to be charged to the client for disbursements. This would include marketing costs, e.g. advertising, travel costs, brochures etc. In some instances it may be appropriate to agree a total maximum sum to be incurred but this leaves it open to dispute when the fee is submitted. It is advisable to be specific about these costs in advance of the money being spent. It is good practice to suggest a total budget but then seek approval for a detailed schedule of costs to avoid doubt. This would include dates of proposed advertisements, sizes, publications etc.

   Where the marketing campaign will take some time it is prudent to provide for such marketing costs to be invoiced to the client on a regular basis, perhaps monthly. Where high costs are to be incurred (e.g. an expensive brochure) it is normal to arrange that the client procure this direct with the company that is to produce it. This avoids any financial liability falling on the agent.

4 Provision for further agreement in writing to be sought should any of these fees change or especially increase. An example would be an additional advertising campaign if the initial one did not secure a disposal. In any such agreement it is important that your client knows at the outset exactly what costs they are to pay and when. The client's approval should be obtained in writing for all fees and costs before any action is taken or costs are incurred.

Any errors in this information might be construed as failing to give material information to a client, thus potentially rendering the contract void. See section 4.9 of the OFT guidance.

A key component of the OFT guidance is that the terms of the agency agreement must be fair and reasonable because the client will make the decision as to who to instruct based on these terms. This is considered to be a transactional decision – see section 3.4 of OFT guidance.

A client falls into the category of a 'consumer' in these regulations as they are potentially consuming your agency services. An agent must not engage in any unfair commercial practices in this relationship. This is before or after a client instructs an agent. So if the client seeks the advice of two potential agents and decides to instruct one but considers that the activities of the other were unfair, an action could still be taken. See section 3.2 of OFT guidance.

Further reading

Murdoch, J. (2009) *Law of Estate Agency*, 5th edn, Estates Gazette, London, chs. 1, 3 and 7.
OFT, 'Guidance on property sales', September 2012 (OFT1364).
RICS (2011) *UK Commercial Real Estate Agency Standards*, RICS, Coventry, section 2.3, 2.4 and 5.2.
www.isurv.com (accessed 02/12/2013).
www.rics.org/uk (accessed 02/12/2013).

1.4 Types of agency – the basis of instruction for disposal

Key terms: sole agency; joint agency; multiple agency; introductory fee

The Estate Agents Act 1979 places a statutory requirement upon an agent to set out and agree their terms of agency with all clients in writing before doing any work on an agency instruction. One aspect of this is that there is a specific form of wording (see below) that should be used. Any changes to this wording must be fully justified.

The wording in the Act relates to the sale of property which is fine for the residential estate agency sector. The commercial property sector is different as other legal interests in property form the majority of agency instructions. The commercial agent will need to amend this wording to allow for a letting, subletting, assignment of an occupational lease, assignment/sale of a long ground lease or surrender. The client may therefore be a vendor, lessor, lessee, assignor etc. The party to whom you have disposed of the property may therefore be a purchaser; tenant; assignee; landlord (in the case of a surrender).

The key types that you must be familiar with are set out below and taken from the RICS UK Commercial Real Estate Agency Standards November 2011. The wording refers to a sale so if the instruction relates to a leasehold transaction the wording would need to be amended appropriately.

Sole selling rights – standard wording

> You will be liable to pay remuneration to us, in addition to any other costs or charges agreed, in each of the following circumstances:
>
> if [*unconditional contracts for the sale of the property are exchanged* or, in Scotland, substitute with *unconditional missives for the sale of the property are concluded*] in the period during which we have sole selling rights, even if the purchaser was not found by us but by another agent or by any other person, including yourself; and
>
> if [*unconditional contracts for the sale of the property are exchanged* or, in Scotland, substitute with *unconditional missives for the sale of the property are concluded*] after the expiry of the period during which we have sole selling rights but to a purchaser who was introduced to you during that period or with whom we had negotiations about the property during that period.

This right may apply whether the basis of instruction is on a sole or joint agency basis. It may appear to be very much in the estate agent's favour; however, it serves as a protection to them. When an agent puts a property openly on the market the advertisements will ask interested parties to contact them for further information and viewings etc. A potential buyer could in some circumstances see this and approach the vendor/client direct to agree a transaction, thereby cutting out the agent. This would be inequitable where such a buyer only became aware of the availability of the property as a result of the marketing actions of the agent. However, a client may not want to pay a fee where they identify the buyer/tenant themselves. It is usual for this right to have a time limit of, say, six months, but that is a matter for the two parties to agree.

Sole agency – standard wording

> You will be liable to pay remuneration to us, in addition to any other costs or charges agreed, if at any time [*unconditional contracts for the sale of the property are exchanged* or, in Scotland, substitute with *unconditional missives for the sale of the property are concluded*]:
>
> with a purchaser introduced by us during the period of our sole agency, or with whom we had negotiations about the property during that period; or with a purchaser introduced by another agent during that period.

This is the normal basis of instruction in the UK where a vendor enters into a contractual agency relationship with the client. The two parties will work together to effect a suitable transaction. It allows a good working relationship to develop between client and agent.

Joint agency or joint sole agency – standard wording

This is in effect the same as sole agency except that there may be two or more agents involved, all having specific instructions to market the property. This is usually the case on a larger instruction where there may be both a local and a national/international demand for the property or part of it where there are a lot of options, for example the building of a new shopping centre. The locally based agent would target the local market, undertaking viewings etc. The inclusion of a national agent may help to attract the national or sometimes international occupiers.

The basis of fees and split of fees on completion of a satisfactory letting will be a matter for the parties involved to agree at the outset. It is often an equal split, but the introducing agent may sometimes receive a larger proportion of the fee.

Multiple agency

If there is no reference to the basis of agency the presumption is that the instruction is on a multiple agency basis. This means that the agent who does the successful introduction receives the fee and no others do. It is not uncommon for some property owners to ask all local agents to market their property on this basis. This generally does not work to anyone's advantage. As an agent it is hard to take such an 'instruction' seriously because you can put in a lot of work for no reward. It is a clear lack of commitment on the part of the client who does not get the job done properly. Such instructions are generally best declined.

Another form of this is an introductory fee where a property owner offers to pay a fee to an agent for the mere introduction of a buyer or tenant. In this case the agent would not normally actively market such property but when talking to companies looking for space might match it to such available property and make an introduction. This may only be an attractive option where you do not have a suitable property on your books to meet the requirement. It is best to register this introduction in writing as soon as is practical and obtain written confirmation.

Further reading

Murdoch, J. (2011) *Law of Estate Agency*, 5th edn, Estates Gazette, London.
RICS (2011) 'UK Commercial Real Estate Agency Standards', RICS, Coventry, section 2.4.
Rodell, A. (2013) *Commercial Property*, College of Law Publishing, Guildford.

1.5 The marketing plan

Key terms: marketing; advertising; board; reception; mail shot

There are many ways of marketing property, depending on the nature of each; the following is intended as a starting point. The crucial issue is to apply any plan effectively to the property. The most important factor is to decide where the market is for the specific property – local, regional, national or international. Plans may include:

Marketing brochure – see Concept 1.6

Board

Size – In the UK an agent may erect a temporary board without planning consent providing it is within a certain maximum size range. Larger boards are costly and need planning consent. Most Local Planning Authorities are agreeable to putting a board up at the same time as the application for consent, but ask first.

Position – Give the contractor clear instructions about where the board is to be erected so that it is most effective. Sometimes it may be best to have a flat board if the building is at a major road T junction. On a normal two-way road a V angle board would be best. This relatively simple method is still one of the most cost effective for a majority of buildings.

Send the contractor a plan and very specific instructions or meet them on site to discuss options. Ensure they take appropriate care not to damage the client's building especially if it is listed. Make sure it is put on the correct building. It is guaranteed that at some point a board will be put on the wrong building resulting in an angry call from the owner.

It must say the correct words, i.e. to let/for sale etc. Boards are normally hired rather than bought and there is an annual charge. Contractors will have the firm's standard style in permitted sizes. A floor mounted board can be put in the ground floor entrance of an office. Most agents have standard to let/for sale etc. posters. These can be put in the windows to supplement a board or where one is not possible.

Clients can be secretive and not want a board. They expect an agent to sell/let the property without publicising it in this way. This does make the job a lot more difficult.

When a deal is done amend the board to say under offer/sold subject to contract etc., then sold/let by, then remove but do so only when definite, as it does not pay to presume a deal will complete. In fact the opposite is the case; the chances are that if a deal is publicised as done before it is then it will fall apart.

Banner

If the building is in a prominent position, e.g. an industrial unit visible from a major road, consider a banner attached to the outside of the building. These are normally canvas, but beware if it is in an exposed windy position. Planning consent must be secured. The increased size makes them more easily visible from a distance.

Advertising

This is costly, however, some is usually justified. With higher-value properties, e.g. a new development, a substantial programme may be needed. But advertising is not a substitute for good agency. Is any really necessary? Do not let costs get out of hand. Agree detailed schedule with the client, but always remember that they have not given the instruction simply for the agent to spend lots of their money on newspaper advertisements. It is the agent's skill that is required.

Most agents arrange this through an advertising agency who will book the space, design the advertisement and ensure that the correct copy is sent to the publications chosen.

National publications

The main national publications are *Estates Gazettes* and *Property Week* but others may be relevant depending on the property, e.g. *Retail Weekly* for retail property.

It is generally very expensive. The agent should decide whether to use colour or simply black and white. It may be effective to take a whole page and include 2–4 properties, thus spreading the cost. Repeating the same advertisement over a few pages can create recognition but will be expensive. The national property press have a programme of focuses – regional/property type throughout the year. This is often the most effective way of using the national property press.

Local publications

Determine the appropriate regional paper and which day(s) it has a property section. The same comments as national adverts apply. Many regional agents opt to take out an annual contract for a block advertisement in a specific location in the paper each week/month.

It is a good idea to check print quality for errors etc. to avoid paying for something that does not appear or if the quality of printing is poor. It is crucial to keep to the schedule agreed with the client. Link ads with other marketing, i.e. make sure marketing brochures are ready. Recover costs regularly from the client. For a large programme in one newspaper ask them to bill the client direct.

Check where and when enquirers saw an advertisement, if that is why they have made contact. The sector does not actively monitor the effectiveness of marketing actions. Keep a copy of all ads placed with cost and apportionment between properties.

Below are some simple suggestions on writing advertisements:

- Does it need a reference on it, i.e. someone's initials?
- Not too much detail – be clear and concise
- State the type of building, location and size
- Does it justify a photo – probably!
- Is there a specific selling point
- Price/rent
- Clients invariably check – send them copies

The aim is that the advertisement should make those reading it decide to ring for further information or look for it on the firm's website.

Marketing suite

This involves the fitting out of an area within a new development/refurbishment. It is usual to do one part of a floor, probably the best and maybe a corner area with good views. It should be carpeted and furnished with a desk and chairs, and separated from the main area by floor standing dividers rather than partitioning that needs to be built.

This area can serve as a good one to sit down with those who view the building. In some instances a computer with floor plans of the scheme or web access may be beneficial. This can be expensive but sometimes it may be possible to persuade an office furniture company to provide the furniture at low cost in the hope that whoever takes the building would instruct them to fit out the whole building. The analogy in the residential sector is a show house.

Reception

This can be very cost effective in a new or refurbished building. It is complimented by a marketing suite and gets the building known in the local, or perhaps a wider market. They are normally held at the building or perhaps at a local hotel if, say, a topping out ceremony.

The cost – price per head is not usually exceptionally high and can be very good value for money. The client should be billed direct.

An invitation list would include:

- other agents;
- appropriate businessmen e.g. local solicitors for their clients;
- media for press release;
- potential occupiers.

The following suggestions may be useful:

1. Circulate around everyone – network.
2. Have all information about the building available, e.g. plans, brochures etc.
3. Ensure good mix of food, wine, soft drinks. Note special diets. Have plenty of soft drinks if those coming must drive. Get quotes from local caterer.
4. Cost is not great but only worthwhile if there is something to show.

5 Do not mix other agents with possible occupiers!
6 Hold an automatic raffle for those who attend or give a present to them. In bigger developments inducements can be offered to encourage people to attend.

Mailing list/mail shot

All firms keep in-house lists onto which all enquiries received should be added in a logical way, perhaps by size/type/geographic location etc. These requirements lists should be updated regularly by phone, letter or (increasingly) email. Many large firms link from a regional office to a company-wide database. The mixing and matching of new instructions to a current requirements list can be a very effective marketing action.

In addition to current requirements most firms will either keep in-house contact lists for potential target groups that include all companies that could/should be interested or buy in a suitable mailing list for each property as necessary. Addresses can be taken from *Yellow Pages* or trade magazines, normally web based. It is best to ring up to get the property person's name. Send details, then follow up – a time-consuming but vital job. Some firms use a mailing house that specialises in producing such up-to-date contact lists. This may be cheaper and easier. In either case the client should pay. Other target groups include:

- public sector;
- local authority lists of vacant space – make sure all properties are on their lists;
- business links organisations for smaller companies;
- any organisation that promotes the region that may have details of available property or is involved in discussing inward investment opportunities with such companies;
- suppliers to major employers setting up or expanding in the area (target their suppliers – this is especially useful in the industrial market);
- surrounding occupiers in the same building/surrounding area/estate/next door;
- recent buyers of similar buildings e.g. for refurbishment/investment;
- agents (those in the same locality and a wider area if appropriate).

The cost of a substantial mailshot would normally be re-charged to the client providing it is included in agreed marketing costs.

Occupier presentations

Following on from the targeted marketing aspect of the mailshot, a follow-up opportunity of meeting a specific potential occupier and giving them a presentation on the scheme may arise. This might initially be to the property people within the company and then perhaps to those who make the decision, possibly the director responsible for the property.

This will normally only be for larger requirements competing with other developments in the region or often with other regions of the country if it is a footloose enquiry or occasionally developments in other countries. It is crucial to determine exactly what the position is and tailor the presentation accordingly.

Press release

This is free advertising, but it needs to identify an angle, such as client expanding and moving to bigger building, therefore vacating space. Is the client's move/action increasing jobs? Refurbishment of derelict old building, perhaps one that is well known or listed with some history to it. It's best to include a photo. Give some background information on both the client and the building. Use paid advertising as a lever to get editorial space.

Internet

See Concept 1.7. All property must be on the firm's website.

Specification guides

In a major new scheme an agent may provide a detailed specification guide for serious potential occupiers to aid their fit-out works and to make the decision in the first instance that this is the best property for them. These will be prepared by the other relevant advisers to the scheme, for example, a building services engineering consultant will produce a specification for all of the services. This may be pivotal for an occupier in deciding whether to take space in the scheme. These guides will be supplemental to the marketing brochures. (See Building information modelling, Concept 3.10.)

Newsletter

When dealing with a major new development especially, where there is a need to market the location as well as the property and where the development of the scheme has a long life span, some developers and their agents will publish a newsletter on a regular basis to update both those taking space in the scheme, potential targeted occupiers and the region or city as a whole, simply to get the message out that this is a development of importance.

TV/radio

Generally expensive but some schemes may be suitable. It would normally be at a local level and probably be used for 'brand awareness'.

Further reading

RICS (2011) *UK Commercial Real Estate Agency Standards*, RICS, Coventry, section 5.6.
www.isurv.com (accessed 02/12/2013).
www.rics.org/uk (accessed 02/12/2013).

1.6 The marketing brochure

Key terms: marketing brochure; letting/sales particulars; desktop; glossy brochure

A marketing brochure is a fundamental part of the marketing of property. This Concept provides an outline of suggested headings of what to include. The ideas must be adapted to the type of property and the legal interest that is offered – freehold, new lease, sublease, assignment or ground leasehold interest. All are different.

The lettings or sales brochure is a crucial part of any property marketing campaign. These may be done in two main formats depending on the value of the property and the level of quality required by the client:

1. Desktop publishing – these are prepared in house but the production is outsourced, normally, using a pre-agreed template. There is therefore no individual design work needed but the final production is a 'glossy brochure'. Almost all low/medium-value properties have this standard of brochure. Most organisations will use this as a minimum.
2. Printed brochure – although much of the content will be prepared by the agent, the bulk of the work is outsourced and done by a design company. Each will be individual and can range from a relatively simple A4 card to a multi-page booklet at significant cost.

The following suggestions may be useful:

- simple A4/A3/booklet style with insertions for different parts, e.g. small/large buildings/blocks;
- agent prepares draft wording/photos;
- design agency does draft/amendments/redraft until finalised;
- content must be correct – if using written material or photos from someone else make sure consent and copyright has been obtained;
- client's approval – print;
- change is difficult without re-run and high cost;
- tend to contain less detail – often do not have prices/rents to allow a longer shelf life;
- insertions should be removed as let/sold;
- electronic files can be amended easily to show changes and sent by email.

Brochures can take a long time to do but they must be correct.

The decision on what to produce will be down to the agent's recommendation and how much the client is prepared to pay. There is a clear move towards digital production and mailing rather than printing paper copies. It is far more efficient to email a pdf of a brochure than post paper copies. There are instances where it may be appropriate to link the identity of the client to your marketing, for example, stating 'On the instructions of....' at the top of the brochure. This is common in disposals by the public sector.

The essential format and contents of marketing particulars is fairly standard but there are variations depending upon the tenure of property that is being marketed. Content will also vary depending upon the nature of each building and the view taken by those doing the marketing. The following is a suggested format for commercial property:

Heading – address, type of property, to let, for sale etc. Be clear, so a prospective occupier can see exactly what you are offering; a colour photograph is crucial.

Situation and description – cover the main factors.

Accommodation – size NIA/GIA/retail frontage etc. Include a detailed breakdown of the various areas if practical and always use the RICS Code of Measuring Practice.

Services – state what services are supplied to the property – gas, electricity etc. and the type of heating system.

Planning – what consent exists and whether there is any potential for a change of use, but then qualify the statement to make it clear that it is subject to obtaining consent.

Car parking – if there are spaces state how many, where they are and the cost; comment on public parking nearby.

Lease terms – what lease terms are on offer including length of lease, repairing liability, rent reviews etc.

Sublease – make it clear what is available within the existing lease.

Assignment of existing lease – state the terms of the lease.

Freehold – be certain it is.

Service charge – if in a multi-occupied building one is payable – what costs are included; is it full recovery or to be capped? The ideal is to state a figure per sq. m/ft but it must be correct based on payments in previous years. Usually this information would be provided in subsequent discussions.

Conditions – any that might apply, e.g. liability to fence a boundary where part of a freehold is to be sold.

Rating – A key cost. Include where known:
RV = rateable value
UBR = uniform business rate
Detail on costs is best left to subsequent discussions.

Floor plans – if available say so. An indicative floor plan can often be incorporated into the marketing details and is very helpful. The availability of computer-aided design (CAD) plans to aid internal office design and layout can help marketing, especially in a new office development. An indicative plan showing the general shape can be helpful.

Rent – asking rent (normally per annum).

Assignment – state existing rent and premium if relevant.

Ground lease – state ground rent if any plus premium sum for the legal interest.

Price – if freehold.

Costs – if the lessor/vendor expects the lessee/buyer to pay its legal and/or surveyors' costs make it clear.

EPC – the rating must be included.

VAT – it is advisable to state that the lessee or buyer is responsible for VAT if payable.

Viewing – what the arrangements are (normally by contacting the agents office).

Plan – a marketing brochure must have a location plan showing where the property is and the extent of what is available where relevant. Usually an Ordnance Survey plan is used at an appropriate scale e.g. 1/1250 site plan, 1/10,000 or greater as a location plan. Where it is a retail property the norm is a GOAD plan.

A good clear plan is very important. Interested parties must be able to find the property from the details. Make it easy for them. Clearly mark the property on the plan and include a scale and north sign.

Disclaimer – All marketing brochures should include a disclaimer in relation to the Misrepresentation Act 1967 and make it clear that the brochure is *not part of a contract*.

Additional notes

Code for Leasing Business Premises – RICS would like to see all marketing brochures making a clear statement that the landlord will negotiate within the Code and that alternative flexible lease terms are available on request. This depends on the view of the client.

Further reading

OFT 'Guidance on property sales', September 2012 (OFT1364).
RICS (2011) *UK Commercial Real Estate Agency Standards*, RICS, Coventry, section 5.6.
www.isurv.com (accessed 02/12/2013).
www.rics.org/uk (accessed 02/12/2013).

1.7 Information technology in marketing

Key terms: IT; cloud computing; social media; virtual reality; internet

We live in a world that is becoming increasingly digital. All surveyors must keep up to date with developments in information technology (IT), as it can benefit our professional lives. Despite the enormous amount of change we have seen over the last 20 years, IT is still in its infancy and the working world will be a very different place again in another 10 or 20 years' time.

Regardless of the computing equipment you feel most comfortable using, we can now undertake tasks that were beyond us in the past. Writing letters and emails, maintaining and using mailing lists, storing data, drafting marketing particulars, mapping, creating presentations for clients and using comparable evidence databases are all commonplace tasks for agents in today's world. It is possible to purchase agency management software to help record the marketing of a property, avoiding the need for paper-based files.

Everything from initial inspection notes to actioning the marketing plan to recording enquiries, viewings and offers can be monitored and managed. Storage of these data is retained within the office in most cases. The advent of cloud computing means that individual firms do not need to have any hard data storage, as it can be provided by a data storage company via the internet. The data storage company will have huge backup capacity gained from the economies of scale of mass storage.

Digital files can be set up to be accessed by all parties involved in the transaction, but with restrictions placed on access depending on individual need. A client should be able to see all data if the agency function is to be wholly transparent. Parties enquiring about a property could see a brochure and other associated marketing information, and have access to direct communication with the agent. A solicitor would only need access to relevant legal information to enable them to finalise the transaction.

The extent of marketing material that can be made available will depend on the nature of the property to be marketed and the costs of what can be done. This may simply be a brochure and EPC for a lower-value property. At the other end of the scale, there might be a virtual 'walkthrough' of a building combined with computer-generated imagery of the surrounding area.

The government is encouraging the property industry to make the whole property transaction process digital including online searches, conveyancing, Land Registry registration, the EPC database, rating information etc.

A website is a must for any surveying firm providing agency services, and it has to be effective and easy to use – a poor website is worse than not having one! The site must be simple to navigate and respond quickly. All properties being marketed should have a brochure (usually as a pdf), an EPC, and a clear location plan with a link through to one of the online mapping facilities with both map and satellite imagery. Indicative floor plans can also be useful and there are software packages that will produce floor plans from inspection notes very quickly. Such floor plans would only give an indication of layout rather than detailed measurements, as accurate architects' floor plans are not always available.

For larger multi-occupied property, the web-based file can include schedules of floor areas updated as sections are let, photos or a video of all floors. Some use 360° panoramic photos allowing a wider image of the property. In addition to a website, there are also the internet portals that firms can pay to upload the properties they are marketing to reach a wider audience (e.g. Rightmove, Costar and Egi).

One significant development is the use of virtual reality (VR), which is defined as:

> A user interface that allows humans to visualise and interact with computer generated environments through human sensory channels in real time.
>
> (IT Construction Forum, 2013)

This technology can be used in conjunction with 3D modelling software to create an interactive 3D walkthrough, allowing the user to navigate a building which can be supplemented by animations, the opening of doors and windows etc. to give a more realistic feeling to the visualisation. This type of VR can also be enhanced with commentary or music. At this level VR can be expensive, but as technology improves it is starting to be used by developers on major new development schemes.

Around the world there are more than one thousand virtual city models developed by various techniques such as detailed aerial photogrammetry. This can be a useful addition to a planning application demonstrating suitable scale and massing within the urban environment. These technologies used together can provide a very powerful tool to both the developer and their agent. All of this information can be made available on an agent's website. In addition a package can be put onto a CD/DVD or USB drive to be given to those enquiring about the property. The laser scanning of an existing building will produce a very accurate 3D image. This can be useful when dealing with listed buildings and especially where alterations are proposed as it can assist in obtaining appropriate consents.

Many firms and marketing portals use mobile phone applications (apps) to enable access to information on property on the market. Some do this using QR codes, which bring up details of the property from the agent's website. Others use a GPS system to provide information of all property within a certain radius. The advent of tablet technology provides a larger screen within a lightweight device and this is much more powerful in terms of app technology. Standard reporting packages can be uploaded allowing the 'mobile surveyor' to carry out an inspection, search for comparables, prepare a marketing report and email it to the client.

Facebook, LinkedIn, Twitter and many more social networks seem to be taking the world by storm. Will it be a short-lived phenomenon that will disappear in a few years or is it the way forward? These are extremely powerful ways of communicating with clients that must be embraced by the property industry.

Further reading

IT Construction Forum (2013) Available at construct-it.org.uk (accessed 02/06/2013).

RICS (2011) *UK Commercial Real Estate Agency Standards, RICS*, Coventry, section 5.6.

Waller, A. and Thompson, R. (2009) 'The role of social media in commercial property'. RICS, available at: www.publicpropertyforum.ca/library/social_media_RICS-26pp.pdf (accessed 02/12/2013).

1.8 Energy performance certificates

Key terms: EPC; energy performance; green retrofit; sustainability; Energy Performance of Buildings Directive

The issue of energy use and sustainability has moved significantly up the ladder of importance within the property industry but still has a long way to go. The cost of energy used in our buildings is increasing at a relatively high rate, to the extent that occupiers are now beginning to consider this as a major cost in the occupation of property. This is an issue that has begun to affect the agent in their marketing of property.

Around 40 per cent of carbon emissions come from energy used in buildings split between commercial and residential. A substantial proportion of the buildings that will be here in 50 years' time already exist so a key issue is the retro-fit of such property to try to improve energy performance. It is relatively straightforward to build more efficient buildings from new – but dealing with the existing stock is a bigger problem.

Green buildings, sustainability and the energy performance of buildings are very topical currently and will remain so. These issues will have an increasing effect on value and the marketability of property as the UK Government raises the levels of energy performance that the property sector must achieve.

This issue is being driven by directives produced by the European Union, which member states are required to apply, as a minimum, in their respective countries. The legislation for the Energy Performance of Buildings Directive (EPBD), which provided for the introduction of an Energy Performance Certificate (EPC), was laid before Parliament in March 2007, and came into force in a phased manner as outlined below.

The key points taken from the Department for Communities and Local Government (DCLG) guidance are:

- the requirement for non-dwellings to have an EPC on construction, sale or rent, was introduced using a phased approach from 6 April 2008;
- the EPC shows the energy efficiency rating (relating to running costs) of a non-dwelling (the rating is shown on an A–G rating scale similar to those used for fridges and other electrical appliances);
- the EPC includes recommendations on how to improve energy efficiency;
- there is no statutory requirement to carry out any of the recommended energy efficiency measures stated;
- the EPC may also include information showing which of these measures would be eligible for finance under the Green Deal scheme, if required;
- EPCs for non-dwellings must be produced by an accredited non-domestic energy assessor, who is a member of a government-approved accreditation scheme;
- the seller or landlord must provide an EPC free of charge to a prospective buyer or tenant at the earliest opportunity;
- a copy of the EPC must also be provided to the successful buyer or the person who takes up the tenancy;
- estate agents and other third parties must ensure that an EPC has been commissioned before they can market a property for sale or rent;
- in addition, all advertisements in the commercial media must clearly show the energy rating of the building (where available);
- EPCs are valid for 10 years and can be reused as required within that period – a new EPC is not required each time there is a change of tenancy, or the property is sold, provided it is no more than 10 years old (where more than one is produced, the most recent EPC is the valid one);
- EPCs have to be displayed in commercial premises larger than 500m^2 that are frequently visited by the public, and where one has previously been produced for the sale, construction or renting out of the building.

The likely effects

The Energy Act 2011 indicated that from 2018, owners and agents will not be able to offer for sale or to let any property with an EPC rating of less than E. The aim is to force owners to improve the two lowest ratings of F and G. There are likely to be exemptions for Listed Buildings and those with no useable heat source that are clearly to be fully refurbished or redeveloped. This change is essentially aimed at buildings in, or ready for, occupation but which are considered to be very inefficient.

Although this rating system is relatively simplistic it can easily be made more demanding for property owners. The minimum rating at which a property can be offered for sale or to let could be moved to E or higher. The rating scale could be enhanced so that A is not the highest by introducing, for example, A1, A2 etc.

Government hopes that owners will improve their property either by direct investment or through the Green Deal whereby the cost of improvement is paid back through reduced energy costs. It should be relatively straightforward to carry out improvements made in the EPC to take the property to the next level. Government may choose to make this mandatory at some point in the future although there is no indication of this at present.

There is some concern that such assessments work well for buildings with a standard form of construction; however, the standard software used is not designed for many older buildings with, for example, solid walls. This is probably a bigger issue in the residential sector as the software used for commercial property is more robust. It is argued that an EPC assessment is a useful measure of comparison rather than a good absolute measure of performance.

It is the owner's responsibility to obtain an EPC. An agent must explain this requirement and most will have contacts with suitable assessors to refer the client to. It must be made clear that an agent cannot market a property until the EPC has been commissioned. However, in practice no advertisements can be made or marketing details sent out without the EPC rating so it is advisable to have it in place when marketing starts or very shortly after.

At the time of writing, business in general does not have this as a high priority and it is to some extent considered to be a box-ticking exercise, but this will change. If the public sector states that it will not acquire additional space below a certain EPC rating, then landlords must take note as the marketplace for such buildings will be limited resulting in a drop in value. This would be bolstered if the larger corporate organisations took the same stance within a corporate social responsibility (CSR) policy. Any such benchmark could easily be increased over time.

Mixed use and multi-occupied buildings will present difficulties. The need for an EPC is essentially based on the heat source in the building, or a part of one. A building with more than one heat source will in general need one for each. For example, a building with shops on the ground floor and residential flats above will probably need a separate EPC for each unit of accommodation, as it is likely that each shop and each residential unit will have individual heating systems. In contrast a multi-occupied office block with one form of central heating system should only need one EPC. In the latter case a tenant seeking to dispose of a leasehold interest and a landlord seeking to let vacant space should be able to use the same one.

An EPC must be done by an accredited assessor. All EPCs must be registered on a central database before they are valid and this register can be checked at www.ndepcregister.com.

Any complaints regarding the availability of an EPC are to be made to and investigated by Trading Standards. The penalty is based on a percentage of the rateable value of the building (12.5% in 2013) and between £500 and £5,000 maximum.

Further reading

DCLG (2012) 'Improving the energy efficiency of our buildings – A guide to energy performance certificates for the construction, sale and let of non-dwellings', December 2012, available at: www.gov.uk/government/organisations/department-for-communities-and-local-government (accessed 03/12/2013).

RICS (2011) *UK Commercial Real Estate Agency Standards*, RICS, Coventry, sections 2.4.3.8 and 3.7.2.1.

1.9 Methods of disposal – private treaty

Key terms: private treaty; invitation to treat; subject to contract; best offer

There are three principal methods of offering property to let or for sale: private treaty, tender (formal and informal) and auction. Part of an agent's role is to advise the seller/landlord which method to follow to achieve the best result for them. It is important that prospective purchasers and tenants understand the difference between these techniques and in particular that they are aware of what stage in the process they are contractually bound to proceed with the lease or purchase.

Private treaty

This is the most common method used in property sales in England and Wales and is generally the cheapest option.

A property is advertised to let or for sale inviting people to make offers to buy or lease the property. Legally the marketing of the property is classed as an invitation to treat (see Chapter 10). The price/rent on marketing details is not an offer to sell/let at that price/rent, it is an invitation to interested parties to negotiate and make an offer. It is important that the asking price/rent is set at a realistic level. If it is too high prospective purchasers may be dissuaded from enquiring about the property. Some may put in a low offer thinking they have managed to negotiate a substantial discount. Setting a low price can attract a lot of interested parties for the agent to negotiate with or go to best offers. The agent should advise on the best course of action depending on the property, the client's needs and the market at that point in time.

The vendor/agent enters into negotiations with interested parties on a 'subject to contract' basis. Where there is significant demand best offers might be invited by a certain date. This will not be contractually binding. It is normal to require all conditions attached to the offer to be clearly stated and the client will reserve the right to accept the highest or any offer and not to accept any, should they choose. The agent needs to be certain that they will receive at least one offer before pursuing this option otherwise this may prejudice future marketing of it.

Once agreement is reached on the principal terms of the deal, heads of terms are normally drawn up by the agent and circulated to all parties and their professional advisers. There is no contract in place at this time – Section 2 of the Law of Property (Miscellaneous Provisions) Act 1989 requires that certain formalities are observed for contracts for the sale of land.

Once the heads of terms are drawn up the buyer and their solicitor will investigate the property further. These investigations and searches generally consist of:

- An investigation of title – checking the seller has the power to sell, the extent of the property and any rights the property has the benefit or burden of.
- A survey, depending on the nature of the property this might be a valuation, Homebuyers Report or full survey.
- Raising enquiries of the seller – this includes matters relating to the condition of the property (flooding, subsidence etc.), relations with neighbours, work carried out to the property and responsibility for boundaries. In most transactions a standard form is used, for commercial property Commercial Property Standard Enquiries (CPSE) and for residential the Sellers Property Information Form.
- Searches – these include the Local Search (planning and building regulation history, public proposals in the vicinity of the property, highways and other public matters), Water and Drainage Search, an environmental report (this is a desktop search which advises whether or not there is considered to be a risk of contamination or flooding) and can also include a Commons Search and Coal Authority Search where appropriate.

During this period the buyer will also be ensuring that they have their finances in place and their solicitor will be agreeing the content of the contract with the seller's solicitor. The contract can be conditional or unconditional. If the transaction is on a leasehold basis then contracts would be based on an agreement for lease, to which a copy of the draft lease would be appended. At this point of the transaction either party can withdraw from the transaction.

The agreement only becomes legally binding when contracts are exchanged. On exchange of the contracts the buyer will normally pay a deposit, which is generally 10 per cent of the purchase price. At exchange the completion date will be agreed. The completion date is the date at which the transaction is complete – the balance of the purchase price is paid, title and possession of the property are transferred to the buyer. Completion can occur on the same day as exchange or there may be a delay between exchange and completion of a week or two. If the property is being sold with vacant possession then on completion the keys of the property will be released to the buyer. If the subject of the transaction is an investment property then on completion the buyer will receive a rent authority letter directing the tenant(s) to make future payments of rent to the new owner.

An overview of the process is set out in Figure 1.9.1:

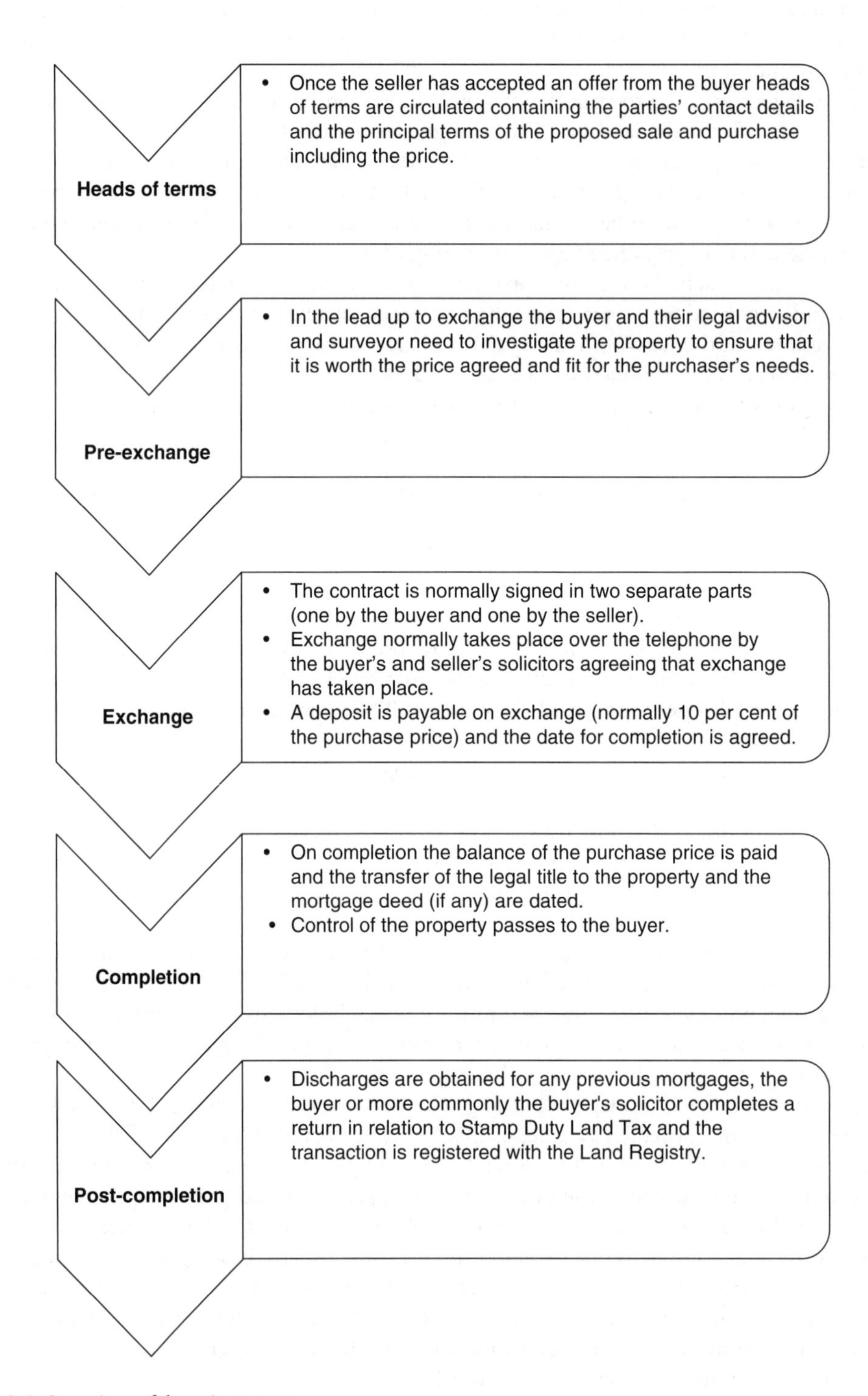

Figure 1.9.1 Overview of the private treaty process.

There are many pitfalls in the process. In a weak market it is not uncommon for a prospective buyer to agree a deal then shortly before exchange try to negotiate a lower price on the basis that the vendor will not want to run the risk of having to find another buyer. This is an unpopular tactic with agents but is quite legal as there is no firm deal until exchange occurs. This is referred to as gazundering.

In a strong market the opposite can occur. Between the point of agreeing a deal (when the property is put 'under offer') and legal exchange of contracts another party may enter the process and offer a higher price. An agent has a legal duty to report all offers to a client. If the vendor chooses to accept this higher offer and instructs solicitors accordingly the first party is considered to have been 'gazumped'. The vendor may choose to create a contract race between the parties such that the first to exchange is the successful buyer. In this case all parties should be made aware that they are in a contract race. In 2007 Home Information Packs were introduced as a way of trying to speed up the conveyancing process and provide some protection for buyers from the risk of being gazumped. HIPs were paid for by the seller and included evidence of title, replies to enquiries and search results. They did not however include a survey. The requirement to have a HIP was suspended in 2010.

Further reading

'Code of Leasing Business Premises', available at: www.leasingbusinesspremises.co.uk (accessed 15/11/2013).

Murdoch, J. (2011) *Law of Estate Agency*, 5th edn, Estates Gazette, London.

RICS (2011) *UK Commercial Real Estate Agency Standards*, RICS, Coventry, section 3.

Rodell, A. (2013) *Commercial Property*, College of Law Publishing, Guildford.

1.10 Methods of disposal – tender

Key terms: offer; exchange of contract; conditional exchange; planning constraints

Informal/non-binding tender

This method of sale is very similar to sale by private treaty except that from the outset of marketing a deadline is given by which offers have to be made. All marketing details, advertising etc. should refer to this date. The agent will need to put together the majority of information about the property prior to marketing so that interested parties are in a position to make a realistic offer by the due date. Offers should be received on or before the date and time in sealed envelopes. The offer should contain all conditions attached to it and the process is then essentially the same as a disposal by private treaty.

All offers must be reported to the seller with a recommendation. The seller then makes a decision that forms the basis of the heads of terms. The transaction then proceeds in the same manner as a sale by private treaty which means that either party can normally pull out any time up to the exchange of contracts.

This is a suitable methodology where you anticipate a reasonable level of interest and are certain, from the start, of receiving some offers. It provides more flexibility than a formal tender for both the seller and the bidders but should give some certainty compared to an ordinary sale by private treaty. It can be effective where the property or site has complications that might require a conditional exchange.

Formal/binding tender

A formal tender is very different to an informal tender as once a seller has accepted an offer in a formal tender a contract is formed and neither party can withdraw from the transaction. Once a decision to sell a property by formal tender has been made, the vendor's solicitors will need to prepare a 'legal pack' containing evidence of title, replies to pre-contract enquiries, search results, form of the draft transfer of the property and any other supporting documentation. The costs of the searches commissioned as part of the legal pack are normally paid by the successful bidder in addition to the purchase price. Any prospective buyers will need to study the legal pack and also consider having a survey/valuation carried out and ensure their finances are in place prior to making an offer.

When making an offer all bidders must include a 10 per cent deposit, normally by way of a guaranteed bank cheque. All offers are to be made in a sealed envelope and may not be opened until the time and date for receipt of offers has passed. The offers should be kept in a safe and normally most are delivered to the agent's office during the last couple of hours before the deadline.

Shortly after the deadline for offers it is normal to meet with the client and their solicitors where the offers are opened and a decision is made. On acceptance of an offer a legally binding contract is created. The parties will normally be required to proceed to completion within a specified period, say 4 weeks. Cheques are returned to the unsuccessful parties.

This is a very useful method but you must ensure there is going to be a reasonable level of interest otherwise you may be wasting time and money. It can produce some very good results. It is often used for sale of development sites or refurbishment opportunities. In such a case it is advisable to explore the planning position prior to marketing and wherever possible agree a planning brief with the local planning authority. This should indicate what uses are likely to be acceptable and any key planning constraints.

Further reading

'Code of Leasing Business Premises', available at: www.leasingbusinesspremises.co.uk (accessed 15/11/2013).
Mackmin, D. (2008) *Valuation and Sale of Residential Property*, 3rd edn, Estates Gazette, London.
Murdoch, J. (2011) *Law of Estate Agency*, 5th edn, Estates Gazette, London.
RICS (2011) *Residential Estate Agency Standards*, 5th edn, RICS, Coventry.
RICS (2011) *UK Commercial Real Estate Agency Standards*, RICS, Coventry, section 3.
Rodell, A. (2013) *Commercial Property*, College of Law Publishing, Guildford.

1.11 Methods of disposal – auction

Key terms: guide price; reserve; auctioneer; bidder

This is a public sale where anyone can make a bid in the auction room, by telephone or sometimes online (this is likely to increase).

The key aim of an auction is to create a legally binding contract on the auction day and in the room. As with sales by formal tender, the buyer will have to have fully researched the property before going to the auction, as if they are the successful bidder they will be contractually obliged to complete the purchase. Also in common with sales by formal tender the seller/their solicitor will prepare a legal pack. The costs of the searches commissioned as part of the legal pack are normally paid by the successful bidder in addition to the purchase price.

They will be held on a specified published date at a time the auctioneer chooses. Many auction houses have regular sales to provide consistency for buyers and sellers. This might, for example, be the first Monday of the month. The larger London-based practices often have sales lasting a full day, or even longer sometimes, whereas for the smaller provincial firms auctions may be an afternoon or sometimes an evening event.

An auction may be held anywhere – sometimes at the property, but often at a hotel, providing it is permissible in the premises proposed (some properties have restrictive covenants against use for an auction). Many properties in the provinces are sold by auction houses in London. The choice between using a regional or national auction house is a constant source of debate. The internet has enabled all to have potentially the same level of marketing coverage. One argument is that the London auctions attract speculators who are looking for a discount on value, whereas the local firms attract the occupier market as well and so achieve nearer to market value. The London firms, however, will attract a larger pool of potential buyers.

A guide price may be given but must not be misleading. It will be a guide only and should be less than the reserve, which is the minimum price at which the auctioneer may sell (if bids do not reach the reserve price then the property will be withdrawn from the auction). The reasoning is that if you indicate a guide price and bidders make offers at that level but the property is withdrawn because it does not make the reserve those bidders will feel they have wasted their time and this is not the environment an auctioneer wants to generate.

The reserve may be *disclosed* or *undisclosed*, although it is usually not disclosed either prior to the day of the auction or at the auction itself.

The property is not strictly for sale until bids have reached the reserve price. This figure is agreed between the auctioneer and the vendor in the days before the auction. It will be influenced by the level of interest shown in the property. There is a strong feeling that reserves must be realistic, especially in a slower market. If a vendor insists on an unrealistic price most auctioneers would decline the instruction. If offered without reserve the property will be sold regardless of the level of offer, providing there is an offer. It is not normally recommended to offer a property without reserve unless it is of very low value, perhaps zero or even negative.

There is a practice known as 'taking bids off the wall' or 'out of the air', under which the seller, his/her agent and the auctioneer can all bid for the property up to the reserve price provided that the conditions of the auction permit this to happen. Surprisingly, this practice is legal (see the Sale of Land by Auction Act 1867) and permitted under RICS Common Auction Conditions Edition 3.

The highest bidder will be the successful buyer providing they bid above the reserve price, if there is one. The auctioneer's final, usually third, knock of the hammer indicates a sale has occurred. The auctioneer will sign the contract on behalf of the vendor. The buyer will be required to sign it immediately bidding has finished on the lot bid for. The purchaser must pay a 10 per cent deposit immediately after the sale and the balance by a specified date.

Why sell by auction? There are three key reasons: certainty for the vendor; quick and clean; public and open. This method avoids the extended period of uncertainty over whether the sale will complete or not providing that there is a successful bidder. It is effective where there is a legal need to show that a sale is fully open and transparent so that all interested parties at that point in time had an equal opportunity to bid for it. This may be where a sale is on behalf of the public sector, a charity, trustees, mortgagees etc. It is considered as evidence that every effort has been made to achieve the best price.

An auctioneer owes the vendor a duty of care. They must:

- do as instructed providing it is legal;
- not act contrary to their client's instructions;
- obtain the best price, i.e. not miss bids;
- be certain the purchaser can buy. Some auction firms require bidders to register with the firm before the sale and to substantiate that they have the funds. Note that the Money Laundering Regulations apply to the sale of property by auction.

The process

- The vendor instructs the auctioneer.
- The auctioneer agrees fees and disbursement costs.
- The vendor is to pay fees if property is sold before or after the auction by private treaty.
- The auctioneer inspects and values the property then reports to client.
- Any issues to be clarified are identified and the best course of action decided.
- The vendor confirms agreement to fees etc. in writing.
- The auctioneer markets – prepares page for auction brochure and website entry, advertising, sale board, targets potential buyers.

- The solicitor provides/makes available legal documentation.
- The auctioneer responds to enquiries.
- The auctioneer arranges viewings.
- The auctioneer manages the day of the auction.

The auctioneer has a duty, where a property is sold before the auction, then all parties that expressed an interest in it must be contacted and advised. They must encourage potential purchasers to check the day before the auction that the property will still be offered. If a bidder arrives on the day of the auction to bid for a property that has already been sold they would justifiably be annoyed, especially if they have travelled some distance.

On the day of the auction, normally many properties will be offered. The auctioneer will seek to create a vibrant atmosphere where properties are selling fast – a full room is best. The auctioneer reads out any last minute amendments before bidding commences. Details will be issued to all those attending – important! The auctioneer seeks an initial bid often stating a possible opening price and bidding begins.

If there are no bids or the reserve is not reached the property will be withdrawn and the next one offered. If this occurs too many times it will affect the atmosphere in the room. The highest bid is the purchaser who signs the contract and pays 10 per cent deposit.

There are a number of firms offering properties for sale on a conditional basis. The successful bidder will secure the right to a pre-agreed period – say 4 weeks – to exchange contracts. They must pay a non-refundable sum, usually a few thousand pounds, and the sale will proceed in the usual way.

It is normal for the vendor to pay the auctioneer's fees; however, some firms auction property on the basis of a 'buyer's premium'. This means that the successful bidder must pay the auctioneer's fee based on a percentage of the price or a fixed or minimum fee in addition to the purchase price. This must be clearly stated prior to the auction so that bidders know exactly how much they must pay.

The sale of property by auction is subject to the laws relating to misdescription; therefore, if a problem arises the purchaser may be able to withdraw from the transaction (see section 10.20 on the CPR/BPR Regulations). All types of property can be and are sold by auction – residential, commercial, rural, retail, leisure, investment, land etc. Auctions are likely to continue to be a popular method of sale. Internet bidding is being looked at seriously. This needs to be in real time and linked by video. It also offers the option to develop an international market place for auctioneers.

The role of an auctioneer is different to that of a normal agent as the auctioneer actually sells the property and will normally have express authority to sign the Memorandum of Sale. In the case of *Greenglade Estates Ltd v Chana and another* (2012) 3 EGLR 99 a third party fraudulently claimed they owned a property and instructed Strettons to offer it for sale at auction. After the auction it became clear that the original vendor did not own the property. The true owner refused to complete the sale and the auctioneer was held liable for damages to the buyer, defined as 'the value of the property less the agreed sale price'. An auctioneer must be certain that a client is the legal owner of any property they are to offer for sale.

Further reading

Estate Gazette Weekly Auction News

Murdoch, J. (2011) *Law of Estate Agency*, 5th edn, Estates Gazette, London.

RICS (2011) *UK Commercial Real Estate Agency Standards*, RICS, Coventry, section 3.

RICS (2011) *Residential Estate Agency Standards*, RICS, Coventry.

RICS guidance notes: 'Common Auction Conditions', available at: www.rics.org (accessed 15/11/2013).

Rodell, A. (2013) *Commercial Property*, College of Law Publishing, Guildford.

1.12 Marketing a property – freehold sale

Key terms: land; development potential; freehold; exchange of contracts

When instructed to offer a property for sale an agent should have reference to the various methods of disposal and make an appropriate recommendation as to which they consider is the most suitable. This will depend on the type of property, the market and to some extent the client; for example, public sector organisations have a responsibility to get best value and to fully explore the marketplace for all possible buyers at that point in time. Auction and tender can be preferred methods but not always.

Many different types of property are offered for sale freehold, including land, buildings for refurbishment, vacant property ready for occupation, investments and a variety of one-off buildings such as an old lighthouse, a garage for parking one car, a redundant toilet block etc.

The client vendor may be the owner that no longer wants it, trustees where someone has died, a receiver where a company has been put into liquidation etc. The property to be offered may be the whole of the client's ownership or part where they have surplus land or buildings. Each will have different requirements.

Land

First it is important to establish whether there is any development potential by considering what the market might want and crucially what planning consents may be available. It is usually best to offer development opportunities subject to planning consent and enter into a conditional contract as discussed in Concept 1.4.

The agent must look into the availability of services to the site to establish whether they are available and whether there are likely to be any significant costs in securing a supply of gas, electricity, telephone, water and drainage.

Ensure the land has road access; that there are no restrictive covenants; look for overhead power lines that might restrict the height of any new development; check whether it is in a conservation area; is all or part of it green belt – if so what are the chances of it being taken out of that category? Is there any government or EU grant assistance for job creation? In short make sure the options for the site are fully understood. Be clear what the boundaries are.

Land with a building requiring a major refurbishment

All of the comments about land above are relevant. In addition if the building is listed it cannot be demolished without listed building consent which is not usually an option. Consider alternative uses as well as its existing use. Be clear about the grade of listing – could part be demolished or perhaps all behind a listed façade?

Building ready for occupation

This might be office, industrial, retail, leisure, mixed use etc. Always check out the potential for alternative use and understand what the market might want so as not to miss out any options.

Investment property

There is a good market for investment grade property but the length of lease remaining and strength of covenant (the tenant) is crucial. Most lenders will not lend on short-term arrangements.

There is a wide range of property of institutional grade that may be of interest to the pension funds etc. There is a good market, at the right price, for investment property with an angle, e.g.

refurbishment to let vacant space, general upgrade, renegotiating leases to increase value etc. The secondary market was quiet during the recession but as the market slowly starts to move so will this part of the market.

General

There are many things to consider including the following: Should the property be sold as a whole or might the client's returns be maximised by splitting the holding into smaller parts? Care needs to be taken to ensure the good bits are not disposed of first leaving a difficult bit that takes a long time to resolve. The ideal is to put together parallel deals on the basis that they will be interdependent and exchange at the same time. Make sure that all parties are committed to undertake necessary works to ensure the split will work, especially in respect of a split of services and responsibility for boundaries. The decision here will rest on suitability for division, availability of planning consent for different uses and the practicality of splitting it.

Where the client wishes to sell only part of their property and retain the remainder, be clear on who will be responsible for services, boundaries and access in order to ensure that the client is not prejudiced. Always be aware that the availability of a freehold property can attract potential purchasers who would not otherwise be in the market for the property in question if it was only available to let. If it proves difficult to secure a sale, always consider the possibility of letting it to create an investment which can either be retained or sold.

Further reading

Murdoch, J. (2009) *Law of Estate Agency*, 5th edn, Estates Gazette, London, ch. 4.
RICS (2011) *UK Commercial Real Estate Agency Standards*, RICS, Coventry, sections 3 and 4.
RICS (2011) *Residential Estate Agency Standards*, RICS, London, ch. 3
Rodell, A. (2013) *Commercial Property*, College of Law, Guildford, ch. 3.

1.13 Marketing a property by way of an assignment

Key terms: assignment; term of lease; occupational lease; premium

This will normally relate to the whole of the demised area for the remainder of the term. When a deal is agreed the whole leasehold interest including the physical lease will transfer to the assignee. The assignor will not be responsible for collection of rent or any other management issues. It is essential to have regard to the issue of Privity of Contract and the Landlord and Tenant Covenants Act 1995, i.e. is it an old or a new lease and if it is a new lease is an authorised guarantee agreement (AGA) required?

When business is good the demand for property is high and it is often acquired without due regard to the long-term implications of that property. Business tends to be driven by short-term needs. It can then become a significant problem a few years later when the economic climate is weak. The agent is then very much in the firing line when the holding costs are high but the property market is weak and it is very difficult, sometimes impossible, to secure a disposal.

Regardless of the tenure of the property a factor to take into account will be the vendor's operational needs for its on-going business. It may be necessary to tie in the disposal with the acquisition of new space. Efforts should be made to avoid the period of time where the company must cover the costs of holding two properties but there will normally be some need to overlap to allow time to relocate.

The first thing to do is obtain a copy of the lease with all additional attachments, e.g. Rent Review Memoranda, Deeds of Variation etc. It is not uncommon to be given incomplete paperwork. Plans will normally be colour coded but copies are invariably only in black and white. Read the paperwork and understand what it means for the lessee. All marketing will be constrained by the terms of the lease.

A factor to bear in mind is that taking over a lease from another occupier is always going to be less attractive than taking out a new lease because then you can have some influence on the terms of the lease.

When disposing of a lease for an occupier it is advisable to involve the landlord at the start of the process which should make it easier, and various consents will be required from them, not least the consent to dispose of the leasehold interest. There are normally statutory requirements placed on landlords to ensure they act in a timely manner. Intransigence is not however uncommon!

Lease details

Confirm the main details of the client's lease (e.g. next rent review, etc.) and any additional papers received to prepare an abstract. This should include as a minimum:

- lease term
- commencement and date of expiry
- rent reviews and break clause
- current passing rent
- repairing liability – full repairing and insuring (FRI) or internal repairing and insuring (IRI)
- rights to sub-let
- user clause.

Any restrictions; for example, is an AGA required on assignment if it is a new lease under the Landlord and Tenant Covenants Act 1996? If yes, are any specific tests to be met by an assignee? Or is the lease contracted out of Security of Tenure Provisions of the Landlord and Tenant Act 1954?

Terms of the lease

In view of the lease terms and market conditions, what can be recommended?

If there is an imminent or outstanding rent review the job will be much more difficult because it will not be possible to say what the rent is going to be for very long into the future. This is an unknown and makes it more difficult for a potential occupier to prepare an accurate business plan.

If the lease is due to expire in the near future, i.e. it's a 'fag-end-lease', establish whether the landlord would be prepared to take a surrender of the surplus area and grant a new lease to any organisation you find that wishes to occupy it.

Negotiating a surrender of the lease, especially in a poor market, is often the best course of action. To assess the potential reverse premium to offer the landlord you would need to take into account and capitalise the costs of rent, rates, service charge up to the termination of the lease or the next break clause. There may also be dilapidations costs to account for.

In a good market the landlord may be keen to take the space back on the expectation of re-letting it at a higher level of rent which then has a knock-on effect for their investment where it is multi occupied. It may be possible to secure a positive premium from the landlord. Some leases have a pre-emption clause providing that the lessee must first offer the landlord a surrender but they do not have to take it.

If the lease contains an AGA the client must be advised that the lease will revert to them if the assignee goes bankrupt. Also there may be some constraints placed on the assignee by the AGA, e.g. their profit must be at least three times the company's total rent bill for the last 3 years.

Rent review and break clause

If there is an outstanding or imminent rent review it is preferable to settle it so that a potential assignee can be provided with clearer costs. If there is a break clause before the lease terminates you must take great care in assigning the lease to avoid the situation where the assignee remains in occupation until after the break but then goes out of business. The lease will generally revert to the previous lessee which must then either occupy it or find another assignee. Depending on the circumstances it may be better to seek a sublet until the break.

Rental valuation

Over-rented – the current rent payable is higher than the market rent in which case the client may need to offer an incentive to induce an assignee to take over the lease. This may be a reverse premium, i.e. a capital sum to reflect this negative value. A rent-free period cannot be offered in an assignment; however, some deals have been done on the basis that the assignor will reimburse the assignee for rent paid on production of proof of payment. This will be for an agreed period. It is an alternative to paying a reverse premium and is especially useful if there are any doubts about the assignee. Avoid a situation where a capital sum is paid following which the assignee company disappears, which has happened.

Profit rent – the market rent is higher than the rent passing in which case it may be possible to negotiate a premium, i.e. a capital sum to reflect the lower level of rent payable until the next rent review or lease renewal. Normally this only occurs in the retail sector.

There are some instances where a tenant has done some approved capital improvements to the property that they have paid for and for which they are not required to rent. It may be possible to secure a premium to reflect this value

Key money – there are examples, mainly in the retail sector, where a capital sum is paid by an assignee which does not reflect rental values but means that a particular retailer wants to trade in that location and is prepared to pay over the odds to do so.

Other restrictions

The lease is as written and is not subject to negotiation in contrast to a company agreeing a completely new lease. If it is on an FRI basis or if there are restrictions on use then those constraints will remain and will influence recommendations and marketing.

You should check that the lease provides for the lessee to assign the lease. This should be subject to landlord's consent. This is normally subject to a test of reasonableness but if not one is statutorily implied by the Landlord & Tenant Act 1927, S19.

The issue of who pays legal and surveyors' costs is a key issue. It is normally the responsibility of the lessee but the state of the market will determine who is to pay. There may be a chain of legal interests in a property all of whom will want their costs paying so be aware that these can mount up significantly!

Further reading

'Code of Leasing Business Premises', available at www.leasingbusinesspremises.co.uk (accessed 27/11/2013).
RICS (2011) *UK Commercial Real Estate Agency Standards*, RICS, London, section 3.
Rodell, A. (2013) *Commercial Property*, College of Law, Guildford, ch. 27.
www.isurv.com (accessed 27/11/2013).
www.rics.org/uk (accessed 27/11/2013).

1.14 Marketing a property by assignment of a long ground lease

Key terms: ground lease; assignment; term and reversion; ground rent

This will normally relate to the whole of the demised area for the remainder of the term. When a deal is agreed the whole leasehold interest including the physical ground lease will transfer to the assignee. Your client will not be responsible for collection of rent or any other management issues. Technically this is the assignment of a ground leasehold interest but is often referred to as the sale of the relevant interest.

As an agent you will want a capital sum for the benefit of the interest being marketed in the same way as a freehold sale. In valuation terms there will often be little or no difference in the value regardless of the interest being freehold or a long ground lease. But sometimes the difference will be substantial. The ground lease may be subject to payment of ground rent with rent reviews, a geared rent or a peppercorn.

Such interests are common with industrial property and less so with other property types. Often found where a local authority or large landowner has permitted a development programme but wished to retain its long-term interest in the land holding and control over use; such an interest will almost always be for a whole building rather than an individual floor.

This will normally be for the whole of the demised area for the remainder of the term. It may be for part of the demised area where a developer whose interest is a long ground lease wants to dispose of individual buildings in a scheme.

The first thing that must be done is to obtain a copy of the ground lease with all additional attachments, e.g. Rent Review Memoranda, Deeds of Variation etc. It is not uncommon to be given incomplete paperwork. Plans will normally be colour coded but copies are invariably only in black and white. Make sure you can make sense of them. You must then read the paperwork and understand what it means for the ground lessee. The whole of your marketing will be constrained by the terms of the ground lease.

A factor to bear in mind is that taking over a ground lease from another occupier is always going to be less attractive than taking out a new ground lease because then you can have some influence on the terms of it. It is also less attractive than acquiring a freehold interest which can be a significant issue in some cases.

When disposing of a ground lease for an occupier it is advisable to involve the landlord, usually freeholder, at the start of the process which should make it easier as various consents will be required from them, not least the consent to dispose of the leasehold interest. There are normally statutory requirements placed on landlords to ensure they act in a timely manner. Intransigence is not uncommon, however!

The crucial issue is to establish what land is included. In almost all cases it will have been granted with a covenant that the ground lessee was to build something – an industrial unit(s), office etc. Occasionally the original demise may have included some buildings. Next you must establish whether there is a ground rent payable or whether this was dealt with by a single premium payment at the start of the ground lease.

Valuation – rental by researching the market for comparable land and property

If there is a ground rent payable you must establish what the market ground rent is at the date you are valuing it. You must then establish a value for whatever building has been built on it. The ground rent must then be deducted from the overall rent before capitalising it. Where there are ground rent reviews you will need to account for this – probably by using the term and reversion method. You should only value to the end of the ground lease because at that point the value of the buildings revert to the freeholder. If there is no ground rent payable you simply value the land and buildings together until expiry of the ground lease.

You should identify specific issues that will help in marketing and be constructive about any weaknesses. The key here is that it is a diminishing asset because at the end of the ground lease the buildings revert to the freeholder. A company looking to acquire the type of property you are dealing with would almost always prefer to acquire a freehold rather than a long ground leasehold interest as it is a cleaner and simpler interest.

If there is an imminent or outstanding ground rent review it would be best to resolve it quickly but as the level of ground rent compared to premium value is often low this may be a relatively minor issue compared to a normal occupational lease. There are, however, instances where the ground rent as a ratio of occupational rent is high, in which case there may be a problem.

If the ground lease is due to expire in the near term the value will be severely limited. Clearly, value will decrease as time passes and expiry of the ground lease approaches.

In general, the user clause in a long ground lease will be reasonably wide as the term is long. However, if the lease was granted by say a local authority for employment generating purposes it is unlikely that consent for a significant change of use would be available. This will inevitably reduce value.

If market conditions have changed significantly it may be possible and more advantageous to negotiate a surrender of the ground lease to allow a comprehensive redevelopment.

Example:

Lease details

- 99 years commencing 49 years ago
- Rent reviews are every 20 years and the last was £2,350pa
- Site area 3,090 sq yds = 27,810 sq ft or 2,583.6 sq m
- Let's say there's an industrial unit of 12,000 sq ft (1,115 sq m) on it
- Market rent £5 psf £53.82 psm gives £60,000pa today
- Yield 9% no adjustment needed
- Market rental value of ground rent today say £5,000pa

Valuation

Term			
Rent	£60,000		
Less GR	£2,350		
Profit rent	£57,650		
YP for 11 years at 9%	6.8052		£392,320
Reversion			
Rent	£60,000		
Less GR say	£5,000		
Profit rent	£55,000		
YP for 39 years at 9%	10.7255		
PV of £1 in 11 yrs at 9%	0.3875	4.1561	£228,587
Value			£620,907
Round up to			£621,000
Compare to freehold value			
Rent	£60,000		
YP in perp. at 9%	11.11		£666,600

Note in practice you would probably drop the yield a little for a freehold valuation. This shows the arithmetic effect. The difference of around £45,600 is the value of the freehold today. At a freehold yield of 8.74 per cent the value would be:

Rent	£60,000	
YP in perp. at 8.75%	11.43	£685,800

Further reading

'Code of Leasing Business Premises', available at: www.leasingbusinesspremises.co.uk (accessed 15/11/2013).
Murdoch, J. (2011) *Law of Estate Agency*, 5th edn, Estates Gazette, London.
RICS (2011) *UK Commercial Real Estate Agency Standards*, RICS, Coventry, section 3.
Rodell, A. (2013) *Commercial Property*, College of Law Publishing, Guildford.
www.isurv.com (accessed 15/11/2013).
www.rics.org/uk (accessed 15/11/2013).

1.15 Marketing a property to let on a new lease

Key terms: new lease; lease terms; negotiating options

The term landlord generally means an owner of property that does not hold it for their own occupation; however, it does include a company that has surplus space in a property that they do own and occupy.

There are a variety of different organisations and people that may be termed as a landlord and each will have their own objectives, which may be highly variable. The main categories are:

a) investor
b) developer
c) public body
d) occupier with surplus space.

It is worth considering each in turn to explore some of the objectives that might influence strategies and therefore the instructions they will give to an agent marketing their property. A lease in this context is an occupational lease, i.e. the party that takes the lease intends to occupy it for their own business purposes. Lease lengths (the term) range from 1 year to 15 years in most cases, although there are some longer ones.

The process

The process detailed below is from the perspective of an agent/consultant advising a client on the letting of a property. The procedure is, however, primarily the same if it is done by an in-house property department. It is written within the framework of the UK market. Having gained an instruction to give advice on marketing the process will normally involve:

a) inspection and measurement
b) rental valuation
c) report to the client
d) marketing plan
e) marketing period – putting the plan into action

f) negotiating with prospective tenants
g) heads of terms agreed
h) instruction of solicitors
i) completion of the transaction.

Once the agent's recommendations and fees are agreed the property will be placed on the market in line with the client's instructions. In theory a landlord offering space to let on a new lease can do so on any terms and at whatever rent they wish but it is not that simple. There are a variety of issues that will influence negotiations.

When acting for major developers or investors they will generally take a medium to longer term view and the agent will be expected to maintain and enhance the asset value when letting empty space. Often rather than taking the first tenant that comes along they would wait for the right occupier to meet tenant mix criteria or quality of covenant.

An investor landlord will always want to maintain rent levels and hopefully improve them. In a difficult market most would prefer to achieve existing rent levels but may need to offer incentives to do so, e.g. rent-free periods etc. In a good market, securing a letting in a multi-occupied building at a new rent level means that all other lettings immediately become reversionary.

In contrast, the aim of most secondary investors is to secure an income, often to meet mortgage repayments. As an agent there will be some pressure to secure a letting to meet this requirement. In some cases any letting may be preferable to an empty property.

The main constraints upon a developer and their agent are any controls on the quality/type of tenant, lease terms and rental levels placed by the bank/financial institution providing finance for the scheme or similar restrictions placed by the investor where it is forward sold. This may create significant constraints for the agent. It is not uncommon to find that minimum lease lengths and/or rental levels must be achieved. This can be a problem where the market weakens during the marketing period.

A developer will normally be keen to secure a pre-letting of space to minimise risk and void costs. In some schemes, especially retail, a developer may be prepared to offer an anchor or larger first tenant a lower level of rent to generate some life in the scheme and attract more occupiers. In a rising market some developers may decide to hang on until completion of the scheme to maximise rental income and hence capital value but this can be a risky strategy.

An imponderable question in larger office blocks is whether to let floor by floor or wait for an occupier for the whole which may never happen of course. Once a single floor is let, the building is immediately no longer available for a single occupier. It is best to tie a few similar lettings together so that a reasonable proportion of the building is let at the same time. It is also preferable to do this on consecutive floors rather than spread lettings around the building in order to be able to meet a large requirement that may arise.

If the available space does not meet current occupier requirements it may be necessary to consider splitting floors. Building regulations must be adhered to. The result may be to reduce the net area as some may need to become common to allow for access and fire exits. The landlord would need to cover service charge costs on these areas. Always try to let the poorer half of a floor first but this may be difficult.

Many larger investors will have a standard form of lease for each property and will not vary from that as it can have an effect on investment value. One example will be the service charge arrangements, because if they are inconsistent, management of it becomes difficult.

In contrast, a public body such as a local authority may have a very different approach where the aim is to provide property for new or small businesses on a flexible basis not normally offered in the private sector. The ultimate aim is employment creation and an opportunity for new businesses to develop and grow in the area.

Some companies own and occupy space for their business rather than renting it and may have surplus accommodation. They may seek to let out this surplus space and the principal aim is generally cost cutting; however, they may well want to ensure that a letting does not compromise the image and identity of the building. This can limit the type of occupiers or the terms under which they can use it, thus making the agent's job more difficult. The result of such a letting will be that the landlord will become a property manager and may need to set up a service charge arrangement.

There are significant costs in holding empty property – local rates, service charges, security, maintenance, insurance etc. This is likely to influence letting strategy. Both RICS and the Law Society have produced standard short form leases to make the letting of lower value property easier. These will normally be used for short-term arrangements and are there to benefit both landlords and tenants. The property industry, led by the RICS, has produced the Code for Leasing Business Premises. The aim is to create a more flexible leasing environment and this is relevant when agreeing a new lease.

Negotiating options

A property transaction is rarely solely about rent or price. There are many factors you can bring to a negotiation in an effort to agree a deal suitable to both parties. The outcome will depend upon the relative positions of the parties involved in the negotiation. Always balance a concession against the level of rent to make sure you do not give away more than you are gaining. The principal options are outlined below:

Rent free

A few months may be appropriate to allow a tenant to move in and fit out. This may be related to the length of a lease, i.e. the longer the lease the longer the rent free period. This need not always be at the start of a lease. It could be spread across a few years or some to be effective after a break clause (see below) if the tenant does not operate it. This concession can be used to offset real costs incurred by the tenant in carrying out essential repair works.

Rising rent

The benefit is that the tenant will start paying some rent from the start. This can be useful if the tenant is unknown or perceived to be weak.

Rent deposit

If you are in doubt about the tenant ask to keep a sum of money on deposit that your client can use if the tenant defaults on rent. Any such monies must be held in a separate bank account.

Guarantor

Again, if you doubt the tenant seek someone to guarantee the lease, e.g. a director if it is a limited company. The creation of a limited company is to limit the liability of the directors, however, so they will normally be reluctant to offer such a guarantee.

Term

This can be any length. The longer the better for the landlord, normally. The client may want a maximum period if they intend to refurbish the building and want all leases to expire at the same time.

Rent reviews

The frequency of these is every 3 or 5 years normally, but they may be any frequency.

Break clause

This can be a useful way of securing a longer lease term but giving the tenant a way out as well. These do have a negative effect on investment value. They can also be difficult to effect depending on how onerous the break option provisions are. They can be subject to a penalty if operated, e.g. the payment of a fixed number of months' rent.

Full repairing and insuring/internal repairing and insuring

Most landlords want an FRI lease, which is sensible in a large, modern and multi-occupied office. But it may be inequitable for an older industrial building in poor condition. If the building is not in a good state of repair or if there is one specific part that is in bad condition, e.g. the roof, it can be beneficial to agree a Schedule of Condition for the property with the understanding that the tenant should yield up in no worse a condition at the end of the lease. Or specific areas of poor repair can be omitted from the tenant's liability, e.g. the roof.

Service charge

This is normally common for all tenants in a multi-occupied building but in an older building a tenant may not be prepared to contribute to an item that is in poor condition, e.g. a heating system. In this case it may be a sensible compromise to agree a fixed or capped service charge.

Security of tenure

In England and Wales, but not Scotland, a tenant has a statutory right (under the Landlord and Tenant Act 1954) to renew a business tenancy at the end of the term. It is possible to contract out of this right at the start of a lease. This is fairly common, especially where the landlord wishes to redevelop/ refurbish in the foreseeable future. If any new letting is to be contracted out it should be made clear at the outset of marketing.

User

The landlord may wish to restrict use, e.g. in a shopping centre where they require a good tenant mix. If the user is too restrictive it can reduce rental value, however. When negotiating a new lease it is advisable to be as open as possible, without prejudicing the landlord's position, in order to maintain the landlord's investment value. The tenant will also gain by giving them maximum flexibility in use and the ability to dispose of the lease if necessary.

Repair/refurbishment

Who is to do any work and who is to pay for it needs to be agreed. Often the tenant prefers to be in control and be given a rent-free period to offset the cost. Sometimes it is preferable for the landlord to do it especially where the works are substantial and they wish to ensure it is done correctly. The key is to make it a condition of the lease that one or the other does it.

Fitting out

In a retail unit it is normal to give the tenant a rent-free period as a minimum to cover the time it takes to do the work. In a slow market the landlord may make a contribution towards the tenant's costs. There are examples of tenants that have taken leases in poor market conditions, where landlords have paid for all fitting out costs and a long rent-free period, with the sole intention of disappearing at the end so no rent is ever paid. Such companies rarely pay any property-based local taxes or any other costs. Beware!

Costs – legal and surveyors'

In the UK it has been standard to ask a tenant to pay the landlord's legal costs but there is no legal requirement to do so. In some cases a tenant or purchaser may be asked to pay surveyors' costs as well. This is common in the public sector and can often be a non-negotiable issue. Sometimes you may be able to negotiate that the tenant is to make a fixed contribution to the landlord's costs. Agree who is to pay what costs – which will often be each party to bear their own costs.

Further reading

'Code of Leasing Business Premises', available at: www.leasingbusinesspremises.co.uk (accessed 15/11/2013).
Murdoch, J. (2011) *Law of Estate Agency*, 5th edn, Estates Gazette, London.
RICS (2011) *UK Commercial Real Estate Agency Standards*, RICS, Coventry, section 5.
Rodell, A. (2013) *Commercial Property*, College of Law Publishing, Guildford.

1.16 Marketing a property by way of a sublease

Key terms: sublease; term of lease; occupational lease

An occupational sublease is the creation of a lesser legal interest out of an existing lease. This may be the whole or part of the demised area for the remainder or part of the term. The lessee will enter into a management role as they must collect rent from the sub-lessee and pay to the lessor. They will remain responsible for all terms under the lease.

An occupational lease is one where an organisation has a lease of property to use and occupy in order to perform the function they are required to undertake – for example, business purposes. Other forms of leasehold interest are not the subject of this concept.

It must be established with the lessee whether it is the whole of the property/demise that is surplus to requirements or whether only part is surplus, in which case they may well wish to remain in occupation of the larger part of the property. If this is the case you may be able to surrender part to the landlord or sublet part.

You should also establish the lessee's future requirements. It is not uncommon for a lessee of a multistorey office building to find that, say one floor, is surplus so they may seek to reduce their costs by subletting a floor. This surplus may, however, be because of a known short-term reason and they may well want to keep their options open to reoccupy the space at a given point in the future. In such a case you would be seeking to dispose of part of your client's demised area for part of the remaining term of the lease.

You must establish exactly what the lessee wishes to achieve which will influence your report and recommendations. This, of course, will be directly restricted by the market conditions at the time and the terms of the lease. The key aim will usually be to minimise your lessee's costs, which will influence negotiations accordingly.

The first thing you must do is obtain a copy of the lease with all additional attachments, e.g. Rent Review Memoranda, Deeds of Variation etc. It is not uncommon to be given incomplete paperwork. Plans will normally be colour coded but copies are invariably only in black and white. Make sure you can make sense of them. You must then read the paperwork and understand what it means for the lessee. The whole of your marketing will be constrained by the terms of the lease. A factor to bear in mind is that taking over a lease from another occupier is always going to be less attractive than taking out a new lease because then you can have some influence on the terms of the lease.

When disposing of a lease for an occupier it is advisable to involve the landlord at the start of the process which should make it easier, and various consents will be required from them, not least the consent to dispose of the leasehold interest. There are normally statutory requirements placed on landlords to ensure they act in a timely manner. Intransigence is not however uncommon!

Where you are acting for a company that has a lease of a whole building and only wishes to dispose of part, say one floor, you will need to consider what areas of the building will become common and any actions your client must take to secure its own position and that of its intended sub tenant. Security can be a concern where a company is to share a building that it has previously occupied on its own. You must make sure the lessee understands that it will need to establish a service charge arrangement that was not in place previously.

A company that occupies a whole building will normally wish to make an effective advertisement out of it but this is inconsistent with sharing it with other occupiers. It will make your marketing efforts more difficult, so identifying the point on inspection should help in your recommendations.

Key factors

Terms of the lease

Establish the main details of the lease – an abstract. This should include as a minimum:

- lease term
- commencement and date of expiry
- rent reviews and break clause
- current passing rent
- repairing liability – FRI or IRI
- rights to sublet
- user clause
- any restrictions, e.g. sometimes any sublettings must be contracted out of the Security of Tenure Provisions of the Landlord and Tenant Act 1954.

Terms of sublease

These need to be flexible but must link with rent reviews and break clause in the lease if any. The maximum term of a sublease is one day less than the client's remaining term. If the lease is due to expire in the near future you should establish whether the landlord would be prepared to take a surrender of the surplus area and grant a new lease to any organisation you find that wishes to occupy it. They may agree but they may not do so if they believe that the building will be more marketable as a whole after your client's lease expires than if they enter into longer term lettings of smaller areas.

Rent review and break clause

If there is an imminent or outstanding rent review your job will be much more difficult because you will not be able to say what the rent is going to be for very long into the future. This is an unknown and makes it more difficult for a potential occupier to prepare an accurate business plan. It is preferable to settle a rent review before marketing a property, however, do not let this influence a higher settlement than necessary.

If there is a break clause ensure that the terms of the sublease do not restrict the lessee's ability to operate the break. It may be best to only sublet up to the time of the break, making allowance for any notice period required.

Rental valuation

Over-rented – the current rent payable is higher than the market rent. You will need to offer some level of incentive to secure a deal, e.g. rent free.
Profit rent – the market rent is higher than the rent passing so the lessee could make a profit.

You must establish what the lease requires. It is normal in the UK for a lease to state that a tenant cannot sublet at less than the passing rent. Some leases require the tenant to sublet at the market rent if it is higher. The rationale here is for the landlord to maximise their investment value. A rental transaction in a multi-occupied building provides the best evidence for future rent reviews and lease renewals. If the evidence shows an increase, all other leases in the building become reversionary.

The Code of Leasing Business Premises encourages landlords to allow flexibility during the lease. One aspect is not to require tenants to sublet at more than the passing rent. The British Property Federation (BPF), which is a landlord organisation, has agreed with many of its larger members that they will follow this protocol. You must establish what the standpoint is of the lessee's landlord.

Repairing liability

If the lease is on a full repairing and insuring basis and the potential sublease only short term this may be a negative. Be clear whether the liability is direct upon the occupying lessee or via a service charge in a multi-occupied building.

Comment re-right to sublet

The rights or any restrictions on the tenant's ability to sublet contained within the lease must be clearly set out. Some leases absolutely prohibit subletting of part or the whole in which case this is not an option. Some leases only permit one subletting and some may stipulate that the sub lessee cannot further alienate the sublease. Both of these are quite restrictive.

User clause

To avoid an incompatible user with your client there may be restrictions. Beware of the effect on rent. There may be some restrictions in the lease that you must comply with. For example it is not uncommon in a building owned by a pension company where surplus space is sublet to find a clause restricting disposal to another company in that same business. There may be restrictions on the sale of alcohol, cooking etc. which will mean that some classes of user will not be permitted, thus restricting marketing efforts. You must establish whether the clause is open, subject to landlord's consent, and whether this is qualified by a test of reasonableness on behalf of the landlord or there is an absolute restriction.

Other restrictions

If the lease is contracted out of the Security of Tenure Provisions of the Landlord and Tenant Act 1954 the lessee can only offer an unsecure interest. Regardless, the lease may say that any subletting must be contracted out so that a sub lessee cannot establish a right to a new lease on expiry. Some clauses only permit one subletting such that the sub lessee does not have the right to further sublet their interest although it is normal to permit an assignment. Finally, do any alterations need to be made to the building to make a sublet practically feasible? Will the landlord's consent be required and will there be any concerns in obtaining it?

Further reading

'Code of Leasing Business Premises', available at: www.leasingbusinesspremises.co.uk (accessed 15/11/2013).
Murdoch, J. (2011) *Law of Estate Agency*, 5th edn, Estates Gazette, London.
RICS (2011) *UK Commercial Real Estate Agency Standards*, RICS, Coventry, section 3.
Rodell, A. (2013) *Commercial Property*, College of Law Publishing, Guildford.
www.isurv.com (accessed 15/11/2013).
www.rics.org/uk (accessed 15/11/2013).

1.17 The marketing process

Key terms: marketing; viewings; for sale; to let; prospective tenant; prospective buyer

Once an agent has received clear written confirmation from the client that recommendations and costs are acceptable, a legally binding contractual agency agreement has been agreed. The agent should put the marketing plan into action as quickly as is practical and sensible.

It is not unknown for a client to have an unrealistic expectation of the speed with which this can be done. Having written the letter of confirmation some clients expect to see advertisements in the newspapers and for sale/to let boards etc. up the next day. Such expectations must be managed.

As well as arranging for all external actions to be put into place, it must be ensured that this new instruction is put onto the internal systems so all involved know it is on the market and can respond to enquiries when needed.

This marketing will hopefully generate some enquiries which must be responded to promptly. While this all sounds quite simplistic it is vital that there is in place an efficient system that ensures these matters are dealt with effectively and that all involved engage with the process. There is nothing worse than receiving no response to an enquiry for information on a property. If this happens an agent is not giving their client an acceptable duty of care and skill and will not enhance the reputation of the organisation.

Ensure that contact details are kept, including telephone numbers and email addresses so that the enquirer can be followed up and the enquiry reported to the client.

The aim of a good agent is to match those looking for property to what they have on the market. The agent must understand an applicant's requirements and know the properties they are dealing with in order to do this effectively. If an agent treats an applicant in an efficient and professional manner they will be happy to deal with the firm and may even instruct it again in the future.

Try to get prospective purchasers/renters to view a few properties to give them a clearer idea of what is available and what they want. It also offers the chance to meet them. When conducting a viewing of a building it is crucial to know it so that any questions can be answered.

Where a company is involved, try to find out some background information on the company especially where they have a substantial requirement. It will help to take a possible deal forward. An agent should also know the market and appreciate what else they will be looking at, in order to present a property in an appropriate light. It is a mistake to criticise other buildings or those involved with them.

Follow up all viewings a few days later to establish whether the applicant wishes to proceed. If so, take the interest forward. If not, is there anything else that will be of interest to them?

A prospective tenant will compare properties in terms of location, specification etc. One of the main criteria for comparison is annual costs usually on a per square metre/foot basis. The three principal costs are rent, rates and service charge. As the agent you need to know what each of these is for buildings you are dealing with. In the UK these costs are quoted on an annual basis but in other countries it may be monthly. An agent must understand what is being quoted and what it includes (see Concept 1.19).

Throughout the marketing period it is very important to report regularly to a client. There is no fixed timescale but even if nothing has happened, a letter or call at least once a month is advisable so the client knows they have not been forgotten. Always try to think of new and fresh ideas to try to generate some interest in a property that is proving difficult to sell/let.

If an agent or a person connected to them has any level of personal interest in the property being marketed this must be made clear and transparent and the client kept fully informed. The client's agreement must be obtained to confirm they are agreeable to the agent continuing with the instruction. This principle applies where an agent offers to supply any form of services to a prospective buyer of property being marketed, and especially where the agent will receive a fee in connection with this. An example of this is where an agent has an in-house financial advisor or links with a lending organisation to which prospective purchasers of property the agent is marketing are referred, and where the agent will receive a fee if a loan agreement is completed.

If an agent or employee of the firm or anyone connected to an employee of the firm instructs an agent to market property on their behalf, this must be transparent on any marketing material. Prospective buyers or tenants must be aware of who they are negotiating with. The agent must also take steps to ensure that other clients of the firm are treated on an equal basis and that all properties are marketed with the same vigour. It is advisable to put in place a procedure whereby another employee of the firm deals with the property concerned – where this is practical.

All offers received for a property must be reported to a client promptly and in writing. An agent must not misrepresent any offers or discriminate in any way. For example where two offers are received for a property, one will take a loan through the agent, for which an additional fee will be received and the other has arranged their own loan, both offers must be reported on the same basis and recommendations made disregarding the loan and fee arrangements.

An agent must not in any way 'ring fence' the marketing of a property or any offers received. 'Ring fencing' is where a property is offered to a limited number of prospective buyers, or perhaps one only, with the sole intention of underselling it. This has occurred where a vulnerable client, unaware of the market, instructs an agent to sell a property and they do so at a knockdown price, perhaps to another friendly client who then subsequently sells it on at a profit. The agent receives two fees and perhaps an additional undeclared fee.

A key change introduced by the Consumer Protection Regulations is that an agent must be entirely upfront about problems with a property being marketed in the same way as positive attributes. An agent is only expected to know things within their line of business but if they do become aware of any problems, amounting to a material factor, these must be fully disclosed.

If for example a building has a major structural defect; is located in a flood plain; is subject to a compulsory purchase order (CPO); has planning proposals that will alter the immediate area and, therefore, value – an agent must disclose these issues. In addition if a previous prospective purchaser withdrew from a purchase following a survey that highlighted problems of repair, this must be disclosed. It is unacceptable to just put the property back on the market at the same price so that the next purchaser gets another survey done to simply establish the same problems.

The withholding of information by an agent will be treated as a 'misleading omission' (see Concept 10.20). The buyer or seller must have all relevant information at the point they make their decision on whether to buy or sell – this is termed a 'transactional decision'. It is the omission of 'material information' that is an offence under the CPR/BPR. The legal principle of *caveat emptor*, buyer beware, is in effect no longer the case.

Further reading

OFT, 'Guidance on property sales', September 2012 (OFT1364).

RICS (2011) *UK Commercial Real Estate Agency Standards*, RICS, Coventry, section 5.6.

www.isurv.com (accessed 03/12/2013).

www.rics.org/uk (accessed 03/12/2013).

1.18 Negotiating

Key terms: duty of care; flexibility; negotiating margin; Data Protection Act

An agent has a duty to a client to get the best deal possible for them. A property transaction is rarely solely about rent or price. There are many factors to bring to a negotiation in an effort to agree a deal suitable to both parties. The outcome will depend upon the relative positions of the parties involved in the negotiation. Where the client agrees, the negotiations for a letting should be conducted within the framework of the Code of Leasing Business Premises which encourages a flexible leasing environment. There are no set rules about how to conduct a property negotiation and most surveyors will have their own style – the following is intended to be a guide. However, the key is to be prepared.

Make every effort to be on time to meetings. If being late is unavoidable contact the other party so they are aware. Being late to a meeting without explanation means the agent is fighting an uphill battle from the start. Turn off mobile phones before the meeting or at least put them on silent. If a phone does go off do not answer it, just stop the call, then turn it off.

Leave a negotiating margin – never start at the lowest acceptable position. When reporting to a client with marketing recommendations it is best to provide a range to allow room for negotiation. In a sale this will generally focus on price but other factors may be relevant. In a letting it is the relation between the lease terms and the rent.

Be realistic in a negotiating stance. There is no point holding out for a price or rent that is wholly unrealistic or unjustifiable in the market place. Flexibility is key. Thinking laterally about how to put a deal together is important. How can the property available match or be made to match the requirements of the party looking for new space? This may mean fitting in with their timescales; the vendor/landlord undertaking some works of alteration to the property; doing a split of space; being flexible on lease terms especially perhaps with a new business that can really only commit to a shorter-term arrangement.

Always take on board the needs and views of the other party in the negotiation. Let the other party to the negotiation have their say, make the case clearly, and stick to it if appropriate. Be prepared to think about arguments put forward and arrange a further meeting.

Treat those in the negotiation with respect and be polite. The other party may be a surveyor but in many instances negotiations will be conducted with the potential occupier direct who will not be a surveyor. They do, however, often have a reasonable knowledge of property in relation to their requirement. If they do not then taking advantage of their position is not a good course of action. It is best to recommend they seek appropriate professional advice.

All offers made/received are subject to the client's instructions. It is not an agent's role to commit a client to any transaction; it is the client's decision based on the advice of their agent. If a deal is provisionally agreed which is outside a client's instructions, make it clear this is the case and that it will be discussed with the client.

In a reasonable market with a good property there may be more than one party interested. This will require the agent to undertake negotiations with two or more parties, so the question arises – is it acceptable to disclose the details of an offer made by one party to any of the others? The answer is no – it is a Data Protection Act issue. An agent must go back to each party to ensure that they have got the best offer out of all interested parties so that the full position can be reported to the client with a clear recommendation.

There may be a temptation to make up or exaggerate an offer to use in negotiations with one or more parties interested in a property. Do not be tempted – it is illegal to create false offers; to misrepresent the detail of an offer; quoting a false high offer to induce a potential buyer to make a higher offer than they would have done otherwise. This is covered in the Consumer Protection Regulations (see section 4.6 of the OFT Guidance). The making of false offers is also covered by the Fraud Act 2006.

The object is:

- to secure a deal;
- to secure the best deal;
- to deliver a duty of care to their client.

Always remember that a deal is a package of terms, not just a rent/price. Do the job thoroughly. Think laterally. A good deal is where both parties feel they have secured the best for their client and that there was something left in their negotiating margin.

Further reading

'Code of Leasing Business Premises', available at: www.leasingbusinesspremises.co.uk (accessed 15/11/2013).
OFT, 'Guidance on property sales', September 2012 (OFT1364).
RICS (2011) *UK Commercial Real Estate Agency Standards*, RICS, Coventry, section 5.7.
www.isurv.com (accessed 15/11/2013).
www.rics.org/uk (accessed 15/11/2013).

1.19 Occupation costs

Key terms: rent; rateable value; uniform business rate; service charge

When marketing commercial property to let – either on a new lease, sublease or by assignment – an agent needs to be aware of the key costs of occupation of each property per annum. Most prospective tenants/lessees will compare the various properties available based on these costs as a total per unit of size either per square metre or per square foot per annum. As far as is practical the agent needs to be aware of how to establish these costs and have a realistic and justifiable idea of what they are likely to be. The agent must also understand how these costs are arrived at and be able to explain how each is calculated. The costs are rent, rates and service charge.

A company looking to take a lease of property will in some way need to prepare a business plan to justify it. Property costs are only part of the process. Other recurring costs will include employee salaries; utility costs (so energy performance of a property will become more important as costs rise), telephone and broadband costs, and many others. There will also be one-off costs, for example relocation from their existing property, new office furniture, redecoration and refitting etc. Some of these you may be able to factor into negotiations.

We will now consider each in turn.

Rent

This will be established by the agent based on a passing rent in the case of an existing lease or the agent's opinion from appropriate comparable evidence. The quoting rent is subject to client approval and may be subject to negotiation either by having a margin to agree a slightly lower level or by way of some form of incentive.

Service charge

This applies in the case of a multi-occupied building such as an office or a shopping centre and relates to the common costs incurred in running the building on a daily basis. In some cases, especially industrial property on an industrial estate in single ownership or offices/shops on a business park, there may well be an estate service charge to cover the costs of landscaping, road maintenance if not adopted, maintenance of estate sign boards etc.

The principle of a service charge is that there are common costs in running the property concerned. The landlord will normally set up a management company or arrangement to look after the property on a daily basis. In an office building this will include maintenance of the structure; common area costs, e.g. heating, lighting, decoration of the entrance and stairs; lift maintenance contract; maintenance contract for the heating system; actual heating costs, e.g. gas for a gas central heating system; and staff costs, e.g. commissionaire, security and caretakers etc.

In a shopping centre more of the costs will be associated with the malls and other common areas. Some centre owners try to put all the costs of marketing the centre through the service charge, which can be contentious. The agent should be aware of any such issues.

Heating is usually the main part of the cost in a service charge so an agent needs to be aware of whether these costs are included in the service charge. In some cases the heating may be separately metered to each office or shop. In an older building that has wall-mounted electricity heating that does not require any form of common heating plant the costs will be paid directly by the tenant.

The managing agent or landlord estimates total expenditure for the year, the tenants pay a proportionate part of this cost on account during the year with rent and actual expenditure then balanced against these costs – this is called reconciliation. The tenants must then pay additional costs if more has been spent, or receive a refund if costs have been less than estimated. The estimate and actual costs should not vary significantly. Costs are usually split on a ratio of floor areas but other methods may be used. The existing lease will state the method or the landlord must confirm it in the case of a new lease.

This process should not be a major problem in an established property because estimates will be based on past costs subject to anticipated work during the forthcoming year. Where the agent is marketing a new building without that past history, costs in similar buildings will need to be considered to provide a realistic estimate. An agent should seek the advice of a suitable management surveyor.

The agent should be able to obtain real costs from either the landlord or tenant client. This will be actual costs from previous years or budget on account costs where reconciliation has not yet been completed.

In general a building with air conditioning will have a higher service charge than one with gas central heating as running costs are more. Such buildings are also likely to command a slightly higher rent, as air conditioning is more desirable.

In some cases the landlord may be prepared to agree a capped service charge – or if it's an existing lease that may have been agreed at the start. This means that the service charge cannot rise above a maximum level for an agreed period. The agent needs to be aware of this.

Local rates

Rates are a local tax based on market rental value. It is invariably a significant element of the cost of occupying commercial property. Rating is a significant discipline within the profession; however, an agent needs to understand the amount payable and the process in outline.

All commercial property is given a rateable value (RV). Each individual hereditament has a separate RV and this may be explained as:

> Hereditament – the defined property comprising the rating *unit*. It does not need to be the whole of a property but can be a defined part of a larger unit, for example, in a multi-let office building or a shopping centre there will be many separate hereditaments.

The RV is the market rental value of the property as at the antecedent (last) valuation date. Previous dates have been 2003 and 2008. All property is re-valued by the Valuation Office Agency and these values, the list, come into effect 2 years after the date of valuation, 2005 and 2008 respectively. This has historically been done on a 5-year rolling programme. A re-valuation was due in 2013 to be effective in 2015 but the government delayed this.

The tax payable is calculated by multiplying the RV by the uniform business rate (UBR). This is set by central government and reviewed annually. Note that this is the maximum sum payable and for a variety of reasons the actual sum payable may be less. The tax is collected by the local authority where the property is located.

RV assessments can be obtained from the Valuation Office Agency (VOA) website (see www.voa.gov.uk). Where an agent is instructed to market a property with an existing RV the process is straightforward. If acting for a tenant it should be possible to obtain copies of the accounts received from the local authority to confirm exactly what was paid. Copies should be kept on file.

If there is not an existing RV the property will be assessed when it is occupied so it is not possible to give an exact assessment during marketing. This situation will arise where the property is a new development or major refurbishment or an existing building where, for example, one floor is to be marketed but which was previously occupied in conjunction with other floors in the building.

In the former case an agent could make an estimate from the RVs in other similar buildings in the same location but it must be made clear that it is purely an estimate. Where there are other RVs of similar accommodation in the same building a clearer estimate can be made but again it must be made quite clear to any prospective tenants that you have only made an assessment.

The rates for the 2013–14 tax year are:

Rates payable = RV × UBR
RV = market rental value at April/2008 (w.e.f. April/2010)
UBR for 2013/14
RV up to £17,999 – 46.2p/£
RV £18,000 or over – 47.1p/£

Further reading

RICS (2011) *UK Commercial Real Estate Agency Standards*, RICS, Coventry, section 8.

1.20 Heads of terms

Key terms: lease terms; landlord; lessee/tenant; subject to contract

When a deal has been agreed in principle it is advisable to prepare a clear and full letter of heads of terms to be confirmed between the landlord and tenant/vendor and purchaser. Heads of terms are a summary of the agreement between the parties and are used to instruct lawyers to produce the formal lease or sale deeds. This must always be subject to contract, i.e. not legally binding on either party. It is the solicitor's job to finalise the legally binding contract; however, the preparation of a good heads of terms letter makes that job easier. It should mean that the solicitor can put together a lease/sale contract that more accurately reflects the intentions of the parties.

The following is a suggested format for a letting and must be subject to contract. If the transaction is a sale, lease assignment, sublet or long ground leasehold, the heads of terms should reflect it. Suggested format for a new leasing arrangement is shown below:

Table 1.20.1 Suggested format for a new leasing arrangement

Property	The full address including postcode.
Lessor/landlord	Name and address.
Lessee/tenant	The name and trading/registered address. Make sure the identity of the tenant does not change before leases are to be signed. It is not unknown for an occupier to negotiate a lease in the name of a company which is of good covenant strength, and then to try to put the lease in the name of a company with much less covenant strength within its group but without any main company guarantees. In any event any financial enquiries e.g. references, credit ratings etc. will be not be relevant.
Solicitors	Names and full addresses for both parties.
Description and accommodation	Refer to the marketing details, although they should not be used as part of a binding contract. This should include a clear location and site plan at an appropriate scale (perhaps 1/1250) and if part of a multi-occupied building, a floor plan showing the demised area (that to be included in the lease) with common areas.
Parking	If any is included in the lease then how many? Where are they (include a plan)? And is the cost included within the rent or an extra?
Term	Length of the lease and is there to be a break clause? If so are any penalties attached to it? Is this a tenant only break or will the landlord have the benefit of it as well?
Rent review	How frequent and on what basis? It is not normal to go into detail about all of the rent review machinery as this can be very long and detailed in UK leases. It is normally dealt with by the solicitor; however, there is potential here for cross-referral work for your rent review specialists to advise on the detail of the rent review clauses.
Rent	State the rent payable and frequency of payment – usually quarterly in advance but the retail sector is pushing for monthly rents to ease cash flow.
Rent free/rising rent etc.	State the periods agreed and any conditions attached, e.g. if the tenant is to undertake certain works in lieu of this.
Guarantor	Is there to be one? If so give full identification details and address. Take up references and a credit check etc.
Commencement	An indication of when the parties hope to start the lease gives the solicitor a date to aim towards.
FRI/IRI	State the basis of the tenant's repairing liability.
Service charge	In a multi-occupied building a sensible landlord and managing agent will set up a standard form service charge and all leases would normally be on this basis. If nothing has been set up, liaise with the landlord to agree what is to be done in conjunction with your management colleagues. If it is not your area of expertise refer to someone who will act for the landlord on a consultancy basis. This is a further opportunity for cross-referral work.
Standard of fitting at start of lease	This is important because a lease normally requires a tenant to give the property back in the same condition as it was at the start of the lease. If it is not made clear what condition it was in at commencement of the lease then negotiations at the end become more difficult for all concerned.
Services	State what is available within the building and specifically to the demised areas.
Use	It should be clearly stated what the building can be used for. If planning consent is required for the proposed use it will normally be the tenant's responsibility to secure it. Write a timescale into the deal so that the prospective tenant cannot prolong the process. This should contain a cut-off date if planning is refused, or if the occupier does not proceed in an acceptable time scale. If there are to be any restrictions on use this should be clearly stated. This is common in a shopping centre to retain a good tenant mix.

VAT	If VAT is to be payable it is normally the responsibility of the tenant. This can be a very real cost to a tenant that cannot reclaim it, e.g. a bank.
Repair/refurbishment	Make it clear who is to do what, by when and at whose expense.
Hours of access	This is only an issue in a multi-occupied building. State the position. Provision will normally be needed for access outside normal opening hours in an office, e.g. by use of a swipe card system, key pad entry or through a security firm.
Rating – local taxation	State who is to pay the sum to be taxed (RV) and the rate of tax (UBR).
Security of tenure	If the parties are to opt out the solicitors should be requested to deal with the appropriate correspondence.
Authorised guarantee agreement	If there is to be the requirement for an AGA when the tenant assigns the lease make it clear. State if the assignee is to be subject to any specific tests, e.g. minimum profit levels.
Costs	Make it clear who is to pay whose costs and when.

When the landlord and tenant have agreed the heads of terms send copies to the solicitors to prepare legal documentation.

Further reading

'Code of Leasing Business Premises', available at: www.leasingbusinesspremises.co.uk (accessed 15/11/2013).
RICS (2011) *UK Commercial Real Estate Agency Standards*, RICS, Coventry, section 5.8.
www.isurv.com (accessed 15/11/2013).

1.21 Money laundering

Key terms: money laundering; dirty money; clean money; legislation

The growth in organised crime, terrorism, tax evasion and other illegal activities has resulted in a significant amount of illegal money finding its way into the world's economy. Most countries have taken some steps to make it more difficult for this to happen. It remains, however, a major and growing problem.

The problem of money laundering is very much an international one and needs to be tackled on an international scale. There is a belief in government, both in the UK and especially in the European Parliament, that property is a popular way to launder money. It is impossible to be certain of the amounts of money involved, but it is thought that the sums are substantial each year.

The European Union (EU) has issued various directives to its member states, which have embodied these into appropriate legislation within their own legal systems. These regulations are reviewed regularly, resulting in a further tightening of controls.

The process

Placement

This is the initial introduction/placing of illegally obtained money. In the context of estate agency it would be the buying of a property.

Layering

The process of moving the money around, through a string of legal sources to create a difficult audit trail.

Integration

The ultimate aim is to give the money a sufficiently detailed history and provenance so that it becomes accepted and unquestioned. The 'dirty money' has therefore been converted into clean money.

Definition

Money laundering is the turning of money obtained by criminal or improper means into money with a known and established provenance. It is the conversion of dirty money into clean money. It is the concealment of the identity of illegally obtained money so that it appears to have been obtained from a legitimate source.

The relevant law in the UK is outlined below:

- Criminal Justice Act 1993
- Terrorism Act 2000; Anti-Terrorism, Crime and Security Act 2001
- Money Laundering Regulations 2007 (MLR)
- The Proceeds of Crime Act 2002 (POCA)

What is the aim of the legislation? To reduce money-laundering opportunities by ensuring that business has effective anti-money laundering systems and controls.

POCA came into effect on 24 February 2003 and redefined money laundering and the offences linked to it. The principal offence of money laundering in the Act applies to everyone, even if the business does not fall under the Money Laundering Regulations. The effect of this is that everyone has a responsibility to be vigilant throughout all property transactions. It this responsibility that can often be overlooked – an agent must be vigilant of the actions of all parties to a transaction, including unsuccessful purchasers, not only their client.

There are three principal offences:

1 *Assisting*
An agent can be prosecuted for assisting a money launderer by becoming involved in an arrangement and by knowingly helping a suspect deal to go through. It is crucial therefore that an agent gives due consideration to all instructions received from a client.
Penalty – up to 14 years in prison and/or a fine.

2 *Tipping off*
Providing information to someone that is likely to prejudice an investigation into money laundering, knowing or suspecting that a disclosure to a money laundering reporting officer (MLRO) or to the authorities, has been or might be made.
Penalty – up to 5 years in prison and/or a fine.

3 *Failure to report*
If you fail to report circumstances where you know or suspect, or have reasonable grounds to suspect, that another person is engaged in money laundering you may be prosecuted.
Penalty – up to 5 years in prison and/or a fine.

Money Laundering Regulations 2007

These regulations placed further requirements on the business sector, which includes estate agents. A firm must put in place a system of controls that achieves the following:

1 Set up identification procedures in respect of those wishing to establish a business relationship with you, whether for a one-off transaction or on a continuous basis.
2 Set up record keeping procedures.
3 Set up internal recording procedures including appointing an MLRO. It is this person that has the ultimate responsibility of deciding whether or not to report suspicious activity.
4 Establish procedures of internal control and communication as may be appropriate for the purposes of forestalling and preventing money laundering.
5 Take appropriate measures to ensure that employees are aware of the regulations and other pieces of legislation that govern money laundering. They must be given adequate training in how to recognise and deal with transactions that may be related to money laundering.

Regulation

Estate agency is now part of the regulated sector. In general in the UK it is advised that an estate agent should not take cash from a client. All monies for fees etc. should normally be paid through the banking system.

All firms must have a MLRO. This person is responsible for dealing with all issues relating to potential money laundering identified by the firm. Where an agent considers that the actions of a client or other party they have dealt with are suspicious, these suspicions must be reported to the MLRO who will investigate and decide whether to report the matter or not. The role of the MLRO is an important one.

Reporting suspicions

In the UK this used to be to the National Criminal Intelligence Service (NCIS) and has now moved to the Serious Organised Crime Agency (SOCA). The normal way to make a report is online and this is known as a suspicious active report (SAR). The following are some of the things to be aware of:

- Upfront fees – it is not normal for a potential client to want to pay fees in advance.
- Offers of lots of work – the promise of lots of work especially in a short space of time can be tempting but is not usual. A client will normally want to develop a relationship with an advisor over a period of time. If the offer of such deals seems too good to be real it probably is not!
- Request for over valuation – never do this under any circumstances, always give advice based on the market.
- Foreign/unknown lending institution – there are many overseas lending institutions that are perfectly acceptable, however, some are not. There are certain parts of the world where the financial sector is not regulated in the way it is in the UK. If a client or potential buyer is using funds from an unknown source, check it out.
- Change of money source – there is little good reason for a business to move money around different banks unless it is part of a layering process.
- Purchaser registered abroad – especially in less regulated countries. The UK has seen a lot of inward investment especially in London which has come from reputable sources, so simply because a client or buyer is from outside the UK it does not mean the money is not legitimate but an agent must take extra care under such circumstances.
- Client has very bullish view – if what you hear simply does not make sense it may be that the party concerned just wants to place money quickly and if they over pay but get the money into the system and legitimise it a small loss is not important to them.
- Unrealistic timescales – the property market is relatively slow as due diligence on a property takes time. Where a buyer wants to proceed too fast or without doing the usual checks, there will be a reason and usually not a good one.

- Large sums of cash – there is no reason for anyone to produce cash and even less reason for any agent to accept it. If someone tries to pay in cash: where has it come from, has the taxman ever seen it?
- Behaviour out of context – where a known client has bought property in a certain price range in the past but suddenly wants to buy an investment property of substantially higher value an agent must establish why and where the additional cash has come from.

Further reading

RICS (2011) *UK Commercial Real Estate Agency Standards*, RICS, Coventry, sections 9.6 and 10.5.
RICS (2011) 'Money laundering guidance for members of RICS and other organisations', available at http://www.rics.org/uk/regulation/regulation-uk/guidance-for-regulated-firms/money-laundering (accessed 05/06/14).
www.isurv.com (accessed 03/12/2013).
www.rics.org/uk (accessed 03/12/2013).

1.22 Safety and security in agency

Key terms: health and safety; accompanied viewings; inspection of empty buildings

Health and safety in the workplace is a key factor in employment in the UK. The base legislation is the Health and Safety at Work Act 1974 supplemented by additional regulation, e.g. the Construction (Design and Management) Regulations 2007. The law places responsibilities on both employer and employee together with an expectation of common sense. There are actions an employer must take and there are guidelines the employee, in this case an estate agent, must follow.

The employer

Where surveyors are required, as part of their work, to operate outside the office they must undertake a risk assessment of those activities and put in place appropriate measures to ensure the safety of employees. An employer owes a duty of care to employees. In the estate agency sector this will include: travelling outside the office either driving, on foot or by public transport in whatever form; surveying buildings that may be either empty or occupied (each has its own risks); attending meetings – the main risk being accompanied viewings in buildings on the market.

In addition employers are required to ensure that all employees are provided with appropriate training in safety procedures and of course the surveyor is expected to engage in this process, especially when new to a job and as a trainee estate agent. Where an employee is in any way disabled or has special needs, the employer must take appropriate action to ensure safety. The employee must of course declare any such issues. The Equality Act places responsibility on the employer.

An employer must ensure that all necessary safety equipment is provided or available; for example, personal protective equipment (PPE) including boots, hard hat and high visibility jacket where visits to construction sites are required. There are occasions when specialist precautions need to be taken – for example protective glasses and other clothing when undertaking an inspection of some industrial property where there are hazardous processes. The estate agent must make full use of such equipment.

The risks

The key risks are mentioned above and some practical suggestions will be made for each. As ever it is crucial to remember that prevention is better than cure. Travelling to an inspection or meeting on foot or by public transport should not normally create any undue issues; however, all urban conurbations have areas where it may be sensible not to walk alone and especially not in a smart suit. Check out the location of any properties you need to visit to see whether there are any problems and act accordingly. If driving, do so legally; especially, don't text, and only use your mobile phone with a proper hands-free facility. If you use your own car be certain it is insured for use in your employment as not all policies cover this.

When inspecting an empty building there are many potential hazards. Be certain the building is safe to enter before doing so. If you are alone, ensure you are not followed into the building. If there is rubbish lying around watch out for nails you could stand on; pits or holes in the floor covered by rubbish; wooden flooring that might be rotten and unsafe; bird droppings that can carry disease; asbestos that may have been disturbed; exposed bare electricity cables etc.

Never use a lift in an empty building even if you think it works. If you do and it stops part way between floors such that you cannot get out, there may be no means of calling for assistance as any emergency facility will probably have been discontinued. Mobile phones will not always work in a lift shaft. You could be in for a long and uncomfortable wait for rescue – then there's the embarrassment factor afterwards!

Take care entering a room where the door could shut behind you but cannot be opened from the inside. This is most likely where the building has a basement. These are just a few things you may come across but there are many others and you must be alert when doing an inspection.

Wherever possible it is best to do an accompanied viewing of buildings that are on the market and it is often a job given to trainee surveyors. If the person is known to, and trusted by the agent it is unlikely there will be a problem. Where those wishing to view a property are unknown there are some sensible precautions that should be taken. These are outlined in the general guidance below. Very occasionally a request for a viewing may come from a known party that can be a problem, in which case the employer should have a specific course of action in place.

Occasionally and without warning the agent may find themselves in an awkward position with someone who is simply unpleasant or tries to commit you to matters that are inappropriate, i.e. make the agent agree to possible terms of a deal by the use of pressure or intimidation etc. If this happens agree to nothing and consider calling the office for assistance or end the viewing. It is unlikely such a party will ultimately do a deal and in any event they would probably not make a good tenant.

General security guidelines

The RICS and NAEA (National Association of Estate Agents) have agreed the following general guidance when an estate agent is asked by a client to undertake accompanied viewings of property being marketed:

- record the name, address and phone number of the potential buyer or tenant in a book or on a system that is available to all staff;
- redial the number to check the authenticity of the person before leaving for the appointment;
- obtain as much information as possible from the buyer or tenant (e.g. employment details, car registration), and ask whether he or she has property particulars so you can check those against your mailing list;
- whenever possible, arrange to meet the buyer or tenant at your office and not at the property;
- do not travel to the property in the same vehicle as the potential buyer or tenant when you have no prior knowledge of that person, but if you need to travel in the same car you should take someone else with you;
- enter the property after the potential buyer or tenant, and keep yourself between the exit and that person so if you do have to leave in a hurry, you have as few obstacles as possible in your way;
- advise your colleagues of your anticipated time of return to the office;
- implement a call-back system so that you can report back to the office soon after the completion of the viewing;
- make sure other members of staff are aware of a distress code, so that you can report to the office by phone and alert help without further compromising your safety;
- never give out your home phone number;

- take a second person with you if there is the slightest feeling of risk;
- carry a personal alarm and a mobile phone, and pre-programme it with an emergency number;
- if in doubt, ask buyers for additional means of identification, e.g. driver's licence.

(from RICS, 2011: section 8.7.1)

Further reading

RICS (2011) *UK Commercial Real Estate Agency Standards*, RICS, Coventry, section 8.

2 Building surveying

Stuart Eve, Minnie Fraser and Cara Hatcher

2.1 Building surveying in an estate management context

Key terms: building surveying; common defects; defects diagnosis; construction components

In terms of building surveying in the estate management context, there is a need for a surveyor to understand the fundamentals of building construction, defect analysis and remedial advice, costing and effect on value.

A building survey is a structural survey of the property including a description of defects and potential problems that may be hidden. It gives a recommendation of remedial action and outlines the issue with a property. A building survey does not include a market valuation of the property (RICS, 2013).

Common defects

There are several inherent or common defects noted in certain types of property built in particular eras:

- *Georgian properties*
  - structural movement due to insufficient or no foundations
  - timber issues – dry rot, wet rot, woodworm
  - maintenance costs high due to age of fabric
- *Victorian/Edwardian properties*
  - structural movement due to insufficient or no foundations
  - timber issues – dry rot, wet rot, woodworm
  - nail rot to roof covering
  - failure of roof covering (usually slate if original)
  - rising and penetrating damp due to insufficient or no damp-proof course
- *1920s/1930s properties*
  - timber issues – dry rot, wet rot, woodworm
  - nail rot to roof covering
  - failure of roof covering – terracotta tile or slate
  - rising and penetrating damp due to insufficient or no damp-proof course
  - electrical failure due to age of the system
- *1950s properties*
  - structural movement due to land built on – post-war influx of development
  - non-traditional housing – some not suitable for mortgage due to severe inherent defects
  - nail rot to roof covering
  - failure of roof covering – terracotta tile, slate and asbestos

   – rising and penetrating damp due to insufficient or no damp-proof course
   – electrical failure due to age of the system

- *1960s/1970s properties*
   – structural movement due to land built on – post-war influx of development
   – failure of roof covering due to lack of lateral restraints to support roof
   – failure of windows due to being made of metal which corrodes over time
   – use of asbestos

- *1980s/1990s properties*
   – fewer issues as more modern forms of construction but dated electrical systems
   – failure of double-glazed units due to age
   – some asbestos materials still commonly used in doors and ceilings for example

- *Modern properties*
   – under 10 years old covered by insurance documentation or an NHBC certificate
   – need to check property conforms to the relevant building regulations.

Construction components and types

- *Foundations*
   – none
   – ash or stone chippings/rubble
   – concrete strip or raft
   – brick or stone footings.

These should not be visible unless there is a problem – be aware of underpinning, whereby the original foundations have failed and remedial work needs to be undertaken to rectify this. In serious cases, this can blight (severely reduce) the value of the property.

- *Walls*
   – solid stone or solid brick – pre-1910 properties
   – cavity – brick and block or stone and block – post-1920s properties
   – timber frame – timber framing covered, externally, by blockwork, brickwork, render, pebble dash etc.

- *Floors*
   – ground floors – solid of concrete or stone construction or suspended timber
   – first floors – suspended timber or chipboard
   – blocks of flats – can be suspended or concrete to all floors

- *Roofs*
   – flat – lead or asphalt covered
   – pitched (and hipped) – slated, tiles, thatch, stone, asbestos

- *Services*
   – water – mains or private
   – gas – mains or LPG
   – electricity – mains and infrequently generator
   – rainage – mains, septic tank, cesspool or private sewerage treatment.

Defect costing

It is difficult to categorically provide costing for works; however, it is a surveyor's role to estimate the costs and, importantly, the affect upon value.

When dealing with common defects, it is easier to establish costs but in more severe cases, the surveyor can withhold their valuation figure pending detailed costing for major works such as underpinning, structural movement or repairs to a property in need of total re-modernisation.

In a residential mortgage valuation report, the surveyor has the option to highlight significant repair work required and will suggest a retention is adopted for the lender. These commonly include:

- electrical repairs – dependent on various costing provided by electrical contractors;
- damp and timber treatment – anywhere in the region of £1000 and in excess of £10,000 dependent on the severity of the issue and the source of dampness;
- structural movement.

Further reading

Hollis, M. (2005) *Surveying Buildings*, 5th edn, RICS Books, Coventry.

RICS (2013) 'Guidance note: building surveyor services', available at: www.rics.org/uk/knowledge/professional-guidance/guidance-notes/building-surveyor-services/ (accessed 28/10/2013).

2.2 Building pathology

Key terms: building pathology; defects; intrusive; investigations; diagnosis; survey; damp; condensation; cracking; distortion; boroscope; thermal imaging; insulation; remedy; pathological

The definition of 'pathology' is '*the science of the causes and effects of diseases*' (Oxford Dictionaries Online, 2013). Applying this term to buildings appears incongruous; however, an analogy can be drawn between the medical and surveying professions.

The fundamental process of assessing a particular defect within a building, whether it be water ingress or distress caused by subsidence is exactly the same as medical pathology, hence the term 'building pathology'. If we swap the word 'diseases' with 'defects' in the quotation above this then defines much of the work undertaken by professionals in order to determine the cause and effect of building defects.

Building pathology involves the careful examination and analysis of surface and sometimes subsurface *symptoms* in order to arrive at the correct *diagnosis*, synonymous with the medical profession. Sometimes a defect can be successfully diagnosed merely by undertaking a visual and non-intrusive inspection. However, in many cases further investigations, involving technology or, more drastically, opening up the structure are necessary to better understand the nature of defects.

The first stage in building pathology would usually be an instruction from a building owner or occupier to undertake a survey. This could either involve a survey prior to purchasing a property or inspecting a specific defect and would usually be undertaken by a building surveyor or structural engineer (in the case of structural movement), although architects and contractors may also undertake this work at varying levels.

Defects could be any or all of the following:

- leaking roof;
- damp (rising or penetrating);
- condensation;
- cracking to walls;
- distortion to floors.

Where the cause of the defect cannot be identified using a visual or non-intrusive method of survey, further investigations would be required. These further investigations might include those shown in Table 2.2.1.

Table 2.2.1 is by no means an exhaustive list of methods available but it gives a broad understanding of the nature of typical defects and further investigations.

A case study illustrating these further investigations is provided below. The case study involves a listed warehouse that had been converted into flats. The external walls had been insulated internally using insulated plasterboard fixed to a metal frame against the internal face of the walls. Dampness had been noted around windows in a number of flats and a pathological approach was required. Tests were undertaken to identify condensation risk. Air temperature and relative humidity were recorded and used to calculate the 'dew point' (temperature at which water vapour will condense on a cold surface). A damp meter such as the one shown in Figure 2.2.1 was used to undertake this analysis.

Table 2.2.1 Defects and further investigations

Defect	*Further investigations*
Leaking roof	• Open up roof structure, either internally or externally. • Thermal imaging (will identify changes in temperature due to moisture saturation) • Deep wall probes to test dampness within structure
Damp	• Detailed moisture testing using samples from within wall structure to create a moisture profile • Deep wall probes • Salts testing • Thermal imaging • Open up structure • Test plumbing (hot/cold water and central heating) for leaks
Condensation	• Monitor temperature, humidity and dew point over a period of time using data-loggers • Further understand use of property (washing/drying clothes, heating, ventilation) • Thermal imaging
Cracks to walls	• Detailed level survey (horizontal deformation) • Distortion survey (vertical deformation) • Detailed monitoring of movement over an extended period • Trial pits to understand soil conditions • Drill holes to use boroscope (inspect within cavity) • Open up structure
Distortion to floors	• Detailed level survey • Open up and access floor voids for detailed inspection • Use of microdrill (drills into damp timber to assess extent of rot)

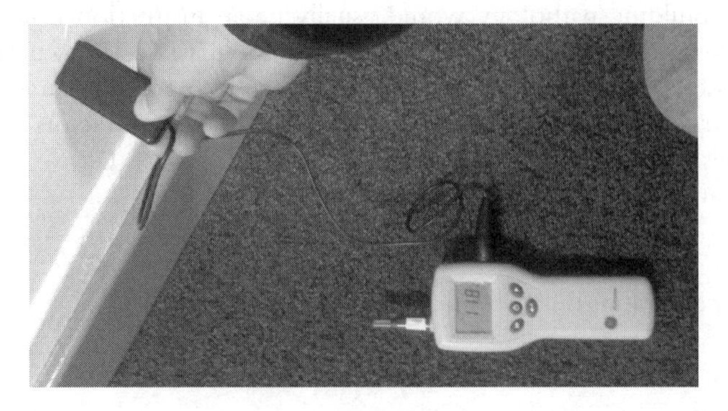

Figure 2.2.1 Damp meter with condensation analysis capability.

There was a relatively high risk of condensation around window openings (where slight mould growth and spoiled finishes were present) and thermal imaging was employed for further analysis (Figures 2.2.2 and 2.2.3). The thermal imaging indicated cold spots to the bottom corners of window openings, illustrated by blue shading in Figure 2.2.3.

Further intrusive investigations were undertaken to understand the nature of the construction as there were no drawings available. This included first using a boroscope (flexible hose with camera), which proved inconclusive, followed by opening up of the structure. It is imperative to understand the nature of the building structure and fabric in order to understand what the defect might be and specify suitable repairs.

The wall was opened up to reveal that although the main walls were insulated (Figure 2.2.4), there was no insulation provided to window jambs (Figure 2.2.5). Plasterboard was applied directly against the solid brick jamb meaning 'cold bridging' could occur. This would result in the wall at this point being much colder than elsewhere, as illustrated in the thermal image (Figure 2.2.3). The temperature at this location will fall below the dew point as a result and condensation will occur.

A pathological approach led to an accurate diagnosis in this case. A lack of thermal insulation around windows meant that cold bridging resulted in condensation in specific locations. A remedy (or cure) could then be confidently specified with full evidence available (Figures 2.2.6 and 2.2.7).

Figure 2.2.2 Dampness to window jamb.

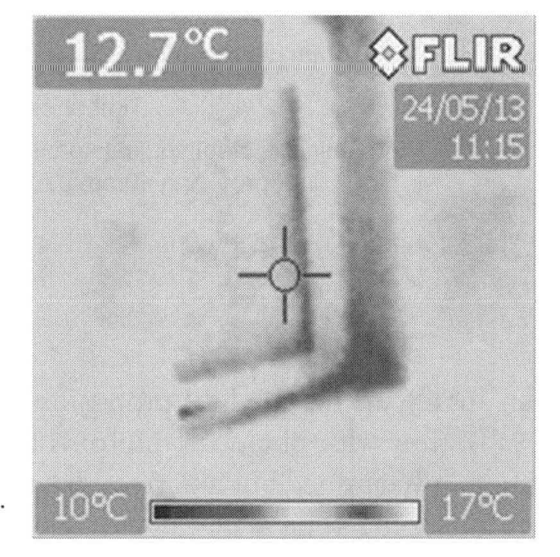

Figure 2.2.3 Thermal image indicating cold spots around windows.

Figure 2.2.4 Wall construction revealed.

Figure 2.2.5 No insulation to jamb.

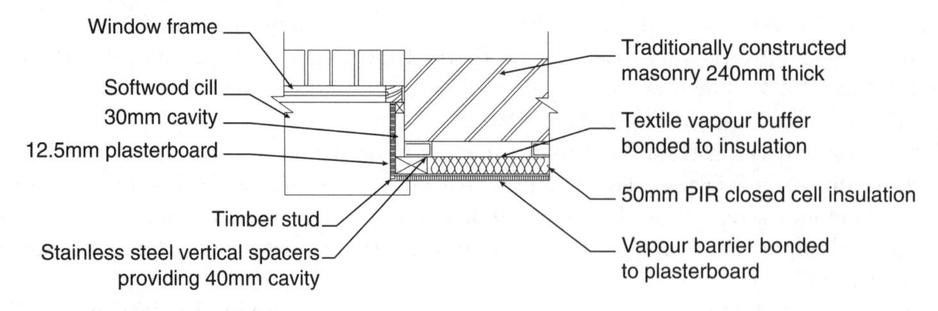

Figure 2.2.6 Window detail as existing.
© Alan Gardner Associates.

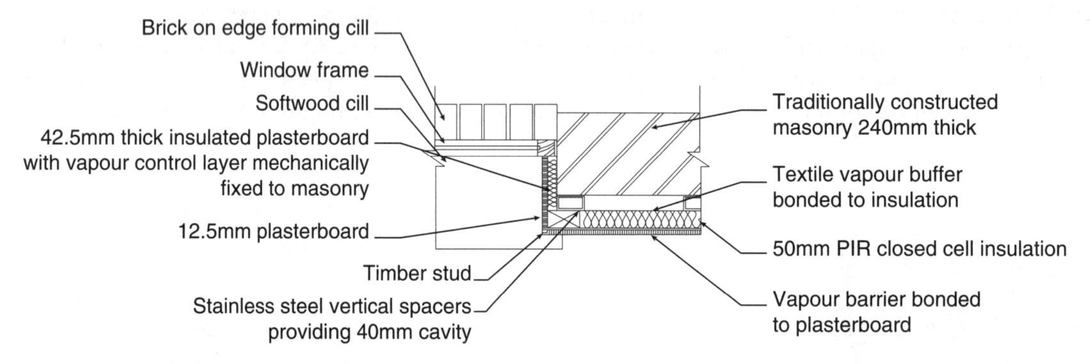

Figure 2.2.7 Proposed remedial works.
© Alan Gardner Associates.

The analogy with the medical profession is an accurate one. The patient (or building) was showing signs of distress with obvious symptoms (damaged finishes and mould growth) visible. The building was carefully examined but a correct diagnosis could not be arrived at without further tests being carried, in much the same way a doctor would recommend X-rays or exploratory surgery. A diagnosis and prognosis was given to the building owner and a choice was made in terms of a remedy or cure. The plasterboard would be removed to all affected window jambs and insulation would be introduced, increasing the surface temperature and therefore eliminating the risk of condensation.

Further reading

Hollis, M. (2005) *Surveying Buildings*, RICS Books, London.

Marshall, D., Worthing, D., Heath, R. and Dann, N. (2014) *Understanding Housing Defects*, 4th edn, Routledge, Abingdon.

Oxford Dictionaries Online (2013) Definition of Pathology in English, available at: www.oxforddictionaries.com/definition/english/pathology (accessed 27/11/2013).

Oxley, R. (2003) *Survey and Repair of Traditional Buildings*, Donhead Publishing, Sparkford.

2.3 Building surveys

Key terms: valuation surveys; acquisition surveys; maintenance surveys; diagnostic surveys; partial surveys; schedules of condition; dilapidations survey

In order to successfully carry out a building survey, the building surveyor requires a good knowledge of building construction, building materials and building pathology. There are a number of different types of survey that are carried out by surveyors for different purposes. There are also a large number of different types of building that may be the subject of the survey as well as different types of clients with their own requirements.

Valuation surveys

Open market valuations (OMV) and mortgage valuations are usually carried out by general practice surveyors. The survey involves an inspection of the building, an assessment of the condition, and costings for necessary repairs. After this, comparisons are made with similar buildings in the area that have recently been sold and their selling prices.

Mortgage valuations are usually instructed by the mortgage lender, but paid for by the mortgage applicant. Often this leads to the usually erroneous opinion that the report will provide information that is of use to the applicant and may account for the fact that approximately 80 per cent of house buyers in the UK do not procure a full survey. The mortgage valuation carried out for the lender does not usually include a survey of the property; a cursory inspection maybe, but not a survey. The purpose is to inform the lender whether or not the property is worth the money they are lending, so that they can be sure that they can recoup their money if it becomes necessary to repossess the property and sell it. Valuations are often carried out by owners of portfolios of property as part of their asset management. These do not usually require detailed inspection or survey of the properties unless there is an additional purpose to the exercise such as maintenance planning.

The type of valuation that is more often carried out by building surveyors is the replacement valuation for insurance purposes, often called a fire insurance valuation (FIV). This involves inspecting the building to discover its construction so that an estimate can be made of how much it would cost to rebuild it if it was so badly damaged that it needed to be demolished and rebuilt. For buildings of standard construction, this is normally done on a square metre basis using up-to-date building price information from pricing books and the Building Cost Information Service (BCIS).

Acquisition Surveys

This type of survey is carried out where a property is going to be acquired either for occupation or for investment purposes. Usually this type of survey is carried out on behalf of the prospective buyer of the property. In the UK sellers of houses have been required to provide information including a home condition report, however, this requirement was suspended in 2010 and now sellers are only required to provide an energy performance certificate (EPC).

There are three types of domestic survey marketed by the RICS; these are the RICS Condition Report, the RICS Homebuyer Report and the building survey. Many house purchasers opt for the Homebuyer Report otherwise known as the Homebuyer's Survey and Valuation (HSV). This is carried out to a standard format set out by the RICS, but is only recommended for properties that are in reasonably good condition, and not for older or dilapidated properties. This type of survey is routinely carried out by general practice surveyors, although building surveyors may occasionally do the inspection and appraisal of the building, property valuation is not usually part of the building surveyor's role. The RICS condition report is a simpler report that does not include a market or insurance reinstatement valuation for the property.

The building survey (formerly referred to as the structural survey or full structural survey) is most typically carried out by a building surveyor. This involves a much more detailed inspection of the building and the report includes advice and recommendations for repairs and future maintenance as well as costings for any pressing repairs. The RICS particularly recommend the building survey for:

- listed buildings;
- older properties;
- buildings constructed in an unusual way, however, old they are;
- properties that the client intends to renovate or alter in any way;
- properties that have had extensive alterations.

(www.rics.org/usefulguides)

The RICS information for prospective clients regarding building surveys states that a building survey can include details of:

- major and minor defects and what they could mean;
- the possible cost of repairs;
- results of damp testing on walls;
- damage to timbers including woodworm and rot;
- the condition of damp-proofing, insulation and drainage (though drains aren't tested);
- technical information on the construction of the property and the materials used;
- the location;
- recommendations for any further special inspections.

(www.rics.org/usefulguides)

Maintenance surveys

Maintenance survey reports will usually contain specific recommendations for repair of elements where defects or wear and tear has been identified and general recommendations for preventative maintenance such as cyclical redecoration of external joinery. The report will cost each item with a projection of when the work will be required and this will allow the client to plan their maintenance budget.

Diagnostic surveys

Where there is a specific problem or defect, the client may instruct a building surveyor to inspect and report on the problem, recommending the necessary remedial action. This type of survey may be associated with a claim on the client's buildings insurance and, occasionally, the report may be commissioned by the insurance company directly. It is often necessary for investigation, testing and/or ongoing monitoring of the problem to be carried out before a diagnosis can be arrived at. The report will describe the investigations carried out and interpret the results before making the appropriate recommendation, which is almost always accompanied by costings for the recommended work. Clearly, this type of survey will only involve surveying the parts of the building that are the subject of the specific defect or problem, unless the client has specifically asked for more.

Partial surveys

Like the diagnostic survey, clients may occasionally want a report that only covers certain parts or aspects of a building. For example, the client may want to acquire a building with a view to refurbishing it. In this case, there will be no interest in the decorations, finishes, fixtures and fittings and the client will commission a 'structure only' survey. Another example may be the client who has had continual problems with the roofs on their estate and commissions surveys of all the roofs to discover the extent of the problems.

Corporate clients whose buildings may be workplaces or buildings that are accessed by the general public need to ensure that their buildings are not going to cause them legal problems. Every time a new set of legislation is introduced, building owners and managers may find they have new duties or new requirements for their buildings. New legislation often results in a flurry of related work for building surveyors. For example, the introduction of the Disability Discrimination Act resulted in building surveyors being instructed to carry out access audits of commercial premises and workplaces. Examples include:

- access audits;
- appraisal of fire precautions and means of escape;
- facilities and spatial layouts of care/nursing homes;
- health and safety risk assessments in workplaces.

Some surveyors may be asked to provide energy audits of buildings – it should be noted that in the UK, an energy assessor must have the relevant qualification to become accredited in order to produce EPCs.

Schedules of condition

A schedule of condition is quite simply a record of the condition of a building at a particular point in time. The schedule often contains written descriptions of the construction and the condition, but nowadays it will always contain photographs. Although the preparation of the schedule does not require the same level of professional knowledge and skill that a diagnostic or building survey does, many clients will ask a building surveyor to do, or arrange this work as the schedule is usually a document of great importance. Any missing details in the schedule could be costly to the client.

Schedules of condition are usually prepared at the commencement of leases or before works are carried out to adjoining buildings. In fact any client carrying out an operation that could possibly cause vibration or damage to nearby buildings should be advised to commission schedules of condition. This will protect them from claims from building owners that every building defect has been caused by their operation.

Dilapidations surveys

Where a commercial property has a FRI lease, the tenant is usually liable to put the building into a reasonable state of repair at the end, or termination of the lease or to pay the cost of the repairs where this does not exceed the diminution in the value of the landlord's reversionary interest. It is necessary to be able to interpret and appraise the duties of both landlord and tenant according to the lease covenants and other circumstances.

The schedule of dilapidations is normally prepared at the end of the lease term or when it has been decided to terminate it. This is called a terminal schedule of dilapidations. Occasionally, the landlord may decide to issue an interim schedule of dilapidations; this may be prompted by the tenant allowing the property to fall into disrepair. A building surveyor is normally instructed to prepare the schedule by the landlord. Before carrying out the survey, the surveyor will need to assess the repairing liability in the lease covenants and examine the schedule of condition if there is one. The inspection will be then carried out to discover and evaluate the wants of repair, which will be fully costed.

The tenant will usually employ another Building Surveyor to act and negotiate on their behalf. Having also assessed the repairing liability in the lease covenants and examined the schedule of condition appended to the lease if there is one, the tenant's surveyor will then take the schedules with them to the property and compare what they find.

Further reading

Glover, P. (2006) *Building Surveys*, 6th edn, Butterworth-Heinemann, Oxford.
Hollis, M. (2005) *Surveying Buildings*, 5th edn, RICS Books, Coventry.
RICS (2004) 'Building Surveys of Residential Property – RICS Guidance note', RICS Books, Coventry.
RICS (2005) *Building Surveys and Inspections of Commercial and Industrial Property* – RICS Guidance Note, 3rd edn, RICS Books, Coventry.

2.4 Dampness in buildings

Key terms: rising dampness; penetrating dampness; condensation; leaks

The vast majority of defects in buildings are caused by moisture or movement, or a combination of both. Water is all around us; it is a necessity of life for all organisms, including insects, moulds and rots that can cause damage to the fabric of buildings. All porous materials will absorb the moisture vapour in the air, depending on their porosity and the relative humidity of the air. The effects of water on building materials and entire buildings are of great importance and the building surveyor needs to have a sound understanding of them. This importance is borne out by the large proportion of successful negligence claims against surveyors relating to defects caused directly or indirectly by moisture.

Dampness does tend to affect older buildings due to the technology of their construction – solid walls, lack of thermal insulation and no damp-proof course (DPC). However, problems with moisture do occasionally affect newer buildings, poor workmanship and the strange things that occupiers do being two possible problems, but the increasing requirement for airtightness to achieve energy efficiency also has an influence. It is also very important to understand that an old building with solid walls and no DPC will not necessarily be damp, many are quite acceptably dry. However, many do suffer some dampness and occasionally people's attempts at bringing old buildings up to date will cause the very problems they are trying to avoid.

The building surveyor needs to know where to look, how to investigate and how to assess the severity of the problem before deciding on the best way of dealing with the problem. The first thing to be discovered is where moisture is coming from; there are five possibilities:

- groundwater rising up through porous materials – rising dampness;
- rain water (or snow) penetrating through the external envelope;
- leaking services or spillages;
- moisture from the air – condensation;
- flooding.

Water in buildings can cause a number of different problems including:

- loss of strength in timber, resulting in warping, twisting and general deflection under load;
- susceptibility to insect and fungal attack;
- spalling and frost damage to masonry – spalling is where the face of the material, such as brick or stonework breaks off and is often caused due to the formation of ice crystals in a porous material as freezing water expands very forcefully;
- corrosion of ferrous metal components;
- staining of finishes;
- mould growth;
- moisture movement.

Rising dampness is classified as water from the ground rising through capillary action through the building fabric. It often affects older buildings that do not have DPCs; however, the lack of a DPC does not necessarily mean that there will be rising dampness. Also, buildings with DPCs can suffer from rising dampness if the DPC is bridged or otherwise compromised. Unfortunately, there has been a tendency in the past for surveyors to diagnose rising dampness in old buildings at the first sign of wetness at low level in walls. Many such surveyors also passed the buck by recommending further inspection by a 'specialist'. Often, clients (usually home owners) would misinterpret this recommendation and go to the first 'damp-proofing specialist' in the telephone book – a company whose main form of income is installing remedial DPCs. It is not surprising therefore that unnecessary damp-proofing work is often carried out, sometimes causing damage to the building and almost always costing the client a large sum of money! This situation should never arise where a competent surveyor has inspected, investigated and made a recommendation.

It is important for a building surveyor to understand the construction of the building and the materials it has been constructed with. Rising dampness becomes a problem when it is severe enough to cause noticeable discolouration and wetness that can support timber decay. There are a number of influencing factors including the materials, the construction technology, the climate and the internal environment. For this reason, problems may manifest themselves in different ways in different places according to the climate and local construction methods and materials.

As an example, the process of a typical case of rising dampness in the UK is illustrated in Figure 2.4.1. An old solid brickwork wall with no DPC is built on a site with a variable water table that rises quite high in winter, above the base of the brickwork foundations. The groundwater soaks into the porous brickwork and wicks up into the wall. The groundwater contains salts in solution which are deposited on the face of the wall as the water evaporates. These salts tend to be hygroscopic and attract moisture from the air, which exacerbates the dampness on the surface of the wall. The level of dampness will be worse in winter when the water table rises.

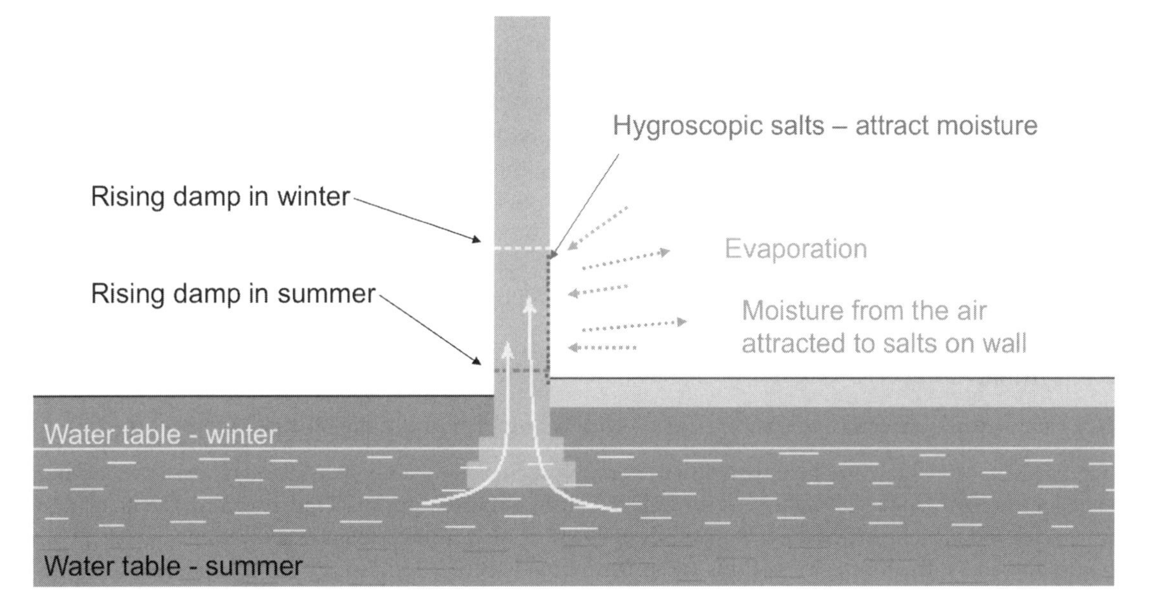

Figure 2.4.1 The process of rising dampness.

Source: Minnie Fraser 2007.

Although the above example is described as typical, it is as well to understand that there are very many different circumstances in which rising dampness may occur. Some water may rise into the walls or through a solid floor of an old building without causing a problem or noticeable dampness, particularly if the building is well ventilated and well heated. Older buildings tended to be very well ventilated, with air gaps around window and door frames and often large open fireplaces which allowed plenty of air movement facilitating evaporation of any water. The fact that the materials of construction are porous also means that they are 'breathable', preventing any water from being trapped in the construction. Problems often arise, however, when older buildings are adapted or modernised. The old draughty sash windows are replaced with uPVC replacements with neoprene seals preventing any air leakage and fireplaces are blocked up. Large houses are divided up into flats with a larger number of people occupying the space – each person breathing out water vapour and raising the relative humidity of the air in each room. These changes would reduce the rate of evaporation of water entering from the ground and a problem may become evident where there was none previously.

Penetrating dampness is classified as water travelling through the external envelope of the building. Moisture can enter the building in a number of ways. It is more common in solid walls as part of the purpose of the cavity in a cavity wall is to prevent penetrating dampness. Solid walls may be of insufficient thickness, particularly if they are in an exposed location. The following problems can contribute to penetrating dampness in solid walls:

- defective rendering;
- eroded pointing;
- defects or cracks in sills;
- defective rainwater disposal;
- unprotected joints around windows, doors, air bricks etc.

Defective gutters and downpipes are a very common cause of penetrating dampness. There may not even be an obvious defect, gutters may fall the wrong way, be blocked with debris or vegetation or simply be of insufficient size.

Other forms of dampness may be caused by leaking services, spillages or condensation. Common sources of dampness are leaking radiator valves and joints in supply pipes; often where they come up through sinks/ basins etc. and join the taps. Spillages can also cause dampness if they happen often enough; for example, frequent floor washing can cause dampness in adjacent walls. Condensation can be a serious problem that can cause black mould growth and damage to decorative finishes or in more severe cases, timber decay and failure of timber structures.

Further reading

Burkinshaw, R. and Parrett, M. (2004) *Diagnosing Damp*, RICS Books, Coventry.

2.5 Timber defects

Key terms: natural defects; conversion defects; weathering; dampness; fungal attack; insect attack

Timber is a very widely used construction material; it is chosen for its easy availability, ease of working, excellent strength-to-weight ratio and nowadays its low carbon footprint. Timber can be categorised by its species and which part of the tree trunk it is taken from. The two main classifications of species are hardwood and softwood. Hardwood is taken from broad-leaved deciduous trees such as oak, ash, elm, sycamore, chestnut or teak. Softwood is taken from coniferous trees such as pines, firs and spruces.

As timber is such an excellent construction material, it has been used very extensively in buildings since the earliest days of building construction. It is used for both structural and non structural elements. For example, structural:

- structural frame;
- roof structure;
- floors;
- staircases;
- lintels and beams;
- load-bearing stud walls.

Non-structural:

- windows and doors;
- frames and sills;
- stud partitions;
- fascias;
- skirtings, architraves, dado rails and other decorative features;
- balustrades;
- weatherboarding/cladding/roof shingles;
- fitted furniture.

The main defects affecting timber in buildings are as follows:

- Natural defects – these are defects that occur as a result of the natural variations in the timber such as knots and wide growth rings that adversely affect the strength of the timber.
- Conversion defects – defects that occur when the tree is converted into usable timber, i.e. when it is cut and dried. Timber can distort by bending, twisting and bowing. It can split, crack or honeycomb, splits radiating from the centre are called star shakes. Sections cut from the edge of the tree can be misshapen or have bark still attached (wane).
- Weathering – weathering is an imprecise term describing the effects on timber of being exposed to the elements. The combined effect of sunlight, water and wind result in loss of colour (it goes grey), surface roughening, and cracking and splitting.
- Dampness – dampness causes swelling and loss of strength and can result in deflection, twisting and warping.
- Fungal attack – including dry rot and wet rot.
- Insect attack – including common furniture beetle ('woodworm'), death watch beetle, powder post beetle, wood boring weevil and house longhorn beetle.

Fungal and insect attack are much more likely to occur in damp timber, so buildings that have problems with dampness are much more likely to suffer from timber decay. The effects and consequences of timber decay can be serious and in severe cases can lead to structural failure. Remedial work tends to be disruptive and expensive and therefore it is vital that the surveyor is vigilant during inspections and knows the circumstances in which these sort of defects are most likely to occur.

Further reading

Bravery, A. F., Berry, R. W., Carey, J. K. and Cooper, D. E. (2003) *Recognising Wood Rot and Insect Damage in Buildings*, BRE, Watford.

2.6 Movement in buildings

Key terms: structural movement; subsidence; settlement; thermal and moisture movement; steel corrosion; sulphate attack

Foundation movement

Structural movement is one of the greatest worries of the building owner or prospective purchaser. Assessing and diagnosing structural problems on behalf of building owners or insurance companies is an important part of the work of the building surveyor.

The investigation, evaluation, diagnosis and prognosis of cracking in buildings takes great knowledge and skill. Correct diagnosis and recommendations can save clients a lot of money and heartache – but often surveyors will recommend unnecessary works to be on the safe side. It should be every surveyor's aim to find the extent of a building problem and recommend appropriate action to the client in order to give good quality and professional service. In the case of structural movement, it is often the case that full investigation to find the cause of the movement will require time consuming investigation and monitoring, which some clients do not like. Increasingly, clients, particularly insurance companies, are dispensing with long-term monitoring of structural movement in favour of quick repair and completion. Surveyors should always advise clients of the possible risks attached to the course of action they choose to follow.

Where buildings move due to movement of their foundations, this is usually called settlement or subsidence. Settlement is downward movement of the foundations due to the compression or shrinkage of the soil beneath the building, or consolidation of the ground due to closing up of small voids. Settlement occurs where the weight of the building exceeds the bearing capacity of the soil, so it is as a direct result of the weight of the building acting on the soil. Subsidence is the downward movement of the foundations due to erosion or shear failure of the soil or collapse of large voids within the ground. Subsidence does not happen purely as a result of the weight of the building acting on the soil, there is always some other factor as well. Movement due to clay shrinkage is usually classed as subsidence.

Generally, subsidence results in more serious defects in buildings than settlement. Although occasionally there may be some doubt over whether movement should be classed as settlement or subsidence, surveyors should be aware of the importance of the distinction between the two. Subsidence is usually classed as an 'insured peril' whereas settlement is generally not covered by building insurance policies.

The weight of a newly constructed building will cause the soil beneath to compress and consolidate, the speed and magnitude of this movement will depend on the soil type as we have already seen, but it is important to note that all new buildings will settle slightly. This small amount of settlement after construction should not cause any problems or even be noticeable in most buildings; however, there can be problems if one part of a building settles more than the rest – this is called differential settlement. Where differential settlement occurs immediately after construction this is usually as a result of a soft spot or inadequately designed foundations for the ground conditions.

Where there is downward movement of the foundations to part of a building, stresses will be set up in the walls. Masonry walls are designed to be in compression, in this state the material is very strong, however it is weak in tension and movement will cause cracking which will follow the path of least resistance, i.e. the weakest part that is under stress. In normal masonry construction, this means that cracking will pass through mortar joints, generally between openings as openings are natural weak points in a wall. The patterns of stresses and cracks will be different in buildings of different construction, such as timber frame, panellised or modular buildings that have been manufactured off site. In these buildings, usually the weak points will be at the junctions between framed panels. The patterns of cracking may also depend on the type and construction of the foundations. The other major symptom of settlement and subsidence is tilting of walls. Where a wall is no longer vertical, it is described as being 'out of plumb'.

A large number of subsidence problems linked to shrinkable clay are also linked to trees in close proximity to buildings. Although in normal circumstances the layer of volume change to the soil does not exceed 1.5 m, where there is a tree taking water from the soil, the depth of desiccation can extend beyond 6 m below ground level. This can have a very significant effect on a building that is close by. The amount of water required by a tree tends to increase as the tree grows and gradually stabilises as it reaches maturity. The effect of trees is felt more in the summer when trees require more water and there tends to be less water available; some differential settlement may be experienced to a building whereby the cracks open up in the summer and close again in the winter. These problems become more severe if there is a very dry summer when there is no rain so the desiccation of the soil due to the tree increases. There is always a great increase in subsidence claims when there is a dry summer.

Thermal and moisture movement

As with other forms of movement in buildings, thermal and moisture movement tend to cause cracking to walls where the cracking will follow the path of least resistance and cracks are generally perpendicular to the stresses that cause them. In both cases, the movements are relatively small and tend not to cause serious damage although there are circumstances where serious damage can happen.

All materials will expand as temperature rises and contract as temperature falls and the magnitude of this dimensional change is dependent on their coefficient of thermal expansion. Concrete has a higher coefficient than brickwork; aluminium, lead and uPVC have very high coefficients and timber has a relatively low coefficient (although slightly higher across the grain than along the grain). Porous materials will expand as they take up moisture; this does not have to be as a result of coming into contact with liquid water, changes in humidity can also cause movement. As a general rule, the more porous the material, the greater the movement, so dense heartwood will not move as much as the lighter sapwood of the same timber. Timber experiences considerably more moisture movement across the grain than along the grain, which can cause problems, particularly in external joinery. Masonry is porous and will expand and contract with changes in humidity as well as with changes in temperature.

A well-designed building should not suffer any problems with thermal or moisture movement because these effects are now well known and the incorporation of movement joints is standard practice. However, older buildings or buildings that have not been well designed may suffer from insufficient allowance for these movements. Long masonry walls without movement joints will develop cracking that is generally vertical stepping up through the mortar joints between openings. The cracks are narrow, although there can be some ratcheting over many years of cyclical opening and closing due to changes in temperature.

Steel corrosion

Steel is a material that has been used for many purposes in building construction over the years. Generally, it is an excellent building material as it is ductile and has excellent tensile strength. For this reason, it is used to provide tensile strength where this would otherwise be lacking in construction, for example as reinforcement in concrete beams, panels and slabs and as lintels supporting masonry over openings in walls and as a structural framework in large span and multistorey buildings. Steel is also the most common material used for wall ties in cavity walls.

Unprotected mild steel will corrode when exposed to oxygen in the air and water. The resulting oxidised product of corrosion – iron oxide – occupies a volume many times that of the parent metal causing expansion. This expansion is very forceful and will break apart material surrounding the corroding metal such as concrete or masonry.

The type and design of steel lintels has changed over the years; many lintels for external use are now combined lintels – that is, they combine the function of cavity tray and lintel. Other types of steel lintel include the box lintel with a flange for the outer leaf to sit on. Most have a generous coat of zinc galvanising, although some are powder coated and some are made of stainless steel. Stainless steel is not often used as it is more expensive than mild steel and is not as strong; however it does have the advantage of corrosion resistance.

Other components commonly made of steel in buildings include the following:

- Bed joint reinforcement in brickwork – used occasionally to increase strength, particularly at returns or junctions.
- Brackets for fixings to external walls for items such as rainwater downpipes, soil pipes, trellises etc.
- Cramps for fixing window and door frames into openings.
- Window frames and French window frames. Steel was popular for window and door frames in the 1950s and 1960s, companies such as Crittall making large numbers of frames for domestic, commercial and industrial properties. Many of these frames are still in place despite their very poor thermal performance.

If the conditions are right for corrosion of the steel, then corrosion will cause expansion and cracking of the surrounding masonry. In the case of window and door frames, the expansion may cause glass to crack and opening sections to bind in their frames as well as pushing up masonry and opening cracks in adjacent mortar joints.

Sulphate attack

Ordinary Portland cement (OPC) used in concrete, renders and mortar naturally contains tri-calcium aluminate, which will react with sulphate salts in the presence of water to produce ettringite or thaumasite crystals or sometimes gypsum. The formation of these long, needle-like crystals pushes the material they form in apart, making it more permeable and causing it to expand. This process is called sulphate attack and affects materials that contain OPC and either also contain sulphate salts or are exposed to sulphate salts as well as being wet for extended periods. This includes the following:

- mortar joints in exposed brickwork;
- ground-bearing concrete slabs with no damp-proof membrane (DPM);
- render coatings to exposed external walls.

It is worth noting that lime also contains tri-calcium aluminate and lime mortars and renders may be similarly affected. Chimneys are particularly susceptible to sulphate attack where they are in exposed locations because the coal smoke contains sulphates that are deposited and build up in the brickwork. The mortar joints on the side of the chimney that are exposed to the prevailing wind will be affected, causing bending over of the chimney and cracking of the mortar joints. Cracking is most noticeable in the bed joints and can sometimes be mistaken for wall tie corrosion (and vice versa). However, sulphate attack is more general and will not confine itself to the joints that contain wall ties. Also, as the cracking is as a result of expansion of the material itself, the cracks tend to be finer and do not open wide.

Where sulphate attack affects render, the render will expand, debonding from the wall, bowing outward and eventually cracking and spalling off. Once sulphate attack has occurred it cannot be reversed; badly affected walls will need to be taken down and rebuilt. Where damage is not so severe, progression of sulphate attack can be slowed by protecting the wall from moisture if that is possible. This could be done for example by providing new copings with damp-proof courses to parapet walls or garden walls.

Further reading

Driscoll, R. and Skinner, H. (2007) *Subsidence Damage to Domestic Buildings: A guide to good technical practice*, BRE, Watford.

Hollis, M. (2005) *Surveying Buildings*, RICS Books, Coventry, pp. 485–538.

2.7 Concrete defects

Key terms: insufficient cover; cracking; corrosion of reinforcement; carbonation; chloride attack; alkali–silica reaction; sulphate attack; high alumina cement

Concrete is strong, durable and can be moulded into all sorts of interesting shapes. With careful design it can be designed to span fairly large distances and form the structure of tall buildings. It is a material that is used all over the world in many different applications and is routinely used for foundations, frames, floors, roof decks and walls.

Concrete has a fairly high embodied energy, but new and innovative forms of concrete are being developed to employ recycled aggregate, and cement replacement materials such as ground granulated blast furnace slag (GGBS) or other pozzolanic agents. Other innovations include the use of non-ferrous reinforcement either as bars, mesh or fibres. As these are new developments, there is little data so far on performance. Concrete defects include:

- defects affecting new concrete such as insufficient cover and poor quality;
- cracking;
- spalling and corrosion of reinforcement;
- carbonation;
- chloride attack;
- alkali–silica reaction;
- 'mundic' or pyrites;
- sulphate attack;
- high alumina cement.

Insufficient cover to reinforcement can lead to corrosion of reinforcement particularly if carbonation occurs. Cover should be at least 20 mm in the most mild environment and up to 60 mm in more severe environments. Problems with insufficient cover to reinforcement are most likely to happen at slab edges, beam edges, construction joints and beam to column junctions.

Almost all concrete has some cracking; cracks appear for a number of reasons including drying shrinkage and thermal movement. The severity depends on the width and depth of the crack as well as its position in relation to reinforcement. Cracks in older concrete may be as a result of corrosion of reinforcement. Other causes of cracking to new concrete include plastic settlement, which usually occurs over reinforcement or at changes of depth of the section. Cracks that are exposed to the weather may allow reinforcement to start corroding as they allow water penetration more easily below the surface.

Some small cracks in new concrete will close up of their own accord, whereas larger more serious cracking may need to be repaired. Brushing dry cement into fine cracks and then spraying with water is often an effective treatment. It is not possible to tell the seriousness of cracking from visual inspection alone, some form of testing such as radar should be used. Core testing can be used but the crack must be filled with resin first as the process of coring may make the crack worse or allow it to close.

Carbonation and corrosion of reinforcement can occur. Concrete is an excellent construction material, however, due to its poor tensile strength, it requires reinforcement where it is to be used for structural purposes or in relatively slender sections. Reinforcement to concrete usually takes the form of steel bars or mesh although sheet claddings, tiles and pipes are reinforced with fibres of various materials (including asbestos in the past). Corrosion of steel requires water and oxygen, but in addition, it requires the reinforcement to be susceptible.

In normal circumstances, the alkalinity of concrete is enough to protect the steel reinforcement from corrosion as it causes a dense layer of oxide to form on the surface of the steel, which protects it. However, carbonation of the concrete occurs where carbon dioxide in the atmosphere dissolves in water to form carbonic acid which then reacts with the alkalis in the concrete reducing its alkalinity. Where the pH falls below 10, carbonation has taken place, and if this carbonation reaches the reinforcement corrosion of the steel is likely to begin. The rate of carbonation is dependent on several factors, the major one being the quality of the concrete; carbonation will be faster where concrete is more permeable and of course oxygen and water will reach the reinforcement more easily if this is the case. The depth of cover to the reinforcement will also play a part in the amount of time it takes for the depth of carbonation to reach it. In typical concrete in moderate exposure, carbonation proceeds at the rate of about 1 mm per year.

Once the reinforcement is no longer protected by alkalinity and oxygen and water are reaching it, it will start to corrode and the products of corrosion will start to build up on the surface, putting pressure on the surrounding concrete. Eventually, cracks will form and the concrete will spall away from the reinforcement. Some rust staining may appear at the surface, bleeding through the cracks before the concrete spalls. Once the concrete has cracked, then there will be greater access to the reinforcement by air and water and the process of corrosion will accelerate. Testing for carbonation can be carried out by spraying a solution of phenolphthalein onto a freshly exposed surface of concrete to find the depth of carbonation.

The presence of chlorides in concrete has the effect of increasing the risk of corrosion of the reinforcement even where the concrete is still very alkaline. This can occur as a result of chlorides being present in the concrete as it is mixed; for example, in the 1970s in the UK, an admixture of calcium chloride was often added to concrete in the UK. In some places, chlorides occur naturally as sand from marine locations has been used to mix concrete – Hong Kong is an example of a place where this practice has been used commonly. Buildings in marine locations may absorb some salts from their environment as will concrete close to areas where salt is spread routinely to prevent ice formation.

An alkali–silica reaction (ASR), also known as concrete cancer (because it can grow gradually over many years), sometimes occurs between the cement and the aggregate in concrete. The product of this reaction is a calcium silicate hydrate gel that absorbs water and expands. This can cause:

- expansion of the concrete;
- cracking;
- differential expansion between different pours of concrete;
- reduced elastic modulus;
- increased tensile strains in reinforcement.

Like sulphate attack, the result is expansion and may also manifest itself as surface crazing or delamination as well as overall expansion of the concrete element. ASR affects stronger concrete that has a high cement content.

In Cornwall and parts of Devon in the UK, waste from tin and lead mines (mundic or pyrite) was commonly used as aggregate in concrete for in situ construction or blocks. The pyrite (FeS_2) oxidises and sulphuric acid results which reacts with the alkaline cement. The result is deterioration and loss of strength which can in some cases require removal of concrete or demolition of buildings. This has caused immense problems in Devon and Cornwall with many properties becoming unmortgageable. Petrographic testing is now required for new mortgages on suspect properties.

High alumina cement (HAC) was used extensively between 1950 and 1976 instead of portland cement as it cures very quickly. However, the mineralogical 'conversion' can cause reduction in concrete strength. There were some high profile-collapses in the 1970s including the roof of a school swimming pool. This led to the banning of the use of HAC in the UK in 1976 – hence all buildings constructed with HAC in the UK are over 30 years old. Conversion is no longer a problem as it occurred quite quickly and therefore the process is complete in existing HAC buildings. Most problems with HAC are now due to carbonation and sulphate attack.

Sulphate attack occurs because OPC used in concrete and mortar naturally contains tri-calcium aluminate, which will react with sulphate salts in the presence of water to produce ettringite or thaumasite crystals or sometimes gypsum. The formation of these crystals pushes the material they form in apart, making it more permeable and causing it to expand. This process affects materials that contain OPC and either also contain sulphate salts or are exposed to sulphate salts as well as being wet for extended periods. This includes the following:

- mortar joints in exposed brickwork;
- ground-bearing concrete slabs with no DPM;
- render coatings to exposed external walls.

It is worth noting that lime also contains tri-calcium aluminate and lime mortars and renders may be similarly affected.

Further reading

Hollis, M. (2005) *Surveying Buildings*, RICS Books, Coventry.

2.8 Structural frames and floors

Key terms: modern steel frames; cast iron frames; wrought iron frames; concrete frames

Defects in structural frames are often hidden from view, but as with other defects, the surveyor can predict where defects are likely if they have a good understanding of the materials used in construction, the technology of how they are fixed together and the most likely agents of damage. As with traditional load-bearing construction, the majority of defects in older buildings are caused by moisture or ground movement. Defects affecting modern steel frames are not very common compared to defects affecting other parts of building, unless the frame is exposed to the weather.

Mild steel has been used for the structure of framed buildings since the early twentieth century. Hot rolled steel sections are the most common steel used for multistorey building due to its great strength, good strength–weight ratio, ductility and ease of connection. Cold-formed steel sections tend to be used for secondary structure such as purlins and cladding support framework. Fixings to cold formed steel are usually self-tapping screws or bolts.

Fixings to hot rolled mild steel frames are usually bolts on site and welds in the factory – although site welding is done occasionally, it is avoided due to its hazardous nature. Great care must be taken with bolt fixings; if bolts are over-tightened the thread can be stripped, but if bolts are insufficiently tightened then additional stress is put on the bolts which can shear as a result. Adequately tightened bolt fixings are vital to the structural performance of structural frames which require rigid joints for their ability to span large distances as the rigid joints ensure that bending is shared by the beams and columns, reducing maximum bending moment at the centre of beam spans. Often where the main structural frame is hot rolled steel, smaller members can be cold formed steel (pressed from sheet steel) – usually the framing that carries lightweight cladding. Hot rolled steel is much heavier and stronger as it has greater thickness.

The BRE and CIRIA carried out a survey of building failures and found that failure of steel frames was twice as likely to be due to faulty design rather than faulty erection on site; most failures occurred within 4 years of completion of the building. It is considered that the use of experienced personnel and checking of design concepts are the best ways of avoiding failure. In their research BRE/CIRIA found that the most likely causes of failure were design that did not take sufficient account of the local conditions or structural behaviour of the frame.

It is highly probable that many design faults are discovered during construction as the steel erectors notice any lack of robustness and may have difficulty in levelling the frame. Typical defects include:

- inadequate design – sizing of columns, beams or bracing;
- overstressing members during erection;
- mechanical damage – particularly to slender or cold formed sections;
- incomplete fire protection;
- incomplete paint coatings;
- corrosion of steel;
- failure of bolt fixings;
- failure of welds;
- inadequate allowance for initial deflection or differential thermal movement between frame and claddings or concrete lift/stair enclosures;
- stressing of frame due to foundation movement.

Most of these defects can be avoided by good management of the design process and the employment of well-trained and experienced site operatives. The report recommends that designers can decrease the risk through a process of peer review where a series of 'what if' questions are asked. Overstressing during construction can be due to poor programming of the steel erection.

Large diameter hollow sections, cold formed sections or thin tie members or tension cables are all susceptible to mechanical damage during construction. Corrosion may be due to leakage or condensation to internal steel or inadequate coatings to external steel. It was also noted that there is a history of failure of welded joints; welding requires good quality control and therefore site welding is to be discouraged.

There are various defects that can affect iron and steel frames. There are some differences in the typical defects according to the materials as they have different properties. For example cast iron is strong in compression but brittle and much less resistant to impact damage than wrought iron. Cast iron was generally joined with bolt fixings as it is not possible to weld; however, occasionally lugs might be missing due to miscasting and these were then 'burned on' by a process of heating the lug and the cast iron component that it was to be fixed to and then pouring molten iron into the joint. The idea was that this would lead to a 'monolithic joint'; however, the quality of these joints was dubious and there is evidence that the practice of 'burning on' contributed to the weaknesses that caused the Tay Bridge disaster in 1879 in which 75 people were killed when the bridge collapsed as the train they were travelling in crossed it.

Wrought iron is stronger in tension than cast iron, but weaker in compression. When subjected to excessive compressive stress, wrought iron can delaminate; this also happens when wrought iron corrodes. Typical defects in iron and steel frames include the following:

- Structural movement or overloading may cause deflection, deformation and distress at joints – in wrought iron deformation may include delamination.
- Impact damage – in cast iron this may cause cracking or even shattering due to the crystalline structure of the material. Wrought iron and steel are more ductile and will dent and bend.
- Water ingress and/or condensation may cause corrosion – due to its laminar structure, wrought iron will delaminate and expand more quickly than cast iron or mild steel although all will be damaged by corrosion with some expansion.

- Corrosion tends not to significantly affect the strength of the frame until it is advanced or where joints are badly corroded. The problems are more often caused by the expansion that corrosion causes.
- The symptoms of structural movement are usually more noticeable in claddings or other elements connected to the frame – cracking and deformation of joints and junctions.

Concrete is a very popular material for frame structures although it has a high embodied energy and in situ cast concrete is very time consuming to construct. It has a number of great advantages including the ability to form complex and interesting shapes, excellent strength, thermal mass and acoustic qualities. In situ cast concrete is generally viewed as not being a 'sustainable' material due to the high embodied energy and the amount of waste generated although the concrete lobby argue that concrete buildings have a long life and lower running costs due to the beneficial thermal mass. Concrete technology is improving all the time with the use of cement replacement such as GGBS and recycled aggregates to make it more environmentally friendly as well as increasing use of reusable formwork to reduce waste on site. Defects in concrete frames include:

- cracking and spalling;
- carbonation and related corrosion leading to cracking and spalling;
- insufficient cover to reinforcement;
- chemical attack – sulphates, chlorides, ASR;
- loss of strength in HAC (see Concept 2.7).

Technologies used currently and historically for floor construction include suspended timber – in older traditional buildings, particularly from the Georgian and Victorian periods, the upper floors were designed with expected loadings in mind. As a result, in domestic buildings, the upper floors designed as bedrooms often had smaller joists than the lower floors used for reception rooms. The top or attic floor was often used only for servants' bedrooms – with very light loadings. Main bedrooms would have had heavier loadings with reception rooms heavier still. Problems can occur where these buildings have a change of use with heavier floor loadings, for example office use. Common timber floor defects include the following:

- timber decay
  - joist ends built into solid walls
  - inadequate ventilation to sub-floor voids
  - dampness encourages decay – insect and fungal.
- excessive deflection
  - inadequate design
  - overloading
  - notching
  - long spans lacking strutting.

Timber decay occurs as a result of dampness; this could be penetration through a solid wall that has become porous due to weathering and age. Excessive deflection occurs as a result of lack of stiffness, e.g. if joists are not deep enough or there is a lack of strutting or lateral restraint. Clearly if floor loadings are increased as described above, then deflection will also increase.

Solid floors include: flag stones or quarry tiles laid on a bed of ash or rammed earth – common in historic buildings up to the end of the Victorian period; and concrete ground-bearing slabs which became common after the Second World War. In newer buildings floors will normally have a DPM and insulation, although arrangements vary. For heavy duty industrial applications, the mix of concrete is so strong that it is virtually waterproof; a DPM is still used as a slip plane to prevent uneven cracking. Problems can arise where historic buildings are made more airtight with closer fitting windows and blocked up fireplaces. This reduces the opportunities for water rising through the floor to evaporate, thus dampness manifests itself. Solid floors may also be cold and attract condensation. The temptation would be to lift the flag stones and lay a DPM and insulation beneath them to prevent dampness passing through and coldness that would encourage condensation; however, this would then cause dampness to rise up in the walls as there is nowhere else for it to go. Typical problems with solid floors include:

- dampness (see Concept 2.4);
- cold and condensation;
- settlement of hardcore that causes loss of support to floor slabs, and consequent cracking and settlement of the slab;
- sulphate attack (see Concept 2.7);
- alkali–silica reaction (see Concept 2.7);
- clay heave;
- swelling of hardcore.

A number of these problems are linked to the hardcore used beneath concrete ground-bearing slabs. If the layer of hardcore is too deep or insufficiently compacted, then settlement may occur with consequent opening up of the joint between the floor and the base of the skirting board plus cracking and breaking of the slab where the settlement is differential. Some hardcore material, for example red colliery shale or reclaimed brick, contains soluble sulphates that can cause sulphate attack to the slab where there is no DPM or where the DPM is damaged, particularly in wet conditions. Other materials are prone to swelling in wet conditions, pushing outwards on walls below the DPC or ground beams, depending on the construction. These materials include:

- steel slag or blast-furnace slag;
- old crushed concrete, which may expand as a result of sulphates in the soil;
- material containing clay;
- material containing pyrites.

Further reading

Beckman, P. and Bowles, R. (2004) *Structural Aspects of Building Conservation*, 2nd edn, Elsevier, Oxford.
Hollis, M. and Gibson, C. (2005) *Surveying Buildings*, 5th edn, RICS books, Coventry.

2.9 Roofs and cladding

Key terms: pitched roofs; slate roofs; clay tiles; concrete tiles; flat roofs; cladding

Roofs and cladding form the main parts of the external envelope of buildings so they are required to keep the building weather tight, control passage of light, heat and ventilation while remaining durable and looking good. As with other parts of a building, it is necessary for the surveyor to understand the technology of the construction, the materials used and how they interact with each other and the environment. One of the great challenges for the parts of the building that form the external envelope is to deal with the temperature differentials that occur between the exterior and interior of the building at various times of day and at different times of year.

Pitched roofs are generally considered to be more durable than flat roofs, particularly in climates such as the UK where there is plenty of rain and sometimes snow. Life cycle costs are generally cheaper with pitched roof forms although the initial capital costs are usually greater. There are various types of pitched roof forms from the basic domestic 'lean-to' monopitch roof to large span portal frames with sheet claddings. Traditional pitched roof coverings include natural slate and clay tiles, while in recent years concrete and artificial slate have become popular, although slate and clay are still used. The main causes of defects in pitched roofs are attributable to complex roof layouts, poor construction, poor design/detailing and lack of maintenance.

Slate roof coverings are prone to the following defects:

- defective mortar fillets;
- problems with torching and bedding;
- split/broken or delaminated slates;
- slipped slates/nail sickness;
- water penetration;
- verge failure;
- problems with ridges and hips.

Concrete tiles are prone to similar problems to clay tiles, although additional problems include:

- overloading of the roof structure;
- loss of surface;
- loss of colour.

Clay tiles are prone to the following defects:

- delamination;
- pitted, cracked or broken tiles;
- slipped tiles;
- efflorescence/salt crystallisation;
- failure of mortar pointing to verges;
- failure of mortar bedding to ridge and hip tiles.

Defective perimeter gutters can cause problems to walls that are repeatedly wetted and leaking valley or parapet gutters allow water to pass easily inside the building. Penetrating dampness through parapet walls can cause decay to timber roof members.

One of the main functional requirements of a roof is that it must protect the building occupants and the building structure from the external environment. To this end the roof must have structural integrity and weatherproofing; these functions are achieved in the elements of the flat roof:

- Structure – depending on the size of the building and the span, steel frame, concrete frame or timber joists are all common.
- Deck – again depending on the type of building, decks are typically concrete, profiled steel or timber (usually plywood), although some buildings may have woodwool slab decks.
- Covering – needs to be more impervious than pitched roof coverings due to the slower runoff.

Types of covering:

- built-up mineral felt;
- single ply membranes;
- mastic asphalt;
- soft metal coverings such as lead, copper or zinc.

Flat roofs are either cold roofs or warm roofs. Cold roofs are quite common in older buildings and generally in house extensions. Because the insulation is below much of the structure, there is an increased danger of condensation and therefore adequate cross-ventilation is very important, particularly with timber structures and decks. The use of a vapour barrier on the warm side of the insulation is recommended, but this is easily damaged due to its position just above the ceiling.

Warm roofs are the most common form of construction in new buildings – as the insulation is above the structure, the danger of condensation is reduced. A vapour barrier is still necessary to prevent condensation between the insulation and the covering. With inverted warm roofs the insulation is above the weatherproof covering and a separate vapour barrier is not required. The insulation must be carefully chosen and installed as it will be open to the weather – it must not absorb water and must not be lifted off by the wind. While the membrane is protected from UV radiation and mechanical damage by the insulation and paving above it, if there is a problem with rainwater ingress, then investigation and remedy often require stripping the roof back which is very expensive and disruptive. This type of roof is excellent when it is installed properly, but occasionally the membrane can be damaged by following trades or if the roof is not kept very clean during construction. Any grit or small objects can pierce the membrane if caught between it and the insulation due to the loading above and the movement of foot traffic over the roof.

The most common problem with flat roofs is water ingress. This can be particularly difficult to diagnose as water will often track along under the covering and appear in the building underneath some distance from the point of entry. Frequently problems arise due to poor design:

- inadequately sized members;
- insufficient members (too widely spaced);
- inappropriate selection of covering;
- inadequate falls;
- inadequate detailing – upstands/outlets/movement joints etc.;
- inadequate rainwater disposal;
- inadequate consideration of roof use (roof-top plant and traffic etc.).

Other problems occur due to poor construction:

- poor workmanship and/or supervision;
- lack of coordination/cooperation;
- inappropriate selection/combination of materials (including fixings);
- unfamiliarity with products and processes;
- damage by following trades;
- working in unsuitable weather conditions.

Cladding materials include:

- masonry;
- pre-cast concrete;
- metal;
- glass;
- glass-reinforced plastics;
- cement-based composites;
- timber.

As claddings have to cope with the weather and differentials of temperature and humidity between the exterior and interior of the building, they have to be well designed and fixed to the building. With all the various forces and environmental changes that claddings have to cope with, it is hardly surprising that there are a wide variety of defects that can affect them depending on materials, design, sizes and climate. Typical causes of defects include the following:

- failure of joints;
- movement:
  - thermal and moisture movement;
  - creep;
  - differential movement;
- failure of fixings;
- deterioration of cladding materials;
- impact damage.

Further reading

BRE (1978) *Digest 217: Wall cladding defects and their diagnosis*, BRE, Watford.

Harrison, H. W. and de Vekey, R. C. (1998) *Walls, Windows and Doors: Performance, diagnosis, maintenance, repair and the avoidance of defects*, BRE, Watford.

Harrison, H. W., Trotman, P. M. and Saunders, G. K. (2009) *Roofs and Roofing: Performance, diagnosis, maintenance, repair and the avoidance of defects*, 3rd edn, BRE, Watford.

Hollis, M. and Gibson, C. (2005) *Surveying Buildings*, 5th edn, RICS Books, Coventry.

2.10 Asbestos in buildings

Key terms: white asbestos; blue asbestos; brown asbestos; health hazards; Control of Asbestos Regulations 2012; management survey; refurbishment/demolition survey

Asbestos is a silicate of magnesium and occurs as a glassy rock, which can be split into very thin fibrous crystals. There are two main groups of asbestos, which have differing crystalline structures.

- Serpentine – white asbestos – fibres are longer and curly.
- Amphibole – blue and brown asbestos – fibres are shorter and straight, brown asbestos being more brittle.

Asbestos is commonly known by its colour; however, this can often be misleading as the material is usually incorporated into another material such as asbestos cement or vinyl that may contain pigment or a matrix of another colour.

- Chrysotile – white asbestos – least hazardous – examples: asbestos cement, Artex.
- Amosite – brown asbestos – examples: asbestos insulation board, soffit board.
- Crocidolite – blue asbestos – most hazardous – examples: pipe lagging, sprayed 'limpet' insulation.

The properties of asbestos are the reasons why it is so popular for building materials:

- Heat resistant – asbestos can withstand a temperature of 900°C without change.
- Mechanical strength – the fibres are strong in tension and are good to add as reinforcement to cement products.

- Chemical resistance – it has good resistance to alkalis, neutral salts and organic solvents. The varieties of asbestos used for building products also have a good resistance to acids.
- It is easily woven into materials to form heat resistant textiles and ropes.
- There are abundant natural sources of asbestos so it is a low cost material.
- It is water resistant.
- It is weather resistant.

Ninety per cent of consumption of asbestos in the UK in 1989 was for construction materials; this has fallen due to the high profile of the health hazards and now new regulations that prohibit its use. As a result of its durability and cheapness, it has been very widely used as a building material and there is a large amount of it distributed around the built environment.

The health hazards of asbestos have been known for some considerable amount of time and in recent years more has become known about asbestos and its effects on health. The amphibole forms of asbestos, crocidolite (blue) and amosite (brown) are more dangerous than chrysotile (white) asbestos. This is not to say that chrysotile is innocuous though, as stressed by the HSE in their publication *Managing Asbestos in Buildings*:

> The carcinogenic risk from chrysotile (white asbestos) has been evaluated by the International Agency for Research on Cancer and it is considered to be a category 1 human carcinogen. HSE's view is that there is sufficient evidence that chrysotile causes cancer in humans but that there is some uncertainty as to the scale of the risk.

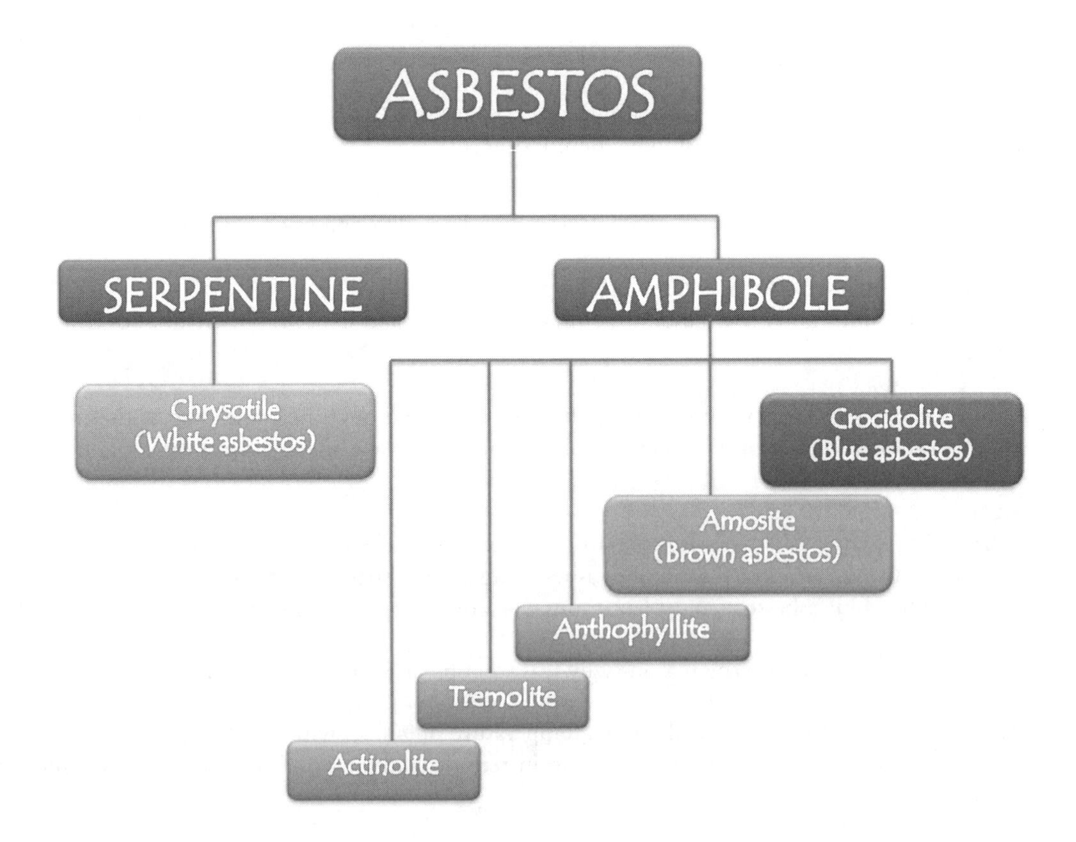

Figure 2.10.1 The asbestos family tree.

The major health problems with any form of asbestos are caused by breathing in dust containing the fibres, which become lodged in the lungs. The HSE website states that on average four plumbers, 20 tradesmen, six electricians and eight joiners die every week in the UK as a result of asbestos-related diseases. Due to the long latency period of these diseases the numbers dying are expected to continue rising and peak in about 2015.

All surveyors that undertake inspections and surveys of buildings should be able to recognise materials that may contain asbestos and warn their clients accordingly. Asbestos-containing materials (ACMs) may be present in any building built or refurbished before the year 2000. Building surveyors should be aware that asbestos may be hidden behind finishes, panelling, inside ducts or voids. To further aid recognition, you are recommended to look at the photographs in the asbestos image gallery on the HSE website at www.hse.gov.uk/asbestos/gallery.htm.

Asbestos has often been used in external claddings and roof coverings. Fibre reinforced cement is a very popular cladding material as it is light, strong and weather resistant. This material is still manufactured and used, but the fibres used to reinforce the material are no longer asbestos. It is not possible to tell whether the material contains asbestos just by looking at it, but unless the cladding is demonstrably new, it is very likely to contain chrysotile although both crocidolite and amosite have been known to be used for this purpose so their presence cannot be ruled out without testing. The most common form is the corrugated sheets that are often seen as roof coverings or cladding to industrial or agricultural buildings.

Other roof coverings include artificial slates that can look quite convincingly like natural slate once it has lichen or moss growing on it. Other artificial slates may originally have been bright colours such as red, orange or even green or blue in some cases. Quite often these slates will have a small metal hook protruding through the lower edge to prevent wind uplift. Soffit boards to the underside of the eaves and occasionally fascia boards are sometimes made from asbestos board, either amosite or chrysotile.

There are various types of asbestos board panels used as external wall claddings and spandrel panels in cladding systems on commercial, institutional and residential buildings. Often these are difficult to recognise as there is such a variety of coatings and finishes, but most are either chrysotile or amosite.

Linings, panelling and ceilings can contain asbestos. Again there is a great variety of internal linings that can be composed of ACMs. This can include wallboards, laminate panels, applied ceiling tiles, suspended ceiling tiles etc. The most common finishes to contain asbestos are vinyl floor coverings – in sheet or tile forms – and textured paint coatings such as 'Artex'. ACMs are commonly found in conjunction with services due to the heat and fire resistant nature of the material. Crocidolite is commonly found as an insulating material around hot water/steam pipes and tanks or calorifiers; this is usually considered to be the most hazardous of commonly found ACMs. There are plenty of other examples of ACM to be found in the plant room such as asbestos board to mount electrical meters and switchgear, textile rope gaskets around doors to boilers, textile movement joints in metal air conditioning ducting, textile sleeves to electrical wiring to hot devices and asbestos tape repairing flue pipework etc. Chrysotile is usually the fibre reinforcing asbestos cement products that have been used extensively in the past for flues, gutters, downpipes, hopper heads and sometimes cisterns and tanks.

The incombustible properties of asbestos make it a good material for fire protection purposes and it has been used for protecting vulnerable parts of buildings over many years. For this purpose, the forms used are usually asbestos insulation board (often referred to as AIB) or sprayed on coating or 'limpet'. Structural steelwork is very vulnerable to distortion and failure in a fire and so it usually has some sort of fire protective coating or encapsulation.

The Control of Asbestos Regulations 2012 relate to non-domestic premises and confer duties onto those responsible for the running and maintenance of the buildings. The regulations require duty holders to:

- take steps to find and assess materials likely to contain asbestos;
- presume that materials contain asbestos unless there is strong evidence to the contrary;

- make and keep an up-to-date record of the location and condition of ACMs;
- assess the risk of anyone being exposed to these materials;
- prepare a plan to manage that risk and to put the plan into effect;
- review and monitor the plan to ensure effectiveness;
- provide information to people who may work with ACMs.

The first of these duties involves finding out whether there are any ACMs in the building; this either requires the carrying out of a sampling survey or the making of assumptions that materials contain asbestos unless there is strong evidence to the contrary. In practice, it is usually more cost effective to carry out a survey than to make a presumption about materials containing asbestos as this requires so much care and management.

There are two types of survey carried out in the UK – the first is the *management survey*, where no major works are planned in the building and the circumstances and design of the building are straightforward. The aim is to:

- ensure that nobody is harmed by the continuing presence of ACM in the property;
- ensure that the ACMs remain in good condition;
- ensure that nobody disturbs an ACM accidentally.

In order to do achieve this, the survey should identify all ACMs that might be disturbed in the normal running and maintenance of the building and assess their condition. This is the standard type of survey carried out and usually involves sampling and analysis to positively identify ACMs, although it can be acceptable to make a presumption that a material contains or does not contain asbestos. A presumption that a material does not contain asbestos would only be made where there is incontrovertible evidence!

The second type of survey is the *refurbishment/demolition survey*, where there are plans to carry out works or demolish the building. It aims to

- ensure nobody will be harmed by work on the ACM;
- ensure that any work involving ACMs will be done in the right way by the right contractor.

In order to achieve this, every ACM present in the building must be found and positively identified prior to commencement of work. This usually involves opening up and destructive testing.

The HSE recommend that surveys be carried out in accordance with HSE guidance HSG264 *Asbestos: The Survey Guide* by accredited specialist surveyors, who have been accredited by the UK accreditation service UKAS. This is not an area of work normally carried out by building surveyors unless they have had specific training in asbestos surveys and the relevant accreditation from UKAS. Standard professional indemnity insurance for building surveying practice does not cover asbestos surveys and surveyors contravene the RICS code of conduct if they carry out work that they are not insured for.

Further reading

Hollis, M. and Gibson, C. (2005) *Surveying Buildings*, 5th edn, RICS Books, Coventry.

HSE (2012) *Managing Asbestos in Buildings*, available at: http://www.hse.gov.uk/pubns/asbindex.htm (accessed 02/06/2013).

HSE (2012) *Asbestos: The Survey Guide*, available at: http://www.hse.gov.uk/pubns/books/hsg264.htm (accessed 02/06/2013).

3 Commercial property

Andy Dunhill, Dom Fearon, John Holmes and Becky Thomson

3.1 Commercial property

Key terms: UK economy; retail; offices; industrial; leisure; health care; owner-occupation; investor; developer; rental value; capital value

According to the British Property Foundation's *Property Data Report 2012*, the UK's commercial property value in 2011 reached £717 billion. Close to 1 million people are employed in commercial property activities. The commercial property sector directly contributed about £45 billion to the UK's GDP in 2011 – more than the country's IT industry, and comparable with the engineering sector.

Commercial property is, generally speaking, property used for the sole purpose of carrying out or performing a business activity. This would exclude residential property and/or primary industries such as mining and agriculture. The commercial property sector contributes a large part of the UK economy and provides a platform for the majority of the country's commercial enterprises. It is a sector that plays a crucial role by providing places in which people can work, shop and enjoy leisure activities. The three main types of commercial property are *retail* (including supermarkets, shopping centres and retail warehouses); *offices* (including business parks); and *industrial* property (including factories and warehouses). Also included to a lesser degree is leisure property (including hotels, bars and restaurants) and health care property (including hospitals, doctors' surgeries and care homes).

Commercial property 'activity' covers those whose main business is the construction and the management and care of buildings. The letting, buying and selling of property is also significant. Investment and fund management is a small but high-value-added part of the industry, and currently the largest in Europe. While the residential property sector is approximately eight times larger than the commercial property sector, the majority of residential properties are owned by private householders and therefore less desirable to large investors.

About a third of all commercial property is acquired by owner-occupiers who need buildings from which to run their business. Some of these occupiers want to buy the property outright by acquiring the freehold of the property, while others may prefer to acquire a leasehold interest in the property for a temporary period to better suit their business requirements. The other two-thirds of all commercial property is generally bought by property investors with the intention of letting out to tenants. This has the benefit of providing the investor with a regular income (by way of rent) that has the potential to increase over time and also a profit from any increase in the *capital* value of the property over the same time period. Many occupiers and most investors, other than institutions, buy commercial property using a combination of debt and their own capital. The proportion of commercial property that is rented grew over the last decade. This is because many businesses have become increasingly reluctant to commit the capital and management time required of owner occupation.

A small proportion of commercial property will be held by property developers whose main purpose is to make gain from the development of existing land and/or buildings to a more profitable use – for example, converting redundant riverside warehousing into prestigious residential apartments. In recent years, due to the economic downturn, much speculative commercial property development has declined due to a fall in demand for the end product but also, due to the risk involved, funding from banks has not been forthcoming.

Commercial property can be owned by a variety of different people and organisations. According to the *Property Data Report 2012*, direct ownership of commercial property in the UK up to 2011 comprised the following:

Table 3.1.1 Ownership of UK commercial property

Ownership of commercial property	*(%)*
UK investors (insurance companies and pension funds)	23
Overseas investors	23
Collective investment schemes	18
UK REITs and listed property companies	15
UK unlisted property companies	9
Private investors	5
Traditional estates/charities	4
Others	3

Further reading

British Property Foundation (2012) *Property Data Report 2012*, British Property Foundation, London.
Rodell, A. (2012) *Commercial Property*, College of Law Publishing, Guildford.

3.2 Private investors

Key terms: investment; security of capital; assurance of income; location; physical condition; covenant; lease terms; yield; risk and reward

> *'Investment'* – Using a capital sum to acquire an asset which will, hopefully, produce an acceptable flow of income and/or appreciate in capital value.
>
> 'The Glossary of Property Terms' (2004), *Estates Gazette*

With regard to commercial property investment, most investors would agree that in a perfect world all investments would provide:

- total security of capital;
- complete liquidity of capital;
- absolute assurance of income;
- absolute regularity of income.

Total security of capital: The investor is guaranteed there is no risk of the large capital invested being lost or its value diminishing in real terms. Ideally, the value of the investment would increase over time.

Complete liquidity of capital: The investor will need assurance that the investment can be easily sold at short notice and/or capable of division into smaller parts should the investor need to recover the capital.

Absolute assurance of income: The investor is confident that the rent received (income) is always forthcoming, will never decrease and would increase over time.

Absolute regularity of income: The investor is guaranteed the full rent is received on the correct due date without the need for demand being made.

Any property that does not possess some or all of these qualities is deemed to pose more risk to an investor. When it comes to choosing which property to invest in, there are many types, e.g. commercial, residential or agricultural, but also they can be classified by quality such as prime, secondary or tertiary. All property is heterogeneous and encompasses a vast array of types with varying qualities and weaknesses. This makes property a difficult asset type to benchmark and define for investment purposes. The quality of a property depends largely on four main factors: *location*, *physical condition*, *covenant* (tenant) *strength* and *lease terms*. Clearly, a property scoring highly in all four areas would be considered a prime property investment.

While physical qualities of a property are important, investors tend to focus more on the quality and security of the cash flow the investment produces. It is the precise terms of the lease that dictates and controls the cash flow. The parties to, and the terms of, the lease will greatly affect the value and attractiveness of the property. This means two almost identical properties in terms of location and physical condition can have very different values depending on the quality of the tenant and the terms of the lease. Three main factors in a lease that will affect a property's value are the *quality of the tenant*, *length of the lease* and, if the tenant has any, *rights to determine the lease* early through a break clause.

In their choice of property investment, investors will often assess the quality of the investment by considering the *yield* or return they will receive from the capital invested. Most investment involves the forgoing of a large capital sum now, in return for an income, with property, via a rent. The annual income return from the investment expressed as a percentage of its market value is termed the yield.

For example:

$$\frac{\text{Annual income} \times 100}{\text{Capital paid}} = \text{Yield} \qquad \frac{\text{£17,000 p.a.} \times 100}{\text{£215,000}} = 7.9\%\ \text{yield}$$

The percentage yield an investor requires from an investment will reflect the characteristics and risks associated with the property. The basic concept is the greater the risk involved, the higher the yield required, and the more secure the investment, the lower the yield required. The ideal investment would therefore represent the lowest yield, as there is no risk or uncertainty connected with the ownership of the investment. The yield provides investors with a common basis of comparison between investments. An investor may have the choice between a prime property providing a 5 per cent yield and a secondary investment providing 15 per cent. The *higher* yield therefore does *not* necessarily represent the best investment as the question that must be asked is does the yield represent good value and a balance between risk and reward.

Further reading

Blackledge, E. (2009) *Introducing Property Valuation*, Routledge, London.

3.3 Private finance initiatives

Key terms: public–private partnerships; public sector procurement; special purpose vehicle; consortium

Private finance initiatives (PFIs) use private money for major public sector capital projects. They were first introduced in Australia in the late 1980s, in order to fund toll road and railway projects. A PFI contract is a way for the public sector to obtain new buildings using private investment which is then repaid to the private sector by rental and maintenance charges over a period of time. The first UK project was the Severn River crossing, which started operating in April 1992.

There were two principal issues driving the move to PFI: the transfer of the risk involved in the procurement of public infrastructure and the use of private finance to keep the cost of infrastructure off public sector borrowing requirement (PSBR). The need for risk transfer was evident from a series of humiliating public sector building procurement exercises, a typical example being the British Library in London which was originally estimated at £150 million but after a series of design changes and delays over 21 years the final bill was £500 million.

Since 1992 there have been more than 720 PFI contracts in the UK. Projects such as highways, hospitals, schools, prisons and government offices have been undertaken to provide more than £100 billion of development in the UK. In a PFI contract the public and private sectors collaborate. Private firms operating in a consortium agree to design, build, finance and manage a public sector facility and in return the public sector client agrees to pay annual charges and/or allows the private sector consortium to reap any profits that can be made over the life of the project (30 years or more). Figure 3.3.1 shows how PFIs are structured.

In theory, as the consortia are responsible for building and maintaining the facility over a protracted period they should produce a building design using life cycle cost analysis, high quality, low maintenance materials and a tried and tested design. The risk in the procurement process is effectively transferred to the consortia and, because the payments from the client only start after the building has been delivered there is an imperative to build to programme and the overruns of previous years have become a thing of the past. While the construction process is reduced, the procurement process under PFI is expensive and protracted. This process requires a considerable input from the client, the potential occupiers of the facility, the design team and the lawyers. Once the output specification has been established and put out to tender, interested consortia will produce designs and costs, and establish a rent, normally index-linked over the period of the contract. The considerable amount of design work is carried out 'at risk' of not getting the project; it is not unknown for five consortia to be bidding for a £25 million capital project and each generating £1 million of costs with a one in five chance of success.

Another problem has been the lack of flexibility inherent in the PFI contract – when a local authority has to cut services, as in the current recession, the PFI library or leisure centre is sacrosanct and everything else has to be closed as the PFI payments must continue whether the building is open or closed. Another example of PFI inflexibility has been the ill-fated Fire Control project, which aimed to replace 46 local emergency call centres with nine regional centres; the project was scrapped by the

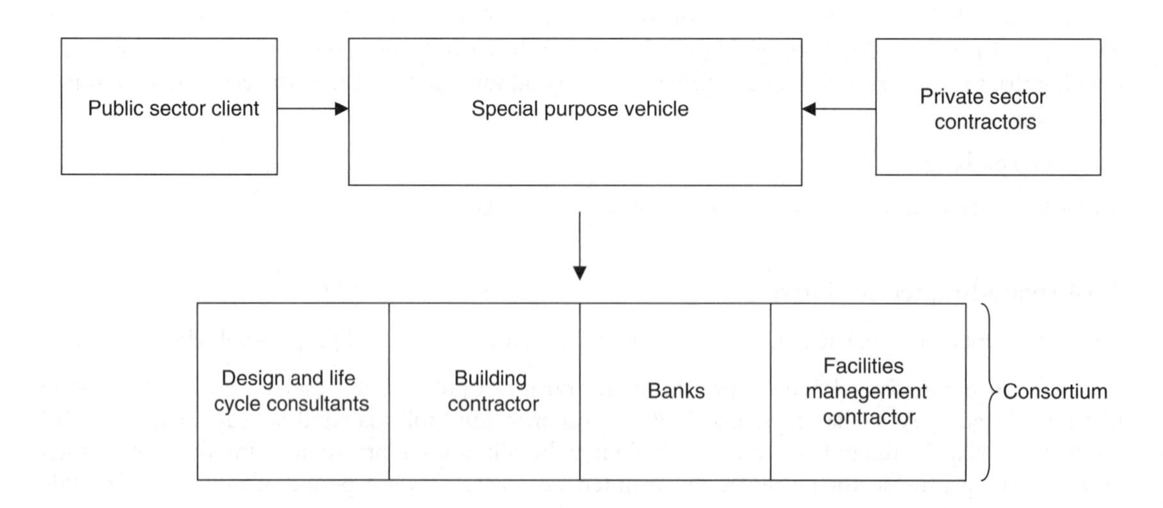

Figure 3.3.1 PFI structure.

new Coalition Government as part of the 'localism' agenda and in the face of union and community opposition, but this was not before the regional control centres had been built and equipped to specification. The nature of the PFI contract means that in 2011 the government was paying £1.4 million per month in rent for empty buildings in addition to writing off £140 million in procurement costs and consultants fees.

Supporters claim that the advantages of PFIs are:

- the contractor no longer 'builds and disappears';
- the client drives the project;
- competition reduces inefficiency;
- there is improved flow of information;
- there are incentives to complete on time, within budget and to expected quality.

Detractors claim that the disadvantages of PFIs are:

- transaction costs are high so only a few firms bid;
- bid costs are often more than £1 million;
- ongoing rental payments can be a burden on public finances (Jowsey, 2011).

The debt created by PFI can have a significant impact on the finances of public bodies. By 2012 there were 719 PFI projects in the UK with a capital value of about £60 billion, but the amount of payments to be made over the next 30 years was more than £300 billion. Annual payments to the private owners of the PFI schemes are due to peak at £10 billion in 2017 and are already stretching constricted public sector budgets, to the extent that public services could be suffering.

A National Audit Office study in 2003 endorsed the view that PFI projects represent good value for taxpayers' money, but some commentators have criticised PFI for allowing excessive profits for private companies at the expense of the taxpayer. There is also evidence that some PFI projects have been poorly specified and carried out. Supporters of PFI, however, claim that risk is successfully transferred from public to private sectors as a result of PFI. The Nation Audit Office again reported on PFI in 2011 and noted that:

> Government Departments and the Treasury have built up considerable experience of PFI and developed guidance, support and project assurance systems to assist in procurement. This has included standard contracts and specialist units to assist individual (inexperienced) authorities embarking on PFI negotiations.

An example of a successful PFI is new office accommodation for the Treasury at 1 Horse Guards Road, Whitehall. This was opened by Alan Greenspan, former Chairman of the United States Federal Reserve Board on 25 September 2002. The increased space available in the new building (known as 1HGR) will enable all Treasury staff to work in the same building for the first time in over 50 years. This was achieved on budget and ahead of time in a successful innovative value for money PFI project.

GCHQ Cheltenham is another successful example of a PFI. This 100,000 m^2 building is a major feat in all respects. As the largest PFI project in Europe, the GCHQ's new building will be home to over 4,500 staff, relocating them from 50 different buildings on two existing sites in Cheltenham. Since 2007 many sources of private capital have dried up; nevertheless, PFI remains the UK government's preferred method for public sector procurement.

Further reading

Jowsey, E. (2011) *Real Estate Economics*, Palgrave Macmillan, Basingstoke, Ch. 21.
National Audit Office (2011) 'Lessons from PFI and other projects', available at: www.nao.org.uk/report/lessons-from-pfi-and-other-projects/.

3.4 Office market

Key terms: office design; office location; low carbon office design; business parks

During the twentieth century office design and office work was transformed beyond recognition. At the beginning of the century offices were often elaborate Victorian edifices with cellular office spaces and coal fires. They were clustered together for easy day-to-day communication, with shipping offices along the dockside and insurance businesses famously around Lloyd's Coffee House in London. By the end of the century offices had been transformed by the modern movement in architecture, the maxim 'form follows function' led to the tower blocks that characterise New York, the design being copied throughout the world. In the UK the modern era was completed by 1 Canada Square, the poster building for Canary Wharf, London completed in 1991 just in time for a recession which kept it empty for years and eventually bankrupted the developers, Olympia and York.

In the twenty-first century office design is being affected by the need for environmental sustainability and, more importantly, the changes brought about by the integration of telecommunication and computers, which has created demand for an entirely new office type, the call centre. In the modern economy offices can be readily classified into two types, characterised by their location: city centre and out of town business parks.

Business parks are an evolution of office design. They are out of town office developments to take advantage of cheap land and good road links. Occupying firms appreciate lower rent, good access and a homogenous selection of buildings with a variety of floor plate sizes. Employees appreciate a new working environment, ample free parking and an attractive parkland setting which may offer recreation facilities for lunch breaks and evenings. The idea of the out of town business park originated in the west coast of America where high-tech and IT start-ups favoured the 'campus style' low rise design that imitated the universities their employees had come from. In the UK, Stockley Park, which opened in 1986, provided the successful model which all business park developers hope to emulate. Stockley Park is handily located just off the M4 motorway to the north of Heathrow airport; it is served by West Drayton rail station and has an adjoining golf course and on-site leisure facilities such as pubs and restaurants. The site is set out at a low density with woodland and lakes providing an attractive setting. The superb accommodation, attractive setting and proximity to Heathrow attracted global blue chip tenants such as Apple, Canon, M&S, IBM and Mitsubishi.

The tranquil parkland environment allows developers to produce environmentally friendly buildings. The low rise offices may be built to a narrow floor plate to allow natural daylight and the absence of traffic noise and pollution allows offices to be built with natural ventilation and openable windows which employees prefer (in reality most developers go for the safe option of air conditioning).

Now all major UK cities have their version of an out of town business park. Many are very successful but at their worst they can provide a bleak experience for employees. The out of town location may be isolated and provide very limited recreational facilities for lunch breaks, the 'ample free parking' referred to earlier may be overwhelmed by demand which leads to full car parks and all the service roads lined with the overflow. The out of town location usually translates into a lack of public transport which means most employees commute by car and as motoring costs increase and salaries stagnate the business park may become an unattractive option.

City centre offices at the nexus of public transport facilities have maximum accessibility and this will drive demand. Where there is demand, the office form can be extended vertically to the maximum height the market and planning regime will allow. The higher the offices the more expensive the build cost becomes and the ratio of gross to net floor space becomes less favourable due to the need for lifts and emergency exit stairs. The modernist slab office block is very efficient but as developing countries produce their iconic buildings, less efficient but more spectacular designs have emerged such as the Taipei 101 building in Taiwan and the Bahrain World Trade Centre.

In the UK the design of offices is largely dictated by the British Council of Offices (BCO) specification, which gives guidance on design of modular sizes, building services, lift specifications, sanitary facilities and finishes. The BCO is a subscription membership group of property developers, designers, investors and occupiers who have come together to research, develop and disseminate good practice in office design. It has its origins in the 1990s when likeminded groups came together to establish an 'investment' standard for offices, partly to prevent wasteful over specification and promote an overarching minimum standard that property developers, financiers and the pension funds and investors would accept as of acceptable quality for prime investments. The current BCO specification dates from the 2009 edition and will be superseded in 2014. The 2009 specification laid great emphasis on environmental sustainability, especially heating and cooling strategies to minimise environmental impact.

Current office design shows a marked departure from the modernist glass box that dominated the latter half of the twentieth century. Current designs will normally feature solar shading, typically brise soleil or a mesh to reduce solar gain and the subsequent cooling demand. Cooling systems are generally low energy designs such as displacement ventilation or chilled beams. As local planning authorities demand an element of renewable energy, hybrid gas and biomass heating systems are being designed, or air source heat pumps, to both heat and cool the building.

Office occupiers have become increasingly concerned to gain the maximum benefit from the space they occupy. In 2005 the Commission for Architecture and the Built Environment (CABE) and BCO published an influential report entitled 'The impact of office design on business performance'; it drew

Figure 3.4.1 Brise soleil on the south elevation of the Museum of London.

Source: John Holmes.

Figure 3.4.2 Mesh solar shading perforated by oval feature windows.

Source: John Holmes.

particularly on the architectural firm DEGW's framework to encompass the business objectives of an organisation:

- Efficiency: making economic use of real estate and driving down occupancy costs (getting the most from the money).
- Effectiveness: using space to support the way that people work, improving output and quality (getting the most from the people).
- Expression: communicating messages both to the inhabitants of the building and to those who visit it, to influence the way they think about the organisation (getting the most from the brand).

Since the publication of the report the principles have been manifest in the office market and office design. As regards efficiency, offices are no longer 9–5 operations; offices are being used more intensively by increasing staff density, open plan working and hot desking. The effectiveness of the space is enhanced by improving staff satisfaction to reduce absenteeism by ensuring the building services work to the optimum to enhance staff comfort as regards temperature, air quality, noise and lighting. Space must be used effectively to facilitate staff tasks, open plan for communication and team working, and individual spaces available when quiet concentrated study is needed, as well as a variety of meeting rooms for project work. Finally expression is the ability of the organisation to transmit its brand values to staff and visitors through the space it occupies. Traditionally, banks have expressed their reliability and gravitas by occupying impressive neoclassical offices or ultra modern HQs in the city. Expression is very important when it comes to attracting high quality staff in competitive employment sectors. For example, the Google HQ under construction in London will feature a 'quirky' fit-out to provide a vibrant workspace for their creative staff.

Offices are places where may people spend lots of time; they can be bleak and dismal, where each day is a struggle to work effectively in a space that is too hot/cold, stuffy/draughty, overlit/underlit, noisy/sepulchral. Or, with some thought and clever design they can enhance productivity, reduce absenteeism and make staff feel valued by their employer.

Further reading

BCO (2009) 'British Council for Offices Specification 2009', available at: bco.org.uk, (accessed 29/10/2013).

BCO (2013) 'Occupier density study', available at: bco.org.uk (accessed 29/10/2013).

CABE/BCO (2005) 'The impact of office design on business performance', available at: bco.org.uk (accessed 29/10/2013).

3.5 Industrial market

Key terms: Planning Use Classes Order 2010; logistics; multi-channel retail; logistics centres; warehouses; distribution centres; industrial estates

From a property perspective, industrial property is property (land and buildings) that can be used for a variety of purposes, with the following being the most common uses:

- production;
- manufacture;
- storage;
- distribution;
- energy production;
- waste disposal;
- industrial land, e.g. mining/quarrying.

As a general rule, agriculture and forestry are not included in the industrial property sector. From a UK planning perspective, industrial property use is generally classified under the Use Classes Order 2010 as:

- *B1 Business* – light industry appropriate in a residential area;
- *B2 General Industrial* – use for industrial process other than one falling within Class B1 (excluding incineration processes, chemical treatment or landfill or hazardous waste);
- *B8 Storage or Distribution* – this class includes open air storage;
- *Sui Generis* – certain uses do not fall within any use class and are considered *sui generis*. An example would be a vehicle/metal scrap yard.

Although industrial property is such a large commercial property sector and its buildings can range in size from a small business workshop of 500 sq ft (46.5 sq m) up to a large warehouse of 350,000 sq ft (32,516 sq m), the sector generally includes the following types of properties:

- logistics centres;
- warehouses;
- distribution centres;
- industrial estates.

A *logistics centre* is the hub of a specific area where all the activities relating to transport, logistics and goods distribution, both for national and international transit, are carried out, on a commercial basis, by various operators. The operators may be either owners or tenants of the buildings or facilities (warehouses, distribution centres, storage areas, offices, truck services, etc.) built there. The whole concept of 'logistics' has had a significant impact on the sector with the emergence of multi-channel retail and the development of multi-channel logistics to support it. The growth of online retail is transforming the way consumers shop, not least by generating multiple channels through which they can research products and make purchases. At the same time, it has added significant difficulty to the logistics operations of retailers. Instead of the relatively simple process of replenishing high street stores, multi-channel retailing is giving rise to multiple channels of distribution to service a huge range of destination points, as consumers can choose to shop at stores, order goods for home delivery or click-and-collect, and return purchased items that they do not want. At present, many companies are struggling with what this change will mean for their business, their logistics operations and their warehouse requirements.

Warehouses are used for storage of goods by manufacturers, importers, exporters, wholesalers, transport businesses, customs, etc. They are usually large open space buildings in industrial areas of cities and towns. They usually have loading bays to load and unload goods from trucks. Sometimes warehouses are designed for the loading and unloading of goods directly from railways, airports or seaports. They often have cranes and/or use forklifts for moving goods, which are usually placed on standard pallets loaded into pallet racks. Stored goods can include any raw materials, packing materials, spare parts, components, or finished goods associated with manufacturing and production.

Distribution centres are similar to warehouses in terms of being generally large in size and used for storage; however, distribution centres are the foundation of a supply network, as they allow a single location to stock a vast number of products. They will be stocked with products (goods) to be redistributed to retailers, to wholesalers, or directly to consumers. A typical retail distribution network operates with centres set up throughout a commercial market, with each centre serving a number of stores. Large distribution centres for companies such as Tesco serve 50–125 stores. Suppliers transport truckloads of products to the distribution centre, which stores the product until needed by the retail location and then distributes the required quantity of goods.

Industrial estates, also known as industrial parks or trading estates, are areas zoned and planned for the purpose of industrial development. They are usually located on the edge of, or outside the main residential area of a city, and normally provided with good transportation access, including road and rail. Examples would include the large number of industrial estates located along the River Thames in the Thames Gateway area of London or in the Team Valley area of Gateshead. The idea of setting land aside through this type of zoning is based on the ability to attract new business by providing an integrated infrastructure in one location – also to set aside industrial uses from urban and residential areas to try to reduce the environmental impact of the industrial uses.

Within the UK commercial property sector, there are difficulties with industrial property investment including problems due to inflation, the strength of the currency affecting manufacturing industry and the economic downturn. These make this sector challenging for investors and developers.

Further reading

JonesLangLasalle (2013) 'The impact of multi-channel retail on logistics', available at: www.joneslanglasalle.co.uk/UnitedKingdom/EN-GB/Pages/ResearchDetails.aspx?itemID=10883 (accessed 25/11/2013).

3.6 Retail market

Key terms: high street; supermarkets; retail parks; shopping malls

Traditional high streets are the product of a way of life that has been fundamentally superseded. From 1900 to the 1960s shopping was a daily event based on 'corner shops' and local high streets. The population was densely packed in terraced houses or semis around traditional high streets providing a captive market for an inefficient retailing offer based on independent traders. Before supermarkets, all groceries were bought over the counter at the butcher, the baker, the fishmonger in a daily excursion by stay-at-home mums. It is this lifestyle that sustained the volume and size of shops on our local high streets; thus, it is unsurprising that when society has changed so radically the traditional shopping high street is struggling to survive and has become the focus of government initiatives to 'save the high street'.

The transformation of retailing has taken place in a number of waves, first supermarkets, initially quite small and on the high street and then larger and larger super stores on the outskirts of town. In the 1960s and 1970s many city centres where 'comprehensively redeveloped' to incorporate modern shopping malls. These malls were very successful and had the advantage of reinforcing the dominance of the city or town centre as the focus of retail activity. The next wave has been out of town retail parks, often resisted by local planning authorities as they rightly predicted that they would undermine the traditional high street. The initial demand for out of town, or more accurately edge of town retailing came from bulky goods retailers, specifically carpets, furniture and white goods. These retailers originally took space in redundant warehouse or factory units on the arterial roads. The cheap space, ease of deliveries and parking outweighed the disadvantage of a non-central location. Retailing from these 'sheds' became more formalised as developers built retail units, with large shared car parks on the orbital roads of large cities and towns. These developments were very successful and attracted carpet and white goods retailers and subsequently leisure uses such as cinemas and restaurants that shared the car parking which would be underutilised in the evening. There was a significant growth of edge of town retailing during the 1970s with a consequent loss of spending power in the traditional town centres. These retail parks have evolved (and smartened up) as high street retailers, in addition to the bulky goods retailers, have joined the edge of town movement, so we now find M&S, Next and Boots taking advantage of the easy free car parking and modern, relatively maintenance-free shops to operate from.

The past 20 years has been characterised by what has been known as the 'space race' from the major supermarket groups. Supermarkets were classified under three basic headings by the Office of Fair Trading report on the groceries market:

One stop shop	Over 1,400 m^2
Mid-range	280–1400 m^2
Convenience stores	Up to 280 m^2

It is the one stop shop and the convenience stores that have been the focus of activity for the major grocery retailers in the UK. The space race alluded to earlier was a period of fierce competition when supermarkets competed for sites and sought to build ever larger hypermarkets, in the range of 10,000–14,000 m^2. These stores offered clothes, white goods, consumer electronics, DVDs and books, undermining local high streets. The attraction of large floor plates was the opportunity to offer a wide range of goods, with higher profit margins than groceries. Mid-range supermarkets were altered to increase sales space by reducing storage space (the shortfall accommodated by more just in time stock deliveries) where possible, mezzanine floors were installed to increase sales area or café facilities. The main grocery retailers, Tesco, Asda, Sainsbury and Morrison's were competing to increase floor space, retail sales and profits. The firms also innovated bank and insurance spin-offs to generate turnover and profits without needing any further floor space.

The current retailing scene is now very different as all operators have moved into online sales and the public are using the internet to purchase clothes and household goods. Currently 9.5 per cent of all retail sales (other than petrol) is online. The large supermarket chains are now finding themselves over-supplied with space, Tesco in particular are installing Giraffe restaurants and Harris and Hoole coffee shops in the stores and are negotiating with gym operators to take space. To maintain their growth in floor space they have been opening convenience stores on high streets and in petrol stations. This was originally a reaction to the difficulty in gaining planning permission for large edge of town stores but has proven to be a sound business model, bringing their high-value, high-profit-margin foods to city centre and high street stores.

Meanwhile on the traditional high street, independent traders have been squeezed by the supermarkets and the internet. Vacancy rates have increased; in 2013 overall high street vacancy rates averaged 14 per cent but within that figure there are significant winners and losers. Winners, with vacancy rates below 10 per cent, are generally characterised by a wealthy hinterland or historic centres which can attract tourists – for example, York, Harrogate, Cambridge, Salisbury, Whitstable. By contrast, the losers with vacancy rates of over 25 per cent are mid-range towns and cities in declining industrial areas – Grimsby, Wolverhampton, Blackburn, Stockton.

Boarded-up shops create a bleak shopping environment; charity shops, which used to be a disguise for low demand, are now considered prime tenants and are benefiting from the recent interest in all things 'vintage'. The government has recognised the crisis and has set up a number of financial initiatives to regenerate ailing high streets, but the logic is inescapable, there are too many shops and high streets need to be repopulated by converting shops to homes and social facilities. It will be a long and painful process that needs to be managed sympathetically to maintain the heart of the community.

Further reading

Competition Commission (2008) 'Supply of groceries in the UK', available at: www.competition-commission.org.uk/assets/competitioncommission/docs/pdf/non-inquiry/rep_pub/reports/2008/fulltext/538 (accessed 05/12/2013).

Portas, M. (2011) 'The Portas Review: An independent review into the future of our high streets', available at: www.gov.uk/government/uploads/system/uploads/attachment_data/file/6292/2081646.pdf (accessed 21/11/2013).

3.7 Leisure market

Key terms: Town and Country Planning (Use Classes) Order 1987; 'boutique' hotel; serviced apartment; Licensing Act 2003; shopping centre development; food courts; roadside restaurants

Leisure property is one of the key sectors within the commercial property market. It has its own use class (Class D2) under the Town and Country Planning (Use Classes) Order 1987 and includes many properties such as hotels, restaurants and public houses, outdoor sporting and recreational properties such as golf courses, as well as holiday centres and caravan sites. Leisure property differs from other commercial property in that the value of the property arises from and very much depends on the value of the business that is being carried on within it. Since the success of a leisure business will often depend on client loyalty, the business or property reputation, or 'goodwill', assumes a greater importance for leisure property than in other property sectors.

The hotel sector, like many other sectors within the leisure industry, is a cyclical and volatile market compared with retail and can be easily upset by events outside its particular sector such as currency fluctuations, tourism related events such as terrorism, bad weather and economic recession. Generally, city centre and provincial hotels can be categorised into very simple bands of hotels. The luxury hotels (i.e. 5 star) were traditionally limited to mainly London but more recently a number of 5 star hotels have been constructed in city centres outside London and a number of refurbished hotels are attaining 5 star status. The bulk of the business hotels in London and the provinces is in the 3/4 star market.

A recent growth has been in the budget hotels sector, e.g. Premier Inn, Travelodge, Holiday Inn Express, which nominally have a 2-star category. They are popular with the business sector and younger generation for associated social/leisure events. They often provide a more comfortable and functional bedroom and en suite facility than is normally associated with the traditional 2 star market. The 'boutique' hotel with examples such as Malmaison has now established itself in the market and provides a very high quality of individual accommodation with personalised service and bar/lounge/brasserie facilities. The latest sector that is starting to grow is the serviced apartment or 'aparthotel'. An apartment hotel uses a hotel style booking system and offers a complete apartment fitted with most things the average home would require.

The history of the public house dates back many centuries, to before Roman times. The true beginnings of the public house probably date back to when the first licences were granted, in the mid sixteenth century. The need for a licence to sell alcohol in these premises together with licences for other leisure businesses (e.g. casino activities and hot food takeaways after 11 p.m.) still remains today under the Licensing Act 2003. Over time, this has resulted in the emergence of four main categories of public houses. The first two are brewery owned, or the pub company will have a substantial shareholding owned by a brewer, or a brewer will have a supply agreement with the owner.

Managed houses. The owning company has the right to nominate all products sold through the outlet including beers, ciders, wines, spirits and all sundry supplies. All outgoings are borne by the operating company, including the salary of the manager who is an employee of that company.

Tenanted houses. These have identical ownership to managed houses, the public house is let to a tenant. The terms of occupation, historically, have been a short-term agreement with a rent or a turnover lease agreement. There will be specific purchasing and stocking obligations, particularly with regard to beer purchases. Unlike the manager, the tenant has a financial stake in the business.

Free houses. Independently owned and not tied to any specified brewer or supplier in respect of liquor supplies. Such properties can be either individually owned or form part of the assets of a multiple company.

Loan tied free houses. Similar to the above category. To a greater or lesser extent the owner, in return for financial support, or other capital sum, undertakes to purchase minimum purchasing obligations from a nominated supplier in the form of a brewer.

With regard to restaurants, this sector, like the hotel sector, has a wide variety of offerings for the public. The location and format of restaurants can be quite diverse, with footfall not always being as important as to e.g., the retail sector. If we ignore fast-food restaurants such as McDonalds, who favour high street locations, most major suburban and provincial centres do demonstrate what is known as the 'crater effect'. In essence, it can be proved that restaurants have tended to move outwards from prime city-centre positions to the periphery and destination locations. This may only be 100 metres or so, away to the periphery of a shopping centre or to locations close to travel termini, cinemas, theatres, leisure and social/drinking circuits. These positions are normally prominent, with good parking facilities, servicing and access. For these reasons, planning consent is generally slightly easier to obtain in these locations than in the prime position.

Developers of the older retail developments in town centres tended to locate restaurants out on a limb in secondary areas of malls. Developers now identify that the public need good catering facilities to enable them to stay for a longer period within the development. The popularity and emergence of the 'coffee culture' provided by Starbucks, Costa Coffee and Caffè Nero has also helped in this regard. Food courts are another concept that have been used with mixed success within shopping centre developments. This is an ever changing field and property professionals need to remain up to date with current criteria and concepts.

Roadside restaurants have also become common in the UK following a long history of success in the United States. Trunk road operators such as Little Chef, Happy Eater, Pizza Hut and Kentucky Fried Chicken, to name but a few, are constantly looking for new sites to expand. Sites range in size from 0.2 acre to approximately 1 acre as a general rule and may be drive-through or drive-to operations, of freestanding sites or linked to a leisure/retail park. Module size can be anything from 1,500 sq ft to 10,000 sq ft but most modules are between 3,000 and 6,000 sq ft. Car parking is an important factor, as is access from trunk roads. Planning consents can be more difficult as the Police and Highways Authority play a major part in consultation and decision making. If in a more residential location in the suburbs, then the likelihood of objections from local residents will be higher.

Further reading

Marshall, H. and Williamson, H. (1996) *Law and Valuation of Leisure Property*, 2nd edn, Estates Gazette, London.

3.8 The health care market

Key terms: ageing population; domiciliary care; fair maintainable operating profit (FMOP)

Due to improvements in mortality rates and an historic decline in fertility rates, the population of the UK is ageing (Office for National Statistics, 2012). Consequently, over the last decade this has had an impact on the levels of private investment in elderly care facilities, and seen the emergence of a key specialist property market in the UK in health care.

There has been a separation in recent years between the care home market and the market for domiciliary care where care is provided to clients in their homes. The majority of individuals requiring domiciliary care are referred through local social services, who then contract out the work to independent providers (see ukhca.co.uk); 89 per cent of publicly funded homecare is now provided by the independent sector, compared to 5 per cent in 1993 (Community Care Statistics, 2013), marking a move away from care homes and towards care in the home.

Many care providers offer both services, and there has been increasing encouragement from the government to assist people who wish to remain in their own homes rather than opting for a care home at an earlier stage. This has potentially placed additional pressure on care home operators as they are required to provide intensifying levels of care and deal with an increase in admissions to hospital.

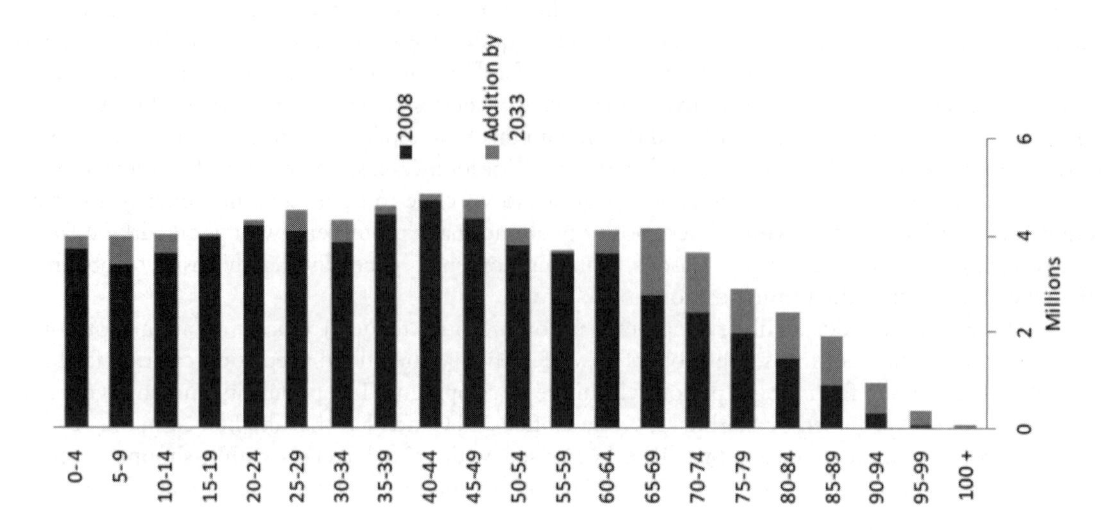

Figure 3.8.1 UK population.

Source: Government Actuary's Department (2010).

In terms of health care properties as an investment, the operational care home business and the property within which it operates are intrinsically linked, so the value of a residential elderly care home is inherent in its profitability. Therefore, market practice is to value a care home by analysing its profit and loss accounts to ascertain the financial position of the business. The valuer then uses their expertise to establish the fair maintainable operating profit (FMOP) that any competent operator could achieve from the business, and calculate the value of the investment based on this figure.

The main income source for care homes comes from weekly fees, which comprise a contribution from the relevant local authority (based on an individual's personal income), and, if necessary, a 'top-up' paid by the individual or their family. The security of this fee income is affected by a number of different external factors, and this subsequently impacts on the value to investors.

Consequently, factors such as local demographics, government policy and property market activity all impact on the value of care properties. For example, in an area with an older population and a high level of personal wealth, a care home operator could expect to achieve a high level of weekly fee in comparison to other areas of the UK. A larger proportion of this would also be a top-up payable by the individual or their family. A good example of this sort of area would be Harrogate in North Yorkshire where care home investments can achieve prime investment yields. In contrast, a poorer area of the UK with a large elderly population would see good occupancy rates for care home operators, but lower fee levels – the majority of which would be covered by local social services.

Changes to the local government fee structure for care homes and the reassessment of individuals for benefit entitlement are also likely to impact on investor sentiment. From 2016, proposed government funding reforms will deliver a new cap of £72,000 on eligible care costs (Department of Health, 2013). The concern from the sector will be how the shortfall between government funding and the cost of care will be covered.

In recent times, local authorities have failed to raise fees by a sufficient amount to cover increasing care costs. In 2010/2011, average baseline fee rates increased by 0.7 per cent, compared with estimated cost increases of 2.1 per cent. In the financial year 2011/2012, the funding gap widened with average local authority increases of just 0.3 per cent compared with estimated care home cost increases of 2.8 per cent. In 2012/13, baseline fee rate increases were slightly higher as local authorities responded to the threat of judicial reviews by providers, but the average increase of 1.4 per cent in England (1.6 per cent across the UK) was still less than estimated cost increases of 2.5 per cent in the year (Bupa, 2012).

However, the health care property market remains the focus of many investors and is subject to expansion into other areas such as care villages.

Further reading

Bupa (2012) 'Bridging the gap: Ensuring local authority fee levels reflect the real costs of caring for older people', available at: www.bupa.com/media/479673/bridging_the_gap_final.pdf (accessed 25/11/2013).

Community Care Statistics, available at: statistics.gov.uk (accessed 26/11/2013).

Department of Health (2013) 'New fairer capped funding system to help everyone plan for the cost of care', available at: https://www.gov.uk/government/news/new-fairer-capped-funding-system-to-help-everyone-plan-for-the-cost-of-care (accessed: 26/11/2013).

Laing, W. (2008) 'Calculating a fair market price for care', Joseph Rowntree Foundation, available at: www.jrf.org.uk/sites/files/jrf/2252-care-financial-costs.pdf (accessed 26/11/2013).

Office of National Statistics, available at: ONS.gov.uk (accessed 26/11/2013).

UK Homecare Association, available at: UKHCA.co.uk (accessed 26/11/2013).

3.9 Student accommodation

Key terms: higher education; student accommodation; residential investment

In the UK, participation in higher education has been on an upward trend for decades; in 2011/12 there were 1.8 million full-time students in the UK, compared to 1.45 million in 2007/8. This growth in numbers has generated additional demand for student accommodation and in parallel with this growth in demand in absolute numbers there is also an increase in demand for high quality accommodation.

In the past, universities held student accommodation as part of their institutional ethos, especially those who ran a college system. The standard of accommodation is very variable; students attending Trinity College Cambridge may be billeted in Lord Byron's old room and dine in halls, students in the ex-polytechnic may have a room in a high rise block built in the 1960s. Universities attempt to 'guarantee' all their first year and international students a place in halls, the national average of provision being 65 per cent and in London 30 per cent. Thus, there is a shortage of student halls to meet demand from first years, and additional supply may persuade students to stay in halls rather than moving into private rented accommodation in their second and final years.

In the current higher education environment student accommodation has become a key determinant in student's selection of university and university management has become more focused on providing modern accommodation to attract students. The new fees regime, at £9,000 per year plus approximately £4,000–6,000 for accommodation has made a university education an expensive product and students are demanding high quality in all aspects of the experience. The standard 'modern' university student accommodation used to be a six-bedroom 'flat', sharing a kitchen and bathroom facilities. This is deemed to be a bare minimum standard and to accommodate new expectations there has been a surge of student halls building across the sector.

This demand has created a variety of providers in addition to the traditional university halls. Two models have emerged, joint ventures and independent providers. Joint ventures between developers and universities have become popular in the past decade. The developer finances, builds and manages the facilities and the university adds the accommodation to their 'portfolio' of accommodation without any capital outlay. For the university this is a zero-cost means of increasing their accommodation offer, the developer consortium is provided with a low risk investment and income stream, provided their accommodation is attractive and the university can attract students. The nature of the joint venture will be negotiated between the developer and the university as regards design, rents, length of nomination agreement and share of income. The largest provider in the sector is Unite which has approximately

25 per cent of the market and 40,000 beds let. They have a vigorous expansion programme recently funded by a £380 million bond issue, with a 3.4 per cent return, which was three times oversubscribed.

Independent providers operate in major university cities without a specific link to a university. Nido is a typical example of this type of operation, with three blocks of halls in London; the Spitalfields block, for example, accommodates 1,200 students over 33 floors, as well as en suite rooms and wi-fi which has become standard. Facilities include 'chill out spaces with giant flat screen televisions and an ultra-modern gym and spa'. It is no coincidence that they have three premises in London – London has 290,000 students, larger than the next five student cities put together and the highest concentration of overseas students, who are most likely to live in university halls.

Student accommodation has become a new investment vehicle; annual investment in the UK has increased from £400,000 in 2009 to £2.7 billion in 2012. With an undersupply in the market there are prospects for growth and lots of opportunity to replace elderly halls built during the 1970s higher education boom. Figure 3.9.1 shows halls built in the 1970s – not old enough to be interesting, not good enough to be attractive.

Figure 3.9.2 shows modern halls under construction – 1000 en suite bedrooms above a supermarket – rooms have fast wi-fi, the accommodation includes fitness facilities and multi-use games area. There is also a multi-screen cinema on site, as well as bars and restaurants.

Figure 3.9.1 1970s student accommodation.

Source: John Holmes.

Figure 3.9.2 Modern student accommodation.

Source: John Holmes.

Further reading

Hubbard, P. (2009) 'Geographies of studentification and purpose-built student accommodation: Leading separate lives?' *Environment and planning*, A 41(8): 1903.

3.10 Building information modelling and commercial property

Key terms: BIM; 3D modelling; facilities management; service charges

This technology is used as standard in construction for new developments in both public and private sectors. It is now a requirement for £5 million-plus public sector contracts, as identified by government strategy in May 2011. Quantity surveyors are embracing it and more recently building surveyors are applying it on schemes relating to existing buildings. It can be defined as:

> The process of Building Information Modelling is the action or the creation and maintenance of a database of information relating to a building. This data is digital and is recorded in such as way as to serve as a useful tool for the future life of the building. For example, details of structure and specifications of windows would make the task of replacing windows quicker and easier. The data can be used to inform maintenance decisions potentially saving time and money over the full lifecycle of a building.
>
> The BIM process can be conducted at different levels, depending on the specifications of the client, represented by either a project or facilities manager. It is important to determine what level of detail is required from the outset as this will inform provision of digital storage space and to some extent what system the model will be created in. Broadly the level of detail can be equated to cost, the more detailed and complete a building information model the higher the cost and the greater the potential benefits and cost savings in the long run.
>
> (isurv, 2013)

A BIM model comprises two distinct sets of information:

1 A 3D visual model drawn in, for example, Autodesk Revit Architecture.
2 Detailed information on the construction and design of the building. It will comprise a full specification of all component parts of the building fabric.

The original design team use BIM design software to create a multidisciplinary model. That model can then be used by others to assist their own needs. Examples can be seen at www.bentley.com, www.autodesk.co.uk.

There is also a range of additional BIM software that allows the end user to manage the building. Some examples are: www.kykloud.com, www.fmsystems.com, www.artra.co.uk, www.archibus.com.
BIM will be an effective tool to assess and, therefore, improve the energy efficiency of buildings. This will become more important as the energy efficiency requirement of property is increased in 2018 and after. The essential aim is to create one model in which all information about a building is stored for the duration of its life. Copyright might be an issue. The developer/investor should specify the use of a BIM model.

New build versus existing property

It is estimated that around 60 per cent of the property that will be here in 2050 has already been built and most will not have a BIM model. In the current economic climate there is limited new construction which is why BIM has been used over the past few years.

New build – it is relatively easy to create and apply a BIM model in a new build as the visual aspects of it will be generated from the design team plans and the information behind that will come from the construction process.

Existing buildings – the required information must be obtained from onsite survey and inspection or laser surveys. This is potentially viable where a major refurbishment is planned. Where existing floor plans exist it may be possible to use them, together with further site inspection to create the visual model. It will assist in change of use and adaptation of buildings.

Potential uses for the commercial property surveyor

- *Development appraisal*
  This facility should be able to aid both the planning and the construction processes especially where variations or changes are made during the construction process. The risk factor during construction will be reduced, thus decreasing the need for contingencies and ensuring work is done on time. This will assist the valuation process done for acquiring funding for the scheme.
- *Agency*
  The information and visual aspects of BIM should assist in marketing during the development process to obtain a pre-let – to assist with layout or specific design requirements of occupiers etc. The visuals would be best enhanced using specific 3D modelling software such as 3Dstudiomax.
- *Property management*
  The information should enable more accurate estimation of service charges throughout the life of the building. The management of this process could be added in to the model by additional compatible software to include all key lease information, rents etc. (see www.fmsystems.com, which Jones Lang Lasalle support). Detailed construction information and models will make future internal alterations easier. In an existing building, records could be kept of problem areas, e.g. the existence of asbestos.
- *Facilities management (referred to as 6D BIM)*
  The model will enable the facilities manager to deal with maintenance and operational issues, alterations and refurbishment of buildings etc. in a more effective way, taking into account factors such as life-cycle costing. They should be involved from the beginning of the process.
- *Investment sale*
  The provision of all information relating to a building being held in one model should make the 'ready for sale' state more easily achieved. This will make property more liquid as an investment, reducing the process of due diligence and avoiding the need to input all lease information about the building each time it is transferred. As a result there should be less uncertainty and more transparency.

Further reading

isurv (2013) 'What is BIM?' available at:
www.isurv.com/site/scripts/documents_info.aspx?documentID=7216&categoryID=1246 (accessed 21/11/2013).

4 Construction

Graham Capper, Barry Gledson, Richard Humphrey, Eric Johansen, Ernie Jowsey, Mark Kirk, Cara Hatcher and John Weirs

4.1 Building Cost Information Service

Key terms: RICS; reinstatement valuation; insurance valuation; tendering; indices

The Building Cost Information Service (BCIS) is a part of the Royal Institution of Chartered Surveyors (RICS) and deals with five areas of surveying and construction:

- construction;
- maintenance;
- rebuilding;
- consultancy;
- insurance.

The BCIS and RICS offer advice and guidance for clients on the above areas of resource and offer statistical information on indices in the relevant surveying and construction sectors. The BCIS is subdivided into several indices that are used to determine the cost involved with building various schemes, including the following:

- BCIS cost indices;
- BCIS tender price indices;
- BCIS regional price indices;
- BCIS output indices;
- BCIS trade indices;
- construction new orders;
- construction output;
- house prices;
- housing starts and completions;
- resource indices;
- retail prices.

BCIS software is actively used in real estate in the form of valuation for reinstatement or insurance valuation purposes. It is calculation of relevant constructional data prices and rates in conjunction with the size, style and type of property. Reinstatement valuations are an integral part of any residential valuation and are also, when required, actively used in commercial valuations.

The purpose of a reinstatement valuation is to provide a valuation figure based on the rebuilding of a property if required. This is based on the square metre of the building and a figure is applied to determine the cost of rebuilding. The reinstatement figure is as important as the valuation of the property, as the purchaser will rely upon this to insure their property. If the figure is incorrect, it may well not be sufficient to rebuild if ever required.

Table 4.1.1 looks at specific pricings that materially affect the construction market. Interestingly, from an estate management context, retail prices are anticipated to increase from 2013/2014 year on year until 2017/2018. The BCIS have also predicted that building costs (as shown in Table 4.1.2) continue to increase steadily.

Cost is a major driver for all construction projects and the BCIS offers information on tendering, indices and reinstatement issues that can be utilised by a client, contractor, developer and valuer in order to provide comprehensive advice on a range of costing issues that might arise when undertaking cost analysis of a project.

Table 4.1.1 Construction costs

Annual % change	*2Q2011 to 2Q2012*	*2Q2012 to 2Q2013*	*2Q2013 to 2Q2014*	*2Q2014 to 2Q2015*	*2Q2015 to 2Q2016*	*2Q2016 to 2Q2017*	*2Q2017 to 2Q2018*
Tender prices	+2.7%	+2.2%	+2.6%	+3.3%	+3.6%	+7.4%	+8.3%
Building costs	+1.6%	+1.3%	+1.9%	+2.8%	+3.0%	+3.5%	+3.7%
Nationally agreed wage awards	+1.1%	+1.6%	+1.9%	+2.7%	+3.3%	+3.8%	+3.9%
Materials prices	+1.2%	0	+1.1%	+2.6%	+2.9%	+3.2%	+3.5%
Retail prices	+3.2%	+3.1%	+2.6%	+3.3%	+3.2%	+3.1%	+3.3%
Construction new work output	−11.3%	−1.9%	+1.6%	+3.7%	+5.4%	+6.6%	+7.3%

Source: BCIS (2013).

Table 4.1.2 Building costs 2013–2018

	Forecast	
Year on Year	*Jul 2013*	*Sep 2013*
2Q2013 to 2Q2014	+1.9%	+1.9%
2Q2014 to 2Q2015	+2.8%	+2.8%
2Q2015 to 2Q2016	+3.0%	+3.0%
2Q2016 to 2Q2017	+3.8%	+3.5%
2Q2017 to 2Q2018	+3.7%	+3.7%

Source: BCIS (2013).

Further reading

BCIS (2013) BCIS, available at: www.bcis.co.uk/ (accessed 13/11/2013).

RICS (2013) 'BCIS overview', available at: www.rics.org/uk/knowledge/bcis/, (accessed 13/11/2013).

4.2 Building control in England and Wales

Key terms: Building Act 1984; services; fittings; elements; material change of use; consequential improvements

Why do we need regulations for building work?

Regulations for building work were initially introduced in the 1980s in order to protect the health and safety of the public after a link was discovered between poor construction and spread of disease. The problems identified as causing this were:

- damp;
- poor ventilation;
- poor lighting (natural and artificial);
- lack of adequate sanitary facilities;
- clean water;
- refuse disposal.

Although the above are still important today, the current building regulations are a major tool in implementing the Government's mandate for reducing carbon emissions from buildings.

The principal piece of legislation that covers building regulations is the Building Act 1984: there are 135 sections and seven schedules arranged in five different parts. The Act gives powers to the Secretary of State to make regulations. The Act also for the first time introduced a set of approved documents that provide technical guidance for designers when submitting plans for a development to building control.

Part 1 of the Act is the Building Regulations 1985; the current version is the 2010 edition. This is quite a small document with only a few pages of legal text – revised in 1994, 1997, 1999, 2000 and 2010 and to be revised again and again.

The principles on which these new regulations were based in 1984 were:

- maximum self-regulation;
- minimal government interference;
- simplicity of operation;
- totally self-financing.

The 1985 Regulations contained 20 administrative, functional and performance requirements and there were 12 approved documents with 377 pages. However, the current 2010 Regulations have 18 approved documents with 877 pages, which contradicts the ethos of the act.

The Building Act 1984 also introduced the provision of building works to be undertaken by approved inspectors. Prior to this piece of new legislation all building work as certified by the regulations was to be administered by the local authority. The aim of this was to allow competition from the private sector which would reduce costs and increase efficiency.

Further reduction of local authority control of the regulations is the introduction of self-certification schemes as detailed in Part 5 of the Building Regulations. Self-certification allows competent persons to assess specific work carried out to the relevant standards. These include the following:

Gas Safe	Work on gas appliances
OFTEC	Work on oil fired appliances and oil storage tanks
HETAS	Work on solid fuel appliance
FENSA	Replacement windows and doors
NICEIC	Specific work to electrical safety in dwellings

Part 2, Regulation 3, of the Building Regulations 2010 states the meaning of building work that needs to be controlled by a relevant body, i.e. local authority, approved inspector or competent person. The following are defined as building work:

- erection/extension;
- provision or extension of a service or fitting;
- material alteration of a building, service or fitting;
- material change of use, as defined by regulation 5;
- insertion of insulating material into the cavity;
- underpinning;
- work required by Regulation 22 (relating to a change of energy status);
- work required by Regulation 23 (relating to thermal elements);
- work required by Regulation 28 (consequential improvements to energy performance).

The above states terminology which is not plain or clear to a person not familiar with the regulations, these are:

Service	These include fixed heat producing appliances, baths, toilets, sinks, wash basins, controlled heating systems and domestic electrical installations
Fitting	These include external windows and doors
Element	These include floors, walls and roofs

Material change of use is defined in Regulation 5 and is in connection with changing the use class of a build, i.e. dwelling house to a flat, etc.

Consequential improvements relates to proposed works to existing buildings (other than dwellings) with a total useful floor area of over $1000m^2$, and including:

- an extension;
- installation of new fixed building services (other than renewable energy generators);
- increasing the capacity of fixed building services (other than renewable energy generators).

Where such works are proposed, additional consequential improvements will be required to make the whole building comply with Part L of the Building Regulations to the extent that such improvements are technically, functionally and economically feasible. In most circumstances, this means that the payback period for the consequential improvements required does not exceed 15 years (or less if the expected life of the building is less than 15 years).

Consequential improvements could include:

- upgrading heating, cooling or air handling systems;
- upgrading lighting systems;
- installing energy metering;
- upgrading thermal elements;
- replacing windows;
- on-site energy generation;
- applying measures proposed in a recommendations report accompanying an Energy Performance Certificate.

Schedule 2, Regulation 9 of the Building Regulations states areas of work that are exempt from the regulations:

- Class 1:
  – any building subject to Explosives Act 1875 and 1923;
  – any building falling under provisions of Nuclear Installations Act 1965;
  – any Scheduled Monument under section 1 of the Ancient Monuments and Archaeological Areas Act 1979.
- Class 2: buildings not frequented by people;
- Class 3: greenhouses and agricultural buildings;
- Class 4: temporary buildings (28 days);
- Class 5: ancillary buildings – site cabins;
- Class 6: small detached buildings:
  – detached single storey, floor area less than 30 m^2, no sleeping accommodation and is a building;
  – no point of which is less than 1m from the boundary; or
  – it is constructed of non-combustible material;
  – a detached building designed and intended to shelter people from the effects of nuclear, chemical or conventional weapons, and not used for any other purpose;
  – a detached building having a floor area that does not exceed 15 square metres, which contains no sleeping accommodation.
- Class 7: extensions:
  – a conservatory, porch, covered yard or covered way; or
  – a carport open on at least two sides;
  – where the floor area of that extension does not exceed 30 m^2, provided that in the case of a conservatory or porch which is wholly or partly glazed, the glazing satisfies the requirements of Part K (Glazing).

Further reading

HMSO (1991) *The Building Act 1984*, The Stationary Office, London.

HMSO (2010) *Statutory Instruments No. 2214 Building and Buildings, England and Wales: The Building Regulations 2010*, The Stationary Office, London.

4.3 Construction firms

Key terms: small firms; subcontracting; linked processes; economies of scale; diseconomies

The construction sector is a key sector for the UK economy and comprises a wide range of products, services and technologies including:

- the construction contracting industry – providing 2,030,000 jobs in 2013 in 234,000 businesses and contributing £63 billion to the UK economy;
- the provision of construction related professional services – 580,000 jobs in 30,000 businesses and £14 billion;
- construction related products and materials – 310,000 jobs in 18,000 businesses and £13 billion;

In total, construction contributes almost £90 billion (about 6.7 per cent of GDP) to the UK economy from 280,000 businesses and almost 3 million jobs which is about 10 per cent of total UK employment (DBIS, 2013).

Firms in the construction industry vary in size from the large civil engineering/building contractors, often undertaking such types of construction work as roads, bridges, office blocks, shopping centres and speculative housing estates, to one-man labour-only contractors such as bricklayers, plumbers, carpenters and jobbing maintenance firms.

Large construction firms can obtain the advantages of large-scale production that economists call economies of scale. These include technical economies of specialised equipment and linked processes; commercial economies in buying materials; specialisation in management, such as their own surveyors and legal experts; the ability to raise finance on cheaper terms; and the spread of risks through diversification into many products, including international contracts. For small or even medium-sized contractors, spreading risks is difficult since a single contract may account for a large proportion of their work.

Nevertheless, the small firm predominates: those with less than twenty-five workers (including the self-employed) covering 96.8 per cent of all firms. However, such small firms account in value for only 29 per cent of total work. Large construction firms with more than 600 employees are only 0.06 per cent of the total number of firms but they account for almost 25 per cent of the value of construction work done (Jowsey, 2011).

This predominance of the small firm is more marked in the construction industry than in other industries such as manufacturing, a phenomenon common to other developed economies. The reasons are to be found on both the demand and supply sides.

Demand in construction is characterised by the dominance of relatively small contracts resulting from the individuality of the product (for example, a single house or extension) and also the importance of minor repair work. In addition there are the fluctuations in demand, which induce firms to remain small and flexible rather than burden themselves with high overheads resulting from specialist equipment that is under-used.

Supply factors in construction also favour small firms. Since production is on site rather than in a factory, the difficulties of supervision are greater. With a large firm carrying out work on many sites, the difficulties of management control are magnified. As a result, management diseconomies of scale may soon outweigh technical, commercial and other economies.

Vertical disintegration in construction through the employment of specialist firms results from the consecutive nature of the operations and the lack of continuity of the work for specialised equipment. This leads to comparatively small firms contracting for pile-driving foundations, roofing, electrical work, heating and ventilation systems, lift installation, and so on. Such subcontracting has advantages to the main contractor, because (apart from saving on overheads) it makes estimating easier and reduces on-site supervision. On the other hand, it can increase the problem of coordinating the various construction operations. For a typical large building project of £20–25 million, the main contractor may be managing around 70 subcontracts of which a large proportion are small – less than £50,000 (DBIS, 2013).

As with other types of production, small businesses exist in construction because the owner is prepared to accept a lower return in order to be his own boss. Entry to the construction industry is fairly easy, as many jobs require little capital equipment (if specialist tools and equipment are necessary, they can be obtained by hiring or subcontracting). Like most other businesses, however, builders find difficulty in obtaining capital to expand beyond a certain size, for their sources are, to all intents and purposes, limited to merchants' trade credit, personal savings, bank overdrafts and ploughed-back profit. Many small builders try to operate on too limited a cash flow, even though progress payments are received on an architect's certificate. Fluctuations in demand can prove financially crippling even for medium and large firms. As a result, the rate of bankruptcies in the construction industry tends to be much greater than that of other industries.

Further reading

DBIS (2013) 'UK construction: An economic analysis of the sector', Department for Business Innovation and Skills, available at: www.gov.uk/government/uploads/system/uploads/attachment_data/file/210060/bis-13-958-uk-construction-an-economic-analysis-of-sector.pdf (accessed 21/10/2013).

Jowsey, E. (2011) *Real Estate Economics*, Palgrave Macmillan, Basingstoke, ch. 13.

4.4 Competitive tendering

Key terms: estimating; tendering; selective tendering; measured term tendering; serial tendering; target cost tendering; successful bids

Competitive tendering is the process used to obtain prices from two or more companies to carry out work or provide goods and services. It involves estimating the cost of supplying all the resources required and submitting the final bid (tender sum) to the client. The vast majority of construction companies rely on competitive tendering as a way of winning work (Harris *et al.*, 2006), while it is the method most commonly used by construction clients for contractor selection (Winch, 2010). Competitive tendering is an expensive process, both in terms of time and money, and is estimated to account for approximately 10 per cent of a contractor's turnover (KPMG, as quoted in Winch, 2010).

Estimating versus tendering

The terms estimating and tendering are often used interchangeably; however, it is important to differentiate between the two.

- *Estimating*: the process of preparing a reasonably accurate calculation of the cost of carrying out the work. A company's estimating team, using the tender documents provided by the client, will consider all labour, plant, materials and timescales required for each section of the work. From this, the estimator can calculate a reasonably accurate cost of the work.
- *Tendering*: the offer to carry out the work for a stated price. Once an accurate estimate is produced, the company will hold an *adjudication meeting*. Senior management will consider the estimate as a whole and agree the percentage 'mark-up' to be added to cover company overheads and profit. The percentage should ensure the contractor's final tender sum is not only profitable but also sufficiently competitive to give them the opportunity of winning the work. All construction projects carry an element of risk, be it production, physical or economic risk and it is during the adjudication process where the contractor will give due consideration to the risks in undertaking the project.

There are several methods of competitive tendering commonly used in the construction industry, including the following.

Open tendering: companies are invited to tender for a project by responding to an advertisement placed by a client. Anyone can request the tender documents but, due to the cost of producing them, a deposit may be requested prior to them being sent out which would be refundable on the submission of their tender. This prevents companies requesting the tender documents out of interest only with little intention of submitting a bid. Open tendering is commonly used for public sector work where, under European legislation, the opportunity to tender must be made widely available. With this method, a contractor's competency to carry out the work is established after tendering but before the contract is awarded.

Selective tendering: with this form of tendering, the client will draw up a list of preferred bidders. The select list is normally drawn up after a pre-selection process whereby the suitability of the contractors is established. Pre-selection will consider factors including financial stability, experience of carrying out similar projects, management structure, and their health and safety record. Therefore, contractor competency is established prior to the tender process.

1 Single stage selective tendering:
  As the name suggests, there is only one tendering stage. Pre-selected contractors are invited to bid for the work and, after due consideration of all tenders, the successful contractor is appointed.
2 Two stage selective tendering:
  This type of tendering is used to encourage early contractor involvement. During the first stage, all selected contractors bid for the work as in single stage tendering. The client will then choose a preferred contractor to go through to the second stage which will involve some cost negotiation but, importantly, early integration into the design team.

Measured term tendering: where a client has regular work over a specific period, (eg. maintenance work), a measured term form of tendering can be used. A schedule of rates is submitted by the preferred bidders, which is used to select the successful contractor. These rates are then applied to the work as it is carried out.

Serial tendering: this form of tendering is often used where a client has a series of projects of a similar nature. A schedule of rates is submitted based on a sample project in the knowledge that a number of similar projects will be carried out.

Target cost tendering: the contractors bid is prepared using basic project information provided by the client, with the successful tender forming the basis of the target cost (sometimes referred to as a 'guaranteed maximum price'). The contractor will be paid the actual cost of carrying out the work providing it does not exceed the original tendered target cost.

Successful bids: while it is the lowest priced tender that is often the main criteria used by the client for contractor selection, it should be noted that, on occasion, cost might not always be the client's main concern. Speed of construction may be the primary concern, particularly where their business depends upon the final building being operational as early as possible. In this instance, providing the tender falls within an acceptable cost range, the contractor indicating the earliest completion date may be successful in winning the tender.

Further reading

Harris, F., McCaffer R. and Edum-Fotwe, F. (2006) *Modern Construction Management*, 6th edn, Blackwell Publishing, Oxford.

Winch, G. M. (2010) *Managing Construction Projects*, 2nd edn, Wiley Blackwell, Oxford.

4.5 Design and build

Key terms: design and build (D&B); employer's requirements (ERs); contractor's proposal (CP); single-stage tendering; multi-stage tendering; novation

The design and build (D&B) form of procurement is often the preferred option for clients who favour both a *shorter overall project period* – that is the period from project inception to final completion, as opposed to just the on-site construction period – as well as *greater certainty over the planned project costs*, and are happy to give greater importance to these project objectives *over the aesthetic considerations*.

When preparing to enter into a D&B arrangement, a set of highly prescriptive employer's requirements (ERs) must be prepared by the client design team which must then be responded to or 'met' in the form of the contractor's proposals (CPs) prepared by the bidding contractors.

One simplistic theory when using a D&B form of procurement is that the quality of the design will suffer due to the lower priority allocated to this aspect of the project, but in reality clients requiring physical assets built to service a particular business need are often better off in engaging the services of contractors who specialise in constructing that particular type of development – industrial, residential, commercial etc. – as in many instances the nature of these particular structures means that the contractor often has partially prepared 'off the shelf' design solutions ready to go.

On one end of the spectrum D&B contracts are often the procurement route of choice awarded when fairly simple structures with a clear and predetermined business need such as new build offices, warehouses and schools are required, as often the nature of these products is that they are so similar in many aspects of their design (structural spans, amount of space to be allocated to each predetermined area etc.) that they become *variations on a theme* and the fabled 80/20 standardised design/bespoke design ratio comes into play.

D&B can also be the choice of clients on more complicated projects who are particularly risk averse and want one party to take ownership of project risk and have determined that, because of the nature of the project, the contractor would be best placed to manage that risk. Other client considerations include a wish to avoid playing the part of referee between contractor and designer whenever a dispute arises, or a wish to avoid the additional expenditure of appointing an external project manager to act as a client's representative to perform this duty – although a project manager can also be appointed regardless to manage a D&B contractor.

The 'design' part of the D&B process can be done in a number of ways. Some major contractors have *in-house design expertise* that can match or indeed better any external design practices. This expertise is not limited to any particular design discipline, and it is not unusual to find contractors who can offer a full range of internal design provision: architectural, civil, structural and building services design. Alternatively companies may have only one or two of these resources – for example, the civil and structural design services are 'in house', with the remainder of the design team appointments made up of external partners to complement the existing team. These partners could come from designers who have pre-existing contractual arrangements with the client or they could be *subcontract design companies* that are employed directly by the main contractor either through *competitive tendering* or existing company-to-company *partnering* arrangements.

Companies with this type of partial design provision and frequent subcontract design partner agreements can have arrived at this business model, either as the result of clear strategic planning, or by organic growth, through the purchasing and absorption of other companies, or by employing multi-skilled specialist resources.

For companies who do not have in-house design provision – or who do have it but do not want to use it on a particular project, then novation of existing project designers often takes place. This can occur either on a traditionally tendered project (single stage tendering) or on higher value–higher risk projects that employ multi-stage tendering strategies such as 2-stage tendering. In essence, this is when a client will have engaged a design team to produce initial, fairly comprehensive tender documentation and would like a fairly early appointment of a major contractor in order to benefit from their expertise in construction.

Design novation is an arrangement that occurs when members (all or some) of the original design team having fulfilled their original contractual obligations to the client by producing initial project documentation up to the level meeting a predetermined RIBA Plan of Work stage, and at a specific period of the project – typically at the end of the period where building contractors have competitively tendered for the project – would then be re-employed on the project, this time by entering into a direct contractual relationship with the successful bidder – the building contractor – rather than in a new contract direct with the client, during the construction phase of the works.

It is important to note that under this arrangement, the designers then effectively become design subcontractors to the main contractor and will have to complete their design duties under the management of their main contractor who may well want to make amendments to the design such as material substitutions. These could be for several reasons such as ease of procurement (shorter lead times), improvements in constructability, or more commercial reasons such as reductions in areas of project cost.

A frequent area of concern in this arrangement is when design team members continue to act as client consultants rather than as design subcontractors and continue to bypass the main contractor in dialogue with the client, particularly in aspects where they consider the quality of the design to be at risk. This is a key challenge for the main contractor to manage in novated design and build contracts.

Further reading

Hackett, M., Robinson, I. and Statham, G. (2007) *The Aqua Group Guide to Procurement, Tendering and Contract Administration*, Blackwell, Oxford.

A full and comprehensive explanation of all available procurement strategies is provided in this highly recommended text.

RIBA (2013) 'RIBA Plan of Work 2013' Available at: www.architecture.com/TheRIBA/AboutUs/Professionalsupport/RIBAOutlinePlanofWork2013.aspx (accessed 1/11/2013).

4.6 Modern methods of construction (off-site manufacture)

Key terms: modern methods of construction (MMC); off-site manufacture (OSM); modular construction; lean construction; volumised and panelised construction

The sub-optimal performance of the construction industry has been identified repeatedly in multiple UK construction industry reviews from the 1940s until the present day. Recurring criticisms have been that the industry is fragmented, adversarial, unwilling to measure performance and ineffective in resource utilisation.

One early attempt to improve the efficiency and effectiveness of the industry was the move after the Second World War to develop 'system building' based on manufacturing. These were standardised on- or off-site systems that were used to deliver repetitive buildings from single dwellings to high rise flats to schools. Unfortunately the quality of many of these buildings was considered to have failed and many were demolished. Later attempts to use timber-framed housing also encountered technical quality problems and were mainly abandoned. For many years the industry mistrusted these types of approach.

One of the more prominent of the later industry reviews was *Rethinking Construction* (Egan, 1998), which highlighted client concerns of inefficiency and waste and advocated greater industry integration. This report focused on additional areas requiring attention that included 'low industry profitability;

low investment in research and development (R&D); inadequate training, and low levels of client satisfaction' (Egan, 1998: 7) Several drivers for change were identified including the need to focus on a 'quality driven agenda' (Egan, 1998: 4) and provided several recommendations for 'improving the project process' (Egan, 1998: 18), such as applying lean thinking in construction and innovating in design and assembly.

These reports led to several movements seeking to reform poor performance and promote construction 'best practice'. A notable response to the recommendations in *Rethinking Construction* was the move to reconsider manufacturing-based approaches through the modern methods of construction (MMC) initiative which was heavily promoted in the early 2000s, particularly by organisations within the residential sector, as a solution for reducing waste and time spent on site, improving site safety and quality, and addressing industry skills shortages.

MMC is a general term for construction technologies that include elements of off-site prefabrication or assembly. There are many other names given for technology that utilises these principles, including (but not limited to) off-site manufacture (OSM), pre-fabrication, and modular construction.

There are two broad categories of MMC:

- panelised construction elements – factory built flat panel units ready for on-site assembly and incorporation into or onto a structure. Typical examples would include external cladding panels that are made up of several different components already joined together prior to site delivery.
- volumised construction elements – factory built units either of a smaller scale for incorporation into a structure or larger-scale units that make up the structure itself.

Smaller-scale units may typically be complete self-contained bathrooms or kitchen 'pods' used in multiple repeat units such as those found in residential type (hotel or large-scale apartment block) projects. Benefits of having these units constructed off site include achieving consistency in the finished quality of the product, due to the greater quality control procedures that factory conditions enable, and reductions in on-site construction time and labour as multiple 'wet trades' operatives such as plasterers, decorators, plumbers etc. are moved off site.

Larger-scale volume units may typically be semi-completed sections of buildings that are constructed off site, transported to site and positioned on foundations and quickly joined together on site. These modular structures reap the benefits of standardisation in design and construction and are repeatable. They are typically used to rapidly provide a new facility for an established business in a new geographical location. An ideal example of this is how fast food restaurants are able to have new facilities constructed in the space of a few days due to a 'plug and play' system of construction. There are also several subcategories to MMC that include hybrid systems that integrate aspects of both volumetric and panelised systems.

At present the use of off-site components is increasing in the industry and things such as 'bathroom pods' are commonly seen. However, the use of panelised or volumised larger-scale units is still in its infancy. Still, there is much interest and expectation within the industry about the possibilities for this to improve effectiveness and efficiency in the same manner that the car industry has seen the use of Design for Manufacture and Assembly (DFMA) to deliver different types and brands of cars using up to 85 per cent of common components.

Further reading

Cain, C.T. (2003) *Building Down Barriers: A guide to construction best practice*, Spon, London.

Constructing Excellence (2013) 'Modern methods of construction', available at: www.constructingexcellence.org.uk//resources/az/view.jsp?id=718 (accessed 4/11/2013).

Egan, J. (1998) *Rethinking Construction: The report of the Construction Task Force to the Deputy Prime Minister, John Prescott, on the scope for improving the quality and efficiency of UK construction*, HMSO, London.

4.7 Managing construction

Key terms: construction management; integration; communication; design/construction interfaces

This is about the 'production' of the built asset. It is difficult to consider as a stand-alone issue because its success is so heavily influenced by other parts of the process. Certainly the following have a major effect on what happens on site:

- The procurement system – other sections in this book deal with aspects of procurement but for the construction manager the key issue is that procurement and the form of contract used tends to control the times at which the parties get involved and to some extent the way they work with each other.
- The timing of involvement of the production side of the supply chain is particularly important – construction is one of the few industries that separates design and production. There are clear advantages to having those who build talking to the designers as they produce the design and offer advice on how buildable the designs are, what the possible problems might be and alternative solutions.
- The main contractor's approach to the appointment of supply chain companies (subcontractors) – few main contractors do much work directly these days. They contract the work to a supply chain (usually referred to as trade or subcontractors). Relationships with these are vital but can vary from 'long-term' relationships in which trust has been developed and best practice solutions are expected; to 'cheapest price' tendering where in many cases the people working on site are an unknown quantity and the contract is relied on to force performance.
- The quality of the management of design/construction and other interfaces – management works well if you use the best people; communicate as close to perfection as possible; and do as much as you can up front before the actual work takes place. This applies to design or production. However, that is easy to say but difficult to put into practice in an industry that has been characterised so much in the past by the 'discontinuities' between the parts of the process. In many ways it is like a dysfunctional family and the long list of reports into the industry since just after the Second World War has highlighted the need for improvements in integrated working, communication and the ways the parts interface with each other.

The industry has traditionally measured itself against the delivery of the project to meet time, cost and quality performance. In the later part of the twentieth century the issue of safety was also added to these headline performance measures. What has become obvious is that the management measures that ensure high performance against one of these key indicators are also vital in delivering the others. Of course what is vital to the project is that any performance indicators arise from a clear understanding of what the client requires but again the management measures on the best projects also ensure that this happens.

While it is possible to argue that a case study taken at a moment in time is not necessarily the best guide to good practice because the world is a complex and changing place, it is more true to say that good practice is generally applicable at any time. Thus, if we look at a specific report for a specific project and take some of the lessons learned from that we can see an excellent and long-standing exemplar for the industry. While there has been much failure in construction over time there have also been excellent examples of good practice. The delivery of the 2012 London Olympics, for example, was generally a triumph for the construction industry, yet many of those involved in it said that they felt that it was no different to other projects and what they had done could be applied anywhere.

We can learn much from the way the Olympics delivered an excellent health and safety record because the issue was embedded in good practice which also helped deliver the other performance indicators. Bolt *et al.* (2012) produced a report for the HSE in which the following issues were considered vital to the success of the Olympics:

- From the beginning – 'leadership and resolve' existed from the highest levels.
- Objectives were developed early and clearly by an integrated team which had built up trust throughout all the actors – 'clients, contractors, designers, workers and regulators'. This allowed relationships and trust to develop quickly in which people's ideas were valued.
- Communication was also an early focus and robust systems were introduced to build on and assist the relationships and trust.
- Agreeing the deliverables and being consistent in providing them was seen as vital for success.

Obviously the Olympics were a huge undertaking and there were many more practical systems and technologies that ensured its success, but essentially these were rooted in the above.

Further reading

Bolt, H. M., Haslam, R. A., Gibb, A. G. and Waterson, P. (2012) 'Pre-conditioning for success: Characteristics and factors ensuring a safe build for the Olympic Park', Research Report RR955, prepared by Loughborough University for the Health and Safety Executive. Available at: www.hse.gov.uk/researcH/rrpdf/rr955.pdf (accessed 13/11/2013).

4.8 Planning and organising construction

Key terms: planning; logistics; site organisation; critical path programming

Essentially the construction manager (CM) succeeds by managing people through excellent communication and good planning. This means that there is an organisation that comes together to allow the work to be planned by considering the management of resources, quality and time. It requires data to allow decisions to be made in advance of action and decisions are usually optimal – that is they are the best solution that balances all of the inputs and produces the best output possible given this balance.

Concept 4.7 dealt with aspects of communication and clarity of definition of deliverables. Here we will consider the practical issues that allow those deliverables to be achieved.

Organisation

This is often seen as the formal relationships between those involved in the process and these are usually expressed using an 'organogram' or 'family tree' diagram that can show the contractual relationships between the parties or the managerial relationships between individuals within a group. For the CM the most useful is the managerial version which indicates the relationships and patterns of accountability between staff. However, once this is captured it becomes simply a signpost, the actual management of the communication between these staff and the contractually linked entities (e.g. subcontractors, suppliers or the design team) requires the CM to set up a communication and decision-making process usually based around a schedule of meetings (either physical or virtual).

Logistics

In addition to the organisation of people, the other resources require to be organised. These include:

- the people, in terms of their location and accommodation, to allow the work to be carried out effectively and safely;
- the materials, so that they can be accessed and used easily and so that waste is minimised;
- the plant and machinery which are needed to allow the people and materials to be used effectively.

The way these are organised is called logistics and it is closely linked with planning and scheduling. The key variable for decisions about logistics is the available room on site. A greenfield site is likely to have fewer logistics problems than a tight city-centre site. Much thought is put into ensuring the logistics are planned and prepared and the physical signs of this are usually seen in a 'site plan' (or plans). This is usually a location plan of the site in which the positions of the following are shown:

- temporary accommodation (assuming a decision has been made to put something on site rather than hire something nearby). This includes all the health and welfare facilities required;
- storage areas including where storage containers can be located and where 'lay down' areas for specific types of material are located (such as rebar). Of critical importance is the amount of time and effort that has gone into making decisions about whether any of the materials can be manufactured as components off site so that only assembly is required on site;
- positions and usage of plant – the transport of materials efficiently and effectively is vital to on-site success. Time and effort must be spent on making strategic decisions about how materials will be stored and transported (horizontal and vertical coverage), which is based on factual data. For example; does the project need tower cranes? To work this out historic data on tower crane performance needs to be used alongside a detailed breakdown of the type, weight and number of lifts required.

Planning and scheduling

Planning involves deciding, before the work starts, what activities and resources are needed to deliver the project on time and how they relate to each other. This has to be communicated in a manner understood by everyone involved. It will involve the evaluation of alternatives.

The schedule is the (usually diagrammatic) output of the plan related to time.

The process is hierarchical moving from the tender to the master programme and then to more detailed short term and sometimes weekly (or even daily/hourly) programmes.

Construction planning involves the following activities:

1 Information gathering – asking what is known, establishing what can be found out and measuring and considering what assumptions can be made (including assessing risk).
2 Deriving logic and sequence – sequence is the order in which activities should take place and involves choice between options. Logic clarifies the relationship between activities and usually does not involve choice (logic dictates what has to happen).
3 Decisions on method and resources – most operations can be carried out using a range of possible methods and a range of resources (e.g. do you dig a hole with one operative and a shovel or with a mechanical digger?). The choice of these also affects costs and time and to calculate the options accurate costs and high quality performance data are required. The latter, in particular, is often not available so it is necessary to be circumspect and investigate the quality of decisions based on this.
4 Calculating and assessing time to completion – for tender programmes this should be based on a confident analysis of historic performance records from previous projects and their programmes but is usually calculated from scratch with some reliance on the experience of the planner. Once the project is won (post tender) the project team should produce a master programme (which should involve reviewing the decisions made at tender stage). The shorter-term and weekly programmes are vital to delivering the project but with the pressures on site-staff can sometimes be not particularly well produced. There are innovative approaches to planning arising from lean construction research which challenges this hierarchical 'push'-based traditional approach.

The last important issue is the choice of planning method. The most common way to express a plan by far is the 'bar chart' programme. Most construction students are also taught to use critical path programming and these are very often used in computer planning software (although the actual plan may be issued as a bar chart). Critical path planning is an excellent tool for understanding the activities within a project and how they might link to each other but there has been criticism of its effectiveness for planning in a complex and dynamic environment such as a construction project.

Further reading

Cooke, B. and Williams P. (2009) *Construction Planning, Programming and Control*, 3rd edn, Wiley-Blackwell, Oxford.

4.9 Managing building services

Key terms: services design; performance specification; engineering solutions; commissioning

Services design and performance output (thermal comfort, air flow etc.) is derived from the initial designs made by the architect about functional spaces and the zones of the building. Many engineers and particularly those in building services will tell you that they are the 'invisible hand' in the industry. By this they mean that no one notices their work until something goes wrong. There is some truth in this in that an environmental problem (e.g. the building is cold, the lights do not work properly or the lifts break down) which occurs in the first weeks of the building's use can stay in the mind of the occupiers for years afterwards.

There is a reverse side to this, in that many others within the industry see engineering solutions as a 'black box which cannot be challenged'. Engineering education is seen as heavily mathematics based and involving derivation of solutions from first principles. This can produce a belief from other members of the industry that engineering solutions are opaque, confusing and so specialist that they cannot be challenged. In fact many engineering solutions can be considered as 'catalogue based' where standard solutions are chosen without the need for deep first-principle analysis. That is not to denigrate the abilities of some of the great engineers we have in the industry who have provided world beating solutions to major problems but most projects do not require this. The drive for more sustainable buildings that are more energy efficient has seen some major developments in things such as façade design, and passive and low carbon system design, and building services engineers are at the forefront of these so much that they are now more often being called architectural engineers.

The major change in services engineering design in the recent past has been to move towards 'performance specification' where the performance outputs required from the systems are provided and the services designers (often the subcontractors) are expected to produce solutions that deliver these outputs. Some engineers feel that this produces 'thinner' and cheaper solutions but there is an argument based on practice in the USA that says the UK over-designed in the past and that this is a more efficient approach if we take the built environment as a whole in that it produces what is needed and therefore reduces the waste of over-design.

Managing building services is one of the biggest challenges on any construction project. This is for a number of reasons:

- All of the services work may be bundled together producing the largest value subcontract on the project with one main services contractor in charge. Thus mechanical, electrical, plumbing, lifts etc. may all have one nominal company looking after them but each area may be subcontracted to other companies, making the chain of command and communication quite long.
- Much of construction is managed on a linear basis whether vertically or horizontally. By this we mean that we work from floor to floor, section to section or room to room. Services are systems based and run through the whole building and are often working across the linear sequence of the other works.

- Services systems often have a 'nexus' where the whole system comes together such as plant rooms or control panels. These are often the most vital and important areas for completing the systems but can be in awkward locations and involve equipment that is on long delivery.
- While there are some excellent examples of off-site manufacture from the services sector, as mentioned in Concept 4.6 (such as prefabricated plant rooms and service modules) there is still a lot of work that occurs in the traditional manner.
- All services systems require a process of 'proving' or putting to work which we tend to call 'commissioning'. These are vital to ensure that the systems work properly within the finished building and often require little or no other work to be going on. However, this happens at the very end of the building construction which is difficult to plan. It is usually under the most time pressure as previous problems and delays have to be incorporated, while the time available is compressed against a fixed end date. Thus it becomes even more difficult to give the commissioners the time and the lack of interference that they need.
- Sometimes on the more traditional types of project the input of services contractors to the overall planning of the project can be lacking because of the complexities of multiple subcontracting.

The solution to managing building services are similar to general issues we mentioned in Concept 4.7 using the Olympics as an example. Perhaps we could rephrase these for the services area as:

- Make sure there is a clear and well communicated vision of what the client wants and the key performance indicators arising from this.
- Involve the services main contractor and as many subcontractors as possible as early as possible.
- Make sure the commissioning engineer (and the client's facilities manager) are also involved as early as possible (and allow time for commissioning to occur properly).
- Communicate regularly and often throughout the project and use high quality management systems (see further reading below) to ensure the quality of planning in particular is high.
- Do not allow services designers or contractors to persuade you that their work is some sort of 'magic' and cannot be challenged. While it is specialist much of it is fairly basic in production.

Further reading

www.leanconstruction.org.uk (accessed 02/12/2013).

4.10 Sick building syndrome

Key terms: sick building syndrome (SBS); building-related illness (BRI); indoor air quality; environmental sensitivity

There is an accepted understanding that sick building syndrome (SBS) is the general, non-specific symptoms of malaise that may be experienced within a building (i.e. discomfort). The cause of the symptoms is typically unknown but, importantly, the symptoms cease shortly after the person leaves the building. SBS describes a situation where more people than normal suffer and where the symptoms increase in severity the longer the occupants spend time in the building.

Related to SBS is building-related illness (BRI), a term perhaps more prevalent in the USA. This has a narrower focus with specific symptoms such as a cough or muscle aches experienced within a building. The symptoms are more clinically defined and have identifiable causes. Instead of the symptoms ceasing upon leaving the building, complainants may require prolonged recoveries. Even in the UK there is a distinction in illnesses associated with buildings – those with an identifiable cause, such as legionellosis and conditions from exposure to building components containing known substances such as asbestos – and those with no identifiable cause and described only by a group of symptoms – SBS.

There are parallels between both SBS and BRI and what is known as multiple chemical sensitivity (MCS), although MCS (or environmental sensitivity) is not recognised by the World Health Organization as a valid diagnosis.

The most typically reported symptoms of SBS are:

- headaches;
- mucus membrane irritation – eye, nose or throat;
- dry cough, dry or itchy skin, rashes;
- dizziness or nausea;
- asthma-like symptoms such as tight chest, difficulty breathing;
- difficulty concentrating, irritability;
- lethargy and fatigue;
- sensitivity to odours.

Many aspects of SBS have been researched and as a result SBS can be considered a function of the interaction of one or more aspects of the individual, the building and the internal environment in the building. The following risk factors are considered to be detrimental/contributory to SBS:

- Personal – job related stress; where the building occupant has no control over workspace conditions including heating, lighting and ventilation; where the work is sedentary and repetitive.
- Physical/environmental – where the indoor air quality (IAQ) is low; the workspace is noisy; lighting levels are insufficient; there is a tendency for overheating and low levels of humidity.
- Design – where there is a low floor/ceiling height; where offices are open plan; where there is an over reliance on artificial light and little natural light.
- Organisational – poor maintenance; poor management and a failure to respond to requests; where there is a high turnover rate of staff (churn); uncertainty and job dissatisfaction.

Very often the spur to investigating SBS has been an increase in absences from workplaces, either through sickness or churn. It may also be as a result of a decrease in productivity. The former is perhaps understandable given that good productivity in any situation requires:

- a variety of settings to support different tasks;
- comfortable working conditions; and
- a degree of personal control.

All of these are in marked contrast to the contributory factors to SBS given above.

From the above, it is important to note that some complaints may clearly be due to psychosocial factors. Some may also result from an illness that was contracted either outside the building and/or at an earlier instance. A typical pattern is for a susceptible person, subjected to an initial exposure, becoming a sensitive person. Acute sensitivity (e.g. allergies) may then be triggered by exposure at relatively low levels.

The investigation of SBS requires an assessment of IAQ and an investigation of possible contaminant sources (pollutants) and possible pollutant pathways. The sources may be the result of:

- human activities;
- materials and products, including heating, ventilation and air conditioning (HVAC) systems;
- combustion processes;
- microbiological organisms, e.g. mould, dust mites;
- outdoor air pollution;
- the ground under a building.

In order of preference the pollution sources may be controlled by:

- source control, e.g. removal, replacement by an alternative material, sealing;
- removal by local exhaust ventilation at the point at which the pollutant is generated, e.g. kitchens, bathrooms, rest rooms, copy rooms, printing facilities;
- dilution by ventilation;
- air cleaning, as an adjunct to one or more of the above.

Both of the major environmental assessment tools in use in both the UK (Building Research Establishment Environmental Assessment Method, or BREEAM) and the US (Leadership in Energy and Environmental Design, or LEED) recognise the potential problems caused by contaminants associated with certain materials and seek to minimise those that are odorous, potentially irritating and/or harmful to the comfort and well-being of both occupants and installers.

Further reading

Institute of Medicine of the National Academies (2011) *Climate Change, the Indoor Environment, and Health*, National Academies Press, Washington, DC.

4.11 Sustainable construction

Key terms: environmental impact; assessment; Considerate Constructors Scheme

Sustainable construction is required in order to produce sustainable buildings. Sustainable buildings are buildings that are constructed in a resource-efficient manner with due regard to the environmental impact of the materials and components used and the methods adopted to put these together. The result should be a building that is healthy to live in and can be heated efficiently at reasonable cost. It should make best use of natural resources and be future proofed for potential changes, such as those through climate change. There should be opportunities for alterations along with an appropriate life span and there should be opportunities for recycling and reuse at the end of its expected life. To be fully sustainable a building needs to be in a sustainable community where other wider issues such as transport and ecology have been addressed.

Sustainable construction therefore requires appropriate materials and appropriate methods.

Materials – ratings

An individual material or element could be selected on the basis that it provides the best balance between the following criteria:

- clean or non-polluting;
- healthy;
- renewable;
- abundant;
- natural;
- recyclable;
- energy-efficient;
- locally obtained;
- durable;
- design-efficient.

However, these are very broad criteria and although a matrix could be drawn up to aid selection it might be difficult to objectively agree on the exact metric for each of the issues and the appropriate weighting across the criteria. What is clear is that the choice of materials and building elements for a sustainable building should not simply be done on the basis of selecting insulating materials to ensure that a building can be heated cost effectively.

The Green Guide to Specification (Anderson *et al.*, 2009) is probably the UK's most extensive guide for design professionals and addresses some of the issues referred to above by limiting the criteria and subdividing them into more measurable issues. It provides 'environmental impact', with cost and replacement interval information for a wide range of commonly used building specifications, and does so over a notional 60-year building life. The factors assessed and included in the *Green Guide* are:

- the implications of mineral extraction to derive the basic product;
- the pollution and energy consequences of the manufacturing/production process;
- the toxicity of product and chemicals etc. used in manufacturing process, e.g. global warming potential/ozone depletion potential;
- the waste issues at all stages of the production and construction processes;
- the distribution/transport issues;
- the life-cycle and recycling options at the end of its expected life.

There are a number of green guides that include the above issues and attempt to provide some weighting on the likely impact of all the above. The *Green Guide* addresses the broad environmental issues and each is separately awarded a rating, and a weighting system is then used to calculate a summary score. The resulting guide gives typical wall, floor, roof and other elements that are listed against an environmental rating scale running from A+ (best) through to E (worst).

Materials – embodied energy

In some environmental assessments and rating systems a single issue is used as a standard proxy for environmental impact and a surrogate of all of the above – 'embodied energy' or 'embodied carbon'. Embodied energy is the total energy impact of a material and the summation of energy use in the extraction, manufacturing, processing, transport to the site and assembly of a product or building element. In order to truly represent the overall impact, any maintenance and disposal implications should also be assessed and included in a whole-life calculation.

In the UK there are a number of potential sources for embodied energy data, but the main, authoritative and open source of data is the ICE database at the University of Bath, published by Building Services Research and Information Association (BSRIA). This contains over 400 values of embodied energy with 30 main material classifications broken down into approximately 170 different building materials. For a global perspective, the Ecoinvent Centre is probably the leading exponent of life cycle inventory data.

The data in the ICE database are values calculated on a 'cradle to (factory) gate' basis and omit any allowances for transport to site, waste etc. Although energy data is preferable, embodied carbon has become the common currency for life cycle assessment, i.e. converting the raw energy data (in MJ/kg) into CO_2 (in kg CO_2/kg). This is done in the UK by utilising common conversion values given by Defra. This has been taken a stage further by incorporating the effect of other greenhouse gases as well as CO_2, resulting in 'CO_2 equivalent' (CO_2e).

Table 4.11.1 shows an extract of the embodied energy for a number of common building materials.

Table 4.11.1 Embodied energy

Material	*Embodied energy (MJ/kg)*	*Embodied carbon (kg CO_2e/kg)*
Aggregate	0.083	0.0052
Aluminium	155	9.16
Clay brick	3.00	0.24
Glasswool	28.00	1.63

The coefficients illustrate the relatively high processing in manufacturing a product made from aluminium. In contrast, a natural product such as aggregate requires little energy.

One clear limitation of embodied energy as a proxy is the failure to take into account any large environmental impacts, such as water usage or waste generation that may be inherent in a low embodied-energy product. It does, however, have the appeal of simplicity – one number, and it is questionable whether most busy designers have time for more than one number, be it a CO_2e value or a single rating such as A+.

Processes

There are a number of recognised schemes that encourage construction sites to be managed in an environmentally and socially considerate and accountable manner. One of the best known is the Considerate Constructors Scheme (CCS) operated by the Construction Confederation. CCS monitors visit sites at intervals and awards credits in accordance with a Code of Considerate Practice. The areas included in the assessment are:

- considerate;
- environmentally aware;
- site cleanliness;
- good neighbour;
- respectful;
- safe;
- responsible;
- accountable.

In addition to the CCS scheme a broader assessment may be made of the measures taken to ensure that the site is managed in an environmentally sound manner in terms of resource use, energy consumption and pollution. It might be expected that a contractor operates an environmental management system (EMS) and demonstrates that some or all of the following are addressed:

- the monitoring, reporting and target setting for energy use on site;
- the monitoring, reporting and target setting for transport to and from a site;
- the monitoring, reporting and target setting for water consumption on site;
- the implementation of best practice in respect of air (dust) pollution arising from the site;
- the implementation of best practice in respect of water (ground and surface) pollution arising on the site;

- the use of an environmental materials policy for the sourcing of construction materials to be utilised on site (i.e. not restricted to a material to be part of the construction of a building, such as timber for site use).

Further reading

Anderson, J., Shiers, D. and Steele, K. (2009) *The Green Guide to Specification*, 4th edn, Wiley-Blackwell, Chichester.
BSIRIA (2011) available at: https://www.bsria.co.uk/information-membership/bookshop/publication/embodied-carbon-the-inventory-of-carbon-and-energy-ice/ (accessed 02/06/2013).

4.12 Fraud in construction

Key terms: monopoly profits; cartel; oligopoly; collusion; cover pricing

Fraud in the construction sector could be worth as much as $860 billion (£539 billion) globally, according to a report by Grant Thornton. The total amount of fraud was calculated to be equivalent to 10 per cent of industry revenues, and it could hit $1.5 trillion (£941 billion) by 2025. Across Australia, Canada, India, China, the UK and the US it is evident that fraud in construction is commonplace and in some cases could be described as endemic.

A cartel is a group of firms in oligopoly (see Concept 6.11) who act together (collude) for the purpose of avoiding a competitive market in order to create monopoly profit for themselves. An example of a cartel is OPEC (Organization of Petroleum Exporting Countries), which agrees international oil prices. For the cartel to work effectively the producers must control supply to maintain an artificially high price. Even in the case of a concentrated market, with few firms, the existence of an unreliable firm may undermine the collusive behaviour of the cartel. Theoretically, cartels are very unstable because they are an informal agreement to collude: to sell at a higher price together, rather than compete down the price. It is very unstable because there is no binding agreement that one of the firms will not undersell the other and capture the entire market share.

The cartel practice may involve the use of false invoices. Construction giants Balfour Beatty and Carillion are among those the OFT accuses of taking part. 'Cartel activities harm the economy by distorting competition and keeping prices artificially high', said OFT chief executive John Fingleton. He further added 'businesses have no excuse for not knowing and abiding by the law' (OFT, 2009).

The most recent example of a cartel was between Unilever and Procter & Gamble who were found guilty of price fixing washing powder in eight European countries. The case was conducted by the European Commission after a tip-off from a German company, Henkel. The resulting penalty was a €315 million fine, split between Unilever (€104m) and Procter & Gamble (€211m). An Australian manufacturer has been fined AUS$2 million (£1.2 million) for its role in a cartel to increase the price of ball and roller bearings. The Federal Court imposed the fine on Koyo Australia following an investigation by the Australian Competition and Consumer Commission (ACCC). It came after the company acknowledged liability for a scheme where it arranged with two competitors to raise prices.

Collusion means competing firms come together in an oligopoly system to decide the price. Price fixing can also be implemented by government (especially in the agriculture sector), in which case it is not considered a collusion. Collusion is easier to achieve when there is a relatively small number of firms in the market and a large number of customers, market demand is not too variable and the individual firm's output can be easily monitored by the cartel organisation. Tacit collusion occurs when companies make an informal agreement to fix prices (i.e. they do this without letting their competitors or official bodies know).

The fewer the number of firms in the industry, the easier it is for the members of the cartel to monitor the behaviour of other members. Given that detecting a price cut becomes harder as the number of firms increases, so much the bigger are the gains from price cutting. The greater the number of firms, the more probable it is that one of those firms is a *maverick* firm; that is, a firm known for pursuing an aggressive and independent pricing strategy. In September 2009, a total of 103 UK construction companies were fined almost £130 million for illegally colluding in bids to secure building contracts including schools and hospitals. The biggest fines were £17.9 million to Kier Group, £5.4 million to Carillion and £5.2 million to Balfour Beatty. The OFT conducted a four-year investigation, uncovering about 4000 tenders that were affected by anti-competitive activities. The firms involved effectively operated a 'cartel', having secretly agreed the prices they would submit during a tender process – a practice known as 'cover-pricing'. In its simplest form, companies that do not have the resources or spare capacity to take on a job but still want to remain on tender lists obtain an inflated price from a rival and submit it in the knowledge that their bid will not succeed. The companies involved had driven up the price of the projects by agreeing to submit artificially high quotes. In at least six cases the companies submitting high tenders were compensated for the cost of tendering. The tendering authority, often a local council, is given a false impression of the level of competition and could end up paying artificially high prices. Some estimates of the amounts overpaid were as much as 10 per cent but the OFT did not put a figure on this.

Bidders for contracts were given a high price from the competitor as a cover price, which was then submitted as a genuine tender offer but not intended to win the contract. Thus, the lowest bidder, and subsequent winner of the contract, in reality did not face any competition. The OFT's investigations revealed that the practice of cover pricing was widespread in the construction industry. As any main contractor would probably have experienced, this issue extends down through the supply chain with cover prices submitted by specialist package contractors as well.

Further reading

CIOB (2006) 'Corruption in the UK construction industry', Survey 2006, CIOB, Ascot.

CIOB (2013) 'A report exploring corruption in the UK construction industry', CIOB, Ascot.

Grant Thornton (2013) 'Time for a new direction', available at: http://www.grant-thornton.co.uk/Global/Publication_pdf/Time-for-a-new-direction.pdf (accessed 02/04/2014).

Jowsey, E. (2011) *Real Estate Economics*, Palgrave Macmillan, Basingstoke, ch. 13.

OFT (2009) 'Construction firms fined for illegal bid-rigging', available at: http://www.oft.gov.uk/news-and-updates/press/2009/114-09#.Uz0fFNNOXIU (accessed 03/04/2014).

Sohail, M. and Cavill, S. (2006) 'Corruption in construction projects', in: A. Serpell (ed.) *Proceedings of the CIB W107 Construction in Developing Countries Symposium: 'Construction in Developing Economies: New Issues and Challenges'*, January 18–20, Santiago, Chile.

5 Development

Hannah Furness, Ernie Jowsey and Simon Robson

5.1 Developers

Key terms: trader developer; investor developer; commercial development; residential development; visionary developers

Developers come from a variety of backgrounds and their involvement in the development process ranges from hands-on day-to-day management to a more remote strategic role. They come from a range of sectors including private, public and voluntary, and all have different motivations, but overall developers are in essence entrepreneurs who identify opportunities and are prepared to take risks in order to deliver a completed property development scheme in anticipation of the requirements of the market in return for profit. The risks inherent in the process are that the requirements of the market (the demand) are uncertain. The developer therefore recognises the potential for development by (i) estimating future demand for alternative uses of existing land resources, and (ii) calculating the cost of the building for new uses. The viability of the development project is the developer's assessment of which scheme will produce the maximum net return subject to the constraints and many variables involved in the process including: the availability of finance; planning and building requirements; and legal restrictions in the title or use of the land. The developer therefore carries the risk and uncertainty of the development. In addition, Miles *et al.* (1991) say that developers have a complex and shifting mixture of roles including:

- creator – the identification of an opportunity and the inception and evolution of a development scheme;
- researcher – of the property development market in order to establish that demand exists for the development;
- analyst – assessing the financial viability of the scheme through analysis of the costs and likely value of the scheme;
- promoter – the marketing agent for the development scheme;
- negotiator – for financial resources to fund the development from lending institutions;
- manager – of the professional team and the construction stage of the development process;
- leader – of the overall development process;
- risk manager – responsible for identifying, managing and mitigating risks associated with the successful completion of the development;
- investor – if the development is financed with the developer's equity.

There are two main kinds of private sector developer: trader and investment. *Trader developers* tend to develop speculatively in order to sell completed development schemes to investors in order to fund the next development project. This is usually true of smaller development companies or 'one-man bands'. Such developers often do not have sufficient equity or resources to commence on a new development project without completing and selling another. *Investor developers* on the other hand will tend to retain completed developments to generate rental income and capital growth and build up portfolios of income-generating properties. Larger development companies and multinationals will tend to operate on this basis.

Depending on the size and type of the development company, a developer might deliver a range of commercial or residential development projects, or may specialise; for example, there are solely residential developers (e.g. the house builders Barratts or Belway), or solely commercial, which may themselves specialise in a specific sector, such as retail (e.g. Dransfield, Westfield) or industrial (e.g. Prologis). Developers can also be property managers, maintaining facilities through refurbishment or improvements to ensure that profits are maximised through attracting the best tenants. They play a vital role in controlling expenses and improving efficiency and effectiveness of buildings for all parties involved. However, no matter what the end product, the developer's principal objective is to generate profit.

Notable visionary developers include:

- John Raskob and Al Smith, who wanted to create the world's tallest building and did so (at the time) with the Empire State Building in New York in 1931.
- Walt Disney, who had the idea for the world's first theme park (near Los Angeles in 1955) despite considerable scepticism from the banks and his own company.
- Victor Gruen, a Viennese architect who was responsible for the world's first modern shopping mall at Northland near Detroit in 1951.

Further reading

Jowsey, E. (2011) *Real Estate Economics*, Palgrave Macmillan, Basingstoke, ch. 7.

Miles, M. E., Berens, G. L., Eppil, M. J. and Weiss, M. A. (1991) *Real Estate Development: Principles and practice*, 4th edn, Urban Land Institute, Washington, DC.

Ratcliffe, E., Stubbs, M. and Keeping, M. (2009) *Urban Planning and Real Estate Development*, 3rd edn, Routledge, London and New York, ch. 11.

5.2 Development

Key terms: development process; property development; factors; land; actors; site

'Development' is defined by statute in the Town and Country Planning Act 1990 s55(1), as:

> the carrying out of building, engineering, mining or other operations in, on, over or under land, or the making of any material change in the use of any buildings or other land.

In addition, property development has been defined by Cadman and Topping (1996) as:

> an industry that produces buildings for occupation by bringing together various raw materials of which land is only one. The others are building materials, public services, labour, capital and professional expertise. The completed building may be let or sold to an occupier. If it is let, it can be held as an investment. It is a venture in which losses as well as profits are made.

In essence therefore, development is necessary to ensure the efficient use of land resources. The developer that can put the site into its most profitable use is the one who can make the highest bid for it, and ultimately make a profit from the development.

Development may be the construction of a new building, or the refurbishment or redevelopment of existing structures/buildings. The first questions that the developer will ask is whether the value of the replacement (or new) building will exceed the value of the current building (or land) plus the cost of construction/rebuilding. The developer will therefore have to assess the rationale for a development by assessing the following factors:

1 choice between development projects;
2 estimate the demand for different developments;
3 decide on the quality of the building;
4 calculate the intensity of development on the site;
5 estimate the costs of development;
6 estimate how much can be bid for the site;
7 obtain finance;
8 decide whether to develop alone or in partnership.

Once these considerations have been made, there are a number of other factors and stages of the development process that must be considered. The development process is made up of four principal dimensions: factors, actors, the site, and events.

Factors include things such as the economy, property cycles and markets, government policy and strategy, legislation, and trends such as the environment, population, politics, transport, technology etc. All of which can have a significant impact on the viability or progress of a development project, which ultimately affects profitability.

Actors are the individuals who are involved in the development process. In addition to the developer, the key stakeholders in any development project are: the landowner, the professional team (architect, contractor etc.), the occupier, the investor/lending institution, local planning authority, the client, the community etc. All of these 'actors' in the process have different aims, roles and status in the process and can have influence over it. It is the role of the developer to manage these differing expectations so that all parties feel that their specific objectives are met.

The site is fundamental to the development. Since each site is different in terms of location and condition, this can impact on the viability of the development. The developer will need to consider factors such as the shape, size, aspect and topography of the site, as well as investigate whether there is any contamination or pollution. The load baring capacity of the site will need to be assessed, as will any existing buildings on the site. Its location relative to water sources and courses, areas of historical or scientific interest, and conservations areas will also have to be assessed.

Finally, there are a number of events that link to all aspects of the development process. Put simply, the process is split into three main phases: pre-construction, construction, and post-construction. Preconstruction includes assessment of the viability of the development, and preparation activities such as site acquisition and securing finance and planning permission. The construction phase includes the tendering for a contractor and the construction of the development. And finally the post-construction phase includes the letting or selling of the completed development. There are a number of different illustrations of the development process that have been proposed. One of the best known, proposed by Birrell and Gao (1997), is made of four principal stages and contains 14 core activities, illustrated in Figure 5.2.1. One or all of these activities are subject to change and variation depending on the nature of the specific development and its various elements.

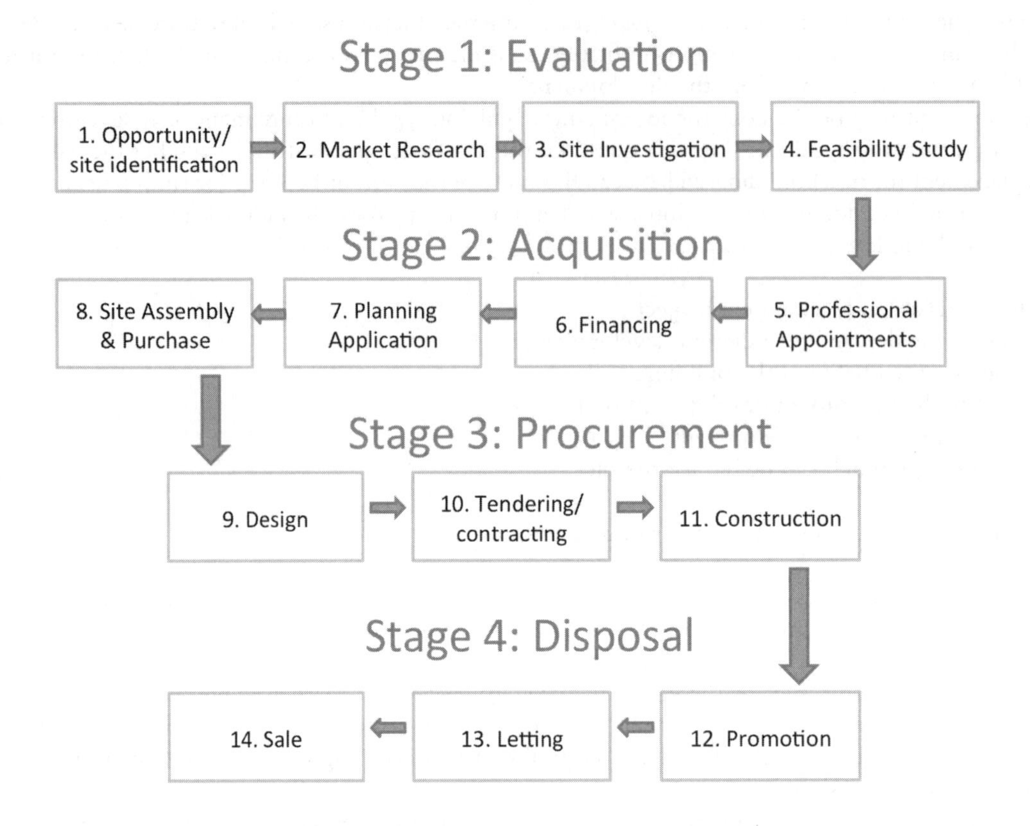

Figure 5.2.1 The 14 phases of development.

Source: adapted from Birrell and Gao (1997).

Further reading

Birrell, G. and Gao, S. (1997) 'The property development process of phases and their degrees of importance'. RICS Cutting Edge Conference, Dublin. Available at: www.rics-foundation.org.uk (accessed 03/04/2014).

Cadman and Topping (1996) *Property Development*, 4th edition, Chapman and Hall, London.

Isaac, D., O'Leary, J. and Daley, M. (2010) *Property Development Appraisal and Finance*, 2nd edn, Palgrave Macmillan, Basingstoke, ch.1.

Jowsey E. (2011) *Real Estate Economics*, Palgrave Macmillan, Basingstoke, ch. 7.

Reed, R. and Sims, S. (2014) *Property Development*, 6th edn, Routledge, Abingdon.

5.3 Development costs

Key terms: land acquisition; site investigation; build costs; finance costs; contingency

There are a number of costs associated with property development, all of which must be carefully considered and managed throughout the process in order to ensure that a profitable development scheme is delivered. The costs of the development therefore must not be greater than the value of the completed development. Although each development project will be different in terms of location, site conditions and scale of development, the main costs of development and construction will usually include:

- land acquisition/assembly – plus Stamp Duty Land Tax and purchasers costs;
- site investigation/works;
- build costs – plus professional fees;
- finance/interest;
- contingency;
- marketing/promotion;
- sale fees;
- letting fees;
- planning fees;
- planning contributions (e.g. s106 obligations; Community Infrastructure Levy).

Land acquisition/assembly

The development site is purchased for a given price, often that which is advertised or suggested by the vendor's agent. In addition to the purchase price the cost of valuer's and legal fees will be in the region of 1.5 per cent of the purchase price depending on the complexity of the deal. Legal fees are usually between 0.25 per cent and 0.5 per cent and valuer/agency fees are normally agreed at 1–2 per cent of the land price. For larger developments, a number of separately owned sites may need to be purchased which can lengthen negotiations and conveyancing which increases the fees payable. In addition, unless a site is in an exempt area, Stamp Duty Land Tax (SDLT) will be payable. Stamp duty rates for non-residential land is divided into four bands: for land priced up to £150,000, no stamp duty is payable; for land priced between £150,000 and £250,000, 1 per cent of the land price is payable; for land priced between £250,000 and £500,000, a duty of 3 per cent is charged; and for land exceeding £500,000, a 4 per cent duty is payable.

Assembling the site and the construction phase both take time, which can expose the developer to a possible fall in profit. For example, if during the site assembly process the government restricts office development, or there is a rise in construction costs. These additional costs can be divided into 'ripening costs' and 'waiting costs'. Ripening costs arise through holding land in anticipation of profitable future development, whereas waiting costs are incurred (even if the land already has planning permission) as construction takes time before any revenue is received.

Site investigation/works

Before construction can begin the site has to be prepared. Site investigations will be carried out in addition to the usual legal searches of title, including ground investigation and land surveys. In addition, costs will be incurred for demolition, clearance, provision of access and provision of utility services. These must all be accurately estimated and in particular the advice of an environmental engineer or surveyor should be sought over potential contamination and the need for remediation. Especially for residential developments, any land contamination must be identified and rectified at the outset in order to reduce risks as far as possible.

Build costs

This is the cost to construct the structure, which is usually expressed as a 'per unit' cost, such as per square metre, which is then multiplied by the gross area to give an overall development cost. The Building Cost Information Service (BCIS) can provide evidence for this purpose if reliable figures are not to hand from previous jobs. Clearly this is an expert matter and will normally be referred to a quantity surveyor (QS). In addition, fees will have to be paid to the various consultants involved in the development. These will usually be subject to variation and negotiation depending on the nature/scale of the project and are estimated at the time of the proposed implementation of the project, but can be assumed to approximate the following percentages of the total build costs: architect 5–7 per cent; QS 2 per cent; structural engineer 2 per cent; services engineer 2 per cent; and project manager 1 per cent.

Finance/interest

The construction phase of the development process is the most expensive for the developer. In addition to the site purchase, the developer is paying out vast sums for building materials and labour. Unless the developer has sufficient equity, in order to purchase the site, pay the fees and duty, and fund the construction phase, the developer will have to borrow. Lending institutions will often charge high rates of interest for the loans due to the risks associated with property development. Interest charges are usually calculated assuming that the borrowing period is the whole development period from start to finish, including the pre and post development stages. The precise draw-down of interest payments month by month will vary according to a number of factors that will influence how much is spent, e.g. weather, contamination, requiring heavy up-front costs etc. might all impact on the time taken to deliver the project; in 2008/9 work deliberately slowed in hope that the market recovered in time for project completion and disposal.

Contingency sum

At the early stages of a project it is usual for a contingency sum to be included in the calculation. This is to cover extra costs and unforeseen risks and is calculated as a percentage of total build costs (including professional fees). For example, the risk of archaeological remains should be considered at the outset but these may turn up unexpectedly. The allowance may be 3 per cent of the running total of costs for a very simple site rising to say 7 per cent for a refurbishment. A figure of 5 per cent is usually used as a standard assumption.

Planning contributions

The local planning authority (LPA) will often grant planning permission for a development scheme with a number of conditions. These may include a planning obligation, whereby the developer will provide or pay for something that is to the benefit of the wider community, or enables the capacity of the site to be increased, e.g. a contribution towards recreational facilities adjacent to a housing scheme; or a contribution towards the upgrade of a highway serving a new business park. These contributions tend to result from negotiations between the developer and the LPA and will vary on a project by project basis.

Other costs

An estimate of the cost of promoting and advertising the buildings for letting or sale has to be included. These costs depend very much on the client but may be estimated at 5 per cent of open market rental value. Letting agent's fees will be in the region of 10 per cent of open market rental value. The cost of agents' and legal fees on selling the investment will be in the region of 1.5 per cent of the net development value.

Further reading

Havard, T. (2014) *Financial Feasibility Studies for Property Development*, Routledge, London, ch. 7.

5.4 Development finance and funding

Key terms: debt; equity; short term; long term; corporate; project

There are numerous ways for funding property developments (Table 5.4.1). In essence, development funding may be classified as debt versus equity, corporate versus project based, or long term versus short term. Each type of development finance package may be categorised as part of this three-way matrix.

Table 5.4.1 The eight main ways of funding property development

	Debt	*Equity*	
Short term	1 Overdraft	5 Sale of assets	*Corporate*
	2 Bridging loan	6 Joint venture	*Project based*
Long term	3 Debenture	7 Ordinary shares	*Corporate*
	4 Mortgage	8 Sale and leaseback	*Project based*

Figure 5.4.1 Property development.

Source: © Simon Robson 2007.

The eight main ways of funding property development are discussed as follows:

1 *Debt finance, corporate based, short term*
An overdraft is available to companies by agreement. These can be called in at any time and are therefore short term. Overdrafts may be used as long-term funds by established companies where call in is very unlikely.

2 *Debt finance, project based, short-term*
Also known as bridging finance, this type of loan is required by a developer to cover the period from the commencement of the development until funding of a more permanent basis is made available enabling the bridging loan to be repaid.

The purpose of a bridging loan is to allow the developer to proceed with the scheme in the short term. The money plus interest typically has to be repaid shortly after completion. This is known as rolling up the interest. Such loans allow expenditure on: the purchase of land and buildings, including adjacent sites; the purchase of leaseholds; site investigation and early site improvement works; access, demolition, security; payment of builders and consultants in stages. Clearing banks are the main sources of bridging finance with loans generally being up to a maximum of 70 per cent of net development value. Some form of security for the loan will be required such as the first legal charge on the site and works in progress. The interest rate is normally at a variable or 'floating' rate of interest based upon 'LIBOR' (the London Inter-Bank Offered Rate).

3 *Debt finance, corporate based, long term*
Loan stock secured against real property(s) is known as a mortgage debenture. They are similar to mortgages but with many stock holders (investors) rather than one 'mortgagee'. Debentures are paper securities or bonds that may be traded on the stock market. 'Gearing' is the ratio of debt to equity for a company. Gearing can increase investment returns but also increases risk. Too much debt and the high gearing may scare off equity investors who fear a lower dividend, which may negatively impact the share price.

4 *Debt finance, project based, long term*
Debt investment generally involves the borrower offering the investor a guaranteed return in fixed terms over a finite period (perhaps 5–30 years). The return is not directly linked to the success of the development being secured in some other way. It provides long-term funding for the developer who does not have to employ so much equity capital and enables the repayment of the short-term bridging loan(s). The property is used as security for the loan and if the mortgagor defaults on interest or capital repayments the lender may 'foreclose', i.e. take over either vacant possession or the income stream. The source of this form of loan is generally a clearing bank with the loan generally being up to 70 per cent of open market value with variable or fixed rates of interest being available.

5 *Equity finance, corporate based, short term*
After paying its interest charges and before declaring a dividend for the equity shareholders any company can announce that it is retaining profits to reinvest in the business. The advantage of using retained profits is that it is often cheaper than borrowing. Profit retention can have effects on dividends and the share price. A smaller dividend can reduce the price, depending on reception by the market.

Property and other assets including recent developments may be sold to raise equity capital for reinvestment into development. Residential builder/developers use sales as normal practice to provide a 'rolling fund' of capital and reduce the overdraft. Commercial 'trader' developers sell on their assets to investors to raise capital either immediately or at the first rent review. Phases of developments may also be sold to release capital and reduce the depth of debt.

6 *Equity finance, project based, short term*
Developers may enter into joint ventures for a variety of reasons including spreading the risk of major projects between two or more partners. Joint ventures can be used to bring in new sources of equity and debt finance. Joint venture companies (JVCs) are where two or more parent companies hold shares in a third subsidiary company they create for the purpose. They operate via shareholders and directors and under company law. Partnerships may also be used to structure joint ventures.

7 *Equity finance, corporate based, long term*
The sale of ordinary shares gives the shareholder an equity share in the company. Ordinary shareholders will receive a dividend, a residual sum from the gross profits, after preference shares and debentures have been paid. Equity shares bear the risk of the company and depend on its success for the dividend. The 'gearing ratio' of a company is the ratio of its total debt to its total equity value.

8 *Equity finance, project based, long term*
One method of obtaining this form of finance is via a 'forward sale' where a fund agrees in advance to buy a property development once completed on agreed terms. The objective of the developer is to reduce their risk and achieve 'take out' of the bridging finance. The objective of the investment institution is to secure new prime investment property with a good return. The investor will pay less than for a completed and fully let development and shoulder some of the development risk.
Another way of securing long-term, project-based equity funding is via a sale and leaseback. This is the sale of the freehold in a development to a fund subject to a long leaseback to the vendor. This vehicle is often used by occupiers who wish to raise funds from their premises, e.g. supermarket operators, hotel chains.

Further reading

Isaac, D., O'Leary, J. and Daley, M. (2010) *Property Development: Appraisal and finance*, Palgrave Macmillan, Basingstoke.

5.5 Site assembly and acquisition

Key terms: site investigation; 'buildability'; land tenure; site purchase; land value

Site investigation

Following the strategic consideration of a development proposal, which will involve an assessment of viability and the likelihood of obtaining planning permission, a site investigation will generally be required. The information needed for a site investigation will be collected from carrying out a desktop study and from a physical inspection of the development site.

Desk research should cover both physical and environmental aspects of risk. A study of historical maps will help identify whether the site has been occupied by polluting activities in say the last 200 years. In addition sites may have been contaminated by activities on surrounding land necessitating the survey to cover an area of one kilometre radius from the boundary.

A site inspection should include a systematic walk identifying all relevant indicators. Samples should be taken of surface soils and water. The position and condition of all boundaries should be noted as legal conveyance plans can be inaccurate. The inspection should also look for visible evidence of site problems such as poor ground conditions, dereliction, contamination and flooding. These may include storage tanks, waste heaps, embankments, presence and condition of any vegetation, discoloured ground, and evidence of past structures or buildings.

To determine the 'buildability' of a site it will normally be necessary to carry out a geotechnical survey. Before carrying out actual work on site, desk enquiries should be made with the British Geological

Survey and the Coal Authority. The survey is directed at the fundamental geology and other factors such as filled ('made') ground, voids, basements, previous foundations or other underground structures. The engineer's report will advise on the bearing capacity of the site and the foundations that will be needed.

If there is an existing building on the site to be refurbished then a complete structural survey will have to be carried out. The report should cover not only the existing state but also the physical feasibility of refurbishment or conversion work.

The developer also needs to know whether the site is liable to flooding. The developer needs to know the location of the site relative to the coast, rivers and minor water courses. What has been the history of past flood events, frequency and severity? Are there any flood defences and if so are they adequate and in good condition? Can the development be carried out in spite of the flood risk? This involves questions of insurance, planning, mortgages, finance and marketing. Can the development be protected by new defences or be designed to be flood resistant?

The presence or absence of utility services such as water, sewerage, electricity, gas and telecommunications affects the development potential of the site. In each case it is necessary to check with the local company to see where the mains are, whether any increase in capacity will be needed and how much it would cost. In addition, the presence of pipelines and cables running over and below the site will add to development cost. Besides bearing capacity, developers need to know whether a site may be contaminated. This will involve a desk survey, physical inspection, site investigation and analysis.

Land tenure

It is ultimately the job of the solicitor to thoroughly check all aspects of the ownership of the site undertaking the 'enquiries before contract'. However, the developer needs to know as early as possible about potential legal pitfalls. When inspecting the site it is often possible to pick up clues as to problems with tenure. Evidence may be boundaries, car parking, tenants, power lines, footpaths, manholes, notices or squatters. The precise nature of the ownership or occupation will then have to be checked.

The basic freehold title to the property will normally be registered at the Land Registry. There may be restrictive covenants placed on the use of the land by a previous vendor. These may be bought out or may be removed on application to the Lands Tribunal.

To carry out a project the developer needs to have complete vacant possession of the site. Leaseholders have the right to remain in occupation until the end of their lease; they may be persuaded to sell their lease to the developer voluntarily though the price may be high. At the end of a lease commercial tenants have security of tenure unless the landlord goes through the relevant statutory process and proves to the court that it needs vacant possession for development.

In addition to leases there may be other minor property rights, which can frustrate projects. These include easements either public or private for sewers or cables, public or private rights of way, rights to light, and rights of support. Removing or rerouting such rights can take time and expense.

Site purchase

A variety of site purchase methods are considered elsewhere. Private treaty is by far the most common method of purchase and has the benefit of privacy. Plenty of time is available for investigation and negotiation. This method may be associated with an informal tender.

The purchaser is able to limit its exposure to risk by using a 'conditional contract' under which a deposit is paid and a price is fixed for the purchase subject to a condition(s). If conditions are met the developer is obliged to complete or lose the deposit. Conditions will normally be planning permission, site assembly, and site investigation. A similar method is an option to purchase whereby a sum of money is paid for an option to buy the site at an agreed price or price formula within a time period.

The purchaser may alternatively agree a delayed purchase (or partnership agreement) with the site owner. Under such a contract the landowner receives an agreed percentage of the sale price of any properties developed when they are sold or let.

A building agreement or licence permits the developer to enter a site and build to agreed plans, normally linked to an agreement to purchase or lease the site. This may happen as part of a joint venture where the proceeds will be shared.

A formal tender is often used by public authorities. In this case the highest bid will be accepted. An informal tender is sometimes carried out by estate agents who will ask for written bids by a certain date.

Definitions of land value

When negotiating the purchase price of a development site it is essential that definitions of land value are clear. These will be largely covered elsewhere – however, it is useful to briefly review the main terms here.

Existing use value (EUV) assumes no development scheme. There is very little risk associated with buying for EUV as no change from the status quo is assumed. It is unlikely that a purchase for EUV will be possible if the potential for development is known.

The *development value* (DV) can be ascertained by undertaking a residual valuation assuming development is to take place. The DV potentially entails maximum risk as it is underpinned by numerous assumptions regarding planning, market conditions etc.

Hope value (HV) is a halfway house between EUV and DV. HV consequently involves a level of risk somewhere between EUV and DV.

Merger value is the extra value that can be realised from assembling two or more adjacent sites to form a (more) viable development. Often referred to in practice as 'marriage' value (see below).

Marriage value is the extra value resulting from marrying freehold and leasehold interests in the same property.

The ability to *compulsorily* acquire interests in property ensures that development for the public good is not held to ransom by site owners. It takes time to get the necessary powers, which are not guaranteed. Such powers may be used to support private development.

Further reading

Keeping, M. and Shiers, D. E. (2004) *Sustainable Property Development: A guide to real estate and the environment*, Blackwell, Oxford.

Ratcliffe, J., Stubbs, M. and Shepherd, M. (2004) *Urban Planning and Real Estate Development*, 2nd edn, Routledge, London.

Reed, R. and Sims, S. (2014) Property Development, 6th edn, Routledge, Abingdon.

5.6 Evaluation and appraisal methods

Key terms: residual; cash flow; internal rate of return; yield; capitalisation; net present value

Individuals' personal objectives in relation to development will vary but in the business world the aim of profit is paramount. The developer is in business to create a profitable scheme without which he will go out of business. The aim of the project is therefore to provide a mix of scale, use type, design and cost that will maximise profit on the site at this time. The two most commonly used methods of determining profitability are residual and cash-flow methods.

Residual method

Central to the residual method is the basic development appraisal equation. The end product of the appraisal is the estimate of capital profit to be made by the developer at the sale of the completed scheme. The equation in its simplest form is:

$$P = V - (C + L)$$

P = Profit

V = Net development value

C = Construction and associated costs

L = Land price and associated costs

Example

Net development value (V)	£10m
Construction and associated costs (C)	£6m
Land price and associated costs (L)	£2m

P = £10m – (£6m + £2m) = £2m

Hence the developer's profit in this example is £2 million. The mechanics of a residual appraisal are relatively straightforward. The key to a 'good' appraisal is the quality of the inputs. The inputs will be considered under the headings of value and cost.

Value

In simple terms the capital value of a commercial development is ascertained by capitalising the open market rental value of a development at the investment yield (normally by multiplying by the years purchase).

Within an active market it is possible to find comparable evidence of rents per square metre of similar local lettings. However, an active market is often a rising one so these get out of date quickly. Within a slack market there are few comparables to be found and the judgement is much harder to make. Where a rent free period or similar has been agreed with the incoming tenant the cost of this must be deducted from the net development value.

The yield used for capitalisation purposes is the one justified by the most appropriate market comparable evidence. Purchaser's costs of fees (legal and agent's) and stamp duty are deducted to give the actual net transaction value.

Costs

Before building can begin the site has to be prepared. Costs for this will include demolition, clearance, provision of access and provision of utility services. These must all be accurately estimated possibly by a QS. The advice of an environmental engineer or surveyor should be sought over potential contamination and the need for remediation.

At this early stage cost estimates will be on a per square metre floorspace basis. Building cost indices can be used or the matter referred to a QS.

Fees will be have to be paid to the various consultants involved in the project including the architect, QS, structural engineer, services engineer and project manager. It is usual for a contingency sum to be included in the calculation. This is to cover extra costs and unforeseen risks.

The developer will generally need to borrow money short term to undertake the project, to buy the site and to pay the builder and consultants. (Development finance is covered in detail elsewhere.) An estimate of the cost of promoting and advertising the buildings for letting or sale has to be included.

In a development appraisal, as opposed to a valuation, it is assumed that the site is purchased for a given price, often that advertised or suggested by the vendor's agent. In addition to the purchase price, the cost of valuer's and legal fees, and stamp duty should be added. In order to purchase the site and pay the fees and duty, the developer will have to borrow. The interest charges are calculated assuming that the borrowing period is the whole development period from start to finish.

Developers' profit is a residual after the total development cost (TDC) has been deducted from the net development value. The rate of the developer's profit is normally expressed as a percentage of TDC.

The same basic methods may be used to estimate the value of the site. This is considered further in the 'residual value' concept.

It is generally acknowledged that the residual approach has a number of flaws. It is cumbersome, often referred to as a 'back of a fag packet' approach. It is poor at handling the timing of costs and returns. It makes inaccurate assumptions in assuming a straight-line build-up of costs. It is poor at handling phased projects unless each phase is the same. It is poor at handling possible changes during the project such as interest rates, revised mix of uses or the sale of sites. It is a poor platform for financial monitoring, i.e. during construction and letting void.

The residual method is widely used in practice because it is simple, quick, widely understood and is thus used in the early stages of development projects.

Cash-flow methods

A cash-flow method will be adopted when more accurate input data becomes available. This is often later on in the project.

In discounted cash flow (DCF) appraisals each cost or return is calculated on a quarterly, monthly or weekly basis and assigned to the correct time slot in a cash-flow chart. The 'period by period' method (PBP) employs a simple allocation of costs and values to time slots. Negative cash flows typically give way to positive ones. Later positive cash flows can reduce interest costs. PBP shows the outstanding debt figure throughout. It identifies the actual break-even point, and is widely used in practice. PBP is, however, very tricky to perform without a computer and does not link well with internal rate of return calculations.

With the net present value method, each individual cash flow is discounted by multiplying by a present value of £1 factor. The results for each period are summed to produce a net present value (NPV).

Internal rate of return

The normal expression of profitability is profit as a percentage of either the total costs or the net development value. In either case time is not accounted for. Where two projects produce the same value any prudent developer will choose the shorter one. A measure of profitability is needed that can compare different profits that are received at different times. This is the IRR (internal rate of return). The IRR is defined as 'That rate of return at which the net present value of the cash flows equals zero' – IRR is thus the true rate of return independent of time and scale.

The residual method is flawed but widely used in practice at the early stages. Cash-flow methods are more accurate but need better data. The PBP cash-flow method is the most widely used method. The NPV method is quicker and easier and relates to the IRR well. The IRR is the most complete and informative method increasingly used since it is an output from all commercial software programmes.

Further reading

Isaac, D., O'Leary, J. and Daley, M. (2010) *Property Development: Appraisal and finance*, Palgrave Macmillan, Basingstoke.
Ratcliffe, J., Stubbs, M. and Shepherd, M. (2004) *Urban Planning and Real Estate Development*, 2nd edn, Routledge, London.

5.7 Intensity of site use

Key terms: diminishing returns; variable factor; fixed factor; marginal physical product; marginal revenue product; marginal cost; economic rent

A developer has to decide how intensively a site should be developed. For example, if the most profitable use of a large suburban site is for housing, to what density should the houses be built? Or, if an office block is to be built in a city centre, how many floors (given no planning restrictions) should it have? In economic terms the issue is: how much capital should be combined with the land?

Since we are dealing with a particular site, we can regard land as a fixed factor. The problem is one of applying units of a variable factor, capital, to a fixed factor, land. As extra units of capital are applied to the fixed site, the law of diminishing returns eventually comes into operation, and the marginal physical product (MPP) of capital falls. This is because building upwards incurs extra costs; for example, more foundations are necessary and lifts and fire escapes have to be provided. So the return on capital eventually decreases, giving a downward-sloping marginal revenue product (MRP) curve (see Figure 5.7.1). Since all capital can be obtained at a given price, the marginal cost (MC) curve is horizontal.

Development of the site will take place up to the point where marginal revenue equals marginal cost: that is, to where the MRP of a unit of capital equals the cost of a unit of capital (OC in Figure 5.7.1) The building reaches its optimum height, i.e. the development is complete, when OX units of capital have been applied to the site.

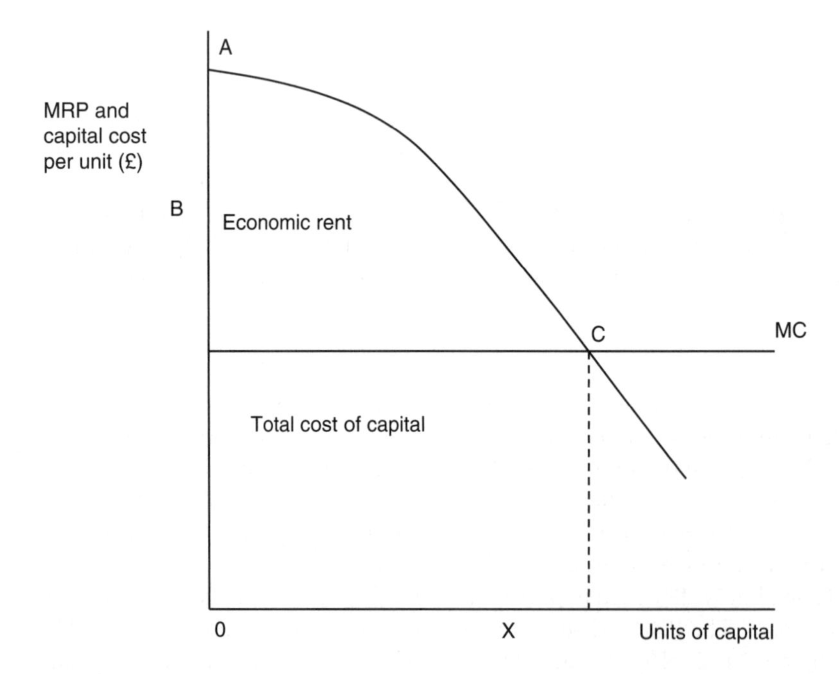

Figure 5.7.1 Applying capital to a fixed site.

Using this analysis we can see that where the marginal revenue product of a site is high, such as in a city centre, the amount of capital applied to the site will be greater and it will be developed more intensively. This can be seen in Figure 5.7.2, which contrasts a city-centre site with a suburban site.

In Figure 5.7.2a the MRP curve is further out because the city-centre site is more productive and more profitable (with higher economic rent ACB) than the suburban site in 5.7.2b (economic rent A' C' B'). As a result more capital (Xa) is applied to the city centre site than that applied to the suburban site (Xb). This means that the city centre site is more intensively developed, and explains why buildings are higher in city centres than in other parts of a city.

Further reading

Harvey, J. and Jowsey, E. (2007) *Modern Economics*. Palgrave Macmillan, Basingstoke, ch. 13.
Jowsey, E. (2011) *Real Estate Economics*, Palgrave Macmillan, Basingstoke, ch. 7.

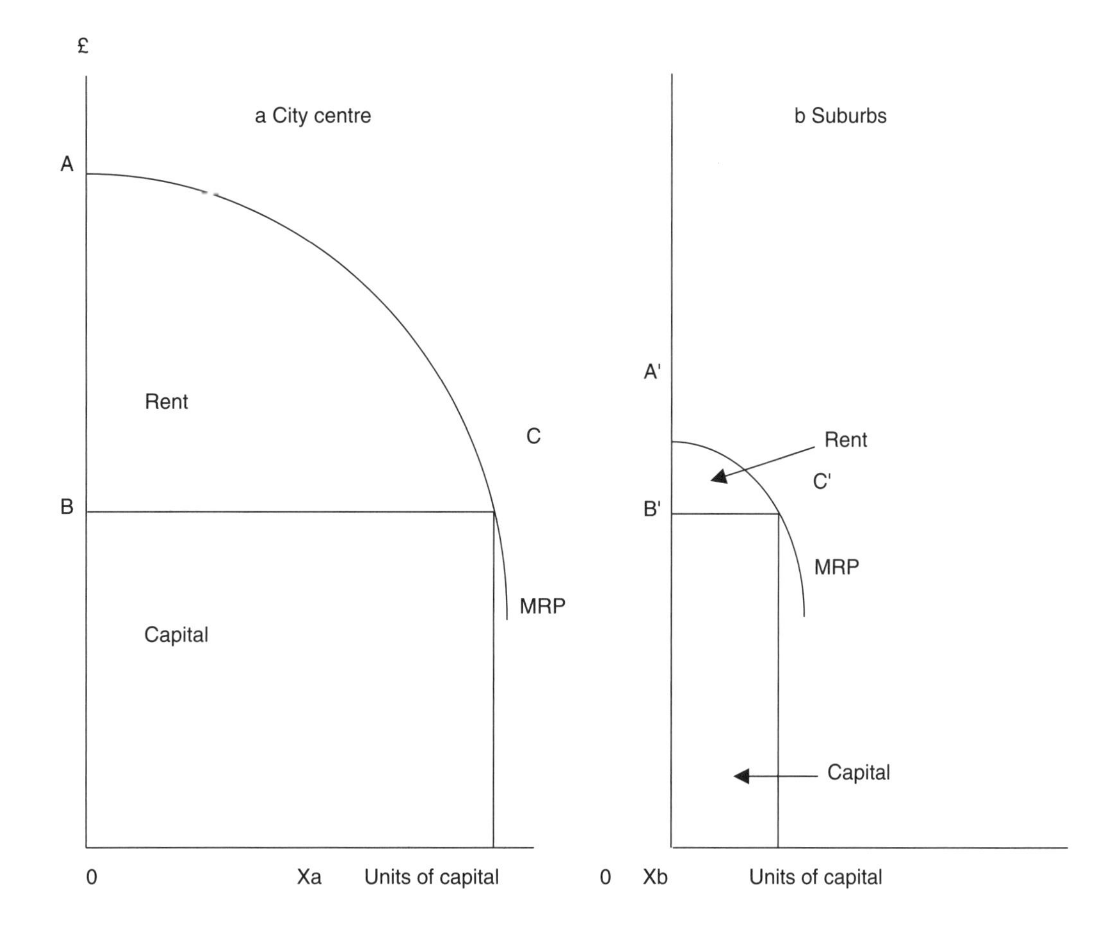

Figure 5.7.2 Intensity of site use.

5.8 Public sector development

Key terms: local authorities; central government; infrastructure; social benefits

'Developer' encompasses a wide range of agencies and organisations, across the public and private sectors, all with their own motivations, aims and objectives. Development undertaken by public sector agencies varies enormously according to the nature and function of the organisation. Public sector development accounted in 2009 for approximately 11 per cent of total development but the percentage varies from year to year. Public development tends to fluctuate with the politics of the government in power, the current requirements of its stabilisation policy, and whether there are any major infrastructure and property development projects under way (such as the Olympics and the proposed HS2 railway).

Local authorities undertake a wide range of development activity across both the commercial and residential sectors. The degree of involvement with capital development projects varies between authorities depending on a number of factors such as location, identified economic and social priorities, land ownership, financial resources and political remit. In addition, national government agencies undertake a wide variety of development initiatives at a national level such as the Highways Agency.

The key difference between the private and public sectors is that the public sector is accountable, and therefore must demonstrate value for money and ensure that all expenditure into capital projects is in the public interest. Therefore the public sector is motivated by social benefits, rather than profit (see Concept 6.8). Typical public sector development projects will include provision of housing, schools, hospitals, and schemes that will generate employment. Sustainable development is also a key objective of public sector development.

Further reading

Ratcliffe, E., Stubbs, M. and Keeping, M. (2009) *Urban Planning and Real Estate Development*, 3rd edn, Routledge, London, ch. 11.

5.9 Redevelopment

Key terms: refurbishment; redevelopment; new development; value of net annual returns; value of cleared site

As a result of changes in the conditions of demand and supply, some structural change of buildings may be necessary. This may take different forms:

- modification of the existing building through refurbishment (for example, new office or shop layouts) or conversion (for example, houses divided into flats, offices converted into residential apartments);
- redevelopment, where existing buildings are demolished and replaced by new ones;
- new development through outward expansion on undeveloped land (for example, suburban housing).

Where land is already developed by having a building on it, fixed capital is embodied in the land. Such capital has no cost in the short period; as a result, redevelopment to a new use, which requires expenditure of further capital, usually occurs only after a considerable period of time. In general terms, redevelopment takes place when the present value of the expected flow of future net returns from the existing use of the land resources becomes less than the capital value of the cleared site. We have therefore to calculate the present value of the land resources in their current use and compare this with the value of the cleared site.

The present value of the most profitable alternative use is obtained by the procedure used for calculating the present value of the current use: (i) the future net annual returns (NARs) in the best alternative use are calculated; and (ii) these NARs are discounted to the present and aggregated to give a capital present value, DD_1 in Figure 5.9.1.

Over time the value of the alternative use (DD_1) rises. This occurs for two main reasons. First, changes in the conditions of demand and supply mean that a new building, being specifically designed for the new use, will earn higher NARs. Thus an old office building will give way to one that is air conditioned and has the structure and space suitable for modern office equipment. Second, any new building would probably have a longer time horizon than the old building (which has already run a part of its life) so that there would be more future NARs to aggregate to obtain its present value.

From the present value of the best alternative use at any one time, we have to deduct: (a) the cost of demolishing and clearing the site (*AB* in Figure 5.9.1); and (b) the total cost of rebuilding for the new use, including ripening costs and normal profit (OA). For simplicity, we have assumed that costs (a) and (b) both remain constant over time. BB_1 represents the sum of (a) and (b).

The present value of the cleared site is thus the difference between DD_1 and BB_1 in any given year. In year O the value of the best alternative use would be only just below that of the chosen current use, for then each was competing for a cleared site. However, once a site has been allocated to a given use and has had a building erected upon it, any new use has the additional handicap of demolishing and rebuilding. So, until year R the value of the cleared site is negative. Eventually it becomes positive and exceeds the present value of existing use and redevelopment takes place at T in Figure 5.9.2.

The economic life of a building is the period of time during which it commands a capital value greater than the capital value of the cleared site. In year T the building becomes economically inefficient because resources (the site) can be switched to a new use having greater value.

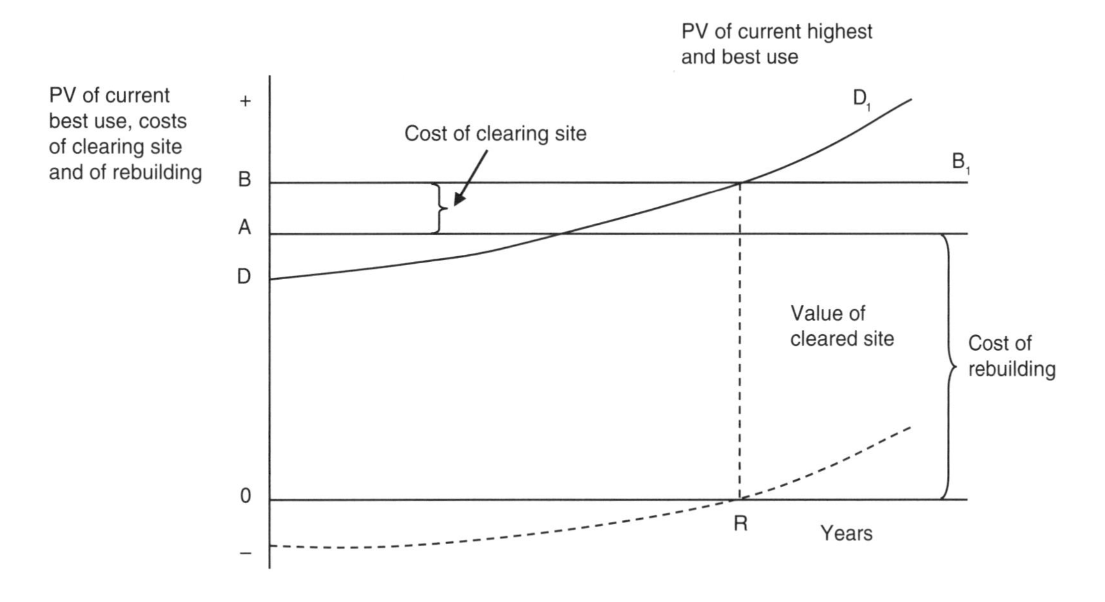

Figure 5.9.1 The value of the cleared site.

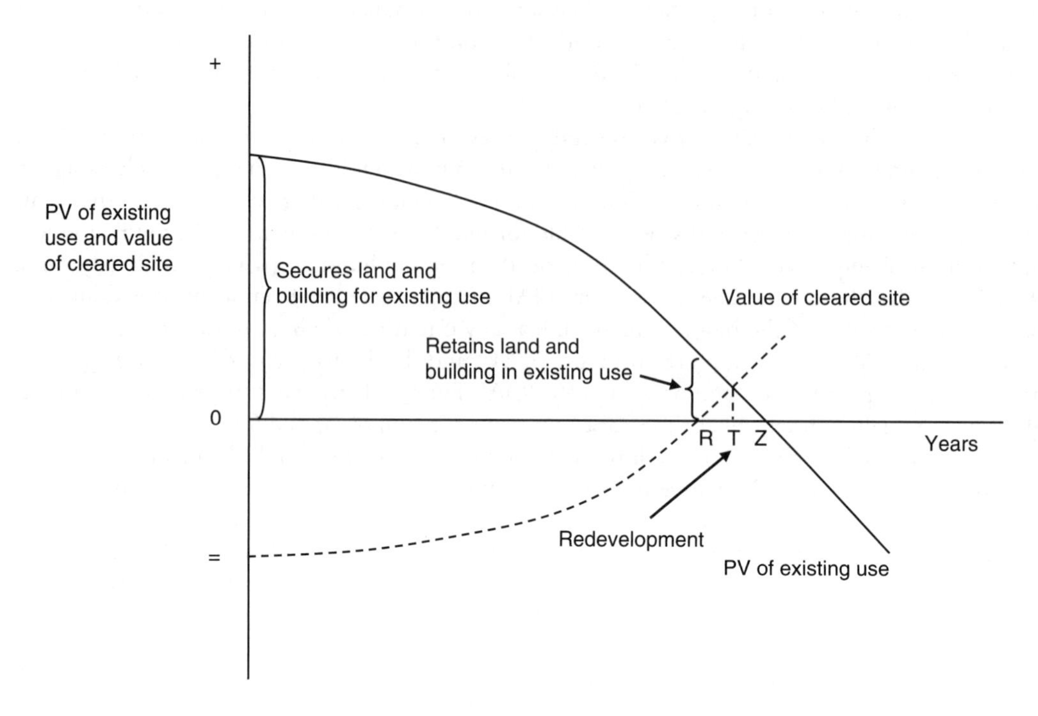

Figure 5.9.2 The timing of redevelopment.

Further reading

Jowsey E. (2011) *Real Estate Economics*, Palgrave Macmillan, Basingstoke, ch. 8.

5.10 Refurbishment

Key terms: renovation; net annual returns; sustainable refurbishment; retrofit

Buildings are maintained, repaired and improved throughout their economic life. At some point a decision must be made as to whether to renovate in order that the building can continue to provide useful services, or to clear it and rebuild or redevelop the site. If the decision is taken to renovate or refurbish the building, the objective must be to raise the present value (PV) of future revenues or NARs by more than the cost of the refurbishment. NARs would increase because future rents in the refurbished building would be higher and because operating costs are likely to be lower, especially if (as is increasingly the case) refurbishment also includes energy-saving and possibly energy-generating measures such as the fitting of solar panels or wind turbines. Renovation or refurbishment would delay clearance and redevelopment as illustrated in Figure 5.10.1.

In many cities the economic life of large old houses has been extended by renovation or conversion into apartments. Inner-city areas can thus maintain the appearance of former grandeur at least from the façades of such buildings, while internal modification and refurbishment provides smaller, more modern accommodation.

Sustainable refurbishment includes insulation and related measures to reduce the energy consumption of buildings, and sometimes installation of renewable energy sources such as solar water heating

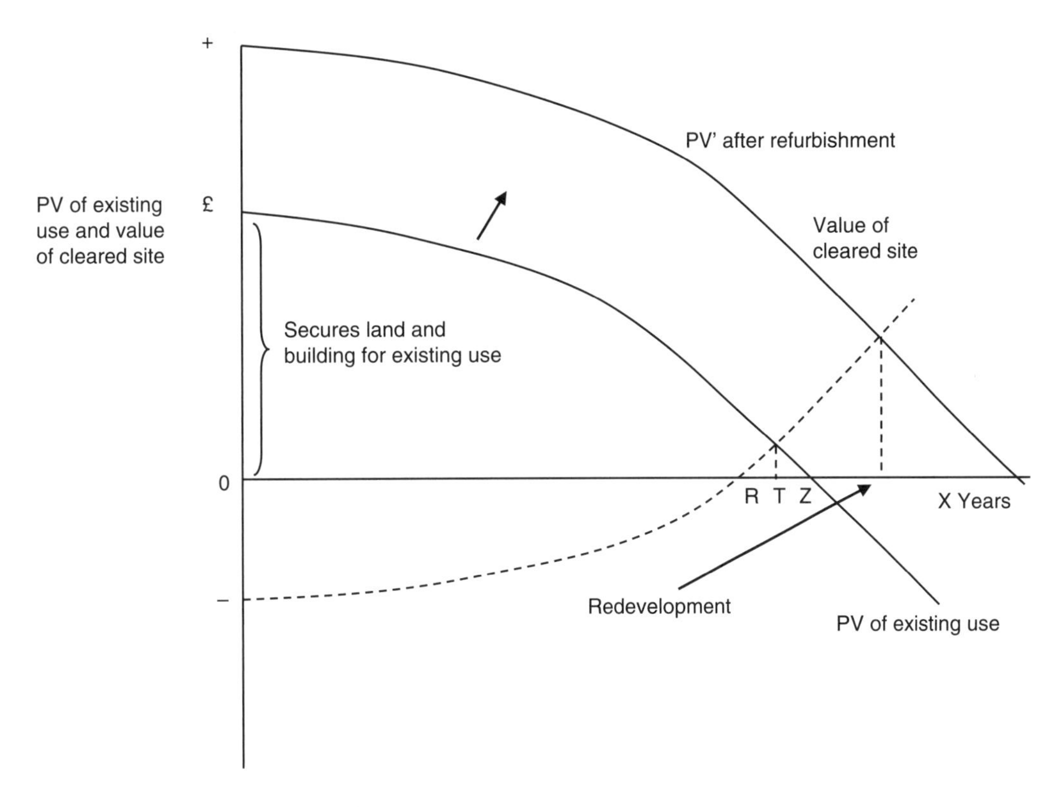

Figure 5.10.1 Refurbishment delays redevelopment.

and solar energy from photovoltaic panels. This process is often termed 'retrofitting' existing buildings. Measures can also be taken to reduce water consumption, to reduce overheating, to improve ventilation and to improve internal comfort. The process of sustainable refurbishment includes minimising the waste of existing components, recycling and using environmentally friendly materials, and minimising energy use, noise and waste during the refurbishment.

The vast majority of existing buildings were constructed when energy standards were low and incompatible with current standards. Much of the existing building stock is likely to be in use for many years to come since demolition and replacement is often unacceptable owing to cost, social disruption or because the building is of architectural or historical interest. Refurbishment of such buildings is necessary in order to make them appropriate for current and future use and to satisfy current standards of energy use.

Concerns about high energy use leading to climate change, overheating in buildings, the need for healthy internal environments and waste and environmental damage associated with materials production, all mean that the importance of sustainably renovating existing buildings is becoming increasingly important.

Further reading

Jowsey, E. (2011) *Real Estate Economics*, Palgrave Macmillan, Basingstoke, ch. 8.

5.11 Residual value

Key terms: development, demand, planning, profit

The residual value is the value of property determined by the residual method. The residual method is used to value property that has the potential for development, redevelopment or refurbishment. When valuing for development purposes a valuer will normally utilise both the comparison method and the residual method. This will depend on the availability of comparable evidence and the nature and complexity of the development proposed.

The basic equation for undertaking a residual land valuation is:

L = V – C – P, where:

L = land price and associated costs

V = net development value

C = building and associated costs

P = developer's profit

So for a site valuation: L = V – C – P

The residual method is essentially the same as the method detailed in the 'Evaluation appraisal methods' concept.

In the first instance it is necessary to determine what the optimum scheme for the site is. This requires an analysis of planning and market considerations. The chosen scheme must have a reasonable prospect of gaining the necessary planning consent. This can be ascertained by an analysis of relevant national and local planning policies. An analysis of the market should be undertaken to establish the potential demand for the ideal alternative forms of development that may be possible. An analysis of existing supply and the development pipeline will also help ensure the identified demand will not be satisfied by an alternative scheme in advance of the subject development project being completed.

The main components of construction and associated costs are considered in the appraisal methods concept. When carrying out a development appraisal an assumption is made regarding the value of the development site and the resultant residual represents the predicted profit for the project. When carrying out a residual valuation an assumption has to be made about profit to enable a land value to be derived.

In determining the level of profit to use in a residual valuation the nature of the development and associated risk must be considered alongside the prevailing practice in the market for the sector. Profit will normally be expressed as a percentage of the total development cost or the gross development value, with the former being more common.

Although the overall concept is fairly simple, there are many variables to consider including:

- what can be built
- what demand there is
- what the levels of supply are
- what the likely build costs are
- method of procurement
- the cost of land remediation

- professional fees
- interest on capital borrowed
- how easy it will be to sell/let the completed building
- the risk of overrunning on costs.

The result is that residual valuations are complex and there is much scope for uncertainty about their outcomes. Nonetheless, they are widely used in the property market for development appraisal, i.e. determining bids and sale prices for land and buildings with development potential.

Further reading

RICS (2008) 'Valuation Paper 12: Valuation of development land', RICS, London.

5.12 Local asset-backed vehicles

Key terms: operational assets; regeneration; development projects; investment LABV; value capture LABV; integrated LABV

Local asset-backed vehicles (LABVs) are joint ownership limited liability special purpose vehicles with the specific purpose of carrying out regeneration or renewal of operational assets. The public sector invests property assets into the vehicle which are then 'value matched' by private sector equity. The partnership then uses these assets as security to raise finance to bring forward further development. The public and private sector are equal equity holders and share profits equally, according to their original equity contribution.

The basic LABV model is illustrated in Figure 5.12.1:

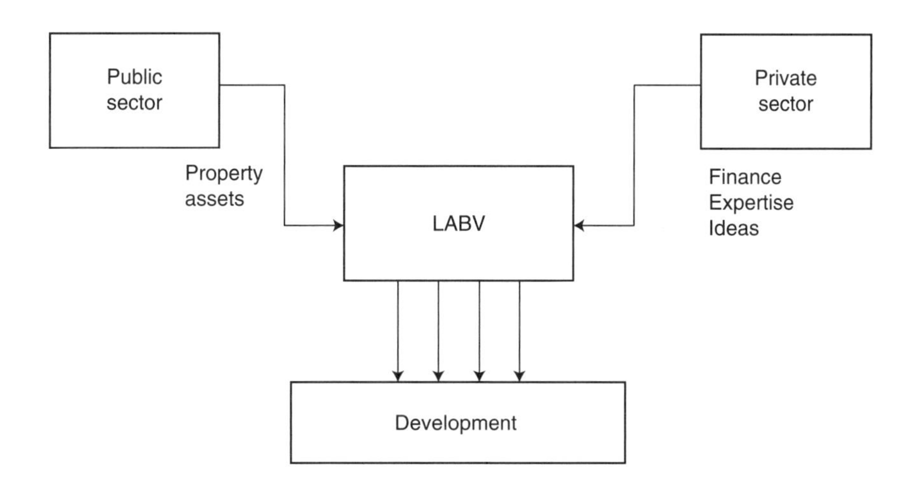

Figure 5.12.1 Local asset-backed vehicles.

Source: adapted from Grace and Ludiman (2007).

LABVs typically run for 20 years, at the end of which the assets are returned to the investing partners through a predetermined exit strategy, e.g. selling the land with planning permission, selling the land once development is completed, or retaining the development as an investment. Before each party enters into a partnership, the long-term goals of the partnership are detailed in the legal documents, which protect the wider social and economic aims of the public sector along with the pre-agreed business plans based on the requirements of the private sector.

Whereas traditional public–private partnership (PPP) methods such as PFI (private finance initiative) have a clear and specific purpose, for example a hospital or school, LABV is a long-term partnership for the regeneration of a town, designated area or a cluster of communities, and must therefore be more flexible as priorities may change over the life of the partnership.

The benefits delivered by such asset-backed vehicles for the public sector, include access to new markets, increased capacity through the possibility of reinvestment of returns into other projects, the sharing of both risks and costs, access to greater resources including staff, private sector expertise and finance, as well as ensuring that development strategies are well focused and coherent. They also offer continued control over the direction of the investment, reduced exposure to deflated asset values, access to gearing and increased flexibility in the vehicle used to deliver infrastructure projects.

The three main types of LABV are:

1 Investment LABV: this can be used where a site requires significant investment to make it marketable, e.g. where major infrastructure, remediation or substantial planning input is required, but otherwise the opportunity is viable. A private sector investor will fund this requirement. Once the works have been carried out and site value is enhanced, the LABV will sell the asset or parts of it on the open market. The risk is relatively low and profit is fairly modest.
2 Value capture LABV: the LABV acts as the developer, ensuring sites are ready for development, carrying out infrastructure works and remediation and obtaining planning permission. This type of LABV provides greater scope for profit but risks are commensurately higher.
3 Integrated LABV: this will deliver most, if not, all of the required development and carries the greatest risks and scope for the highest level of profit. The construction supply chain is procured before the LABV is established. The vehicle potentially carries the risk for land assets, planning infrastructure, some or all construction sales.

(Pinsent Masons, 2011)

The 'patient capital' investment characteristic of LABVs is regarded as an attractive quality of the model as it incentivises the private sector to invest and deliver over the longer term, as returns are subject to the performance of the partnership over 10–20 years time frame. To date, LABV take up in the UK has been relatively modest. There are a number of reasons for this – firstly, local authorities have a responsibility to ensure they get best value from the disposal of public assets and are cautious about 'selling off the family silver' at an unfavourable point in the property cycle or of handing over control of public assets.

Further reading

British Property Federation and the Local Government Association (2012) *Unlocking Growth Through Partnership*, BPF, London.

Grace, G. and Ludiman, A. M. W. (2007) 'Local asset backed vehicles: The potential for exponential growth as the delivery vehicle of choice for physical regeneration', *Journal of Urban Regeneration and Renewal*, 1(4): 341–353.

Pinsent Masons (2013) 'Guide to LABVs', available at: www.out-law.com/en/topics/property/structured-real-estate/local-asset-backed-vehicles (accessed 28/10/2013).

Thompson, B. (2012) *Local Asset Backed Vehicles: A success story or unproven concept?*, RICS, London.

6 Economics

Ernie Jowsey

6.1 Allocation of resources

Key terms: wants; resources; economics; production possibilities curve; opportunity cost

Allocating scarce resources is the overall economic problem. In comparison with all of the things we want, our means of satisfying those wants are very inadequate. On a personal level there are probably many things you would like to have but you cannot afford them. On a global scale, if everyone on earth was to live at the same standard of living as people in the United States, more than three planets would be needed!

So we have unlimited wants and very limited resources, and this is a problem that can never be completely resolved. But we can make the most of what we have – in other words, 'economise'. Each household has to make the most of its limited resources, and the word 'economics' is derived from the Greek word meaning the management of a household. Shoppers have to make decisions as to what to buy in order to obtain maximum satisfaction. Scarcity forces us to economise. We weigh up the various alternatives and select that particular assortment of goods that yields the highest return from our limited resources. Students have to make sure their funds last the term. Businessmen have to decide on quantities of raw materials to buy and labour to hire. Even the richest person in the world must decide how to allocate their time!

There are four aspects of this economic problem:

1. the wants of human beings are without limit;
2. those wants are of varying importance;
3. the means available for satisfying those wants – human time and energy and material resources – are limited;
4. the means can be used in many different ways – that is, they can produce many different goods.

An illustration of this can be seen by considering the problem facing a house builder or property developer when deciding exactly what type of properties to build on a piece of land. Suppose there is a choice of affordable homes or executive houses or a combination of the two. We need to assume there is no interference from the planning authorities. Table 6.1.1 shows the combinations that can be produced.

These production possibilities can be shown in a production possibilities curve (PPC). The straight line PPC in Figure 6.1.1 represents the trade-off between executive homes and affordable homes facing this developer. She cannot build 40 executive homes and 80 affordable homes because she does not have the resources and so that combination lies outside the PPC and is unattainable. Of course the actual combination that is decided upon will be influenced by demand for executive homes and affordable homes in the market and almost certainly by planning conditions. If the decision is made to build 40 executive homes, then the opportunity cost of that decision is 80 affordable homes (that are not built).

Table 6.1.1 Production possibilities for housing

Affordable homes	*Executive homes*
80	0
60	10
20	30
0	40

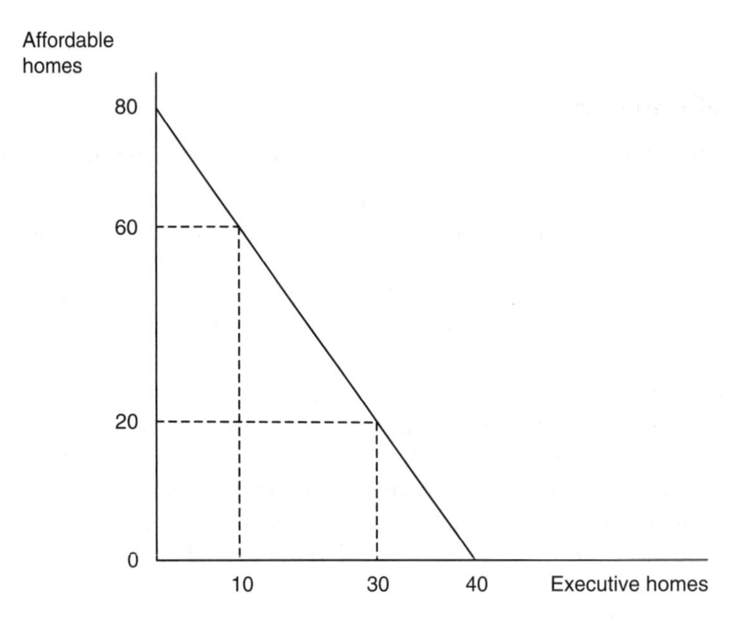

Figure 6.1.1 A (straight line) production possibilities curve.

Governments also have to decide how their resources or funds are allocated. Less spent on roads might mean more can be spent on the health service, but might lead to more road traffic accidents. Better schools and hospitals may be needed but they are competing for the limited revenue that can be raised from taxation. Extra houses, new roads and conservation or wildlife areas – all are claiming a share of the limited land available. A decision to provide more of one type of facility will have an opportunity cost.

Further reading

Jowsey, E. (2011) *Real Estate Economics*, Palgrave Macmillan, Basingstoke, ch. 1.
Harvey, J. and Jowsey, E. (2007) *Modern Economics*, Palgrave Macmillan, Basingstoke, ch. 1.

6.2 Supply and demand

Key terms: demand curve; supply curve; market equilibrium; clearing price; complements; substitutes

Demand in economics is the desire for something plus the ability to pay a certain price in order to possess it. Technically, it is how much of a good people in the market will buy at a given price over a certain period of time. The higher the price, the less people will want to buy and at a very high price their demand is zero. In Figure 6.2.1 there is a demand curve with price rising on the Y-axis and quantity demanded increasing from left to right along the X-axis, the demand curve slopes downwards from left to right.

Demand is determined by price and other factors affecting demand. These other factors are:

- prices of other goods, especially complements and substitutes;
- incomes of consumers;
- tastes of consumers.

If price changes then quantity demanded adjusts along the demand curve. If there is a change in one of the factors affecting demand, such as tastes of consumers, the demand curve will shift to the left or right within Figure 6.2.1.

Supply in economics refers to how much of a good will be offered for sale at a given price over a given period of time. This quantity depends on the price of the good, and the conditions of supply. More will be supplied at a higher price, because producers can make more profits. Supply is determined by price and other factors affecting supply. These other factors include:

- prices of factors of production;
- the level of technology.

If price changes then quantity supplied adjusts along the supply curve. If there is a change in one of the factors affecting supply, such as a change in the price of a factor of production, the supply curve will shift to the left or right within Figure 6.2.1. A supply curve will rise from left to right as in Figure 6.2.1.

The intersection of demand and supply in the market is at price p and quantity q. At price p the quantity demanded and supplied is q. There are no unsold goods and no shortage of goods at the price p (the market 'clears').

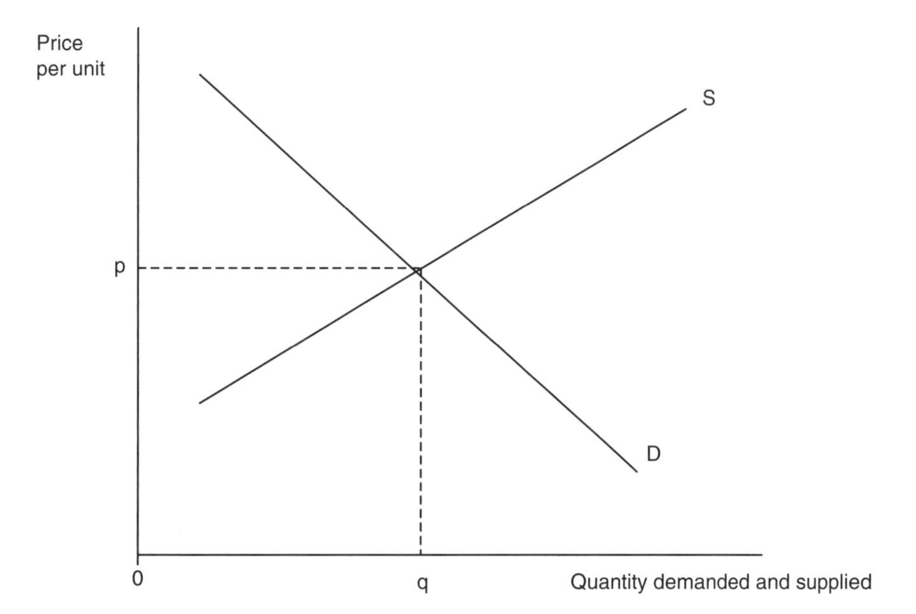

Figure 6.2.1 Equilibrium price and quantity.

In reality, prices are rarely ever at this equilibrium level – house prices, for example, do not usually clear the market and there is often excess supply with downward pressure on prices; or excess demand, with upward pressure on prices.

To summarise: a market can be defined as 'all buyers and sellers of a good or service who influence its price'. The price of a good or service is determined by the interaction of the forces of demand and supply. Since demand and supply curves slope in opposite directions, they intersect at a single point. This is the market or equilibrium price where demand equals supply – the 'market clearing price'.

Further reading

Harvey, J. and Jowsey, E. (2007) *Modern Economics*, Palgrave Macmillan, Basingstoke, ch. 3.
Jowsey E. (2011) *Real Estate Economics*, Palgrave Macmillan, Basingstoke, ch. 3.

6.3 Pareto optimality

Key terms: welfare; distribution of income; potential Pareto improvement; cost–benefit analysis

Efficiency in the use of resources is often called Pareto optimality or the Pareto criteria in honour of the Italian economist Vilfredo Pareto (1848–1923), who first identified this idea.

The Pareto optimal condition is that welfare is maximised when no one can be made better off without somebody else being made worse off. Any improvement in economic efficiency that involves nobody losing will represent an increase in welfare. This avoids the distributional problem which may result from very unequal income distribution. Even a society with a few very rich people and a lot of very poor people can be Pareto optimal if the poor cannot be made better off without taking something from the rich. For this reason Pareto optimality is known as a weak welfare criterion.

Figure 6.3.1 illustrates this. Starting from the initial income position X, with A's income equal to OA and B's equal to OB, movement to M would represent an increase in welfare for both A and B; a

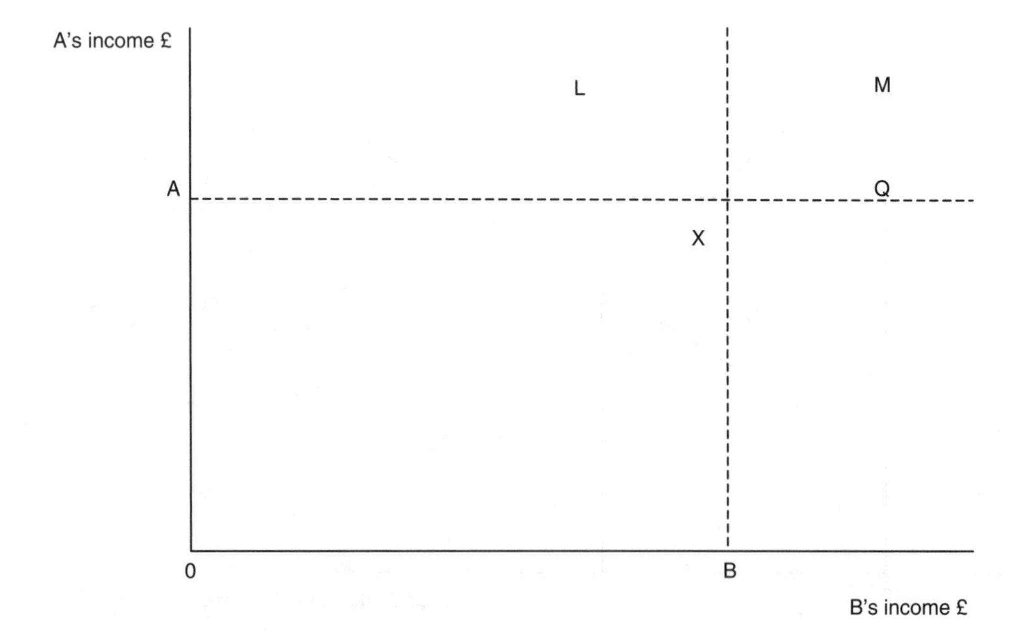

Figure 6.3.1 Pareto improvements.

movement to Q would increase B's welfare without reducing A's. Both M and Q therefore represent Pareto improvements. It is impossible, however, to say whether position L represents an overall gain or loss, since A's income has increased but B's has fallen.

The problem with the strict Pareto-optimality condition is that its application is restricted to those cases where there are gainers but no losers resulting from the reallocation of resources. There may be ways of improving production and overall welfare which make some people much better off but others a little worse off. Applying the strict Pareto criteria would not allow this change. A weaker condition is that of *potential* Pareto improvement if those who gain could compensate those who lose and still be better off. The compensation does not actually have to be paid – it is only necessary to show that it could. This condition forms the basis for decision making in cost–benefit analysis (see Concept 6.8). It can be summarised as:

$$\Sigma_i [B_i - C_i] > 0$$

where Σ_i is the sum over a time period i; B_i is total benefits over a time period i; and C_i is total costs over a time period i. In words: if total benefits minus total costs over the time of the project are greater than zero then the project should go ahead.

Further reading

Jowsey E. (2011) *Real Estate Economics*, Palgrave Macmillan, Basingstoke, ch. 2.

6.4 Economic efficiency

Key terms: limited resources; resource allocation; production possibilities curve; Pareto optimality; competition; perfect competition

Economic efficiency is achieved when society has secured the best allocation of its limited resources, in the sense that the maximum possible satisfaction is obtained. Society's aim is to maximise its welfare in terms of the way it uses its limited resources; and the distribution of income between members of society. Decisions on income distribution are political decisions, so we can leave that to politicians and adopt a fairly limited condition, that of Pareto optimality or Pareto efficiency (see Concept 6.3). This states that: 'welfare is maximised when no one can be made better off without somebody else being made worse off.' Thus, any improvement in economic efficiency that involves nobody losing will represent an increase in welfare.

A PPC (or frontier) can illustrate the concept of economic efficiency. In Figure 6.4.1 an economy's production of food and manufactured goods is shown in a graph of possible outputs. If only food is produced then 40 units can be made – and if only manufactured goods are produced then 50 units can be made. Of course society is likely to want a mixture of both and it can choose any combination on or inside the PPC.

Choosing point X where 20 units of each resource are produced is not Pareto optimal because more of either good could be produced without reducing output of the other. Any point inside the PPC, such as X, is not technically efficient. Both A and B are on the PPC and are technically efficient – as much as possible is being produced. But society may prefer more food to be produced than it is at point B, where only 10 units are made. There are two ways for society to achieve its preferred allocation at point A – by government direction (in a command economy) or by using the price system to allow people to express their preferences. Starting from point B with a lot of manufactured goods and little food would mean the price of food is high and the price of manufactured goods is low. Producers would respond by growing more food and reallocating resources away from manufactured goods. This process would continue until prices stabilised for both food and manufactured goods – where demand

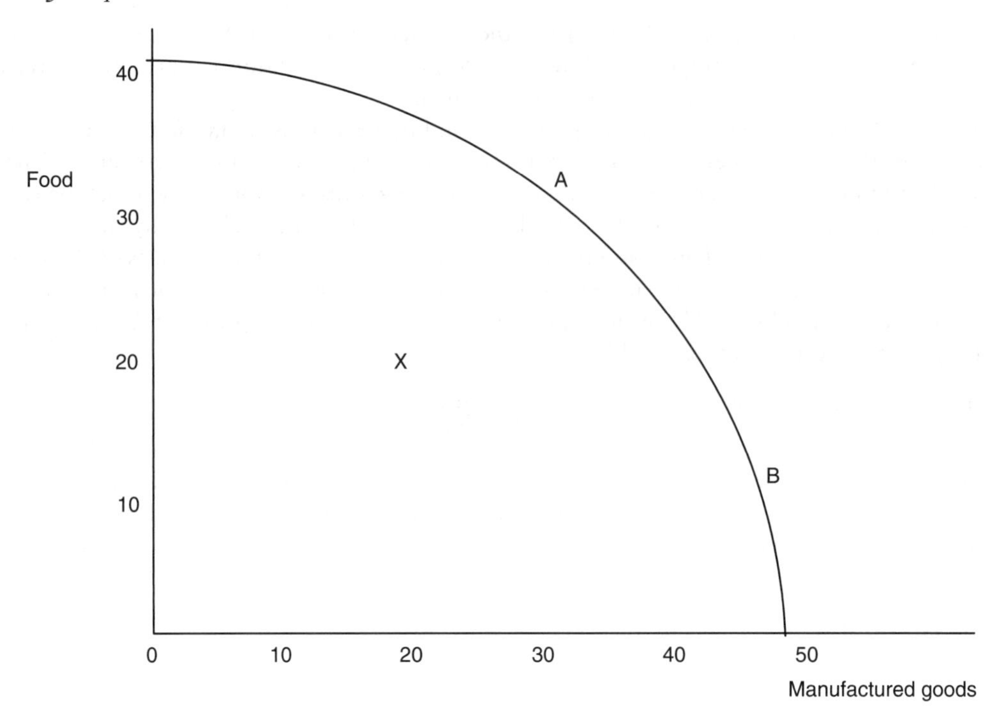

Figure 6.4.1 A production possibility curve.

equals supply for both goods. If this is at point A on the PPC then point A is both technically efficient and economically efficient.

Competition increases efficiency in the economy. Without competition, producers have no incentive to reduce costs (and prices) and to innovate. Perfect competition is a market form where there are many buyers and many sellers of a product and this situation will ensure that competition results in the most efficient methods of production.

The market economy or price system is very efficient but there are instances of market failure that arise from externalities such as pollution, and from the existence of goods and services that cannot be priced, such as defence and street lighting. As a result, the market economy is unlikely to be fully efficient in allocating resources, but allocation by government decisions in a command economy can present even greater problems. At least the market system does start with the advantage that economic decisions are based on prices that reflect consumers' preferences and relative costs. This suggests the compromise of a mixed economy. This uses the price system as the basis for allocating resources, but, recognising its defects, relies on the government to provide, as far as possible, the conditions necessary for its efficient functioning.

Allocation of real estate resources is still mainly through the price mechanism of a market economy. How efficient is the real property market in registering changes in demand and supply through their effect on price? In an efficient market we could expect the market to be in equilibrium most (if not all) of the time. This would mean the market clears or demand equals supply. So if the real property market is efficient there would be no prolonged periods of excess demand or excess supply. Any observer of the cyclical nature of the real estate market in most market economies would recognise that this is not the case – in the upswing or boom there is a period of excess demand and in the downswing or recession there is a period of excess supply.

Further reading

Harvey, J. and Jowsey, E. (2007) *Modern Economics*, Palgrave Macmillan, Basingstoke, chs. 1 and 2.
Jowsey, E. (2011) *Real Estate Economics*, Palgrave Macmillan, Basingstoke, chs. 2 and 3.

6.5 Market, command and mixed economies

Key terms: market economy; command or planned economy; invisible hand; community goods; externalities; economic efficiency

Market economy

The first exchanges of goods and services were quite simple: there was a direct swap of one good for another – a market was established. Eventually money was developed, allowing goods to be priced and sold more easily. The subsistence economy had now become a market economy, where decisions on production follow from people's decisions in the market. In the market economy, emphasis is laid on the freedom of the individual, both as a consumer and as the owner of resources.

Consumers express their choice of goods through the price they are willing to pay for them. The owner of resources used in production (usually his or her own labour) seeks to obtain as large a reward as possible. If consumers want more of the goods than is being supplied at the current price, this is indicated by their 'bidding up' the price. This increases the profits of firms and the earnings of factors producing that good. As a result, resources are attracted into the industry, and supply expands in accordance with consumers' wishes. On the other hand, if consumers do not want a particular good, its price falls, producers make a loss, and resources leave the industry; consumers control the economy.

Prices therefore indicate the wishes of consumers and allocate the community's productive resources accordingly. There is no direction of labour and people are free to work wherever they choose.

Efficiency is achieved through the profit motive: owners of factors of production sell them at the highest possible price, while firms keep production costs as low as they can in order to obtain the highest profit margin. Factor earnings decide who is to receive the goods produced. If firms produce better goods or improve efficiency, or if workers make a greater effort, they receive a higher reward, and have more spending power to obtain goods in the market.

In a market economy, the price system acts like a huge invisible hand or computer, registering people's preferences for different goods, transmitting those preferences to firms, moving resources to produce the goods, and deciding who shall obtain the final products. The driving force is the motivation of individual self-interest.

Of course, the market economy has problems and does not work quite as perfectly as this. In particular:

- community goods, such as defence, may not be provided;
- competition, upon which the efficiency of the market economy depends, may break down;
- imperfect knowledge may impair the working of the price system;
- externalities or 'spillover effects' such as pollution may occur;
- great inequalities of income can arise.

Command economy

In a command economy, the decisions are taken by an all-powerful planning authority. It estimates the assortment of goods which it considers people want and directs resources accordingly. It also decides how the goods produced shall be distributed among the community. Economic efficiency largely depends upon how accurately wants are estimated and resources allocated. The central planning authority can ensure that adequate resources are devoted to community and other goods; and it can use its monopoly powers in the interests of the community, e.g. by securing the advantages of large-scale production, rather than make maximum profits by restricting output. There can be more certainty in production by integrating plans and improving mobility by direction of resources, even of labour. External costs and benefits can be allowed for when deciding what and how much to produce, and uneven distribution of wealth can be considered when planning what to produce and in rewarding the producers.

There are also problems with a command or planned economy, however:

- Estimating the satisfaction derived by individuals from consuming different goods is impossible – although some help can be obtained by introducing a modified pricing system. Often, however, prices are controlled, supplies rationed or queues form for limited stocks.
- Many officials are required to estimate wants and to direct resources and this may lead to bureaucracy, such as excessive form-filling, 'red tape', slowness in coming to decisions and an impersonal approach to consumers. At times, too, officialdom has been accompanied by corruption.
- State ownership of resources, by reducing personal incentives, diminishes effort and initiative.
- Direction of labour may mean that people are dissatisfied with their jobs.
- Officials may play for safety in their policies.
- There is a danger that the state will make it easier to pursue economic objectives by restricting freedom of action.

Mixed economy

Most economies in the world are free market economies regulated by government. They have a mixed economy in which more than half of production is carried out by private enterprise through the market (though subject to varying degrees of government control), while for the rest of the economy the government is directly responsible. The government influences the allocation of the goods and services produced using fiscal and monetary policy, income redistribution and subsidies.

Further reading

Harvey, J. and Jowsey, E. (2007) *Modern Economics*, Palgrave Macmillan, Basingstoke, ch. 2.

6.6 Externalities

Key terms: private costs; social costs; private benefit; social benefit; spillover effects

Resource allocation in the economy is the result of the decisions of consumers and producers who seek to maximise the difference between benefits and incurred costs. These are private benefits and private costs. There may be benefits and costs – externalities – additional to those that are the immediate concern of the parties to a transaction and that are not provided for directly in the market price. These benefits or costs spill over onto others not directly involved. For example, a firm may decide to build a new factory on a derelict site in a depressed district. In doing so it confers some external benefits: tidying up the site and reducing the cost of government unemployment benefit payments. But if the factory is built in a predominantly residential district, it would incur external costs of vehicle movements, noise, etc.

The full social benefits are, therefore, private benefits plus external benefits. The full social costs are private costs plus external costs. If an economically efficient allocation of resources is to be achieved, externalities should be allowed for.

The spatial characteristic of land in particular gives rise to considerable 'spillover' costs and benefits. For example, if the grounds of a previously derelict property are landscaped and much improved, there can be benefits for the whole area. And if planning permission is granted for the construction of several new houses on a greenfield site, there may be additional traffic costs in the form of delays, congestion and pollution for existing residents of the area. This situation is shown in Figure 6.6.1.

On private benefit/costs considerations alone, houses would be built to a plot density of N where the developers benefit (MSB) is equal to his costs (MPC). But more intensive development gives external costs of pollution and congestion. MSC, therefore, exceeds MPC. Plot density should therefore be limited to N_1, in order to achieve optimal development where MSB = MSC (equalising social benefits and social costs).

Of course, some spillover costs are inevitable. Living in cities limits the open space available to people and they suffer continuous traffic noise. But people live in cities because of the advantages and these include external benefits, such as convenient amenities and leisure facilities.

Sometimes externalities are reflected in the market price. People will pay more for a house in an area that has good schools, open spaces, leisure facilities, and road and rail links. Shops where traffic congestion is a serious problem will command a lower rent because trade will be affected. Externalities can also be 'internalised' by private arrangements. Developers often attract prestigious shops to a shopping centre by making prior agreements with key retailers. They offer these anchor tenants a good deal on their rent in order to have the benefit of their reputation and good name in their centre. Houses and flats may be sold subject to leases or covenants in order to secure external benefits (such as satisfactory maintenance) or to avoid external costs (for example, excessive noise, car-parking nuisance etc.).

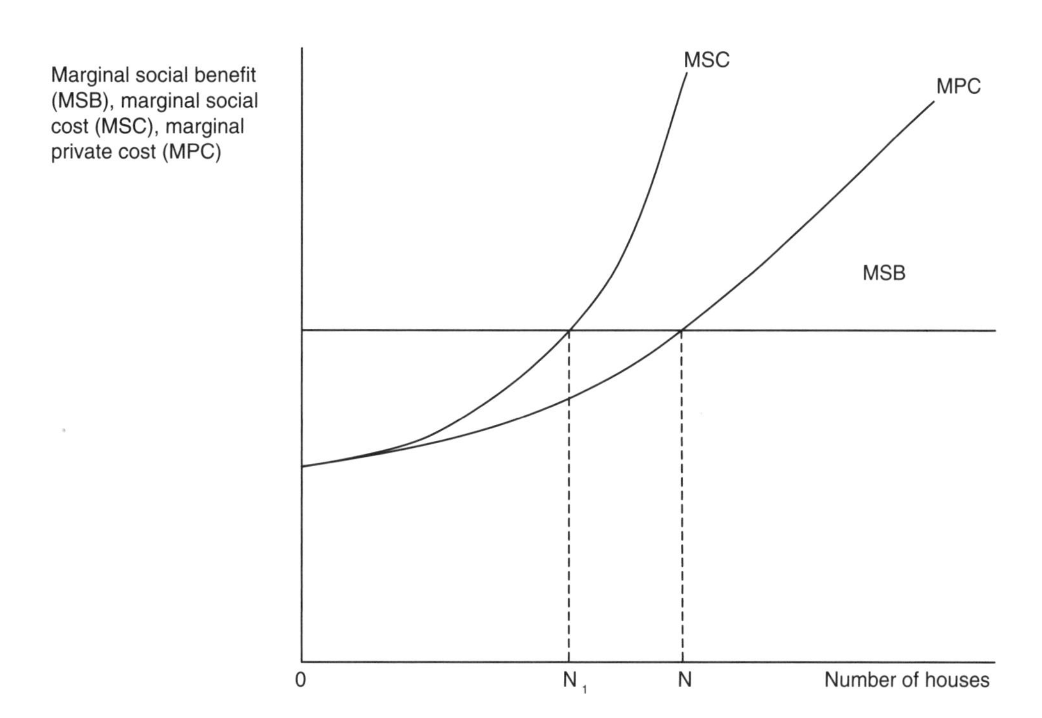

Figure 6.6.1 Additional marginal social costs of development.

Further reading

Harvey, J. and Jowsey, E. (2007) *Modern Economics*, Palgrave Macmillan, Basingstoke, ch. 17.
Jowsey, E. (2011) *Real Estate Economics*, Palgrave Macmillan, Basingstoke, ch. 12.

6.7 Market failure

Key terms: exchange efficiency; technical efficiency; economic efficiency; externalities; imperfect competition; government intervention; community goods

The vast majority of the world's economies work by using market forces to allocate resources. Usually this is done in a very efficient way (see Concept 6.4). If so there will be:

- 'exchange efficiency' so that no overall gain in satisfaction can be obtained by an exchange of goods between persons;
- 'technical efficiency' in production so that no increase in output can be obtained by producers substituting one factor for another or reorganising the scale of production;
- 'economic efficiency' so that out of society's limited resources, the selection of goods and services produced is that which gives the greatest possible satisfaction (so that supply equals demand).

If the market economy is to achieve these types of efficiency, there must be perfect competition (see Concept 6.9), no external benefits or costs, and all economic goods must be priced in the market. Obviously, not all markets are perfectly competitive as there are many examples of imperfect competition, oligopoly and even monopoly in the real world and in many real estate markets. There are also many external benefits and costs that are unpriced because they are outside the market (see Concept 6.6). For example, the design of a new building may be out of character with the surrounding area, imposing 'spillover costs' on other residents. And with many economic goods and services, such as defence, street lighting, common land and many environmental goods and services it is not possible to exclude non-payers and so they are not priced.

So for the reasons mentioned above, the market economy does not always lead to an efficient allocation of resources and so government intervention is necessary. For example, governments have rules and regulations relating to monopoly and imperfect competition; they provide community goods; they try to improve imperfect knowledge; and to 'internalise' or bring in to the market external benefits and costs such as pollution.

In real estate markets there are many examples of externalities and, therefore, the potential for market failure. A firm may decide to build a new business park on a derelict site in a run-down area. This confers external benefits by tidying up the site and providing much needed employment. If the business park was built in a mainly residential district, however, it would incur external costs of extra vehicle movements, noise, air pollution and so on.

Another example is provided when an individual household spends time and money on house improvements, and there are benefits for other houses in the area. If the improvement makes the general area tidier and more appealing, then other households in the vicinity benefit from a positive externality. If several households make improvements the area may be considerably improved and all property values could increase. Of course this issue could suffer from the problem of 'free-riders' – some households doing nothing but still benefitting from the uplift in property values. The benefits of the improvement to the area are non-excludable and if enough households acted as free-riders then the improvement would not materialise (Jowsey, 2011).

In order to try to encourage the households in the area to engage in individual (and thus general) improvement, the local authority could provide improvement grants in run-down areas. This will be Pareto optimal if the private and social benefits are greater than the private costs to the residents plus the cost of the grants. There is evidence to suggest that subsidising housing investments produces significant and sustainable external benefits to urban areas, and of course, they can be used to reduce the external costs of inadequate housing (health and social problems etc.).

These are examples of market failure where government intervention takes place. There is no guarantee of course that the government solution will not have its own adverse effects and there are many examples of 'government failure'. The poor history of twentieth-century rent controls in the UK provides a fairly clear example.

Further reading

Harvey, J. and Jowsey, E. (2007) *Modern Economics*, Palgrave Macmillan, Basingstoke, chs. 17, 18 and 19.
Jowsey E. (2011) *Real Estate Economics*, Palgrave Macmillan, Basingstoke, chs. 3 and 12.

6.8 Cost–benefit analysis

Key terms: Pareto optimal; 'spillover effects' or externalities; discount rate; shadow prices

If those who gain from a property development can *in principle* fully compensate those who lose then a potential Pareto improvement has occurred. In the public sector, cost–benefit analysis may be used to determine whether a development will lead to a potential Pareto improvement. In a cost–benefit analysis (CBA) it is necessary to:

- list all relevant items – including spillover benefits and costs (see Concept 6.6);
- value expected benefits and costs;
- discount future flows of benefits and costs using an appropriate rate of discount;
- assess the net present value of the development – if it is positive then there is potential Pareto improvement.

Difficulties arise at each stage of the CBA and especially in the choice of discount rate (which is often based on market rates of interest by default). Nevertheless, CBA provides a rational technique for appraising developments where market information is lacking.

Government responsibility for roads, bridges, airports, parks, amenity land, new urban areas and housing means that decisions have to be made as regards the allocation of land and land resources. Questions arise such as: is investment in a new motorway justified? Which site should be chosen for a new airport? The difficulty is that, since many public-sector goods are provided free or below market price, indications of the desirability of investment through the price system are either non-existent or defective. CBA is a technique that seeks to bring greater objectivity into decision making. It does this by identifying all the relevant benefits and costs of a particular scheme and quantifying them in money terms so that each can be aggregated and then compared.

CBA is likely to have its main use in the public sector where

(i) price signals are inadequate to guide investment decisions;
(ii) 'spillover' benefits and costs are important because of the magnitude of the schemes; and
(iii) the welfare of unborn generations (intergenerational equity) has to be allowed for – for example, by conservation measures.

CBA can be used in all public investment decisions, it has particular application to the allocation of land resources, where externalities are likely to be considerable. As a result CBA studies have been undertaken for:

(a) the construction of the M1 motorway;
(b) London Underground's Victoria Line;
(c) the siting of the third London Airport;
(d) consideration of a third runway and sixth terminal at Heathrow;
(e) the re-siting of Covent Garden Market.

There are, however, several difficulties in the application of CBA:

Distributional effects: country lovers may lose pleasure through mobile phone masts intruding on the landscape. If they are fully compensated, there is no problem in terms of Pareto optimality. However, the difficulty of identifying such losers means that compensation is not actually paid, and there are thus distributional effects.

Adjusting market prices: where market prices reflect the true opportunity cost to society of employing resources in a particular way (assuming no externalities), they can be used to estimate the cost of a project. However, in the real world it has to be recognised that prices in the market may not accurately reflect opportunity cost. This may be the result of imperfect competition, indirect taxes and subsidies, or controls that interfere with the free operation of the market mechanism.

Pricing non-market goods: market prices may not be available. This occurs with the following goods:

1 community and public goods, where 'free riders' cannot be excluded (examples are street lighting, land, radio programmes) or where it is decided to make no charge (for instance, for public parks or bridges). Here the cost is covered by taxation that is unlikely to reflect true 'willingness to pay';
2 intangible externalities, such as noise and congestion cost, human lives saved, the pleasure derived by passers-by from flowers and trees in private gardens or from a walk in a park.

Since both enter into CBA calculations, it is necessary to ascribe notional prices to them so that benefits and costs can be quantified in money terms. But formulating such 'shadow' or 'surrogate' prices faces formidable difficulties.

Valuing time saved: transport improvements usually result in reducing the time spent in making a journey. What price do we put on this benefit?

The value of human life: such projects as road improvements reduce deaths and accidents; others such as rock quarries (with associated lorry movements) may increase them for people in the vicinity. How can a money value be given to human life?

Spillover effects: the problem of which spillovers should be included is concerned, first, with the difficulty of distinguishing between real changes and distributional effects, and, second, on deciding the cut-off point – how widely effects are considered.

Intangibles: in aggregating costs and benefits, how much weight should be attached to the shadow prices of intangibles compared with true market prices? In so far as there is perfect competition, market prices at factor cost reflect true opportunity costs. In comparison, shadow prices are derived indirectly, and to that extent are somewhat suspect.

Further reading

Harvey, J. and Jowsey, E. (2007) *Modern Economics*, Palgrave Macmillan, Basingstoke, ch. 17.
Jowsey, E. (2011) *Real Estate Economics*, Palgrave Macmillan, Basingstoke, ch. 11.

6.9 Perfect competition

Key terms: many buyers and many sellers; homogeneous products; freedom of entry and exit; normal profits

As the term suggests, perfect competition is the market situation where there is the most possible competition. This is because:

- there is a large number of relatively small sellers and buyers;
- the product they sell is exactly the same (homogeneous);
- there is perfect knowledge of market conditions;
- there is free entry to the industry;
- in the long-run there is perfect mobility of the factors of production.

In Figure 6.9.1 price in the market or industry is set in the graph on the right, where demand equals supply. In the graph on the left, the firm is a price-taker; it can sell as much of this product as it can produce at the market price. Real world real estate examples are hard to find but the market for breeze blocks used in construction may provide one. These blocks are homogeneous and producers can sell a large number at the market price. Without a perceived difference in quality any one producer would find it difficult to raise the price of their product and so the firm's demand curve is perfectly elastic at the market price (P).

It should be apparent that there are no real estate markets where all of these conditions apply and so the real estate industry generally does not operate under conditions of perfect competition. If it did prices would be as low as possible (because of all the competition) and the industry would be very efficient (with costs as low as possible). In fact it is widely believed that the real estate industry is quite inefficient (see Concept 6.4), and the reason for this is that it does not exhibit the conditions listed above as being necessary for perfect competition.

In particular, every property is unique – even if that is only because it is in a slightly different position to other properties – so the condition of homogeneity cannot apply. There may be many buyers and many sellers, as is often the case in residential estate agency, but they most certainly do not sell products that are exactly the same. And often there is not perfect knowledge, for example, of the true value of a property – that is why professional valuers are required to advise on prices.

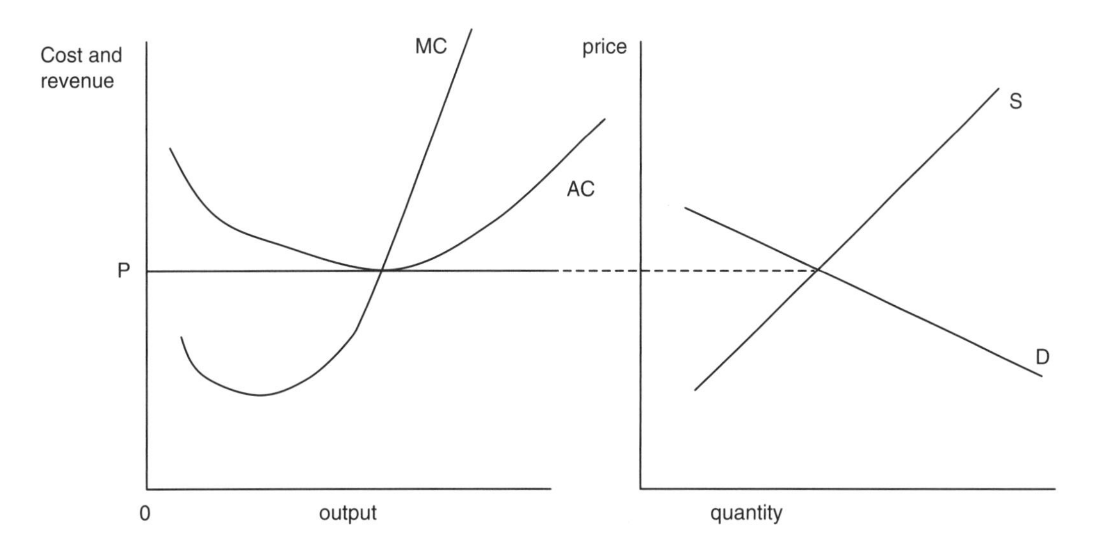

Figure 6.9.1 Long-run equilibrium in perfect competition.

Areas of the real estate industry where there is most competition, such as residential estate agency, general practice surveying firms, property management companies etc. are generally those areas where only normal profits can be made (sufficient profit to keep the firm in the industry). If supernormal profits could be made (much greater than normal profit) then because there is relatively free entry to the industry, new firms would be attracted and provide more competition to reduce those profits. Of course it would be in the interests of the existing firms in the industry to try to restrict entry of new firms by forming professional bodies and trade associations, but this is difficult to do without infringing laws on fair trading.

In a perfect market any price differences would quickly be eliminated and demand would always equal supply at the market price. It is obvious that this situation rarely exists in real estate markets where periodically there are conditions of excess demand (in a property boom) and excess supply (in a recession).

Further reading

Harvey, J. and Jowsey, E. (2007) *Modern Economics*, Palgrave Macmillan, Basingstoke, chs. 1 and 2.
Jowsey, E. (2011) *Real Estate Economics*, Palgrave Macmillan, Basingstoke, chs. 2 and 3.

6.10 Imperfect competition

Key terms: product differentiation; elastic demand; freedom of entry; normal profits; competitive advertising

Between the two extremes of perfect competition and monopoly there is a situation where many firms or perhaps only a few compete imperfectly in the market. The products are not exactly the same and so producers can make claims about them that differentiate them from others. Such a situation is sometimes called 'monopolistic competition' because the slight differentiation of each product gives it a little monopoly power. An example might be an estate agent claiming that they provide the best advertising options or a small builder stressing that the quality of their work is their advantage over others.

If such firms raise their prices, some customers will buy competitor's brands, but some will consider competitor's brands inferior and will only switch after a large rise in price. And if the firm lowers its price it will only attract a limited number of customers from rival producers. Demand will be elastic (with a shallow slope) but not perfectly elastic (horizontal – as in perfect competition).

In the short run, monopoly (or supernormal) profits can be made. The short-run situation is the same as for a monopolist (see Concept 6.12). But these supernormal profits attract other firms and because there are no barriers to the entry of new firms, the extra competition from new producers will reduce the profits of existing firms. Entry of new firms will continue until the supernormal profits in the industry disappear and only normal profits remain in the long run.

In Figure 6.10.1 at the profit maximising output of MC = MR, only normal profits can be made because AC = AR. These are sufficient to keep the firm in the industry but no more. The situation is not as good for consumers as perfect competition, however, because average costs (AC) are not at their lowest point (where the AC curve is cut by marginal cost, MC). This is because extra costs will be incurred by firms in competitive advertising and in finding ways to differentiate their product. For example, estate agents may offer viewing services and expensive photography. Small builders may pay for local advertising or sponsor local events etc.

So, imperfect competition is not as efficient as the situation that prevails with perfect competition because firms' costs are greater and they are not producing at the lowest cost output. But it is a much closer representation of most real world markets where firms compete through advertising and reputation. Most firms involved in the real estate industry, from general practice surveyors to letting agents, property management companies, maintenance firms, security firms and builders operate in this form of market.

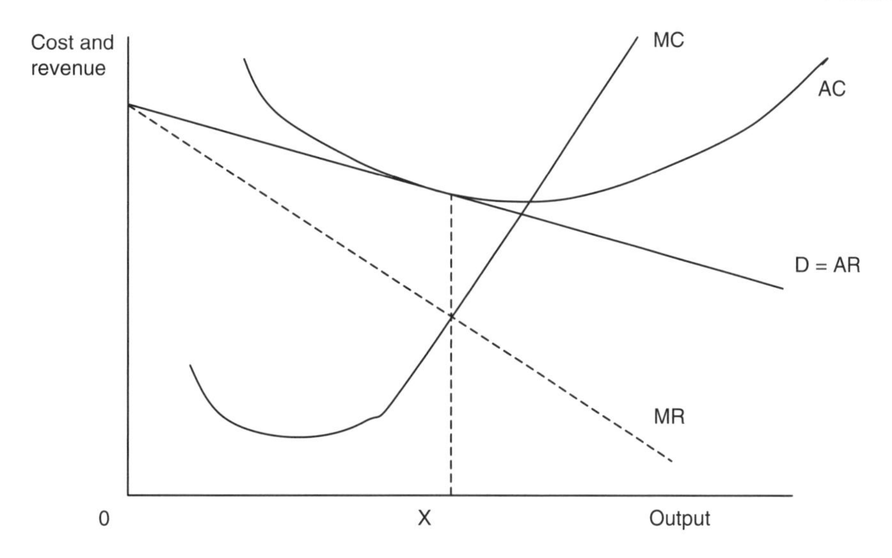

Figure 6.10.1 Long-run equilibrium in imperfect competition.

Further reading

Harvey, J. and Jowsey, E. (2007) *Modern Economics*, Palgrave Macmillan, Basingstoke, chs. 1 and 2.
Jowsey, E. (2011) *Real Estate Economics*, Palgrave Macmillan, Basingstoke, chs. 2 and 3.

6.11 Oligopoly

Key terms: competition among few; price leader; kinked demand curve; price stability; non-price competition; cartel

Where there are only a few firms in an industry, conditions of competition can produce surprising outcomes. This situation is not uncommon in the real estate world where the very large building firms, national and international property companies and many financial institutions experience competition among the few. In oligopoly, pricing and output conforms to no given principles. Sometimes one firm is the 'price leader', setting the price that will maximise its profits and taking its share of the market. Smaller firms then accept this price and try to compete in terms of quality and service.

If the firms are of roughly equal size, then no one firm can set a price without considering its competitors' reactions. Cutting price does not guarantee more sales because competitors may retaliate and reduce their prices. Raising price to try to make more profits may not work if competitors leave their prices unchanged. This means that the firm's demand curve is relatively inelastic for price cuts, but relatively elastic for price increases. This is shown in Figure 6.11.1 where the demand curve is kinked at P. If the original price is P, the firm could expect to move along the relatively elastic section towards D_1 if it raises its price. This would mean losing customers to competitors if it raises its price. Lower its price and the firm could expect to move along the relatively inelastic section of the demand curve towards D.

Under such circumstances the firm will be reluctant to alter its price – it gains little from either a price increase or a decrease. Even if its costs change the profit maximising. output and price may well remain the same because of the broken marginal revenue curve MR, which is the result of the kinked demand curve. So marginal cost could vary from MC to MC_1 without the profit maximising output X (where MC = MR) changing. So price stability or rigidity is likely where a firm in oligopoly is unsure of its competitors' reactions to a change in its own price.

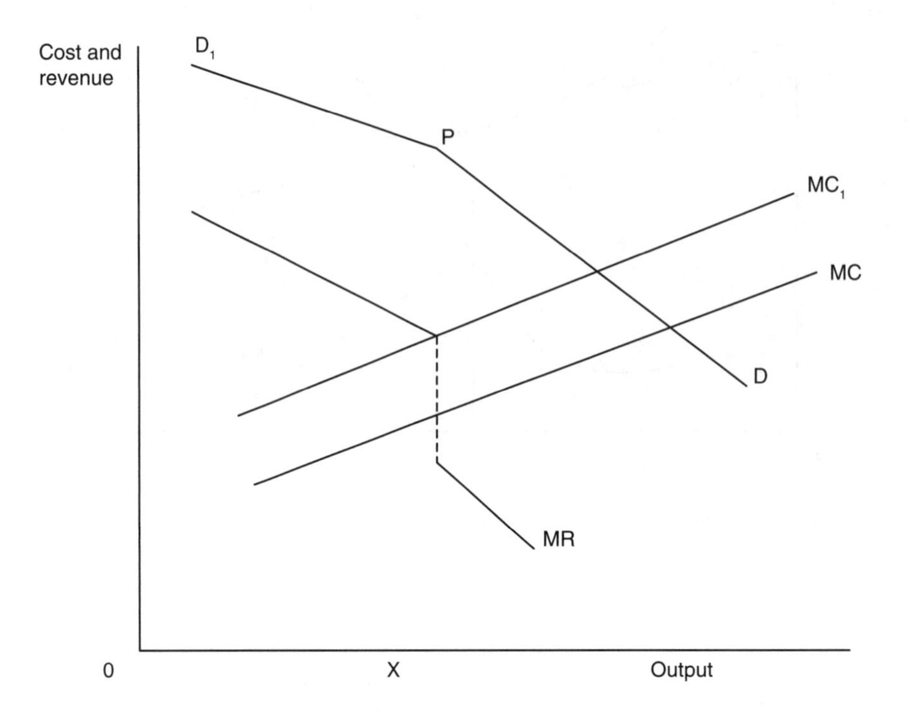

Figure 6.11.1 The kinked demand curve in oligopoly.

If firms in oligopoly are reluctant to compete in terms of price, they may engage in non-price competition through advertising, discounts, loyalty cards etc. Or they may come to a tacit agreement in order to jointly maximise profits. This might involve following the price of the leading firm. In some circumstances illegal anti-competitive agreements may be made between firms, effectively forming a cartel. Instances of this happening are not unknown in the construction industry (see Concept 4.12).

Further reading

Harvey, J. and Jowsey, E. (2007) *Modern Economics*, Palgrave Macmillan, Basingstoke, ch. 16.

6.12 Monopoly

Key terms: monopoly power; barriers to entry; supernormal profits

The term monopoly literally means one seller. It is possible for firms with a large market share to have some degree of monopoly power even if there is more than one seller in the market. In practice pure monopoly – one seller only – is rarely found in real life. But one seller can dominate market supply of a good or service, and competition laws in the UK consider a market share of 25 per cent to constitute monopoly power.

Monopoly power can come from the following sources:

- control of the source of supply by one firm, e.g. mineral spring water, a secret recipe for a product, specialist skills such as fashion design;
- patents, copyrights and trademarks that are allowed in order to encourage innovation;
- legal prohibition of new entrants such as in some utilities and essential services;
- indivisibilities – it may require enormous amounts of capital to enter the industry (for example, investment in prime city centre offices can mean hundreds of millions of pounds are required);
- restrictions on imports.

Essentially conditions in monopoly, where there is a downward sloping demand curve, mean that withholding supply achieves a higher price and more profit for the monopolist. Certain professions achieve this by restricting membership and so increasing remuneration for their members.

In the long run the firm in monopoly can make supernormal profits because there are barriers to the entry of new firms. This means that competition is restricted, allowing the monopolist to produce less than competitive firms would, and to sell it at higher prices. The situation is illustrated in Figure 6.12.1.

In Figure 6.12.1 the monopolist sets the profit maximising output where MC = MR. At this output supernormal profits are made equal to PCAD, because at output 0X average revenue (AR) is greater than AC. Total profit is the difference between AR and AC multiplied by output 0X, which is equal to PCAD.

In real estate markets, sometimes conditions allow an owner to gain monopolistic control. For example, geographical divisions between markets can allow local monopolies for selling agents to develop; and the imperfections of the capital market may exclude all but the biggest investment firms from purchasing certain developments; and because property is fixed in location, certain site owners

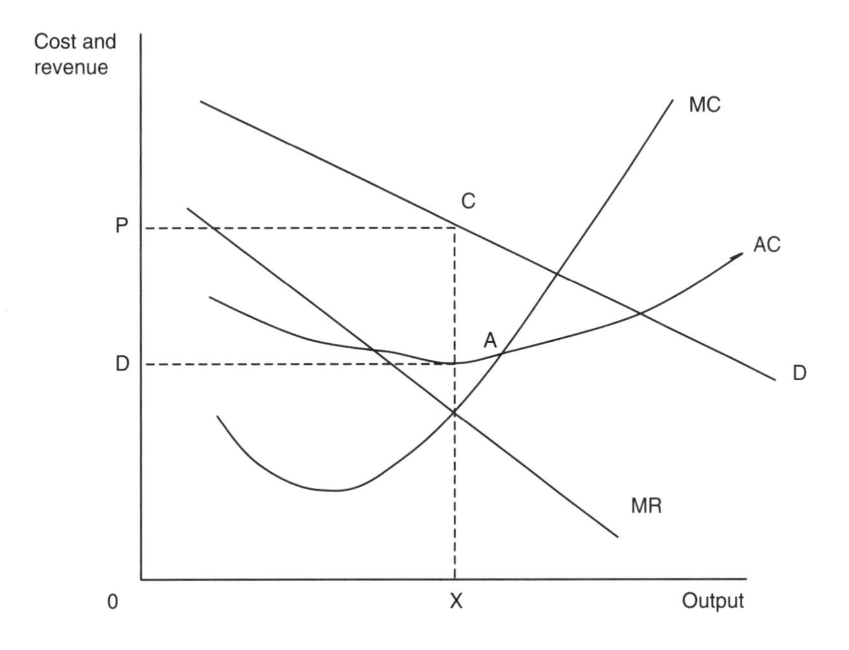

Figure 6.12.1 The long-run situation in monopoly.

can have a strong bargaining position relative to buyers, especially if their land unlocks a development. The owner of such a site can exploit his monopoly power by demanding a price far in excess of that paid for other sites, and so virtually secure all the developer's supernormal profit from the scheme (Jowsey, 2011).

Further reading

Harvey, J. and Jowsey, E. (2007) *Modern Economics*, Palgrave Macmillan, Basingstoke, chs. 1 and 2.
Jowsey, E. (2011) *Real Estate Economics*, Palgrave Macmillan, Basingstoke, ch. 15.

6.13 Economies of scale

Key terms: average costs, long-run average costs; mass manufacturing; minimum efficient scale; industrialised building

Production on a large scale can reduce average costs of output. This is the basis of mass manufacturing which standardises units of output so that they can be produced in large numbers at low individual cost per unit. The advantages of large-scale production are referred to as 'internal economies of scale'. These include:

- technical economies – greater division of labour, more specialised machines, linked processes that reduce costs;
- managerial economies – dividing the functions of management into separate roles such as sales, production, transport etc.;
- commercial economies – bulk buying at lower unit prices, spreading advertising costs, selling by-products;
- financial economies – large firms offer better security and so borrowing costs are lower;
- risk-bearing economies – large firms can cover their own risks and better manage fluctuations in demand than small firms.

The minimum efficient scale (MES) is the long-run output for a business when internal economies of scale have been fully exploited. This will be the lowest point on the long-run average total cost curve and is also known as the output of long-run productive efficiency. The MES will vary from industry to industry depending on the nature of the cost structure in a particular sector of the economy. There are many areas of real estate where significant economies of scale can be found. These include large property agencies with specialised departments (managerial and risk-bearing economies), large property investment companies with low borrowing costs (financial economies) and large house building companies (commercial and technical economies).

When the ratio of fixed to variable costs is very high, there is great potential for reducing the average cost of production. This would be the case when a large investment in a factory is required in order to engage in large-scale production. Industrialised building requires a demand that is both large and continuous. Thus, in Figure 6.13.1, in order to secure the lower long-run average cost curve of industrialised building ($LRAC_{IB}$) resulting from design/construction integration, factory production of components and economies of scale, there must be a demand for at least 300 dwellings per period. To take full advantage of the economies of scale available from industrialised building, however, output would need to be at 700 units, where the $LRAC_{IB}$ is at its minimum point. Long-run average costs for traditional building ($LRACTB_{TB}$) are higher than those for industrial building ($LRACIB_{IB}$).

In the construction industry, demand is in small parcels: 90 per cent of all contracts are for fewer than 100 dwellings. Increasing the size of contracts would only be possible if the variety of buildings was restricted. But private clients, especially house buyers, demand buildings that are individual in some way. Industrialised building, in contrast, is frowned upon as being uniform, a view mainly

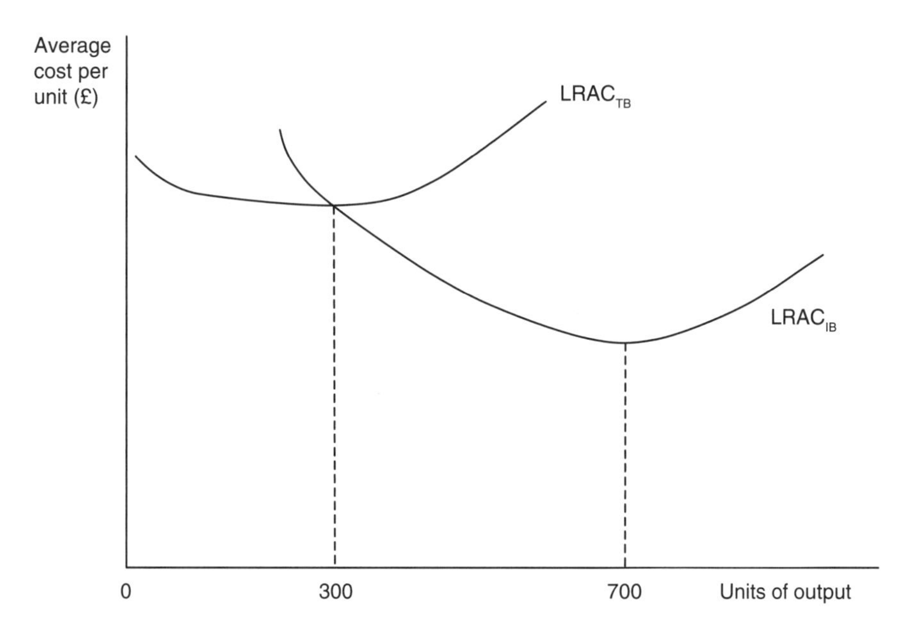

Figure 6.13.1 Economies of scale in industrialised building.

derived from the austere appearance presented by local authority flats that were subject to severe cost restrictions.

Demand for industrialised building therefore has to come from those clients where some uniformity of product is acceptable and where the highest and best use of the site is not impaired by the erection of a factory-built construction. This has meant that in the private sector the method has been most successful for retail parks, factories, hotels, farm buildings, small garages, house extensions and garden sheds.

A twenty-first-century revival of industrialised building could be seen with the entry of Ikea-Skanska into the market via the 'BoKlok' (Swedish for 'Live Smart'). About 800 such homes are built in Sweden each year and the first development in the UK – construction of 119 dwellings consisting of 36 BoKlok apartments, 57 BoKlok houses and 26 LiveSmart@Home dwellings – is sited along the Felling Bypass in Gateshead. The company aims to assemble about 500 affordable dwellings per year in the UK (Jowsey, 2011).

Further reading

Harvey, J. and Jowsey, E. (2007) *Modern Economics*, Palgrave Macmillan, Basingstoke, ch.7.
Jowsey, E. (2011) *Real Estate Economics*, Palgrave Macmillan, Basingstoke, ch.13.

6.14 Mobility of labour

Key terms: efficiency; occupational mobility; geographical mobility; transaction costs

For an economy to be efficient, the factors of production, including labour, should be free to move to where they are most valued – which will be where they are most profitable. So labour should be mobile and free to move to where it commands the highest wage.

If there was perfect geographical mobility of labour, then workers would move from low wage areas to higher wage areas and this would continue until wages were equal in all areas. And if there was perfect occupational mobility of labour then wages would be equal across all occupations. Clearly these situations do not exist and so there must be some geographical and occupational immobility of labour.

The property sector contributes to geographical immobility of labour in the UK. This is because property markets are not the same in all areas. Property prices are higher in London and the southeast than in other parts of the country. This can mean that even if a better paid job is available to someone from the regions, they do not take it because to do so would mean spending much more of their salary on accommodation. This can be the case for owner-occupiers who cannot afford the higher property prices in the south east, and for tenants who cannot afford the higher rents. Council tenants may feel unable to move away from the area that they are in because they will have to go onto a waiting list for rent controlled council accommodation in the new area – and such waiting lists can be very long and so this could take a long time.

The costs of moving can also be considerable, especially for owner-occupiers who will have to pay estate agents fees, solicitors fees, removal costs and stamp duty (if the property they purchase is above the threshold). These transaction costs can amount to several thousand pounds. Stamp duty alone would be £9000 on a house costing £300,000 – not an extravagant property price in London. Since the majority of properties in the UK are owner-occupied (about 67 per cent in 2013) this is a significant obstacle to labour mobility and economic efficiency. It can be argued that an economy with a larger proportion of rented property such as Germany has a more efficient labour market as a result.

Further reading

Harvey, J. and Jowsey, E. (2007) *Modern Economics*, Palgrave Macmillan, Basingstoke, ch. 20.

6.15 Property rights

Key terms: interests; spatially fixed; fee simple absolute; freehold; leasehold

Real estate markets actually deal in 'property rights' often referred to as 'interests'. This is no different, however, from any other market. It is possible, for instance, to go to a Land Rover dealer and buy a car. The car is handed over and the buyer is given the exclusive right to drive it as long as they wish, and, to part with that right by selling it to somebody else. Alternatively, a car could be hired from Avis. Here the right to drive the car would be restricted to a certain period of time and subject to conditions regarding damage etc. These restricted rights are more likely to be clearly defined in a written agreement which, by implication, also excludes Avis from letting anybody else drive the car during the specified period.

The situation is similar for real estate, but here it is not possible to hand over land and buildings because they are spatially fixed. Because of this, it is 'rights' that are dealt in, especially as, accompanying the transaction, there is a written statement defining the exact rights that are being transferred. In short, the real property market deals in the rights relating to real property rather than in the land and buildings themselves.

With real estate the separation of rights is more usual than with personal property. The largest collection of rights that can be held in real property is 'fee simple absolute'; that is, the unencumbered freehold. But all the rights included in the 'fee simple absolute' can be separated and transferred individually to other people. For example, a lease may be granted to a person to erect a building on a plot of land and to enjoy the rights to this building and land for a 99-year period on payment of a specified yearly sum to the lessor. Provided the terms of the lease are fulfilled, the lessor's rights are, for 99 years, now restricted to the receipt of a freehold ground rent. The freehold has then been divided into two interests: the leasehold and the freehold ground rent. In any given land resource different people may have many different rights – for example, a freehold ground rent, a head lease, a sublease, a mortgage, a rent charge, and so on (Jowsey, 2011).

The exact rights transferred can be finely adjusted according to the individual preferences of the seller or buyer, for example, by a restrictive covenant. Such a fine differentiation to meet the individual preferences of sellers and buyers is achieved automatically through the free-market mechanism and is reflected in prices.

Within his rights the 'fee simple absolute' owner can possess, use, abuse and even destroy his real property. He can sell rights in such a way as to restrict their future use (for example, by covenant), or he can bequeath them to distant heirs. Nevertheless, his rights are limited to some extent if there are other people's rights in force, such as ancient rights to 'easements' (see Concept 10.9),or if the land is subject to the legal restraints imposed by planning restrictions, building regulations and similar legislation.

An interest comes into existence simply because a bundle of certain rights is wanted. But no rights would be wanted unless their owner could exclude others from them – there would be no point in paying for the right to use something that others could use for free. Thus the concept of *rights* is essentially a legal one: it presupposes that there is a government authority that will, if necessary, protect the rights vested in the owner. Moreover, being a legal concept, a right must be clearly defined. This implies limitations to the right. Thus a right is merely exclusive, not absolute or unlimited. In fact, different rights are really only differences in 'exclusiveness'.

Where rights are well defined and the costs of negotiating and enforcing contracts are small compared with the benefits of the transaction, an exchange system based on prices works smoothly. Economic theory tends to assume these conditions. At times, however, the failure to define rights unambiguously may lead to economic inefficiency.

Property is durable and so the rights existing in real property have a long timescale. Real property rights, such as stocks and shares, are, therefore, demanded as investment assets and the real property market can now be regarded as a part of the wider investment asset market.

Further reading

Jowsey, E. (2011) *Real Estate Economics*, Palgrave Macmillan, Basingstoke, ch. 3.

6.16 Economic rent

Key terms: land; resources; opportunity cost; transfer earnings; surplus

This is payment made for a resource that is not necessary as an incentive for its production, often regarded as a surplus over cost. Undeveloped land, or 'pure' land, refers solely to natural resources and space. Thus land as a *whole* – that is, the earth' land surface – can be regarded as being fixed in supply. No one had to be paid to make it and so rent paid to the landowner is economic rent. If the land has been improved, however, part of the rent is payment for the improvement.

With man-made commodities, including capital goods, price is a function of demand and supply and because supply is influenced by cost of production, price itself is influenced by cost. But since land as a whole is a fixed supply provided by nature, it has no supply costs and the earnings of 'pure' land are determined solely by demand.

Thus in Figure 6.16.1 POXT represents the earnings of land when demand is D, and P_1OXL when demand is D_1. No matter how small the earnings, the total supply of land is still the same, hence the supply curve S is perfectly inelastic (a vertical straight line at X). This means that its opportunity cost is zero and so all the earnings of land as a whole are an excess over opportunity cost. They represent economic rent, that part of the earnings of a factor that results from it having some limitation of supply – the return arising because supply is not perfectly elastic.

Because the earnings of land are a surplus over opportunity cost does not mean that payments do not have to be made for land. Price still rations scarce supply of land among competing uses. This ensures that, in each location, land is put to its most profitable use according to the preferences of consumers and society.

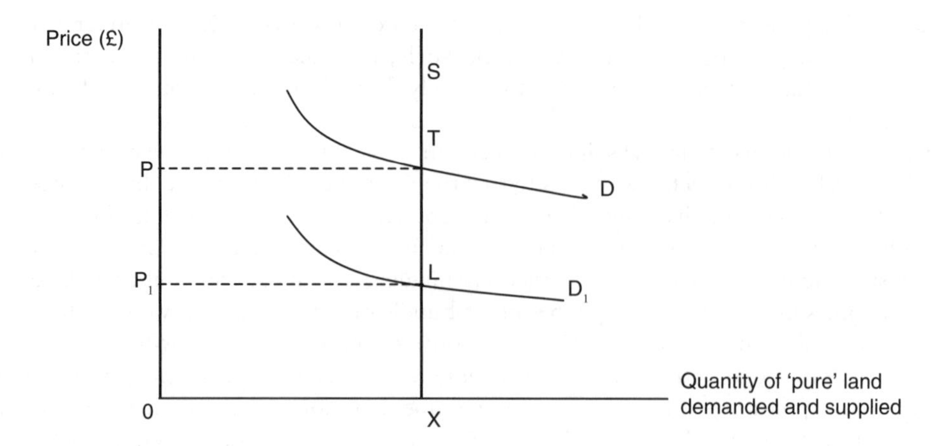

Figure 6.16.1 Economic rent.

So the supply of land can never be regarded as fixed. Additional supplies can always be bid from other uses if the proposed new use has a higher value than the existing use. Furthermore, the productivity of land can usually be increased in response to additional demand by using it more intensively by the addition of capital.

The fact that the earnings of land as a whole are entirely demand-determined is important from the point of view of taxation – land will still be there, no matter how heavily it is taxed. As a result a tax on pure land has no disincentive effect on the supply of land and economic rent could theoretically be taxed away entirely without affecting output.

David Ricardo showed in 1817 that the price of wheat was high not because land commanded a high price or rent but because the demand for wheat was high and this determined the high rent of land. Rents are highest on the most productive land and, as Tim Harford shows in *The Undercover Economist*, coffee shop rents are highest in the best locations such as train stations (Harford, 2006).

The concept of economic rent can be applied to payment made to people with special abilities, such as footballers or rock stars, but it is difficult to distinguish their natural talent (for which they receive economic rent) from their skills acquired by hours of practice and training. Transfer earnings are the amount that a resource could expect to earn in its best alternative use. In the case of land for development, its best alternative use might be for agriculture and that value is at least what must be paid to secure it for development. If it commands a higher price (and it almost certainly will because of high demand and competition among developers to buy the land) then the additional earnings are economic rent.

Further reading

Harford, T. (2006) *The Undercover Economist*, Abacus, London, ch. 1.

Jowsey, E. (2011) *Real Estate Economics*, Palgrave Macmillan, Basingstoke, ch. 4.

6.17 Gross domestic product

Key terms: gross national product; national income; GDP per head; living standards

The title of this concept could be 'national output measures', but many people will come across the terms gross domestic product (GDP) or gross national product (GNP). Fluctuations in the level of economic activity are monitored by quantitative information on the national income. Income is a flow of goods and services over time: if our income rises, we can enjoy more goods and services. The principal figures are published annually in the United Kingdom National Accounts (The Blue Book – see www.statistics.gov.uk). They include several measures of national output and it is important to realise that national output equals national income and national expenditure – these are measures of total production from the economy, but measured at different points in the circular flow of income (see Figure 6.19.1).

For goods to be enjoyed they must first be produced. A nation's income over a period, then, is basically the same as its output over a period. National income is the total money value of all goods and services produced by a country during the year. The GDP is simply the money value of the final output of all resources located within a country. In order to obtain GNP we have to add the balance of net property income from abroad – such as profits made by UK companies in other countries.

Figures for GNP are calculated for income, expenditure and output. Because information is incomplete and derived from a variety of sources, results are not identical and there is a statistical discrepancy.

Very often the term national income is used in economic literature. Technically this is net national product which is equal to GNP minus depreciation (machinery and buildings that have worn out). National income, then, is quite a good representation of the additional output/income/expenditure produced in a twelve-month period.

Property contributes to GDP in a number of ways:

- Rent is a significant part of incomes.
- The value of home ownership is given an imputed rent in the national income accounts.
- Goods and services are produced in space provided by land and buildings.
- The incomes of at least 3 million people in the UK come from working in construction and wider real estate occupations such as facilities management and security.
- Investment in property development is a significant component of national output.

GDP divided by the population size gives GDP per head. This is often used as a measure of living standards and because of economic growth it usually rises over time. This can also be used to compare living standards between countries but caution must be exercised here because so many things that contribute to living standards (such as environmental quality, leisure time, cultural development and health) are not measured by the national accounts.

Further reading

Harvey, J. and Jowsey, E. (2007) *Modern Economics*, Palgrave Macmillan, Basingstoke, ch. 28.

6.18 Economic growth

Key terms: production possibility frontier; full employment output; gross national product; capital goods; consumer goods

If an economy is not producing the maximum that it could and there are unemployed resources, the economy's actual output is below its potential output. In terms of Figure 6.18.1, the economy is producing inside production possibility frontier I, say at point X. Starting from X, output can be increased by measures that utilise unemployed resources. But even full employment of an economy's resources (as would be the case actually on the production-possibility frontier) does not necessarily mean that the economy will grow. Growth occurs over the long run where the potential full-employment output of the economy increases over time.

In terms of Figure 6.18.1, economic growth means that, over time, production possibility frontier I is pushed outwards to II and III. Increases in the productive capacity in the economy over time are usually measured by calculating the rate of change of real GNP per head of the population.

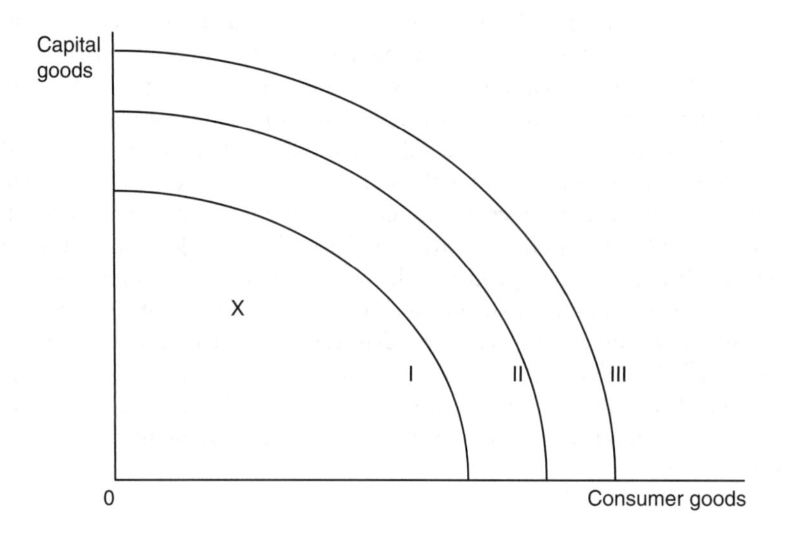

Figure 6.18.1 Production possibility frontier.

Economic growth is the major way of achieving improvements in the standard of living. It means more consumer goods, better living conditions, better quality food and so on. While such improvements occur gradually and almost imperceptibly from year to year, small differences in the annual rate of growth produce large differences in the speed of growth. For instance, a rate of growth of 2.5 per cent per annum will double real GNP in 28 years, whereas a 3 per cent rate doubles GNP in only 24 years. An annual rate of growth of 10 per cent (which has frequently been achieved by China in recent years) doubles GNP in only 10 years! In addition, growth makes it easier for the government to achieve its economic policy objectives. Revenue from taxation increases, allowing government services, e.g. education and health care, to be expanded without raising the rates of tax while still allowing the standard of living of the better-off to show some improvement. Economic growth makes it possible for living standards generally to increase.

There are five basic causes of growth:

- a rise in the productivity of existing factors of production, e.g. education improves labour productivity;
- an increase in the available stock of factors of production, e.g. development of natural resources such as shale gas;
- technological change, e.g. inventions and innovations;
- change in composition of national output, e.g. moving from agricultural production to manufacturing production;
- sustained improvement in the terms of trade – enabling more imports to be bought for a given quantity of exports.

Further reading

Harvey, J. and Jowsey, E. (2007) *Modern Economics*, Palgrave Macmillan, Basingstoke, ch. 34.

6.19 The multiplier

Key terms: national income; injections; leakages/withdrawals; marginal propensity to consume; marginal propensity to withdraw; equilibrium

The 'multiplier' refers to the multiplied effect on incomes of an increase in spending. It is sometimes referred to as the 'Keynesian multiplier', after J. M. Keynes, the economist, or the national income multiplier, where the effect is on the whole economy.

The basic principle behind the multiplier is that one person's spending is another person's income. So when there is a boost to spending in an area, it is multiplied because the shopkeepers or dealerships receive a boost to their income and they will spend at least some of it.

In a simple model of the economy known as the circular flow of income, there is a flow of expenditure from households to firms and a flow of factor payments (income) in return in the other direction. The situation is shown in Figure 6.19.1. There are also 'leakages' out of the circular flow in the form of savings (not spent within the flow), taxes (taken out of the flow by government) and spending on imports (which goes to firms overseas). There are also 'injections' into the circular flow in the form of investment, government spending and exports.

In short, as long as the amount leaking out equals the amount being injected into the economy, the economy will be in equilibrium, remaining the same size. If the leakages exceed the injections then the economy will shrink; and if the injections exceed the leakages the economy will grow. These leakages and injections are subject to the multiplier effect. So if there is an increase in investment (an injection into the economy) the economy will grow by more than the amount of the extra investment, that is, it is multiplied. If there is an increase in savings (a leakage out of the economy) the economy will shrink by more than the amount of the extra savings – it is multiplied.

If we assume that in the national economy on average people spend 60 per cent of their income; and that there is an increase in government spending of £1 million, then the recipients of the £1m will spend £600,000, and the recipients of this will spend £360,000 (600,000 × 0.6) and so on. The final increase in income will be the cumulative amount of all of these spending flows (Figure 6.19.1).

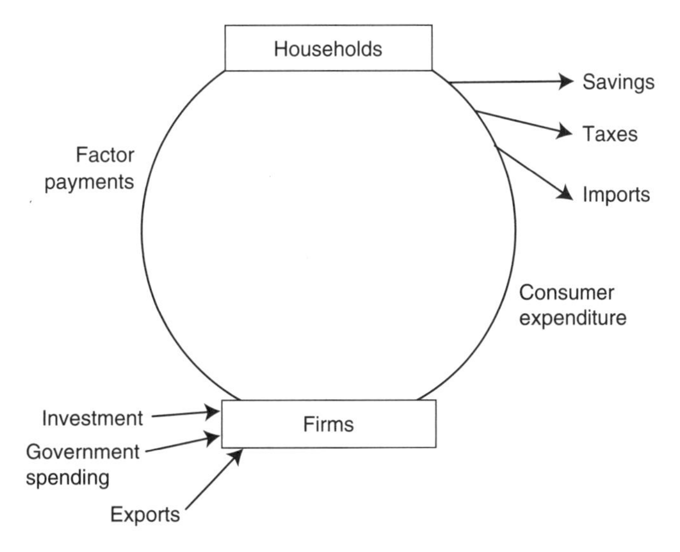

Figure 6.19.1 The circular flow of income.

Technically, if the re-spending was 60 per cent of any injection, the total of the leakages (sometimes called withdrawals) would be 40 per cent or 0.4. The formula for the multiplier is:

$$K = \frac{1}{1 - MPC} \qquad \text{or} \qquad K = \frac{1}{MPW}$$

where K = the multiplier, MPC = the marginal propensity to consume (in this case the 60 per cent expenditure) and MPW = the marginal propensity to withdraw (in this case the 40 per cent leakage). The value of the multiplier would then be $1/1 - 0.6 = 2.5$ or $1/0.4 = 2.5$.

This means that an injection (for example, an increase in investment) of £1 million would increase the circular flow of income or national income by £2.5 million. The extra expenditure on investment is multiplied. Of course this example is very theoretical – in the real world the multiplier does exist and it is easy to see the simple premise that one person's spending is another person's income, but the value of the multiplier for the UK (where there is a lot of spending on imports so a lot leaks out of the economy) is lower at, perhaps 1.6. Nevertheless it is still an important concept with implications for government spending, employment, output and incomes.

Further reading

Harvey, J. and Jowsey, E. (2007) *Modern Economics*, Palgrave Macmillan, Basingstoke, ch. 30.

6.20 Fiscal policy

Key terms: macroeconomy; government spending; taxation; aggregate demand; automatic stabilisers

Fiscal policy is the use of government spending and taxation in order to influence the macroeconomy. Broadly speaking, an increase in government spending will add to aggregate demand and stimulate the economy. A cut in government spending will reduce aggregate demand and slow the expansion of the economy. An increase in taxation will reduce aggregate demand (households and firms have less money to spend); and a reduction in taxation will increase aggregate demand, encouraging more spending and an expansion of the economy.

Some fiscal policy operates automatically – taxation yields increase as incomes increase because many taxes are progressive (see Concept 17.1). Some government spending also automatically stabilises the economy, for example, as the economy slows and unemployment increases, so do benefit payments. This allows for a 'softer landing' than would be the case as the benefits help to maintain aggregate demand.

Some fiscal policy is discretionary – it is specifically initiated by government. An example of this would be government changing the type of taxes levied in order to influence the economy, say by switching from direct taxation such as income tax, to indirect taxes such as VAT in order to try to reduce tax avoidance. Changes to taxation and government spending are announced each year in the budget. Again broadly speaking, aggregate demand will be increased if taxation is less than government spending and vice versa.

Property is directly affected by a number of taxes including stamp duty land tax, council tax and uniform business rates, all of which can be changed by government with subsequent consequences for the real estate sector. Government spending is often directed at the property sector in order to stimulate the overall economy. For example, in the 2013 budget the UK government announced a 'Help to Buy' scheme to help first-time buyers to get onto the property ladder. The scheme provides a government loan of up to 20 per cent of the price of a property if the buyer can raise a 5 per cent deposit. The loan is interest free for 5 years and is available on properties costing up to £600,000.

The government can use fiscal and monetary policy to influence the macroeconomy. Fiscal policy, because it involves altering areas of government spending and taxation, can be used to stimulate or slow down specific parts of the economy and because property is such an important part of the economy, fiscal policy is often employed to affect the property sector.

Further reading

Harvey, J. and Jowsey, E. (2007) *Modern Economics*, Palgrave Macmillan, Basingstoke, ch.31.
Jowsey, E. (2011) *Real Estate Economics*, Palgrave Macmillan, Basingstoke, chs. 20 and 22.

6.21 Property cycles

Key terms: boom; recession; recovery; endogenous causes; exogenous causes; time lags in supply

In all developed economies there are regular cyclical fluctuations in economic activity, which includes output, employment and incomes. The business cycle has distinct phases:

- A boom is the top of the cycle, when GDP peaks – industries become fully utilised and bottlenecks begin to appear, putting upward pressure on prices.
- In a downturn or recession, consumption, investment and employment fall and business expectations become negative. A severe recession could lead to a depression which is characterised by high unemployment, a low level of consumer demand and surplus productive capacity plus low business confidence.
- In a recovery, employment, incomes and consumer spending all begin to increase; business expectations improve and investment increases.

The business cycle is illustrated in Figure 6.21.1. The long-term trend line rises over time because of economic growth.

Property cycles involve fluctuations in the rate of all property returns. In real estate markets, housing, commercial, retail and industrial sectors are subject to periods of booms and slumps. Property

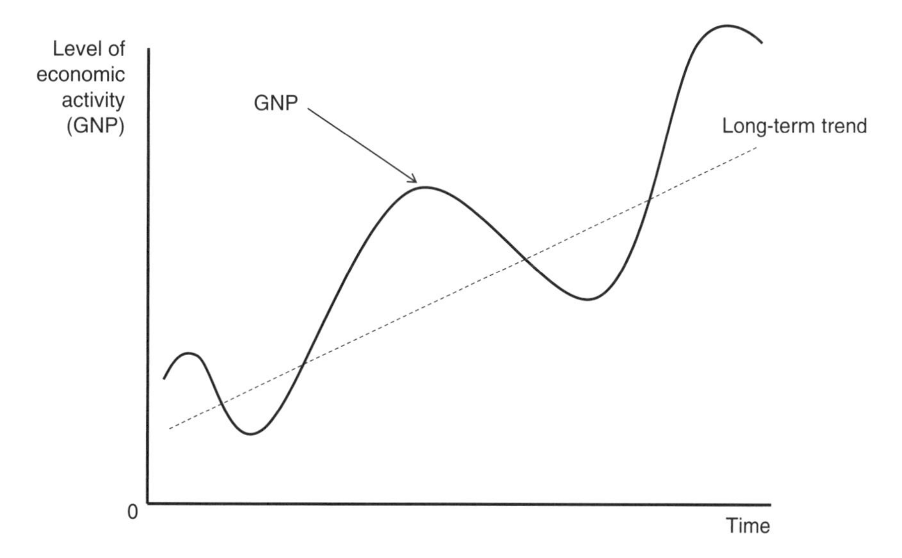

Figure 6.21.1 The business cycle.

cycles take place over irregular periods of time and their impact on prices and activity varies. Most developed countries have a property cycle that is closely linked to GDP and the business cycle.

Property cycles occur for reasons that are endogenous (from within the property sector) and exogenous (influences outside the property sector). Endogenous reasons can arise from time lags in production that lead to periods of excess demand and excess supply which mean the property market is hardly ever in equilibrium. At first, demand increases but there is a delay before new building can commence while planning permission and finance are arranged; further shortages of space lead to rising rents bringing forth more new developments; speculation that rents will go on rising adds to supply; completions then provide excess supply leading to falling rents and less development.

Exogenous influences on the property sector include: fluctuations in income, employment, availability of credit, rates of interest, exchange rates, changes in government policy etc. A simple example is the introduction of energy performance certificates that involve considerable expense for commercial property owners.

A strong economic upturn coinciding with a shortage of available property may be the starting point of a cycle. Rents and capital values increase, and this encourages new development. Further speculative developments are financed by an expansion of bank lending and a building boom results but it takes time for supply to reach the market and so rents and capital values continue to rise. When the new developments are completed the business cycle may have peaked and slowed down, causing falling demand for property and a property slump with falling values, high vacancy levels and widespread bankruptcies in the property sector. It is clear that the economy, the property sector and the financial sector are strongly interlinked (see Figure 6.21.2).

There have been three major boom/slump cycles in the UK in the last 40 years (the 1970s, late 1980s and 2007–9). All were the result of strong demand growth, supply shortages and credit expansion acting together. In each of the three cycles the government and/or the banks pursued expansionary monetary policy (Barber boom 1972/3, Lawson boom 1987/8 and the property bubble created by US and UK banks 2004–7). In each of these 'bubbles' an upswing in the business cycle was added to by relaxation of credit restrictions. Increases in demand meant existing space was quickly occupied

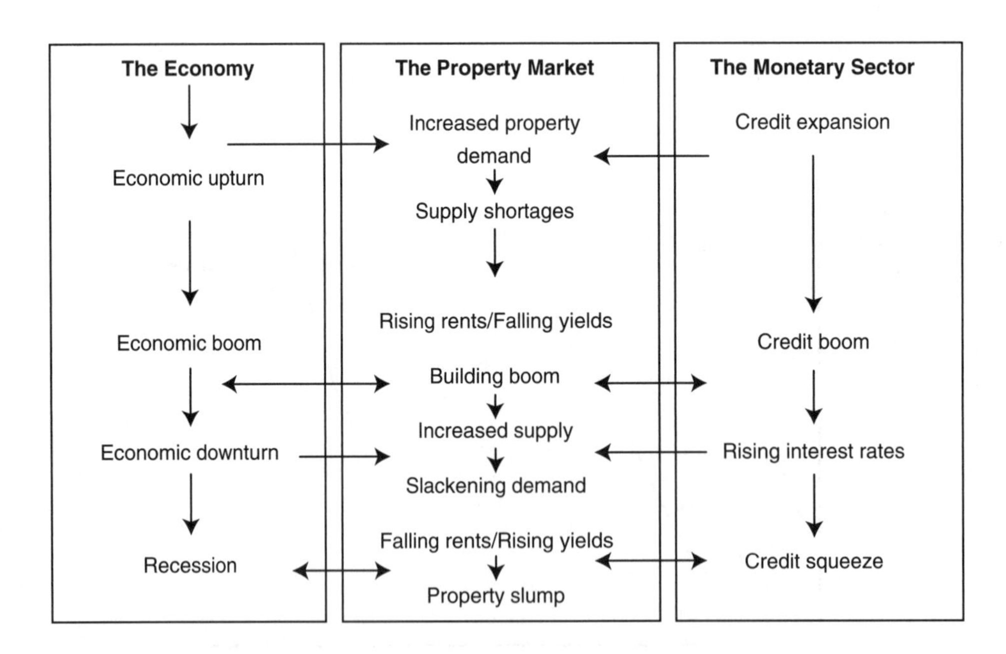

Figure 6.21.2 Links between the property cycle, the economy and the monetary sector.

Source: Jowsey (2011) adapted from Barras (1994).

and rent and capital values soared. The availability of easy credit fuelled a speculative boom. When the economy went into recession, however, over-supply of property caused a fall in rental and capital values (Jowsey, 2011).

Further reading

Barras, R. (1994) 'Property and the economic cycle: Building cycles revisited', *Journal of Property Research*, 11: 183–197.
Jowsey, E. (2011) *Real Estate Economics*, Palgrave Macmillan, Basingstoke, ch. 6.

6.22 Globalisation

Key terms: international trade; migration of people; exchange of technical knowledge; foreign direct investment; global transparency index

Since the 1980s globalisation has increased the free flow of information, people, goods, services and capital across international borders. This has increased the integration of economies and societies and as a result most economies in the world have become more interdependent. Transnational corporations can now locate the production of each component part of a product in the country that can produce it at least cost as long as the quality is satisfactory. Many workers in the USA, Germany, Poland, France, Japan and many other countries now work for foreign-owned firms. Many companies have a global presence, such as Nike, Coca-Cola, MTV, McDonalds and Guinness. Markets are also global thanks to worldwide media so that tastes in such things as designer clothes, music and films are similar in cities everywhere.

Technological progress has made it easier to supply services remotely, to separate production activities and to manage them remotely. International corporations engage in 'offshoring', which involves firms selecting and holding on to the stages in the value chain that they consider to be core, while relocating their less essential activities to foreign countries or to subcontracting firms in foreign countries. This transfer of production, office support, and research and development centres to developing countries has boosted commercial real estate in many of these countries. Many countries have adopted more open economic policies. Economic reform in India and China has enabled their development as low wage producers of many of the world's manufactured goods and international trade in intermediate inputs and services has grown strongly in the last decade.

The increased exposure to alternative markets, more investment choices, and access to global capital markets has led investors to look beyond their own borders, in part because their home markets are too small or too competitive and do not offer many investment opportunities. Investors try to realise higher returns by taking advantage of inefficiencies or unique opportunities in local markets, while owners seek to expand their existing portfolios of assets and build on the success of their domestic businesses. Securitisation and the development of a number of financial instruments has enabled international investors to have liquidity. For the host country, global investment means increased employment, access to international capital to finance growth and compete internationally, and access to global technologies. Real estate investors have become increasingly international in their outlook over the last decade. The opening-up of markets has increasingly led to international development and investment in real estate. With finance and investment capital increasingly mobile, it is possible for investors to achieve higher returns, to widen their investment opportunities, and to diversify internationally as a strategy to reduce risk.

Of course real estate stays where it is, but the reduction of regulatory barriers to international finance flows has led to greater integration of worldwide financial markets and real estate investment flows now take place in and out of most countries with the aid of information technology. Knowledge of international real estate markets is key to successful investment and 'transparent' markets are those where information on security, opportunities and values is good. Jones Lang LaSalle produces a Global Transparency Index which ranks countries according to the transparency of their real estate markets. In 2012 the most transparent markets were the USA, UK and Australia.

The UK has about 8.4 per cent of the global real estate market, largely in London assets. There are no UK exchange controls on inward or outward investment (direct or portfolio), or on the repatriation of income or capital, the holding of currency accounts or the settlement of current trading transactions. And with a few exceptions there are no limitations on foreign ownership of real estate in the UK, although the recent introduction of the Annual Residential Property Tax on properties worth more than £2 million may discourage foreign investors (see Concept 17.8). Property may be acquired or occupied by individuals, trustees or companies. It can be acquired as a freehold interest in the land, on a long lease, on a short lease or on licence. There are, however, some restrictions on how the owner of an interest in real estate uses or develops the land or building in the form of covenants, planning legislation and building regulations.

Further reading

Jowsey, E. (2011) *Real Estate Economics*, Palgrave Macmillan, Basingstoke, ch. 18.

6.23 The credit crunch

Key terms: business cycles; property cycles; speculation; property bubble; globalisation

The term 'credit crunch' technically means a severe shortage of money or credit in the economy, but it has become associated with the extremely severe worldwide recession of 2007–9. In all economies there are regular cyclical fluctuations in output, employment and incomes. Sometimes the cycles are severe and impact upon all sectors of the economy, often having very serious effects on property markets. These business cycles repeat periodically as fluctuations in aggregate economic activity with expansion in many sectors at the same time, followed by contractions in many sectors at the same time – they vary in length from 1 to 12 years.

In a boom, at the top of the cycle, GDP peaks and industries become fully employed and shortages and bottlenecks begin to appear, putting upward pressure on prices. In a downturn or recession, consumption, investment, output and employment fall and business expectations become negative. In a recovery, employment, incomes, output and consumer spending all begin to increase. The catastrophic credit collapse of 2007–9 was triggered by events in the United States when the housing boom in the US was halted by interest rates rising from 1 per cent to 5.35 per cent between 2004 and 2006. Property is a major part of most developed economies and it has sufficient influence on other areas and especially the financial sector to cause significant macroeconomic effects.

House prices rose in real terms in most developed countries except Germany, Japan and Switzerland in the decade up to 2006. The boom in house prices was driven by low interest rates and changing perceptions of relative returns on assets which caused investors to switch from shares to property. Speculation that property prices would continue to rise added to demand and further increased prices. US banks relaxed lending criteria in order to give mortgage loans to people with poor credit histories. These were high-risk loans and have become known as 'sub-prime' (i.e. not the best) lending. These sub-prime mortgages along with other loans, bonds or assets were bundled into portfolios known as collateralised debt obligations (CDOs) and sold to investors around the world, thus providing a transmission mechanism for the crisis to go global when the bubble burst (Jowsey, 2011).

When prices of property reach unsustainable levels relative to incomes or rents or some other economic fundamentals the 'bubble' bursts and subsequent falls in house prices result in many owners finding themselves in a position of negative equity – a mortgage debt higher than the value of the property. Homeowners in the US began to default on their mortgages. Default rates rose to record levels and the effects were felt across the financial system because of the CDOs that had been sold worldwide.

On 9 August 2007 the French bank BNP Paribas told investors they could not take money out of two of its funds because of 'complete evaporation of liquidity in the market'. In other words, the banks had stopped lending to each other. The European Central Bank injected more than 200 billion euros into the banking system to try to improve liquidity. The US Federal Reserve and other banks around the world also began to intervene, because the liquidity crisis threatened to cause a recession.

Banks around the world began to announce losses. UBS lost $37 billion. Merrill Lynch announced a loss of $7.9 billion. The US Federal Reserve coordinated action by five leading central banks around the world to bail out the banks with loans and interest rates were reduced in the US and UK. In April 2008 the IMF warned that losses from the credit crunch could be more than $1 trillion and stock markets worldwide fell significantly.

The two largest US mortgage corporations, Fannie Mae and Freddie Mac, were bailed out by the US government in September 2008 as they could not be allowed to fail because they were owners or guarantors of $5 trillion worth of home loans. In September 2008 Lehman Brothers became the world's biggest corporate bankruptcy ($639 billion).

In October 2008, governments and central banks around the world implemented rescue packages worth hundreds of billions of dollars. The Eurozone, the UK and the US all went into a severe recession, the worst since the 1930s. In the UK, GDP fell by a record 5 per cent in 2009 and the property sector was badly hit with a substantial fall in values and reduced development and construction activity.

The credit crunch recession of 2007–9 was almost certainly the worst worldwide recession ever. It was triggered by a property 'bubble', and is a demonstration of the interlinked nature of property markets, financial markets and a globalised world economy (see Concept 6.21).

Further reading

Jowsey, E. (2011) *Real Estate Economics*, Palgrave Macmillan, Basingstoke, ch. 6.

6.24 Currencies and exchange rates

Key terms: international transactions; currency risk; international property investment

Different countries around the world have their own currencies. When they engage in trade and international transactions these currencies must be exchanged and the price of a currency in terms of another currency is its exchange rate. Thus one pound sterling might be worth 1.58 US dollars (£1.00 = $1.58) and 1.19 euros (£1.00 = €1.19).

Exchange rates are prices and are determined by the demand and supply of currencies. Factors affecting exchange rates are:

- global trade movements
- interest rate expectations
- inflation
- macroeconomic events
- investors' appetite for risks
- economic growth.

Property and land of course stays where it is and does not cross borders, and this means that currency exchange for imports and exports of property is very unlikely to be needed. It is increasingly possible, however, to invest in property overseas (see Concept 8.16) and this necessitates currency exchange. Indeed currency exchange is an additional risk of overseas real estate investment. This is because the exchange rate can change after the purchase price has been agreed – during the time it takes to complete a transaction. For example, a British investor may decide to buy a €10 million office block in Berlin when the exchange rate is £1.00 = €1.19. The cost in pounds is £8,403,361. When the deal completes 6 weeks later, however, the exchange rate has changed to £1.00 = €1.06 and the cost in pounds is now £9,433,962; more than £1 million pounds more!

If the transaction is made at the current exchange rate, this is known as the 'spot rate'. It is possible to secure an exchange rate now for a transaction at a specified future date – this is a 'forward contract'. The forward rate is adjusted for the differential between interest rates in the two different countries over the time to the specified future date. Of course, a forward contract is something of a gamble because it is binding and the actual rate may actually be more favourable when the transaction takes place.

The exchange rate can also change between the time of purchase of the investment and the time of its sale. For example, a UK individual buys a property in Florida for $350,000 when the exchange rate is £1.00 = $1.50 and so the cost in pounds is £233,333. After 3 years the Florida property has risen in value to $500,000 and the owner decides to sell and realise the profits on the investment. The exchange rate, meantime has changed to £1.00 = $1.97 (the pound has strengthened against the dollar) and the sale now produces 500,000/1.97 = £253,807. When transaction costs such as legal fees and taxes are taken into account there would be very little profit left!

Further reading

Jowsey, E. (2011) *Real Estate Economics*, Palgrave Macmillan, Basingstoke, ch. 18.

7 Finance

Ernie Jowsey and Hannah Furness

7.1 Banks

Key terms: central bank; clearing banks; deregulation; base rate; credit creation; monetary control

Banks vary, both in the type of function they perform and in size. They can be classified as:

- The central bank – in the UK this is the Bank of England, which, on behalf of the government, exercises a general control over the banking system (through the Financial Conduct Authority since 2012) and the availability of credit (using monetary policy).
- The clearing banks – these banks, once dominated by the 'Big Four' (Lloyds TSB, Barclays, Natwest and the Midland, now HSBC) now include former building societies, such as the Halifax and Nationwide. Even so, unlike the systems of other countries, such as the USA, which are composed of a large number of unitary small banks, Britain has only a few large banks, each having a network of branches throughout the country.

The basic business of the clearing banks is holding cash balances or deposits and lending on the basis of these holdings. Bank loans create bank deposits and banks are able to lend multiples of their deposits. A bank's main assets are loans to customers, cash and short- and long-term bonds and securities. Its liabilities are customers' deposits.

In relation to real estate, the banks are a source of funds for developers, private buyers of investment properties, businesses purchasing premises and home buyers. Property provides collateral for loans from the banks, making lending against property relatively low risk in normal circumstances. Of course there have been times, notably in the run up to the credit crunch recession of 2008, when the banks over-extended their lending as property prices went up and up – until the bubble burst and the crash led to massive losses and bank failures around the world (see Concept 6.23).

The Bank of England has control over the base rate of interest which influences all interest rates in the economy. This affects the housing market through mortgage interest rates and the property and development market through the direct effect on yields on investments and indirectly through interest rates affecting aggregate demand in the economy.

The Bank of England also influences the quantity of money in the economic system and the availability of credit. This can be done in order to stimulate the economy (as with quantitative easing after the 2008 crash) or by tightening monetary policy by increasing interest rates and contraction of the money supply in order to reduce inflation.

Since deregulation of financial markets in the UK in the 1980s, the clearing banks, building societies and insurance companies have encroached on each other's territories in competition for the mortgage market (worth over £77 billion in the UK in 2010). More competition in the market gave prospective buyers more choice, but contributed to risky lending and eventually the 'credit crunch' recession of 2008–9. Over-optimistic developments were left empty or not completed, and banks had to write off billions in bad debts from property companies.

Further reading

Harvey, J. and Jowsey, E. (2007) *Modern Economics*, Palgrave Macmillan, Basingstoke, ch. 26.

7.2 Bridging loan

Key terms: short-term finance; closed bridge; open bridge; risk

A bridging loan is a type of secured short-term loan, typically taken out for a period of weeks or months while longer-term finance is arranged. Security for the loan is provided by property but bridging loans are normally more expensive than conventional financing to compensate for the additional risk of the loan. Bridging loans typically have a higher interest rate and the lender may also specify a lower loan-to-value ratio. An arrangement fee, interest and possibly an exit fee will be payable. However, they can be arranged quickly with relatively little documentation.

In the residential property market, a house buyer may plan to make a deposit with the proceeds from the sale of a currently owned home, but the sale will not complete until after the date that the new property must be bought. A bridging loan is a short-term loan to bridge the period between someone buying a new property and selling their previous home. This allows the buyer to take equity out of the current home and use it as down payment on the new residence, with the expectation that the current home will complete within a short time frame and the bridging loan will be repaid.

There are two types of bridging loan: closed bridging and open-ended. With a closed bridge the money is borrowed for a short and clearly defined period, perhaps a week or a month and so the risk is quantifiable and limited. With an open bridge, contracts have not been exchanged on the sale property and so there is much greater risk as the buyer may pull out. If there is not yet a buyer at all an open bridge is very risky and expensive.

Bridging loans can also be used by developers to finance a project while planning approval is sought. Because there is no guarantee the project will happen, the loan will be at a high interest rate and from a specialised lending source that will accept the risk. Once the project has full permission, it becomes eligible for larger loans from more conventional sources that are at lower interest and for a longer term.

A bridging loan may be used to purchase a property offered at a discount if the purchaser can complete quickly. The discount offsets the costs of the short-term bridging loan used to complete. Long-term lending may not be available in auction property purchases where the purchaser has only 14–28 days to complete and a bridging loan may be necessary.

In 2012 the UK bridging loan market was growing rapidly in response to the cautious lending policies of the banking sector. The size of bridging loans was also increasing and commercial loans were increasing as a proportion of all bridging loans. Bigger projects are being financed with lower loan to value ratios, meaning that the interest rates on bridging loans have fallen.

Further reading

www.cml.org.uk/cml/publications/newsandviews/121/454 (accessed 15/10/2013).

7.3 Company accounts

Key terms: balance sheet; income statement; cash flow; liquidity; assets; liabilities; ratio analysis

Company annual reports provide data that can be analysed in order to assess the financial health of a company. Within the annual reports are financial statements that are of use to potential investors, existing shareholders, landlords assessing the reliability of the company as a tenant and potential partners considering a joint venture. Other interested parties could include company creditors, employees of the company and the tax authorities. The financial statements almost always include a balance sheet, from which the value of the company can be ascertained; an income statement showing the profits (or losses) over the previous 12 months to the accounting date; and a cash flow statement showing the flow of cash in and out of the business. The statement of cash flows is useful in determining the short-term viability of a company, particularly its ability to pay bills.

The *balance sheet* of the company shows the assets, liabilities and shareholders funds of the company at a particular date. Generally:

total assets = shareholders funds + liabilities

and

shareholders' funds = total assets – liabilities.

The balance sheet provides details under the following headings:

- non-current assets (fixed assets such as investment properties, loans, joint ventures etc.);
- current assets (cash and money on deposit, properties for sale etc.);
- current liabilities (any short-term borrowing, trade debts, tax owing etc.);
- non-current liabilities (longer-term borrowing)

Total assets minus total liabilities gives net assets and these are equal to total equity or shareholders funds (including reserves of capital held by the company) and, hence, the balance sheet balances.

Property companies often have considerable capital reserves in the form of investment properties that rise in value over time. These are not liquid assets, however, and if they were to be sold quickly there would probably be considerable loss of value. As a result the net asset value of the company can be less than the stock market value of the shares in the company. To put that another way the price of the company's shares can be less than the net asset value per share. Such a situation could make the company a target for 'asset stripping' – the process of buying the shares then selling some or all of the assets to make a profit.

The *income statement* of a company (sometimes called a profit and loss account) shows the firm's trading results for the previous 12 months. This is usually presented as revenue and costs under the following headings:

- Revenue (rents, income from sales etc.)
- Costs (maintenance costs, staff costs and overheads).

Then revenue minus costs gives gross profit, from which indirect costs such as interest payments on loans and costs of disposals are deducted to give profit/loss before tax. Then tax is shown and deducted to give profit/loss for the year. Some of this may be retained (ploughed back) by the company (transfer to reserves) and some distributed to shareholders (dividends to shareholders). It is customary to show earnings per share which is total dividends to shareholders divided by the number of shares issued in the company.

Both the balance sheet and the income statement normally show figures to the accounting date plus the figures for the previous year so that comparisons can be made. A visit to a company's website on the internet will quickly enable you to peruse their annual report and accounts; see, for example www.britishland.co.uk.

Ratio analysis can be used to assess the performance of a company using figures in their accounts. For example, assets per share can be calculated by dividing net assets by the number of shares and profit as a percentage of turnover (revenue) can be calculated and compared with other companies' performance. Other commonly used ratios include:

- profit before interest and tax as a percentage of average capital employed;
- interest cover – the number of times any interest payable on loans is covered by profits;
- gearing – the proportion of total funds that is borrowing;
- liquidity ratio – the ratio of current assets to current liabilities.

Further reading

Isaac, D. and O'Leary, J. (2011) *Property Investment*, 2nd edn, Palgrave Macmillan, Basingstoke, ch. 10.

7.4 Debentures

Key terms: fixed interest security; redemption date; convertible debenture

In the UK a debenture is a secured loan raised by a company, usually at a fixed rate of interest and secured against specific assets of the company (often property). The interest must be paid whether the company is making a profit or not. The redemption date, when the loan is to be repaid, is normally specified. Debentures are the most common form of long-term loans that can be taken by a company. Debentures are usually loans that are repayable on a fixed date, but some debentures are irredeemable securities. A debenture holder enjoys no ownership rights of voting on management and policy.

'Mortgage debentures' are secured on specific assets of the company such as land and buildings. A company whose profits are subject to frequent fluctuations is not in a position to raise much of its capital by debentures. They are really only suitable to a company making a fairly stable profit (sufficient to cover the interest payments), and possessing assets that would not depreciate a great deal were the company to go into liquidation. Debenture interest is included in the costs of a company for the purposes of calculating tax and so it reduces taxable profits.

Debenture holders (the lenders) have no control over the company as long as the loan conditions are complied with and the interest is paid. If this is not the case, they can take control of the company because the loan is secured against the company's assets. Debenture holders are paid first in the event of liquidation of the company. Debentures usually have a lower nominal rate of interest compared with unsecured loans. If the debenture has a 'sinking fund' this means that the borrower must pay some of the value of the bond after a specified period of time. This decreases risk for the creditors, as a hedge against inflation (which diminishes the value of the fixed interest payments) or bankruptcy. A sinking fund makes the bond less risky, and therefore gives it a smaller 'coupon' (or interest payment).

Convertible debentures combine the security (to the lender) of a fixed interest loan with the growth potential of equities. Fixed interest is paid but the debenture holder also has an option to convert stock into a specified number of ordinary shares at a given date. If the option to convert is not exercised, the debenture continues as a fixed interest stock. A convertible debenture is less risky than ordinary shares because the fixed interest to be paid limits the fall in value that could occur if the company does badly. If the company does well, the potential for conversion means the convertible debenture will rise in value because it can be converted into shares in the successful company. If a profitable conversion is anticipated, the price of the convertible debenture will reflect the expected share price at conversion.

In the United States debentures are not secured by physical assets or collateral. These debentures are backed only by the general reputation and creditworthiness of the issuer. Both corporations and governments issue this type of bond in order to secure loan capital.

Further reading

Harvey, J. and Jowsey, E. (2007) *Modern Economics*, Palgrave Macmillan, Basingstoke, ch. 6.

7.5 Depreciation

Key terms: asset value; obsolescence; rent voids; yields; refurbishment; redevelopment

Depreciation means a reduction of asset value, often because of physical deterioration of a property, but also because of economic change reducing demand, or technological change making the building less functional. Land does not deteriorate in value in the same way that buildings do and it is possible for buildings to depreciate while the land value increases.

Property is affected by depreciation, which reduces its value and the income it can make from rent. Eventually, when a building has no rental value it is obsolete, meaning redevelopment of the site is required. Redevelopment may take place before this point for economic reasons because the value of the cleared site is greater than the existing buildings.

Buildings (but not land) depreciate and since the value of commercial and industrial building is normally much greater than the value of the land it is on, this can be a substantial cost to companies. Rents can be affected by depreciation and capital value can fall as the property becomes less attractive or fit for purpose. The risk of voids (periods when the property is vacant) also increases. To try to offset the effects of depreciation, the property owner may refurbish or redevelop, but this requires capital outlay and possibly loss of rent during the process.

For office buildings, as new stock is built older buildings gradually become secondary stock and their rents decline compared with the new prime properties. Periodic refurbishment can offset this. Industrial buildings are greatly affected by economic fluctuations and so factory buildings are at more risk of void periods. Factories and warehouses usually have a shorter building life than offices and most shops, and so the impact of depreciation on them is greater. As a result yields on industrial property are normally significantly higher than those on other commercial property.

Traditionally, yields are lowest on agricultural property (where land is a greater component of asset value than buildings); higher on retail premises and higher still on offices (which are thought to be less vulnerable to rent voids); and higher still on industrial property.

Further reading

Jowsey, E. (2011) *Real Estate Economics*, Palgrave Macmillan, Basingstoke, ch. 8.

7.6 Financial gearing

Key terms: equity; debt; loan to value; high gearing; leverage

The principle of financial gearing is to use a smaller equity stake by an investor to supplement a larger loan to enable the purchase of an investment, for example, 25 per cent of an asset's value is used to acquire a loan for the remaining 75 per cent. Gearing enables investors to gain exposure to a number of new investments by spreading available equity to supplement different loans. It is a high-risk and potentially high-return strategy and is applicable to both residential and commercial property investments.

Gearing exaggerates market returns; however, it also increases risk. The capital gearing effect can be explained by the following example:

> Mr Smith has agreed to buy a house for £100,000. In return for paying £25,000 of his own equity, the bank will loan him the remaining £75,000 in order to purchase the property. If the house price was to increase by 10 per cent one year after purchase, then Mr Smith's equity would also increase as illustrated below (see also Figure 7.6.1):

House prices have risen by 10 per cent.
£100,000 house is now worth £110,000.
Bank is still owed £75,000 (assuming an interest-only loan).
Therefore, Mr Smith's equity = £110,000 – £75,000 = £35,000 (40 per cent uplift in equity from £25,000 to £35,000).
Had Mr Smith bought the house wholly with his own money (i.e. 100 per cent equity) then he would have received only 10 per cent in return.

In Figure 7.6.1 the advantage of gearing is illustrated by comparing a low-geared investment of 50 per cent loan to value (LTV) and a highly geared investment of 90 per cent LTV. When the investment increases in value from £100,000 to £110,000, the return on equity is 20 per cent with the 50 per cent LTV loan (£10,000/£50,000); but 100 per cent with the 90 per cent LTV loan (£10,000/£10,000). As long as the investment increases in value by a greater percentage than the rate of interest on the loan this will be the case.

However, while the principle can amplify returns in a rising market, in a falling market it increases volatility of returns. An investment's underlying growth can be supplemented by gearing; however, in a rising market gearing magnifies equity returns, but in a falling market it magnifies market falls.

Companies are said to be 'highly geared' if their gearing ratio (debt/equity) is greater than 50 per cent. If this is the case, in a good market profits are likely to be greater than the interest payable on the companies' borrowing and being highly geared provides better returns for shareholders. Gearing is referred to as leverage in the US because the borrowing is used to lever out greater returns for shareholders. In a poor market, however, profits are likely to be lower than the interest on borrowing and highly geared companies can get into financial difficulties.

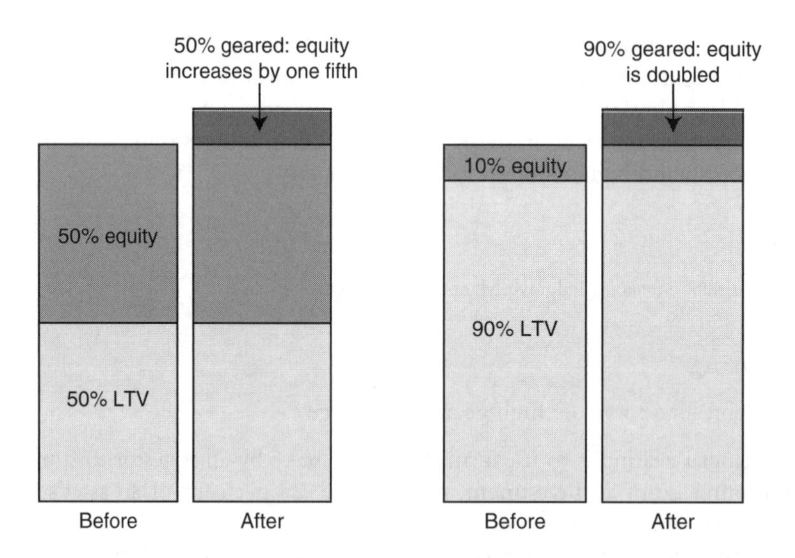

Figure 7.6.1 The capital gearing effect – 50 per cent equity stake versus 10 per cent equity stake – house price rises 10 per cent.

Source: Furness (2013).

Further reading

Isaac, D. (1994) *Property Finance*, Macmillan, Basingstoke, ch. 4.

Isaac, D. and O'Leary, J. (2011) *Property Investment*, Palgrave Macmillan, Basingstoke, ch. 8, ch. 10.

7.7 Liquidity

Key terms: conversion into cash; direct property investment; asset values

The concept of liquidity is important when considering property. All assets can usually be turned into money eventually and so liquidity is largely a matter of degree, often depending upon the organisations that exist to make assets such as pension funds, company shares, government bonds and insurance policies liquid. All assets except money yield either a flow of income or direct satisfaction. Only money is perfectly liquid and can be changed into another asset without delay, extra cost or some form of capital loss.

Of course these non-money assets can be sold at some capital loss, so they do give some liquidity to their owners, meaning that they could keep a lower cash balance for emergencies. Ownership of a house makes it possible to raise cash by re-mortgaging if there is sufficient equity in the property to provide security for the lender.

So a liquid asset is one that can be converted quickly into cash without loss of value. Of course if I own a property I can almost certainly sell it quickly but not at full value. If I need to sell it very quickly I may have to accept a lot less for it than it would be worth if I could afford to wait for the proceeds. And of course property takes time to sell because a valuation is required and the transaction can be complex and require legal searches, arrangement of finance and preparation of legal documents.

The illiquidity of direct property investments – investment in actual buildings rather than in PUTs or REITs (see Concepts 8.8 and 8.9) – means that this form of investment is at a disadvantage compared to more liquid investment assets such as shares in companies. In a recession it is very difficult to sell property because buyers are reluctant to buy when prices are falling. Of course this makes property assets even less liquid and increases the risk that full value or something close to it cannot be achieved. When house prices fall and few buyers want to enter the market, sales fall to almost zero. At such times the market has become illiquid.

Liquidity indicates how quickly an asset can be converted into cash and so liquidity is a desirable trait to investors; so generally the more liquid an asset the lower the return it offers, due to investors bidding up its selling price. The most liquid asset markets have a high turnover of assets and many participants, and the cost of doing business in them is lower.

Property companies with considerable investments in actual buildings, may find that their balance sheets (see Concept 7.3) fluctuate enormously as their asset values change with the cyclical performance of the overall economy.

Further reading

Jowsey, E. (2011) *Real Estate Economics*, Palgrave Macmillan, Basingstoke, ch. 16.

7.8 Freehold ground rent

Key terms: property rights; leases; reversion; government bonds; gilts; yields

A ground rent is created when a freehold piece of land or a building is sold on a long lease or leases. The ground rent provides an income for the landowner. With real property it is not possible to hand over land and buildings in the same way as would be the case with movable goods. Therefore, it is 'rights' that are dealt in and accompanying the transaction there is a written statement defining the exact rights that are being transferred (see Concept 6.15). In short, the real property market deals in the rights relating to real property rather than in the land and buildings themselves. One such property right is a freehold ground rent (FGR).

So, FGRs are the annual payments accompanying long leases. Originally they were mainly charged on land leased for development, but they are now received for flats sold on long leases. Since they are small relative to the full value of the developed site upon which they are secured, the income is certain. Thus, where reversion is distant, FGRs are comparable with irredeemable or very long-dated government stock, although, being less liquid, yields are higher, especially if management is involved (as with

FGRs on flats). Because of inflation, freeholders granting long leases have insisted on provisions for periodic revision of the ground rent or for profit sharing or for some combination of both. Certainty of payment means that FGRs are similar to an investment in gilt-edged securities.

When reversion is less than 50 years, an FGR increasingly assumes equity characteristics, as its reversionary value will tend to rise. Its price will tend to be higher if it has a special value to a particular person, such as the current lessee of the property or a property developer who wishes to assemble land for future development.

The contemporary accepted meaning of ground rent is the rent at which land is let for the purpose of improvement by building. So it is a rent charged in respect of the land only, and not in respect of the buildings on it. It is therefore usually lower than the rent that might be achieved for a building let on the open market and is let for a longer term. Under the terms of a lease agreement, the freeholder (the outright owner of the land or property) grants permission for a leaseholder to take ownership of the property for a specified period of time. This could range from 21 years to 999 years (which is common in central London). During this lease time the leaseholder will pay ground rent to the freeholder. Freeholders lease property primarily for the initial premium paid by the original leaseholder for granting the lease; but in addition ground rent (often a token amount) will be payable over a long term, and this may be an attractive fixed income investment for some types of investor.

Of course, inflation has eroded the value of most ground rents with long leases and non-rising incomes, so their value is small where there is no prospect of a reversion (when the ownership of the property reverts back to the freeholder) within say 150 years.

The Commonhold and Leasehold Reform Act, 2002 and the Landlord and Tenant (Notice of Rent) (England) Regulations 2004 now govern the form of notice that needs to be issued to collect ground rent. Previously there had been a problem with some landlords sending confusing or dishonest demands for payments to tenants.

Further reading

Jowsey, E. (2011) *Real Estate Economics*, Palgrave Macmillan, Basingstoke, ch. 3.

7.9 Reverse yield gap

Key terms: yields; gilts; inflation; nominal rate of interest; real rate of interest; asset bubble

Demand for prime properties, such as city-centre shops and offices from institutional investors, is often very great because such investments appear to have future growth potential. As a result, yields on their freeholds are low and sometimes below the yield on long-dated government securities (gilts). A reverse yield gap is the amount by which the yield on gilts exceeds that on property.

The term is also used to describe the situation where interest charges are greater than yields. This can often occur during times of high inflation when nominal interest charges are high in order to compensate for inflation and maintain a real rate of interest (the real rate of interest equals the nominal rate minus the rate of inflation). For example, a property developer might borrow £1 million at 11 per cent interest in order to finance a development with initial rents amounting to only £90,000 a year. The £90,000 equates to 9 per cent interest and so there is a reverse yield gap of 2 per cent. Of course this is a loss-making situation (interest £110,000 p.a. is greater than the income of £90,000 p.a.), but it may be acceptable if the developer expects the capital value of the property to increase or if rents are likely to increase at review.

Since gilts are considered a risk-free investment (because they are government backed) it would seem to make very little sense to invest in relatively risky property that provides a lower return than this. In fact it indicates that the investor is prepared to accept the initial lower return in the expectation that in the long term the capital value and income return from the property will be much more profitable than the yield on gilts.

Since yield is the return on the initial capital investment, if they are less than the return on gilts this can be a sign that capital values have been driven up too high in a boom and might be a signal that the market is overheating and the asset bubble is about to burst.

Further reading

Jowsey, E. (2011) *Real Estate Economics*, Palgrave Macmillan, Basingstoke, ch. 16.

7.10 Sale and leaseback

Key terms: equity release; operational control; tax benefits

Sale and leasebacks, in which companies sell property assets and then rent them back, became increasingly popular during the years of rising property prices as companies turned their property assets into cash in a rising market. Sale and leaseback is a method of raising funds for companies with property assets. Equity can be released from the property while still occupying the premises on a leasehold basis. This might enable the company to obtain funds for expansion without resorting to borrowing. Property is becoming more important in corporate decision making (see Concept 8.18), and attachment to corporate property ownership has weakened, with the recognition of the often significant funding potential within assets where value has often appreciated significantly over recent years.

Sale and leaseback can offer a profitable company significant tax advantages. For example they could sell their asset to another company in their group and the lease payments would then become an allowable tax deduction.

Sale and leaseback has several advantages:

- it allows a company to gain access to money tied up in valuable business assets;
- capital is released to fund merger and acquisition activity or to reallocate to core activities;
- the asset can still be used even if it is no longer owned;
- tax benefits are realised by offsetting lease costs as an operating expense;
- the seller remains in day-to-day operational control of the property;
- property value risk is transferred to a third party.

Disadvantages include:

- the asset is no longer owned by the company, thereby weakening its balance sheet;
- there may be better ways to gain access to the funds either by refinancing or by securing a loan using that asset as collateral;
- the long-term costs of the premises are likely to be greater than if they remain in ownership of the company.

Developers can raise finance using sale and leaseback arrangements. The freehold of the development with the benefit of planning consent is sold to an institution which then advances the development costs at a fixed rate of interest. If the development is not let within, say, 6 months of completion, the developer receives a balancing payment in return for either entering into a leaseback or providing a leaseback guarantee at an agreed base rent. This balancing item represents the developer's profit and is based on a certain yield to the institution of about 7.5 per cent. If costs rise or completion takes longer than expected, the developer makes up the difference from his balancing item.

Further reading

Jowsey, E. (2011) *Real Estate Economics*, Palgrave Macmillan, Basingstoke, ch. 9.

7.11 Mortgages

Key terms: fixed rate; variable rate; repayment mortgage; interest only mortgage

The term 'mortgage' is from the French language and means 'death pledge'. Thankfully, most long-term loans secured against residential property – or mortgages – are for a fixed term of 10, 20 or (more normally) 25 years and the borrower usually outlasts the mortgage. Owner occupiers comprised about 67 per cent of all UK households in 2013. This was slightly less than at the peak of 2007 because the private rental sector grew in the recession.

Most people cannot afford to buy a property outright and instead require a deposit and a long-term loan or mortgage, which they pay off with interest over a number of years. A mortgage can be a 'repayment mortgage' where the capital sum and interest are repaid over the term of the loan – or an 'interest only' mortgage where the interest is paid but the capital sum remains outstanding and must be repaid in some way at the end of the mortgage term. Repayment mortgages guarantee that the whole loan is repaid by the end of the term, making them a low-risk option. More interest is payable overall on an interest-only mortgage as the borrower pays interest on the whole loan for the whole term. In the past, interest-only mortgages were more commonly linked with endowment policies as a repayment vehicle, but endowments are risky because they are investments related to the performance of the stock market. There's always the possibility that if the stock market doesn't grow enough, the investment might not become big enough to pay off the total amount of the mortgage. Many households are facing shortfalls on their mortgage when it matures because their endowment has underperformed. Because of their poor history, endowments are now rarely recommended for interest-only mortgages.

With a fixed rate mortgage, the interest rate stays the same for a set period of time. This means that for every month during this set period, the mortgage repayments will remain the same. This enables easier budgeting and greater security. This is in contrast to a variable rate mortgage, which will go up or down in relation to the Bank of England base rate, or the lenders' standard variable rate (SVR). A SVR mortgage is a lender's 'default' rate – without any limited-term deals or discounts attached. The term of a fixed rate mortgage usually lasts between 2 and 5 years, but can be longer, before reverting to SVR. In certain circumstances, fixed rate mortgage deals can have higher rates than variable rate mortgages – if you're on a fixed rate and the base rate drops, your monthly repayments don't change but variable rates probably will. In addition, arrangement fees may be payable to set up a fixed rate mortgage – and you're also likely to face early repayment charges if you pull out of a fixed rate deal before the end of the term.

A tracker mortgage interest rate tracks the Bank of England base rate at a set margin (for example, 1 per cent) above or below it. Tracker mortgage deals can last for as little as one year, or as long as the total life of the loan. Once the tracker deal comes to an end, the mortgage reverts to SVR and, usually, this will have a higher rate of interest.

Flexible mortgages let you over and under pay, take payment holidays and make lump-sum withdrawals. This means the mortgage could be repaid early and a saving made on interest.

Other types of flexible mortgages include offset mortgages, where savings are used to offset the amount of mortgage interest that is paid on each month. So, for example, if you have a £200,000 mortgage and £10,000 in savings, you only pay interest on £190,000. Current account mortgages combine current account, savings and mortgage into one so all credit balances offset the mortgage debt.

Flexible deals can be more expensive than conventional ones but they can be particularly worthwhile for higher-rate taxpayers because no tax is paid on interest from savings as they are offset against the mortgage.

The LTV ratio of a mortgage represents the amount of money that is borrowed as a percentage of the value of the property. For example, if you were to buy a £500,000 property and borrowed £400,000 you would have an 80 per cent LTV mortgage. The bigger the LTV the more expensive the mortgage is likely to be. This is because the more you borrow the bigger the risk for the lender – they have more to lose if repayments are not made – so the larger the deposit, the better the mortgage rates that will be offered.

In the past UK lenders would lend 100 per cent or more of a property's value but these types of mortgage are not available anymore. The most lenders will offer now tends to be 95 per cent of a property's value but most people will borrow less than this. The government's 'Help to Buy' mortgage scheme enables borrowers with deposits of just 5 per cent to buy a property, with the government guaranteeing up to 15 per cent of the loan in return for an insurance fee. The government guarantees lenders losses if the property is repossessed and prices fall.

A commercial mortgage is similar to a residential mortgage, except the collateral is a commercial building or other business property. Commercial mortgages are taken on by businesses instead of individual borrowers and so assessment of the creditworthiness of the business can be more complicated than it is with residential mortgages. Businesses may purchase premises using a commercial mortgage which can enable them to own a large asset that is likely to increase in value and repayments can be similar to rental costs. The mortgage interest payments are tax-deductible.

Some commercial mortgages are non-recourse; this means that in the event of default in repayment, the creditor can only seize the collateral, but has no further claim against the borrower for any remaining debt. A commercial mortgage is often supplemented by a personal guarantee from the owner, which makes the debt payable in full if repossession of the mortgaged collateral does not pay off the outstanding balance.

Further reading

Jowsey, E. (2011) *Real Estate Economics*, Palgrave Macmillan, Basingstoke, ch. 5.

7.12 Sources of finance

Key terms: financial devices; liquidity; security; collateral; recourse loan; non-recourse loan

New financial devices are constantly being invented in order to meet changes in market conditions. New development projects often involve finance that is designed to suit both borrower and lender. Short-term finance covers the development period until the completed development is sold. This is mainly obtained from banks – the clearing banks, merchant banks and the UK subsidiaries of foreign banks. All will be concerned about liquidity, security and the profitability of the loan. Banks are increasingly sensitive to being 'locked in' as a lender of project finance where the sale of the completed development is the sole means of repayment. Since the financial crisis of 2008, the banks have put much more emphasis on liquidity so that money lent can be recovered quickly in the event of a property market collapse.

Lenders will look at a number of areas when considering whether they will provide finance for a development project. These include:

- experience as a property developer;
- financial strength of the developer;
- how much equity is brought to the project;
- the location of the proposed development;
- the profit potential of the development;
- builder experience and capacity;
- project management team experience;
- type of development (residential housing, apartments, commercial or mixed use);
- level of pre-sales/pre-leases;
- ability to cover cost overruns;
- exit strategy.

A clearing bank limits the loan to about two-thirds of the cost of the development, usually determined by its own valuer. This may present a difficulty for a housebuilder or minor office developer with few capital resources. A well-established property company has an advantage, because its existing property holdings provide collateral and net revenue from them may cover interest payments on the new loan. Even so, it will limit its collateral as far as possible since uncommitted property can be used to support later borrowing. Where a loan is confined to the specific project it is known as a 'non-recourse' loan; where other assets of the company or parent company can be called upon, it is a 'recourse' loan. usually the final agreement lies between these two – a 'limited recourse' loan.

A large property company may wish to form a subsidiary company to carry out a particular development, but limiting its equity commitment to an initial 5 per cent of the finance required. A bank would advance, say, 80 per cent of 'senior debt'. The balance of 15 per cent – 'mezzanine debt' – could probably be obtained by arranging cover against loss with an insurer who specialises in loans secured against commercial property.

Longer term finance can come from: priority yield schemes, sale and leaseback, profit erosion, forming a joint company and raising equity capital by selling shares. Table 7.12.1 summarises the advantages and disadvantages of different forms of finance.

Table 7.12.1 Advantages and disadvantages of sources of finance

Source of capital	*Method*	*Advantages*	*Disadvantages*
Finance by cash/assets	– paid before acquisition – paid after acquisition – paid by instalment	– certainty – no ongoing liability – retain control of development	– drain on resources – can affect cash flow – restricted to own cash/ assets
Finance by loans	– bank/institutional loan – issuance of shares/bonds/ futures/options/REITs	– steady cash flow – risk sharing – default option/ prepayment option – better control of quality and asset – specificity	– liability/possibility of liquidation – cost of loan (depends on credit rating) – collateral at risk
Finance by sub letting land interests	– sale and leaseback – joint venture (PPP/PFI) – build-operate-transfer (BOT) – build-own-operate-transfer (BOOT) – sale of land promises (e.g. naming rights/roof-top antenna/external wall advertisement)	– no cash drain – ultimate ownership of land retained – own the development with no construction and operation costs – risk sharing – can attract value-added developers and increase profits	– loss of control of land interests – high asset specificity (the investments made to support a particular transaction have a higher value in that transaction than they would have if they were redeployed for any other purpose) – moral hazard (may act differently if risk is reduced) – quality of development affected

Source: adapted from Jowsey (2011).

Further reading

Jowsey E. (2011) *Real Estate Economics*, Palgrave Macmillan, Basingstoke, ch. 9.

8 Investment

Ernie Jowsey and Hannah Furness

8.1 Investors

Key terms: owner-occupiers; private investors; institutional investors; charities and trusts; foreign investors; PUTs; REITs

Real estate investment is carried out by private persons, private trusts and the institutions – insurance companies, pension funds, charities, property companies, property bond funds, property unit trusts (PUTs) and real estate investment trusts (REITs).

Unlike many other types of investment, such as shares and bonds, property is a tangible asset. While property valuations can fluctuate, the property is still there. Many private investors feel that they understand property better than other assets, because they can see it and because they often have the experience of buying their own homes. The considerable sums required for major commercial property investments, however, act as a barrier to entry for private individuals.

Anybody who purchases a property rather than renting is an investor (home ownership confers consumption benefits and investment benefits). The satisfaction or return received should at least equal what could be obtained if, instead, premises were rented and the money invested elsewhere. So the return on investment in property is in competition with the return on other investments.

- Owner-occupiers – for example, shop-owners, farmers and householders – are holding wealth in the form of real property. They enjoy a full equity interest, which includes income or satisfaction from the use of their property, and (normally) a hedge against inflation.
- Other private persons investing in real property usually have only limited funds. Thus, their direct investment tends to be restricted to dwellings and secondary shops. Indirectly, however, they can invest in prime shops and offices by buying property bonds or shares in property companies or unit trusts specialising in quoted property companies.
- Insurance companies and pension funds are major institutional investors in commercial property. Pension funds compete strongly with insurance companies and property companies for first-class properties, because the inflation hedge helps to retain the real value of the accumulated pension funds. The average pension fund had around 20 per cent of its assets in property in the 1970s, a figure that fell to 12 per cent in the 1980s and 6 per cent by the mid-1990s. The smaller pension funds invest in property indirectly through PUTs and REITs (see Concepts 8.8 and 8.9), which give the advantages of property investment without management problems. The larger funds, however, prefer to purchase and manage their own properties and sometimes even participate directly in investment with development companies, property companies and construction firms.
- Charities and trusts require investment income (from which periodic distributions are made) but must also retain the real value of trust funds. Consequently, although they pay no income tax, they cannot invest entirely in high-yielding securities. Unlike most institutional investors, charities

receive little 'new' money for investment each year, so they are constantly reviewing their existing portfolios to see what possible adjustments could best serve their beneficiaries.

- Property investment and development companies tend to be highly geared, their capital consisting of a high proportion of loans to ordinary shares. Properties owned provide the security against borrowing, while interest charges are covered by regular rents. High gearing is beneficial to the shareholders when profits are good, and it makes it easier to retain control. The larger companies tend to specialise in office blocks or prime shop properties, and a few, such as SEGRO, in industrial property. Hammerson specialises in shopping centres. Residential property investment is confined mainly to smaller companies.
- Foreign investors into UK property (particularly in central London) have increased their investment considerably since the fall in property prices through the 2007–9 recession and the fall in the sterling exchange rate. The UK also tends to have longer leases than continental Europe, 15 years compared to six or nine, and investors are attracted by the income security that offers during the recession.
- Property bond funds, PUTs and REITs invest in commercial properties and investors can buy shares in them, taking advantage of their favourable tax treatment (see Concepts 8.8 and 8.9).

Further reading

Jowsey, E. (2011) *Real Estate Economics*, Palgrave Macmillan, Basingstoke, ch. 16.

8.2 The property investment market

Key terms: assets; commercial; office; retail; industrial; residential; PUTs; REITs; income; capital gain

Investment is an exchange of present capital value for future income and capital. Future cash flows from the investment depend on:

- variability of income
- variability of capital value
- security of income
- security of capital value.

Direct property investment is expenditure on the purchase of existing assets – for example, shares of a company, or an interest in an existing (perhaps tenanted) office block. There are several types of real estate investment property including:

- residential property investment
- commercial property investment
- buy-to-let property investment (see Concept 8.11)
- investment in land purchase.

Residential property investment is the least risky form of investment because the purchase of a residential property by the owner is not only an investment decision as the property performs the function of providing somewhere to live and is thus a consumer good as well as an investment good. One of the main reasons for home ownership rather than renting is because generally investment in property is profitable. By combining the consumer good function of shelter with the investment nature of property purchase, investment in residential property provides a safe and usually profitable dual use of funds. In addition, homeowners have an excellent incentive to regard their property purchase as a long-term financial priority because it provides their home and as a result lenders are often willing to advance up to 90 per cent of the purchase price of a residential property, seeing this as low risk lending. In most developed countries loan to value ratios can be in the region of 70–90 per cent as a result.

Commercial property investment involves the purchase of office buildings, retail space, hotels and motels, warehouses, industrial units and workshops, and other commercial properties. For the UK the main commercial property investment sectors are office, retail and industrial. In many other countries investment in residential apartments is also common. The investment usually provides a rental income and the possibility of growth in the capital value of the asset, but commercial property is generally regarded as being more risky than residential and so loan to value ratios are usually in the range of 50–70 per cent.

Real estate investments have certain characteristics that affect their performance:

- They are in a fixed location.
- Heterogeneity – they are all different.
- They tend to be of high unit value (or large acreages) and this can be a barrier to entry.
- Illiquidity – they are among the least liquid of assets.
- Lack of transparency in the market (poor information). Aggregate data has been available in the UK since the 1960s and in the USA since 1978 but in other countries only recently. A further information problem is the need for valuers (and subjective valuations) to measure some of the returns on property that arise from increases in capital value.

It is also possible to invest in property indirectly by buying shares in PUTs and REITs (see Concepts 8.8 and 9.9). These provide tax advantages and many of the benefits of direct property investment with the added advantage of liquidity as they can be traded quickly on the stock market.

The broad objectives of investment are to preserve or enhance the real value of the asset and to receive a flow of income over time. Different interests in real property really represent different bundles of rights, such as freeholds, leaseholds, freehold ground rents and mortgages. Because land resources are durable, rights existing in them have a long time scale and, although there may be management costs, no problem exists in storage. Property rights, such as stocks and shares, are therefore demanded as investment assets. The real property market can be regarded as a part of the wider investment asset market which also includes company shares or equities and company bonds or government bonds (gilts).

Further reading

Jowsey, E. (2011) *Real Estate Economics*, Palgrave Macmillan, Basingstoke, ch. 16.

8.3 Commercial property investment

Key terms: prime property; secondary property; efficient markets; retail; office; industrial; yields

The commercial property investment market has two parts (possibly three) which are not well defined. The primary or prime property sector consists of top quality premises in the best locations. The secondary sector is poorer-quality property in fringe locations. If we were to include a third category it would be tertiary property which is in run-down, possibly declining areas and is difficult to let.

The market for prime properties is dominated by large corporate investors, many of them investing internationally in order to get higher returns and to diversify their investment portfolios. Prime property is often considered an efficient market (see Concept 6.4) because there is a lot of knowledge about the properties, the rents they command and the yields that can be obtained. Because information is so available it is difficult to 'beat the market' and the strategy for prime property investors may be to accept market returns and diversify their portfolios in order to reduce risks. Market returns are often considered to be those reported by Investment Property Databank (IPD).

The secondary property market is less efficient as there is less information about rents, capital values, yields and tenant covenants. As a result there may be opportunities for investors in secondary property (and tertiary property) to 'stock pick' and try to beat the market by investing in properties that may be under-priced. Indeed it is possible to become an expert in certain specialised secondary markets, such as industrial property in south London boroughs, in order to obtain better than market returns.

More than 20 per cent of all UK property investment by value is in central London, and more than 50 per cent is in London and the southeast. The main types of property invested in in the UK are:

1 *Shops (retail sector)*
  Location is all important in determining the yield on shops. Generally speaking the best high street positions are occupied by the multiples, their values being enhanced because they complement each other. The main reasons for the low yield (and high capital value) on prime shops are:
  - the supply of such sites is limited;
  - multiples are willing to pay high rents for these sites;
  - the goods sold have a high income elasticity of demand, thereby ensuring growth in turnover; and
  - institutions (such as pension funds and insurance companies) seek such investments because occupiers have excellent covenants and the rate of rental growth has been the highest of all types of property.

  Even a short distance from the prime shopping location rents fall off considerably, while potential rental growth is not as good. As a result yields on secondary shops are much greater than those on prime shops.

2 *Offices*
  Prime office blocks in the central business districts of cities appeal to institutions with large funds, because they can often be let to a single tenant providing an excellent covenant, thereby reducing risks and management costs. During the recession of 2007–9, amalgamations of firms, and reduced staff requirements because of the introduction of new technology, led to an over-supply of offices, resulting in falling rents, over-renting and reduced prospects for future increases. This has highlighted the problem of obsolescence through technical advances and changes in working practices. Hence, though the institutions are still buying prime modern office blocks, it is the property companies that are acquiring the older properties where there are opportunities to add value by management expertise or even redevelopment.

3 *Industrial factories and warehouses*
  Industrial premises tend to be less popular as investments than other types of property. This is because rents are usually more affected by economic depression and many factories are built for a special purpose, and if they have to be re-let it can be difficult to find a similar tenant. Sometimes expensive modifications will be necessary. Changes in techniques of production and handling goods can also make industrial properties obsolete; for example, greater height may be required for forklift stacking or container unloading in a warehouse. Industrial properties are also liable to more rapid depreciation than other properties, though tax reliefs give some compensation.

The result is that yields on industrial premises are about 8 per cent. Newly built B_1 premises are becoming more popular as investments, however, as they are of simple construction, on the ground floor only, have large clear spaces, roof lighting and office accommodation attached, and are easily adapted to different uses. Furthermore, they can be written off as depreciation for tax purposes, and may carry special tax allowances.

Other commercial property investments would include farms, hotels and leisure premises such as cinemas and golf courses, but they are not major sectors.

In other countries residential investment, often in the form of multi-family apartment blocks are popular investments for institutions and sovereign wealth funds. This has not, traditionally, been the case in the UK where buy-to-let investments (see Concept 8.11) are generally small to medium-sized investors. To some extent there may be cultural resistance in the UK to the idea of large corporate landlords (Isaac and O'Leary, 2011).

Investment in commercial property gives the possibility of rental income plus capital gain; and such investments are tangible assets that appear less risky as a result. Such investments are very illiquid (see Concept 7.7), however, and are subject to considerable market volatility in periodic booms and slumps.

Further reading

Isaac, D. and O'Leary, J. (2011) *Property Investment Markets*, Palgrave Macmillan, Basingstoke, ch. 2.
Jowsey, E. (2011) *Real Estate Economics*, Palgrave Macmillan, Basingstoke, ch. 16.

8.4 Portfolio strategy

Key terms: equities; gilts; diversification; direct property investment; indirect property investment; efficient market theory; passive management; active management

An investment portfolio can include a number of different types of asset, one of which is property. Other investment assets are government bonds (gilts), shares in companies (equities), commodities such as gold and antiques or works of art. Investment portfolios often consist of a mixture of gilts, equities and property so that the portfolio is diversified and the risk of losses from one asset is spread. Institutions such as insurance companies, pension funds, investment trusts, unit trusts and sovereign wealth funds are important investors.

There are two methods of investing in property;

- as a direct investment by buying a physical asset;
- as an indirect investment by purchasing a paper asset backed by property (property company shares, PUTs, REITs, mortgages and loans).

Corporate property investors (such as REITs) aim to maximise returns for shareholders while reducing risk. They take an active role in managing property investments, undertaking refurbishments, partial redevelopment and even complete redevelopments. Even fully let prime properties require asset management.

There are a number of problems with commercial property as an investment:

- an optimal mix of properties from different sectors would include office, retail, industrial and leisure properties with different management issues;
- an optimal mix of properties from different geographical locations would increase management costs;
- often the investments are indivisible – large expensive buildings that are beyond the reach of small investment funds;
- the funding of the investment can be complex;
- there is a need for strategic management decisions on acquisitions, upgrades and disposals of properties in the portfolio.

Efficient market theory suggests that in efficient market share prices reflect all that is known or knowable about a company. There are, however, three levels of market efficiency:

1. Strong form efficiency – where no information can achieve superior returns to the overall market (not even insider dealing). These markets are unlikely to exist in reality.
2. Semi-strong form efficiency – no public information helps investors beat the market, since it is already allowed for in prices (so fundamental analysis can't beat the market) but private information might (e.g. insider dealing). Stock markets are like this.
3. Weak form efficiency – where fundamental research into details about the investments can help the investor do better than the overall market ('beat the market').

If efficient market theory (strong form) is accepted, then it is not possible to outperform the market consistently – so the strategy to employ is 'passive management'. This involves constructing a portfolio with the same structure as the market: a market fund or index fund. It will have a return below that of the market because of management costs. The portfolio should be diversified by sector (office, retail, industrial, leisure, residential, agricultural, forestry, infrastructure etc.) and by geographical location (UK, Europe, global). The portfolio can also be diversified by property size, age and type of business. The IPD produces an annual index that measures returns from the total professionally managed UK investment market and this is often used as the benchmark for market performance.

In a weak form efficient market, the strategy would be to engage in active management of the portfolio and through research and management expertise to achieve above average returns consistently. Active fund managers seek to buy, sell and improve assets in order to add value to portfolios, over and above that possible from a passive strategy. Investors return is related to the risks taken and, generally, higher risk leads to higher returns. Investors need to decide on their preference for a risk and return combination and achieve that in the construction of the portfolio (i.e. use modern portfolio theory; see Concept 8.5).

Further reading

Hoesli, M. and MacGregor, B. (2000) *Property Investment: Principles and practice of portfolio management*, Pearson, Harlow.
Sayce, S., Cooper, R., Smith, J. and Venmore-Rowland, P. (2006) *Real Estate Appraisal: From value to worth*, Blackwell, Oxford.

8.5 Modern portfolio theory

Key terms: unsystematic risk; diversification; correlation; correlation coefficient; efficient frontier

Markowitz's modern portfolio theory (MPT) suggests that unsystematic risk (non-market risk) can be reduced and returns optimised when investment assets that have negative or low correlations are included in an investment portfolio. The purpose of MPT is for a given level of risk to maximise return or for a given return to minimise risk.

It was developed for share dealing, but has some relevance for commercial property portfolios because it is possible to find properties with a low correlation of returns, although it is more difficult to find negative correlations. The correlation coefficient is the degree to which variables change in relation to one another over time; if time series data showed that retail rents in a city always moved in the same direction and with the same order of magnitude as average income then the two variables have a correlation coefficient of 1; in practice anything over 0.6 suggests high correlation. For diversification, low or negative correlation between investments is required.

A beta value (β) indicates the volatility of an investment relative to its group, e.g. if the commercial property index shows capital values falling 10 per cent in one year but within that office values fell

20 per cent, then offices are more volatile with a β of 20/10 = 2. If shop values only fell by 8 per cent over the same period, the β value is 0.8 – lower volatility than the asset class.

In an efficient market prices instantly reflect all public knowledge and so no abnormal returns can consistently be made. It is only possible to achieve high return by taking on high risk, and management of risk therefore becomes important. MPT develops formal ways of diversifying portfolios to achieve chosen risk/return profiles. A more informal approach is 'naïve diversification', i.e. don't put all your eggs in one basket.

Markowitz developed the theory using two assets. First we define the return and risk expectation of each available investment asset. Then if we split our money between two assets, we can work out the risk and return of the resulting portfolio, based on risk and return of the two component assets. We should look at expected return and risk; practically, investors tend to use:

- past returns as a proxy for future expected returns; and
- the standard deviation of past returns as a proxy for future expected risk.

The return of a portfolio with assets A and B depends on:

- return of A, proportion invested in A;
- return of B, proportion invested in B.

For example, two assets A and B provide a range of portfolio options, with returns ranging from 10 per cent up to 18 per cent (Table 8.5.1).

The risk of a portfolio with money distributed between assets A and B depends on:

- risk of A, proportion invested in A;
- risk of B, proportion invested in B;
- correlation coefficient of A with B.

The correlation coefficient measures how closely linked the returns of the two assets are. Correlation coefficients always lie in the range –1 to +1:

- if the coefficient is +1, it means the returns move exactly in step: perfect positive correlation;
- between 0 and +1 means the returns are linked, the nearer to 0 the weaker the link: positive correlation;
- if the coefficient is –1, the returns move exactly oppositely to one another: perfect negative correlation;
- between –1 and 0 means the returns have an inverse link, weaker the nearer the coefficient is to 0: negative correlation.

The lower the correlation between assets' returns, the better the effects of diversification. Ideally we would combine negatively correlated assets, but most investments are tied into the general economy, so can only find low positive correlations; however, these still offer the possibility of managing risk and return by diversification.

Table 8.5.1 A two-asset portfolio

All A	*70% A*	*50% A*	*No A*
No B	30% B	50% B	All B
18%	15.6%	14%	10%

When assets are combined in a portfolio, the expected return of the portfolio is the weighted average of the expected returns of the component assets; the risk is determined by the covariance structure of the component asset returns. The way in which asset returns covary is central to portfolio risk: it provides diversification opportunities.

The risk of the portfolio can be lower than the average risk of the constituent assets. Even adding a risky but negatively correlated asset to a portfolio can reduce portfolio risk. A high-risk portfolio will offer high returns and a low-risk portfolio will offer low returns. For any given return level, an optimal portfolio (one with the lowest risk for that return) can be found.

There will be combinations of assets that are not optimal, meaning that by using different asset combinations it is possible to get extra return for the same risk, or to have less risk for the same return. By eliminating the suboptimal points it is possible to construct the efficient frontier (see Figure 8.5.1).

The choice of the optimal portfolio along the efficient frontier depends on the investor's trade-off between risk and return. Markowitz's theory of diversification was developed for stock markets in order to design a portfolio to manage risk and return. It does have relevance for property, showing:

- how to use property to diversify a mixed asset portfolio;
- how to diversify within a property portfolio.

There are, however, problems applying the theory because of property's data and investment characteristics. Defining 'the property market' and its sectors/regions is difficult, as is getting returns data to identify risk/return for property sectors (office, retail, industrial). It is usual to use IPD data, but it relates only to the 'institutional' part of the market, which is dominated by prime and some good secondary property.

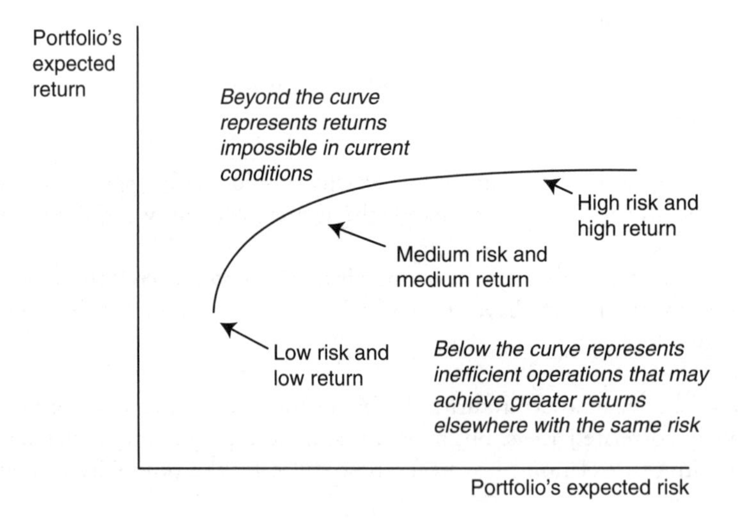

Figure 8.5.1 The efficient frontier.

Further reading

Baum, A. and Hartzell, D. (2012) *Global Property Investment: Strategies, structures, decisions*, Wiley-Blackwell, Oxford, ch. 4.
Markowitz, H. M. (1952) 'Portfolio selection', *The Journal of Finance*, 7(1): 77–91.

8.6 Capital asset pricing model

Key terms: investment portfolio theory; beta value; risk and return; systematic risk; asset specific risk
The capital asset pricing model (CAPM) considers systematic or market risk and asset specific risk. It is a theoretical way of determining the required rate of return for an asset to be added to an already diversified portfolio – given that asset's market risk (see Concept 8.5). Risk is considered to be the sensitivity of the asset return to changes in market returns represented by the quantity beta (β). If the return on the asset is highly sensitive to a change in market returns it will increase the volatility and risk of the portfolio – and so it requires a higher risk premium in its return.

CAPM puts a numerical value (β) on this sensitivity, allowing calculation of this required return. If an asset increases the volatility of the portfolio, it increases the risk and so increased returns are necessary if it is to be included. Stocks with a β of one have the same risk as the market because their returns move in step with the market. Stocks with a β of more than one are more volatile/risky than the market (sometimes called 'aggressive stocks') and stocks with a β of less than one are less volatile/ risky than the market (sometimes called 'defensive stocks').

So the beta value measures the volatility of an investment in relation to a market portfolio – which would be a portfolio of all known assets weighted in terms of market value. A β of 1.0 means that if the market increases by 20 per cent, the expected value of the new investment increases by 20 per cent. A β of 2.0 means that if the market increases by 6 per cent, the expected value of the new investment increases by 12 per cent. A β of 0.5 means that if the market decreases by 10 per cent, the expected value of the new stock decreases by 5 per cent.

The expected return on the market portfolio (E(Rm)) should be higher than the risk-free rate of return (RFR). It comprises the RFR plus an expected risk premium (E(Rp)):

$$E(Rm) = RFR + E(Rp)$$

The risk premium depends on the β:

$$E(Rm) = RFR + \beta \text{ x } [E(Rm) - RFR]$$

If we assume a risk premium of 3 per cent and a risk-free rate (say the rate available on government bonds) of 4 per cent, the expected return on the market is:

$$E(Rm) = 0.04 + 0.03 = 0.07 \text{ or 7 per cent}$$

The return on a risky investment should comprise the risk-free rate plus a risk premium that reflects the systematic risk of the investment relative to the market. If an investment is more risky than the market, it should earn a higher risk premium. β measures the risk and so the return on a risky investment (y) is:

$$E(Ry) = RFR + \beta \times (Rp)$$

If the risk premium is 3 per cent, the risk-free rate is 4 per cent and the asset β is 2, then the expected return on the asset is:

$0.04 + 2 \times 0.03 = 0.10$ or 10 per cent

Relating this to property investments, suppose research gives us estimated betas of:

Offices $\beta = 1.5$

Shops $\beta = 0.8$

Industrials $\beta = 0.7$

Then if the risk-free rate of return is 4 per cent and the risk premium is 3 per cent, the expected return on an asset in the different sectors is:

Offices: $ro = RFR + \beta(Rp) = 0.04 + 1.5(0.03) = 0.085$ or 8.5 per cent

Shops: $rs = RFR + \beta(Rp) = 0.04 + 0.8(0.03) = 0.064$ or 6.4 per cent

Industrial: $ri = RFR + \beta(Rp) = 0.04 + 0.7(0.03) = 0.061$ or 6.1 per cent.

This looks straightforward but the problem with CAPM applied to property is that it is very difficult to calculate returns (which include change in capital value) on an individual property or even a portfolio on a frequent basis. REITs (see Concept 8.8) are actively traded on stock markets, however, so there can be some useful application of the theory with these assets.

Further reading

Baum, A. and Hartzell, D. (2012) *Global Property Investment: Strategies, structures, decisions*, Wiley-Blackwell, Oxford.

Brown, G. R. and Matysiak, G. A. (2000) *Real Estate Investment: A capital market approach*, Prentice Hall, Upper Saddle River, NJ.

8.7 Risk and return

Key terms: risk–return trade off; risk averse; risk taking; volatility; variance; standard deviation; risk-adjusted return

Investors look at both return and risk when evaluating investments. Return is usually expressed as a percentage per annum of the capital cost of the investment. They prefer a high return to a low return, given equal risks; some may choose lower risk in preference to high return. Attitudes to risk-taking vary, but in general, low risk is preferred to high risk, given equal returns. The risk–return trade off is shown in Figure 8.7.1. The risk-free rate of return might be the rate available on government bonds (gilts).

In Figure 8.7.1 point A gives low return with low risk and point B high return and high risk. A risk-averse investor will require a larger increase in return to take on more risk – and a risk taking investor will accept more risk in order to achieve an increase in return than a risk averse investor.

There are many sources of property investment risk:

- risk to income flow (tenant risk);
- risk of increased future costs (e.g. because legislation changes);
- risk of capital value changing (it varies with income flow and outgoings and yields).

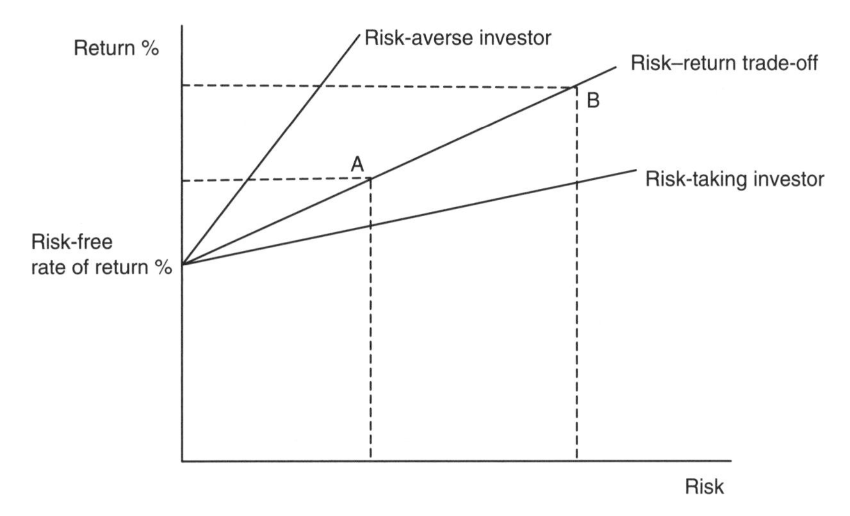

Figure 8.7.1 The risk–return trade-off

The risk profile of an investment is made up of two factors: the risk associated with assets of that class or subclass – systematic risk, usually called the beta factor; and the risk of that particular asset itself – unsystematic risk called alpha. Risk is measured by volatility of returns over a given period and volatility of returns is measured by their variance or standard deviation (which equals the square root of the variance).

A risk-adjusted return (RAR) shows the amount of return per unit (percentage) of risk. So if retail return is 11.3 per cent and risk is 7 per cent (standard deviation), then retail RAR would be 11.3/7 = 1.61 units of return per unit of risk. This can be compared to the RAR of other sectors (Table 8.7.1). And it can be seen from the table that retail investment gives the highest RAR.

Table 8.7.1 Risk-adjusted return

	Risk-adjusted return
All Property	1.2
Retail	1.61
Office	0.87
Industrial	1.24

Further reading

Isaac, D. and O'Leary, J. (2011) *Property Investment Markets*, Palgrave Macmillan, Basingstoke, ch. 9.

8.8 Real estate investment trusts

Key terms: tax efficiency; liquidity; capital appreciation; yield; diversification

Individual investors in property usually have insufficient funds to buy prime property, which has shown the greatest capital growth and least-risk income returns. Real estate investment trusts (REITs) are a way around this problem. Already well established and very popular in many countries overseas such as Australia, the USA, Germany and Japan, REITs were launched in the UK on 1 January 2007. A REIT is a quoted company that owns and manages income-producing property on behalf of shareholders. A REIT can contain commercial and residential property. Most of its taxable income (at least 90 per cent) is distributed to shareholders through dividends, in return for which the company is largely exempt from corporation tax (see Concept 17.3). REITs are designed to offer investors income and capital appreciation from rented property assets in a tax-efficient way, with a return more closely aligned with direct property investment. This is achieved by taking away double taxation (corporation tax plus the tax on dividends) of ordinary property funds. The introduction of UK REITs created an opportunity for a wide range of investors to invest in property as an asset class by creating a more liquid and tax-efficient method.

REITs provide a way for investors to access property assets without having to buy property directly. In the UK, REITs can apply for UK-REIT status, which exempts the company from corporate tax. The introduction of a UK-REIT regime, combined with the traditional strengths of London's capital markets, created opportunities for the growth of the property investment sector. A UK-REIT enjoys all the benefits of any other company, in addition to the tax advantages. UK resident companies listed on a recognised stock exchange are eligible for REIT status. Companies that qualify as REITs are not subject to corporation tax on their qualifying rental income or chargeable gains. A conversion charge applied to companies adopting REIT status, equal to 2 per cent of the market value of their investment properties at the date of conversion, but this was scrapped in 2012. Nine UK property companies converted to REIT status in 2007, including five that were FTSE 100 members at that time: British Land, Hammerson, Land Securities, Liberty International and Slough Estates (now known as SEGRO). The other four were Brixton, Great Portland Estates, Primary Health Properties and Workspace Group (Jowsey, 2011).

REITs receive special tax considerations and typically offer investors high yields, as well as a highly liquid method of investing in real estate. Equity REITs invest in and own properties (thus responsible for the equity or value of their real estate assets). Their revenues come principally from their properties' rents.

REITs offer capital appreciation along with yield, and so provide more diversification to an investment portfolio. REITs are considered to have a low correlation with equities, meaning they do not normally move in the same direction as equity stocks or bonds. Since their introduction in 2007 in the UK, REITs have shown disappointing returns (as might be expected given market conditions) but they have performed much better in the last few years and they allow much greater liquidity than direct investment in property with the same effective tax treatment.

Further reading

Jowsey, E. (2011) *Real Estate Economics*, Palgrave Macmillan, Basingstoke, ch. 16.

8.9 Property unit trusts

Key terms: collective investment schemes; offer price; bid price; diversification

Property unit trusts (PUTs) are collective investment schemes where the underlying properties are held on trust for the participants. They are unit trusts that specialise in property (for example, Cornhill

Property Share specialising in emerging property markets and off-plan residential investments), but in order to avoid management commitments, such unit trusts use their funds to buy shares in property companies or in companies such as hotels that are concerned with property.

PUTs operate on a similar principle to REITs as 'open-ended' collective investment funds that can continually accept new cash and use it to buy more property assets. Closed ended collective investment funds are where investors pool money to buy one or more property assets but after that no further investment takes place. Unit trusts that specialise in property avoid management commitments using funds to buy shares in property companies or companies concerned with properties (e.g. hotel groups). The underlying assets of a PUT are mainly direct property investments; some PUTs also invest in other indirect property vehicles (i.e. other PUTs or property company shares or bonds).

An operator (usually a bank) sets up a Trust, which holds a group of properties. Units in the Trust are bought, often by small pension funds or by private investors, who cannot afford direct property investment but want exposure to property returns. Authorised PUTs (APUTs) have stricter Financial Services Authority (FSA) rules to protect small, 'retail' (i.e. non-professional) investors. Non-authorised PUTs are available only to 'professional' investors such as pension funds.

Investors can 'redeem' their units if they wish at the price set by trustees, based on periodic revaluation (often monthly). The offer price equals the price to buy a unit; and the bid price equals price on sale back to the trust (always lower). The difference between bid and offer prices is called the 'bid–offer spread' and covers costs of administration and 'round trip' investment costs (stamp duty, legal fees etc. on buying and selling the properties).

Some examples of specialist funds are:

- Airport Industrial Property Unit Trust
- Henderson Central London Office Fund
- Standard Life UK Retail Park Trust
- The Residential Property Unit Trust
- The UK Logistics Fund.

By buying units in specialist vehicles, investors can manage their own diversification – or diversify against their direct property holdings.

Further reading

Ball, M., Lizieri, C. and MacGregor, B. D. (1998) *The Economics of Commercial Property Markets*, Routledge, London, ch. 12.

8.10 Active fund management

Key terms: passive management; active management; Investment Property Databank; structure setting; stock selection

Investment assets can be managed either actively or passively. Passive management involves setting up a portfolio of assets with similar characteristics that are traded to maintain these characteristics, such as index funds. These funds require very little management, therefore have low management costs, and have guaranteed market performance due to the minimal risk involved in their management.

Property, however, is heterogeneous, and therefore requires active management in order to increase performance and control risk. Active management assumes that the market contains inefficiencies that can be exploited by research, and enables added value to be realised by outperformance of the market. Active management dominates all investment markets.

The purpose of active property fund management is primarily to increase rental income and therefore create capital value. This creates market evidence for the rental and investment markets, and it can improve the potential market value of an area by, for example, improving the tenant mix and therefore the perception and market performance of an area. The remit of active management is to outperform, and so therefore the fund manager will set a performance target against a benchmark, such as 12-month all property returns published by the Investment Property Databank.

Approaches to active property fund management include structure setting, stock selection, and asset management.

Structure setting involves assessing how much to invest in each category of asset based on performance by reference to data such as the Investment Property Databank Monthly Property Index. This type of management is known as *naïve diversification* – the fund is only concerned with the type of stock, rather than individual properties themselves. It is a strategy adopted by larger funds, but the disadvantage being that one cannot reasonably expect to consistently select properties with above average expected returns in this way and it is more of a speculative and longer-term strategy.

Stock selection focuses on assessment of the individual assets within the portfolio, whereby fund managers will target assets with above average expected performance. This is a high risk strategy for smaller funds whereby the specific risks of particular impacts can have much greater impacts on income. *Traditional management* is interested only in stock selection, and there is little explicit analysis of the structure of the portfolio or regard to outperformance of a benchmark. Properties are selected based purely on their individual over average performance.

Finally, *asset management* adds value to existing stock by improving the asset through means such as lease re-gearing, refurbishment, redevelopment etc. to maximise returns.

Modern management involves a combination of stock and structure analysis to inform asset selection and active management to ensure that the maximum income is being realised by the portfolio. Value is added by positioning against a benchmark, actively managing individual stock, and using pricing models to identify individual assets to buy and sell. At the outset, a clear statement of objectives details the benchmark, expected outperformance (overall return) within a specified timescale, and an indication of anticipated risk. The structure of the portfolio is then analysed before individual assets are considered and an active management strategy is then established for each individual property and how that will contribute to the overall income and value of the portfolio.

Individual properties can be actively managed in a number of ways: lease re-gear; surrender and re-let; refurbish and re-let; attract/improve the quality of tenants; or change the use of a property.

Lease re-gearing is usually carried out towards the end of a tenancy or when a break option is approaching. In order to maintain income, the landlord might offer an incentive, such as a rent-free period, or the forfeiture of a break clause, in order to retain the tenant for a longer term.

Surrender and re-let is used when a tenant wishes to vacate all or part of their demise while still within the term of the lease, for example if a business is struggling financially. In order to do so, the tenant must pay a capital sum to the landlord to compensate for the loss in income and for any empty rates that may be payable for the vacant property. The premium paid to the landlord will greatly depend on the re-letting prospects of the property, however the potential benefits to the landlord are to secure a tenant on superior terms and increased income stream.

Refurbishment is typically carried out when a tenant is vacating and provides an opportunity to upgrade and modernise existing premises in order to command a higher rent. This is most relevant for office properties where there is a high rate of obsolescence. It is the balance of the cost of the refurbishment versus the uplift in rental value and how easily the refurbished space will be re-let, since better quality space is more likely to attract better covenants and longer lease terms.

Tenant mix is particularly important for retail parades and shopping centres, but also relevant to other sectors. The landlord will seek to attract the best-quality tenants to improve the overall quality of the tenant mix in order to maintain strong covenants willing to pay higher premiums. However, higher quality retailers will require significant incentives to relocate, and so there must be a careful balance to ensure that the short-term loss of income from a rent-free period, for example, is balanced by a long-term uplift in income on superior terms.

Changing the use of a property is another way that income and value can be increased. The high capital outlay for converting a property from say an office to a residential property must be balanced by a significant increase in income. There is also the issue of negotiating and securing planning permission which can significantly delay the benefits of the conversion project and therefore change of use is a relatively high-risk management strategy. However, the rewards can be significant: residential values can be up to double those of commercial, depending on location and specification, for example in the West End of London.

Further reading

Ball, M., Lizieri, C. and MacGregor, B. D. (1998) *The Economics of Commercial Property Markets*, Routledge, London, chs. 10 and 11.

Baum, A. (2009) *Commercial Real Estate Investment:A strategic approach*, 2nd edn, Routledge, Abingdon, chs. 2 and 5.

Hoesli, M. and McGregor, B. D. (2000) Property *Investment Principles and Practice of Portfolio Management*, Longman, Harlow, ch. 9.

Isaac, D. and O'Leary, J. (2011) *Property Investment*, 2nd edn, Palgrave Macmillan, Basingstoke, chs. 6, 7 and 9.

8.11 Residential property investment and buy-to-let

Key terms: rental income; capital value; mortgage; landlords; mortgage interest rates

Residential property investment is undertaken to some extent by everyone who buys a house or apartment. The decision to buy confers consumption benefits (from having somewhere to live) and investment benefits as the value of the property increases over time. Since home ownership in the UK has reached about 67 per cent of all households, most people have some experience of residential property investment. On the whole these investments are very satisfactory – as house prices rise over time people are better off in terms of the assets they own. A few of these investors get into difficulties with defaults on mortgage payments and negative equity if they buy shortly before a market downturn; but for most home buyers the investment is a profitable one.

In other parts of the world, investing in apartment blocks as part of a portfolio including offices, retail properties, industrial properties and hotels is quite common. In the UK this has not been the case and most residential investment has been by small investors rather than by institutions. This may be because of the incentives that have been offered to home ownership by successive governments. One area where investors do branch out beyond home ownership is into the buy-to-let market.

Buy-to-let investment is the investment strategy of buying a residential property to be let for profit. The benefits for a buy-to-let landlord can include a stable income from rental receipts, as well as a capital gain if house prices go up over time. Rising house prices in the UK have made buy-to-let a relatively safe way to invest in normal times. The main risk occurs where the landlord takes a loan to buy the property, with the expectation that rental income will meet or exceed the cost of the loan and that the house can be sold later for a higher price. If the landlord cannot meet their mortgage repayments, however, then the bank will 'repossess' the property and sell it to try to get their money back. The broad popularity of buy-to-let investments has made a large number of new (inexperienced) landlords in the UK.

The 1988 Housing Act made investment in residential property more attractive to landlords when it introduced a new type of tenancy giving landlords more control over their properties, and there has been a substantial recovery in the private rented sector since then. The availability of loans for buy-to-let purchasers has also increased the appeal of owning rental property.

Since the mid-1990s in the UK there has been a surge in demand for rental property. Low interest rates and soaring house prices helped fuel the appeal of buy-to-let, together with changes in legislation which made it easier for landlords to deal with sitting tenants. Historically, buyers of property to let were surcharged or forced to borrow at commercial rates and potential rental income was not taken into account for servicing the borrowings. Since the introduction of the Assured Shorthold Tenancy in the mid-1990s, however, rights have changed more in favour of the landlord and mortgage lenders have been more willing to provide finance.

This led to a rapid expansion in the amount of mortgage finance available with schemes specifically designed for amateur and professional landlords that became known as 'buy-to-let mortgages'. Lenders supporting buy-to-let schemes brought their interest rates into line with the rates for owner occupation and took rental income into account for servicing the loan. This policy switch was encouraged because professional letting and property management agents were more involved in the selection of suitable properties for the rental market, in the selection of tenants and in the management of properties. This enhanced the security and creditworthiness of buy-to-let investments. From only 4 per cent of mortgages in 2000, buy-to-let mortgages increased in number to 29 per cent of all mortgages in 2006, according to the Council for Mortgage Lenders website. At the height of the boom in property and buy-to-let investment in 2006–7, some investors bought property more as 'buy-to-leave' – relying solely on the appreciation in value of the property and not renting it out. Clearly if this type of activity became commonplace it would have consequences for communities where there were a number of empty properties, and would almost certainly increase the problem of homelessness.

The credit crunch recession, however, has caused most UK lenders to limit buy-to-let mortgages. The part of the buy-to-let market hardest hit by the downturn of 2006–8 was that for new inner-city flats. These were often purchased with high mortgages taken out using inflated valuations. This risky and highly speculative part of the market ended with the downturn in prices for city-centre apartments from 2006 onward.

Further reading

Jowsey, E. (2011) *Real Estate Economics*, Palgrave Macmillan, Basingstoke, ch. 16.
www.cml.org.uk/cml/consumers/guides/buytolet (accessed 05/12/2013).

8.12 Mortgage-backed securities

Key terms: securitisation; asset-backed security; collateralised debt obligation; sub-prime lending; repossessions; foreclosures

Securitisation is the process of creating securities by pooling together various income-producing financial assets. These securities are then sold to investors. Any asset may be securitised as long as it produces an income. The terms asset-backed security (ABS) and mortgage-backed security (MBS) reflect the assets behind the security.

A MBS is an ABS that is secured by a parcel of several mortgages. The mortgages are held by or sold to a financial institution that packages the loans together into a security that can be sold to other financial institutions or investors. Shares of this security, called tranches are sold to investors who buy them and ultimately collect the dividends in the form of the monthly mortgage payments. These tranches can be further repackaged and sold again as other securities, called 'collateralised debt obligations' (CDOs).

The mortgages of a MBS can be residential or commercial (CMBS). Residential MBS carry a risk of prepayment. This is where the mortgage is repaid early, thus depriving the MBS holder of some of the interest. This is less likely to happen with CMBS because commercial mortgages are more rigid in their terms and conditions.

Of course, if mortgage providers are selling on their mortgages in this way they are not at risk if the borrower defaults on their repayments. There is then a temptation to issue mortgages without due diligence. As a result, many mortgages were issued in 2004–7 in the United States to borrowers with poor credit histories, and possibly even to unemployed borrowers. This has become known as 'sub-prime' lending and has been blamed for the credit crunch crash and recession of 2008–9. A large proportion of the mortgages taken out by sub-prime borrowers were adjustable rate mortgages. These loans had a discounted fixed interest rate for a number of years and then reverted to a higher rate. A homeowner could find the monthly payments doubled after the rates adjusted and defaults and ultimately repossessions ('foreclosures' in the US) increased dramatically. As a result, the average CDO/MBS lost about half of its value between 2006 and 2008 and the most risky sub-prime ones became worthless.

Since then in the US there has been a wave of repossessions in the housing market and the first national decline in house prices since the 1930s. The chronic depletion of capital in banks involved with mortgage lending or MBS has meant a massive reduction in house buying. The Federal National Mortgage Corporation and the Federal Home Loan Mortgage Corporation (known as Fannie Mae and Freddie Mac), the government-sponsored mortgage corporations, made losses of at least $15 billion. This led to a severe recession in the US and because the MBS were traded around the world, banks in many other countries also suffered severe losses and the financial collapse and recession became widespread (see Concept 6.23).

Further reading

Jowsey, E. (2011) *Real Estate Economics*, Palgrave Macmillan, Basingstoke, ch. 6.

8.13 Land banking

Key terms: undeveloped land; planning permission; speculative holding; land hoarding; externalities

In the UK land banking refers to the practice of buying undeveloped land with or without planning permission and retaining it until planning permission is granted or the market conditions are right. Many reputable building companies engage successfully in land banking for future building projects and regard it as necessary in order to achieve continuity in their business. Building companies and developers tend to keep land banks for several years in order to secure housing supply needs. Some of this land will be held speculatively in the hope that planning restrictions will be relaxed in the future. They can use the land as collateral to obtain loans and they need a reserve of land to remain in business; if they ran out of land they would have to buy from competitors at higher prices. Unfortunately, the term land banking has also been used to describe two other situations that are more disreputable.

The first such situation we can describe as 'land hoarding', where a company retains undeveloped land with planning permission in the expectation that its value will rise over time. The practice has been highlighted in recent years in London where housing shortages persist and yet developers are sitting on a large number of sites with planning permission. Speculative hoarding of government released land is regarded as one of the main obstacles to achieving housing affordability in Australia and the government there is monitoring the holding of large parcels of undeveloped land by foreign owners. In the United States, land banking refers to the practice of counties or cities buying up derelict land that is a blight on neighbouring areas.

The second situation is a 'scam' where someone buys a large area of undeveloped land and divides it up into smaller plots to sell on to unsuspecting investors at much more than market value. The schemes claim that the plots have good investment value as big profits will be made in the expectation of future development when planning permission is granted.

The Land Registration Act of 2002 enabled companies to purchase land sites and easily divide them into smaller plots. The company can then offer these plots for sale to individual investors, often claiming that they will make considerable returns. This practice in the UK does not fall under the control of the FSA, although collective investment schemes are regulated. Typically, land bankers buy land for less than £10,000 an acre and divide it into plots of 0.1 acres, selling these for £8000 to £20,000 each, making profits of between £80,000 and £200,000 an acre. The selling companies often go into liquidation leaving the buyers with virtually unsaleable plots. In reality, the land may be totally unsuitable for development and have little hope of ever getting planning permission. The land often ends up abandoned and neglected.

In 2008 the FSA ordered UK Land Investments to be wound up as it constituted an unregulated collective investment scheme. UKLI owed £70.8 million to creditors and 4,500 people had bought 5,000 plots of land. The same owner operated Regents Land and St James's Land, which continued trading. Several other land banking companies have been put into compulsory liquidation by the Insolvency Service Companies Investigation Branch. As a result of increased regulation, many land banking companies – some still selling UK plots – have moved outside the European Union.

Such schemes do more than trick investors out of their money; they lead to considerable costs or externalities for society. The land can no longer be used for farming or forestry because of the multiple ownership. No one is responsible for its maintenance and the plot owners have no interest in its use or management. The land will slowly become derelict and subject to unlawful activities such as fly-tipping. Local authorities find it difficult to enforce clean up or to take action against the landowners because of multiple ownership of small plots. Recombining the plots in order to make the land usable again is extremely difficult unless some form of compulsory purchase can be used.

Further reading

Jowsey, E. (2011) *Real Estate Economics*, Palgrave Macmillan, Basingstoke, ch. 16.

8.14 Property indices

Key terms: Investment Property Databank (IPD); property index; weighting; base year

An index number is a number showing the size of some variable relative to a given base (usually 100). It is a method of expressing the change in a number of variables through the movement of one number. The procedure is: select a base (year) and give it a value of 100 – express all subsequent changes as a movement of this number.

Because the variables differ in relative size/value within the data set, they need to be 'weighted' according to their relative importance, e.g. in the IPD UK Annual Property Index all property has a value of 100; within that, retail has a value of 50.7, office of 29.5 and industrial of 14.5, reflecting the relative importance of these sectors within the UK property investment market. The index number is usually a weighted average of the index numbers for the various components:

- a base weighted index (Laspeyres Index) uses weights from the base period;
- a current weighted index (Paasche Index) uses weights from the current period.

There is a trend to create passively managed mutual funds based on market indices – index funds, and it is believed these beat actively managed funds (see Concept 8.10). Index funds attempt to replicate the holdings of an index and so they remove costs of active management and have a lower 'churn rate' (turnover of stock/securities). Typically, pension funds will set the IPD annual return as a benchmark, that is the widely accepted return available from the market.

Tables 8.14.1 to 8.14.4 show the IPD UK Annual Property Indices for 2010.

Table 8.14.1 IPD indices

Capital value (in portfolio) (%)	*Total return*	*Income return*	*Capital growth*
Retail 50.7	16.0	6.2	9.3
Office 29.5	15.8	6.3	9.0
Industrial 14.5	10.8	7.2	3.3
Residential –	–	–	–
Other 5.3	15.0	6.3	8.2

Table 8.14.2 IPD indices – capital growth of all property 2010

Sector	*Capital growth*	*Weighting (%)*	*Capital growth × weighting*
Retail	9.3	50.7	4.7
Office	9.0	29.5	2.6
Industrial	3.3	14.5	0.5
Other	8.2	5.3	0.4
			Total +8.2%

Table 8.14.3 IPD indices – income growth of all property 2010

Sector	*Income growth*	*Weighting (%)*	*Income growth × weighting*
Retail	6.2	50.7	3.1
Office	6.3	29.5	1.8
Industrial	7.2	14.5	1.0
Other	6.3	5.5	0.3
			Total +6.2%

Table 8.14.4 IPD indices – total return of all property 2010

Sector	*Total return*	*Weighting (%)*	*Income growth × weighting*
Retail	16	50.7	8.1
Office	15.8	29.5	4.6
Industrial	10.8	14.5	1.5
Other	15.0	5.3	0.8
			Total +15.0

To re-base an index

Total return index 1980 = 100 (this is the base year; 2009 = 1219.4; 2010 = 1403.4).
It may be that you think the base year (IPD 1980) is getting a little distant in time. If you want to re-base the index, then for each year:

$$\frac{\text{Old base year}}{\text{Required base year}} \times 100 = \text{new index value}$$

So if 2010 is now the base year (=100) calculate the index value for 1980 and 2009:

$$1980: \quad \frac{100 \text{ (old base year figure)}}{1403.4 \text{ (required base year 2010)}} \times 100 = 7.12$$

$$2009: \quad \frac{1219.4}{1403.4} \times 100 = 86.8$$

$$2010: \quad \frac{1403.4}{1403.4} \times 100 = 100$$

Further reading

See: IPD.com/UK (accessed 27/11/2013).

8.15 Discounting and discount rates

Key terms: present value; compound interest; discount rate; long-term projects

Discounting is the valuation in present day terms of the future costs and benefits arising from a project. It progressively reduces the present value of future costs and benefits. The reduction is greater as distance into the future increases and as the discount rate increases. Discounting is done because people have a positive time preference – they prefer benefits now rather than later, perhaps because of impatience or because they expect to be richer in future. A further justification of discounting comes from the productivity of capital – money makes more money. £1 invested now will be worth £1 plus interest by the end of the year. So equal money amounts have lower present values (PVs) the further into the future they are received.

Money received now produces more money in the future, so an equal amount of received money in the future must have a reduced present value. For example:

– Suppose £1000 is invested for one year at an interest rate (i) of 10%. This will be £1100 at the end of the year.

£1000(1.1) = £1100 or

$PV(1 + i) = B_1$

where PV = present value; i = interest rate; and B_1 = benefits at end of year.
– At the end of 2 years, the £1000 becomes, with compound interest:

£1000 (1.1)(1.1) = £1210

and $PV(1 + i)(1 + i) = B_2$

or $PV(1 + i)^2 = B_2$

And to discount:

$$PV = \frac{B^2}{(1+i)^2}$$

$$\left[\text{Check } £1000 = \frac{£1210}{(1.1)^2}\right]$$

The future value of benefits at end of year 1 can be related to its present value (discounted) by rewriting as:

$$PV = \frac{B_1}{(1+i)}$$

$$\left[\text{Check } £1000 = \frac{£1110}{1.1}\right]$$

A general discounting equation relating benefits in any year (t) to their present value is:

$$PV = \frac{B^t}{(1+r)^t}$$

For an investment yielding benefits over many years (t=0, 1, 2, … n)

$$NPV = \frac{B_0}{(1+r)^0} + \frac{B_1}{(1+r)^1} + \frac{B_2}{(1+r)^2} + \ldots + \frac{B_t}{(1+r)^t}$$

The further into the future we go, the discount factor increases, causing PV to fall; and as the discount rate increases, PV decreases. A high discount rate for returns far in the future significantly decreases present values. This can be appropriate as risk increases the further into the future the income stream goes, but it disadvantages long-term projects such as infrastructure that is expected to last for decades.

Further reading

Isaac, D. and O'Leary, J. (2011) *Property Investment Markets*, Palgrave Macmillan, Basingstoke, ch. 5.

8.16 International property investment

Key terms: diversification; information costs; transaction costs; prime property; transparency; investable stock

International real estate investment increased rapidly in the 1980s with globalisation of commerce and multinational business operations. Financial institutions became landlords for prime commercial property. Growth in international investment practices enabled investors to look outside their own countries for better performing properties. Floating exchange rates from 1971 increased foreign exchange risks and led to more international financial instruments, e.g. currency futures. Telecommunications and computers enabled stock, option and commodity exchanges to trade globally, creating a global financial market (24 hour and instantaneous). Deregulation of financial markets in the 1980s allowed global borrowing and lending and increased diversification of investment. 'Prime' investment quality real estate began to appear after 1980 in cities throughout the world – not just London and New York.

The rapid growth of cross-border investment in commercial property took two forms: direct foreign investment in the private property market to receive a rental income stream and capital appreciation (sometimes as a joint venture with a company in the host country); or indirect investment in publicly traded vehicles, e.g. property company shares, funding of property projects and purchases of securitised debt (Jowsey, 2011).

Investors consider real estate to be an important component of their global portfolios. Real estate is regarded as an illiquid, long-term asset class, more suited to investors without short-term liabilities. The stable income streams from property rents or dividends from listed real estate securities appeal to pension funds and sovereign wealth funds (government investment funds). There is diversification potential, given the low correlation of real estate with other asset classes, such as equities and bonds.

Measuring international real estate investment flows can be difficult because although records are kept for direct international property investment, they are not for special purpose vehicles and some corporate acquisitions are recorded as domestic transactions. Corporate sector real estate investment takes place for operational or strategic reasons. The main problems with international property investment are:

- information costs – these can be considerable because of different legal systems and variations in methods of valuation between countries (Jones Lang Lasalle compiles a Global Transparency Index to combat this);
- cultural barriers and possible restrictions or regulations concerning foreign ownership;
- management and maintenance costs;
- adverse currency movements that affect investment performance;
- international variations in transaction costs that can be considerable;
- political instability affecting rents and capital values;
- market inefficiency – especially in smaller emerging markets.

International property investment may offer higher returns than those available in the national market of the investor and there may not be sufficient opportunities for high quality real estate investment in the home market. International real estate investment also takes place in order to diversify investment portfolios although there is now evidence that the correlation between many international real estate markets is surprisingly high and international property returns move in the same direction, at least in developed countries. This is because real estate is affected by much the same fundamental economic variables that affect GNP. The near simultaneous downturns in the office markets of major financial centres such as New York, London and Toronto in the early 1990s and in 2007 confirm this. There is also a very close correlation between changes in nominal rental values for offices in major North American and European cities.

The largest global real estate investors are pension funds, insurance companies and sovereign wealth funds. Real estate is estimated to be 50 per cent of the total value of world assets but that is not investable stock (stock that is of sufficient quality to become the focus of institutional investment). DTZ in their 2012 *Money into Property* global report estimated total investable stock at $21 trillion (US dollars), of which $8.6 trillion is owner-occupied and $12.3 trillion is invested in. The individual countries with the greatest investable stock of real estate in terms of world share are the US with 30 per cent, Japan with 17 per cent and the UK with 11 per cent. The sectoral breakdown of investable real estate assets is: 34.8 per cent offices; 25.1 per cent apartments; 23.9 per cent retail; 14.3 per cent industrial; 1.9 per cent hotels.

Further reading

DTZ (2012) 'Money into Property', available at: http://www.dtz.com/Global/Research/Money+into+Property+2012+UK (accessed 03/06/2014).

Jowsey, E. (2011) *Real Estate Economics*, Palgrave Macmillan, Basingstoke, ch. 18.

8.17 Transparency index

Key terms: transparent markets; market fundamentals data; environmental sustainability

International property markets vary in the amount of accurate information that is available to prospective investors and corporations seeking property for business. Some markets are 'opaque' in the sense that there is very little useful information about investable property in that country. As might be expected, the most transparent markets with easily available accurate information are to be found in developed countries. Jones Lang LaSalle (JLL) and LaSalle Investment Management produce a Global Real Estate Transparency Index, which is a unique biennial survey covering 97 markets worldwide. It aims to help real estate players understand important differences when transacting, owning and operating in foreign markets.

The 2012 listing of the most transparent countries is shown in Table 8.17.1.

Table 8.17.1 2012 Transparency Index

2012 Ranking	*Country*
1	United States
2	United Kingdom
3	Australia
4	Netherlands
5	New Zealand
6	Canada
7	France
8	Finland
9	Sweden
10	Switzerland

Source: joneslanglasalle.com

The transparency index is compiled by assessing the following areas in each country:

- direct property indices
- listed real estate securities indices
- unlisted fund indices
- market fundamentals data
- financial disclosure
- corporate governance
- regulation

- land and property registration
- eminent domain
- sales transactions
- occupier services
- debt regulation
- valuations.

JLL show that transparency improved in more than 90 per cent of countries between 2010 and 2012. Improving market fundamentals data and performance measurement, combined with better governance of listed vehicles, have led to transparency progress over the 2 years. Much improved performance data have been made available in Brazil, emerging Asia, Mexico, and central and eastern Europe (CEE). Market fundamentals data have been enhanced in most markets (joneslanglasalle.com). There is still room for improvement of course, but there is growing recognition in many emerging economies that the current lack of performance indicators and accurate market information hinders inward investment and holds back development of competitive domestic real estate sectors.

The issue of environmental sustainability is gradually moving to the forefront of real estate investor and corporate occupier concerns. JLL have launched a separate Real Estate Sustainability Transparency Index for a subset of 28 countries, covering such issues as energy benchmarking and Green Building rating systems. The United Kingdom, Australia and France are the top-scoring countries in this new index.

Further reading

joneslanglasalle.com
Jowsey, E. (2011) *Real Estate Economics*, Palgrave Macmillan, Basingstoke, ch. 18.

8.18 Corporate real estate asset management

Key terms: strategic management; cost effectiveness; sale and leaseback; corporate relocation; sustainability

The aim of corporate real estate asset management is to add value to a company through strategic management of real estate assets. Property is both an investment asset and an operational asset. Strategic management of real estate is crucial for both investors and occupiers of property. The value of any business can be enhanced through strategic management of the property that supports the business.

Many corporations are not aware of the importance of property even though it is the second highest cost after labour. Commercial property comprises 30–40 per cent of total costs in the balance sheets of most UK companies (Bootle and Kalyan, 2002) and 75 per cent of corporate borrowing in the UK is secured against property. In the private sector, if the importance of property in the total asset valuation of the corporation is underestimated by the owners, the result might be a hostile takeover bid and possible asset stripping. In the public sector, lack of awareness of property assets can lead to missed opportunities, e.g. a hospital on a prime city-centre site could realise the value of assets to relocate to a larger suburban site with easier access and parking.

The corporate property manager should prepare a corporate plan for the property that flows from and feeds into the corporate business strategy (Table 8.18.1). The property manager must:

- manage property for maximum value;
- understand occupier's objectives;
- understand activities the property supports;
- develop extra plans for capital value, rents received, operating costs and security of income.

Table 8.18.1 Alternative real estate strategies

Occupancy cost minimisation	*Cost effectiveness*
Flexibility	Facilities and workplaces are adaptable
HR objectives	Acknowledging that the workplace impacts on productivity
Promoting marketing message	Using the buildings and environment as a marketing (image) tool
Promote sales	Locate for high footfall
Facilitate and control production, operations, service delivery	Locations should link with suppliers and customers
Facilitate managerial process	Physical workspace should be designed to enhance knowledge working
Capture the real estate value creation of business	e.g. A major retailer acting as anchor tenant for a shopping centre

Source: adapted from Haynes and Nunnington (2010).

A major decision to be made is whether business premises should be freehold or leasehold. Owning the freehold has been the historic choice of many organisations – for cultural reasons (desire to have control) – for financial reasons (capital growth) – and for business reasons (specialist premises might be difficult to obtain under a standard lease). The main arguments for leasing are:

- to finance growth opportunities;
- to have a lower level of debt when leasing is reported as off balance sheet financing;
- to obtain lower rental costs where capital allowances are passed on by landlords;
- to increase efficiency by treating property as a cost and managing it more efficiently.

As a result of these perceived advantages there is a growing trend towards 'sale and leaseback' (Table 8.18.2). This is where the owner of a property sells that property to a third party and simultaneously takes a lease on that property from the third party. Examples include:

- 2004 – 38 off-licences from Thresher Group for sale at auction by Allsopps on a sale-and-lease-back basis;
- 2005 – Debenhams £495m sale and leaseback to British Land;
- 2009 – Unite Group £21.5m sale and leaseback of student accommodation to M&G Secured Property Income Fund.

Table 8.18.2 Sale and leaseback

Advantages	*Disadvantages*
Frees up capital to fund merger and acquisition activity or to invest in core business	Potential loss of capital allowances
Tax benefits by offsetting lease costs as an operating expense	Exposure to rent increases and upward only rent reviews (in UK)
Seller remains in operational control of property	Partial loss of control of asset
Improvement in balance sheet through exchange of fixed assets for cash	Loss of any capital growth potential
One-off profit and loss account benefit of profit realised over book value	Uncertainty at end of lease term
Transfers property value risk to a third party	Loss of an asset that can be used to secure further borrowing

Source: adapted from KPMG and Haynes and Nunnington (2010).

Another part of corporate real estate asset management is the corporate location (or relocation) decision. Strategic decisions to be made include:

- macro location (country)
- macro location (city)
- micro location (options within selected city)
- micro location (characteristics of selected location)
- the building specification
- the building configuration
- specific operational requirements
- sustainability or green building considerations.

Further reading

Bootle, R. and Kalyan, S. (2002) *Property in Business – A Waste of Space?*, The Royal Institution of Chartered Surveyors, London.

Haynes, B. P. and Nunnington, N. (2010) *Corporate Real Estate Asset Management*, EG Books, Oxford.

www.kpmg.com/global/en/industry/real-estate/ (accessed 27/11/2013).

9 Land management

Dom Fearon and Ernie Jowsey

9.1 Archaeological sites

Key terms: PPG 16; site evaluation; preservation *in situ*; Sites and Monuments Record

The term 'archaeological site' covers a wide range of land – from an area where a single historical object was found, to the remains of monuments such as Stonehenge.

As archaeological sites can be damaged by any ground disturbance, managing them on cultivated land is very difficult. If their location is known, they can be managed to prevent damage occurring. However, although some archaeological sites are easily recognisable, such as large monuments, many cannot be seen as they are buried below the level of normal ploughing, and regular cultivation can damage any hidden remains.

Indicators of an archaeological site on arable land include:

- objects brought to the surface by ploughing – such as pottery, burnt clay, flint tools or metalwork;
- patches of stony ground and building materials, which indicate the presence of disturbed walls or roads;
- darker or lighter patches in a field, which indicate buried features – such as ditches;
- differences in crop growth caused by buried archaeological features – crops grow better in pits and ditches because the soil is deeper and wetter, whereas reduced soil depth over patches of stone, together with restricted water, can lead to stunted growth and early ripening.

Prevention of damage to archaeological sites requires careful site management. The best way to protect a ploughed archaeological site is to remove it from cultivation as permanent grass or long-term, non-rotational set-aside. If removing the land from cultivation is not viable, there are measures that can be taken to minimise damage from agricultural practices. If possible, the landowner should:

- avoid tilling;
- use minimum cultivation techniques;
- maintain current plough depth to avoid new damage;
- avoid sub-soiling, pan busting, stone cleaning or new drainage operations;
- avoid growing potatoes, sugar beet, short rotation coppice or turf;
- avoid any harvesting operations that involve rutting, soil removal, significant soil compaction or soil erosion.

Planning Policy Guidance 16 (PPG 16) advised that archaeological remains are a finite and irreplaceable resource and that their presence should be a material consideration in applications for new development. It accepted that development will affect archaeological deposits and that this effect must be mitigated. PPG 16 stresses the importance of the evaluation of a site for its archaeological potential in advance of development in order to inform future management decisions. This evaluation may involve non-intrusive methods such as a desk-based study and/or a more direct method such as trial trenching.

Following the results of the initial evaluation, PPG 16 offered two solutions for preserving any significant archaeological deposits found to be on a development site. The first, and explicitly preferred, method involves preservation *in situ* whereby the archaeology is left untouched beneath a new development through methods such as adaptation of foundation design and architectural layout of the proposed new development, or by raising the level of the development with made ground so that its foundations do not reach the archaeological level. Where nationally important remains are encountered, this method of preservation is strongly preferred.

If preservation *in situ* is not feasible then PPG 16 permits preservation by record. This involves archaeological fieldwork to excavate and record finds and features (thereby destroying them). This may involve a full excavation, further trenching in specific areas or an archaeological watching brief that involves an archaeologist monitoring ground works for the new development and recording any finds or features revealed as construction continues.

All forms of archaeological investigation undertaken through PPG 16 are funded by the developer. The work is intended to be undertaken in advance of any planning consent being granted but often happens to satisfy a planning condition placed on an application for development.

Because of the potential for destruction of significant remains, PPG 16 prefers evaluation to take place in advance of any planning decision being made. A developer tenders for the work to be done and chooses an archaeological organisation to retain. The work is monitored by a curator, normally the county archaeologist, who is nominated by the local planning authority (LPA) as an adviser and who also identifies sites where archaeology might be threatened by development. Following submission of a satisfactory site report and demonstration that a site's archaeological potential has been properly safeguarded and/or recorded, the curator will usually advise that development can continue.

A Sites and Monuments Record or SMR is kept by curators; this is a database of known archaeological sites and is often used to inform decisions on archaeological potential. Areas of archaeological potential are often drawn on GIS maps which can indicate any potentially damaging development automatically.

PPG 16 has resulted in an explosion in archaeological fieldwork in the UK. Developer funding has led to dozens of archaeological organisations competing for work along with archaeological consultants working for developers to oversee projects. This has contributed to the growing professionalisation of archaeology.

Further reading

www.gov.uk/sites-of-special-scientific-interest-and-historical-monuments (accessed 05/12/2013).

9.2 Coastal and marine heritage

Key terms: coastal heritage; marine planning; English Heritage; Marine Management Organisation

Coastal and marine heritage includes: submerged remains of prehistoric landscapes, such as historic shipwrecks; the archaeology and built heritage of major ports and minor harbours; and coastal defence systems. Since the 2002 National Heritage Act, English Heritage (EH) has responsibilities for archaeological remains below mean low water within UK territorial waters adjacent to England. Scottish National Heritage does this for Scotland and Natural Resources Wales does this for Wales.

EH aims to secure the preservation of ancient monuments in, on or under the seabed, and to promote the public's enjoyment of and knowledge of ancient monuments in, on or under the seabed.

Natural erosion, flood and coastal erosion and commercial developments all pose risks for coastal heritage assets. English Heritage provides advice on the implications of these factors for the historic environment at a strategic and scheme-specific level. The Marine Management Organisation (MMO) is another important organisation, established by the Marine and Coastal Access Act 2009 with responsibility for preparing marine plans for the English inshore and offshore waters. It has a duty to take decisions on proposed developments in the plan area in accordance with the Marine Policy Statement (MPS) and marine plans, unless relevant considerations indicate otherwise. The MMO must:

- design a planning process suitable to deliver marine plans;
- work to integrate and balance all the current marine and future activities into a comprehensive plan;
- deliver a marine plan for each marine plan area;
- monitor and review plans on a regular basis.

The MMO is responsible for marine planning around the UK coastline which currently contributes about £50 billion to the economy and has the potential to contribute significantly more. The MMO administers a range of statutory controls (other than oil and gas licensing) that apply to marine projects, including:

- seabed and foreshore construction;
- coastal flood risk management works;
- navigation and aggregate dredging;
- the disposal of waste materials in the sea.

This is done on behalf of the Secretary of State Defra, and covers the UK marine area, not subject to devolved administration.

Further reading

www.helm.org.uk/managing-and-protecting/coastal-and-marine-heritage (accessed 05/12/2013).

9.3 Farm buildings

Key terms: barn conversion; local planning authority; permitted development rights; Town and Country Planning (General Permitted Development) Order 1995; Town and Country Planning (Use Classes) Order 1987; English Heritage

Farm buildings no longer required for agricultural purposes can often provide a landowner with a variety of potential reuse and/or development opportunities. While this is usually seen as a positive, landowners should be aware that if they wish to continue using and occupying surrounding areas for agricultural or business purposes, this can cause some conflict between the existing and proposed uses. The obvious choice, and one that has become very popular over the last few decades, is a 'barn conversion' for residential purposes that can be sold or let separately from the main farm steading or used as a supplementary income, e.g. holiday let, bed and breakfast accommodation or camping barn. More recently, in addition to residential use, we have seen larger rural estates convert redundant buildings and whole farm steadings to office space and/or small, light industrial business space for local rural businesses. These are just a few examples of reuse, but there is clearly more potential, especially in light of the current government's intention to relax certain planning policies in a bid to simplify and speed up the current planning process.

Critics may argue this form of development can be harmful to the countryside and rural villages, and there is usually much debate and many hurdles to be crossed in terms of planning approval before these developments can go ahead, if at all. The regulations concerning change of use for farm buildings are further discussed below. In its favour, reuse is inherently sustainable. Some farm buildings represent a historical investment in materials and energy, and contribute to environmentally benign and sustainable rural development. The concept of reuse is not a new one. Farm buildings have often been adapted over a long period to accommodate developing farming practices and technologies. Some have a greater capacity to accommodate change or a new use than others, and a small number are such historically or architecturally significant elements of our heritage that they should be conserved with minimal or no intervention. This is traditionally where the role of the LPA takes its part in the process.

Traditionally, under planning legislation, agricultural buildings had certain permitted development rights. These rights are basically a right to make certain changes to a building without the need to apply for planning permission. These derive from a general planning permission granted from Parliament in The Town and Country Planning (General Permitted Development) Order 1995 (SI 1995/418) (the 1995 Order), rather than from permission granted by the LPA. Schedule 2 of the Order sets out the scope of permitted development rights. It is worth noting here many traditional farm buildings are listed – meaning listed building consent will most likely also be required. Again, traditionally under planning legislation, The Town and Country Planning (Use Classes) Order 1987 puts uses of land and buildings into various categories known as 'use classes'. The categories give an indication of the types of use that may fall within each use class.

More recently changes to permitted development rights (PDRs) came into effect in May 2013, offering an easier process for changing the use of farm buildings in England. Changes to the PDRs came into force on 30 May 2013 and allow the change of use of small farm buildings in England without formal approval subject to the following:

- The changes apply to agricultural buildings with a floor space of up to 500 sq m, which were in sole agricultural use as of 3 July 2013.
- They can be changed to an alternative use without the need for detailed planning permission for that change of use.
- Any building converted under the new rules after this date must have been in agricultural use for a minimum of 10 years to be eligible.
- Note that the PDRs apply only to change of use – changes that physically alter the building may still need building consent or planning permission.
- Also, for some changes of use to buildings larger than 150 sq m, the LPA will need to refer proposals to statutory consultees such as the Highways Authority and or the local Environmental Health Department.

PDR change of use includes conversion of farm buildings into shops, restaurants, cafés, offices, light industrial units and hotels without detailed planning permission being needed. They also allow some farm buildings already in alternative uses to be changed to a further alternative use. Please note, at present these changes do *not* include changes to residential use although the government opened a consultation, which ended on 15 October 2013, looking at the reuse of existing agricultural buildings for a dwelling house. These specific proposals are for up to an additional three dwellings with an upper limit of 150 sq m for each dwelling.

When considering the conversion of traditional farm buildings there is good guidance provided to landowners and developers via English Heritage, which includes their document *The Conversion of Traditional Farm Buildings*. To understand the character and significance of the farm building and its landscape setting English Heritage have provided a summary of good practice for conversion proposals comprising some of the following actions:

- A thorough understanding of the farm building's historical, structural and spatial attributes is needed to inform the possible future use of the farm building and subsequent design work.
- Try to understand as much as possible about the way the building is constructed and its condition before undertaking significant works of repair/alteration.
- A comprehensive measured survey in plan section and elevation together with an accurate survey of condition is essential before embarking on the works.

Further reading

English Heritage (2006) *The Conversion of Traditional Farm Buildings*, available at: www.english-heritage.org.uk/publications/conversion-of-traditional-farm-buildings (accessed 22/11/2013).

www.gov.uk/planning-permissions-for-farms (accessed 22/11/2013).

www.gov.uk/government/consultations/greater-flexibilities-for-change-of-use (accessed 22/11/2013).

9.4 Fishing and fishing rights

Key terms: riparian rights; Land Registry; Salmon and Freshwater Fisheries Act 1975; national byelaws; regional byelaws

Fishing rights: the law in England and Wales

Many estates or sizeable landholdings are likely to include at least one river or watercourse and it is therefore important to understand the legal framework that governs them.

There are laws that govern fishing in England and Wales. Like many other activities these laws have developed over time, and there are different laws for tidal waters (e.g. the sea) and non-tidal waters (e.g. lakes and rivers).

- For tidal waters, members of the public have a right to fish in the sea below the mean high water mark of tidal waters. Anyone can fish either from the bank, a pier or by boat assuming there is suitable public access. You may, however, require a licence to fish salmon or sea trout in a tidal estuary. Fishing at sea is subject to controls over the manner in which fish can be caught, the minimum size of the fish and the size of the mesh in the nets. In addition, European Union quotas are applicable to commercial fishing.
- For non-tidal waters, the arrangement is a little more complicated. There are two parts of property in a river that can be owned: these comprise the river bed and the fishing rights. The right to fish in a river, commonly called 'fishing rights' is a class of legal interest called a *profit à prendre* and is often the most valuable part of the river. It is not an interest in the land as such, like being the owner of it, or having a tenancy of it, but rather the right to do something with a certain part of the land; in this case, the water. The fishing rights can be bought, sold and leased independently of the land or together with it, depending on what the owner wants to do. The fishing rights can also be registered with their own, independent title at the Land Registry.
- The owner of the land adjoining one side of a natural river or stream owns the exclusive fishing rights (often called *riparian rights*) on their side of the bank. These rights extend up to the middle of the water. If there is an island in the river then it is also owned by the riparian owner whose side of the river it falls in, or, if both, it is again split along the imaginary middle line between the two banks. If the river lies wholly in their ownership, then they own all of it. However, as this can sometimes be a presumption, express wording to the contrary will negate it. This can mean that by agreement, one owner can own all of the bed of the whole river and the adjoining owner's interest may start on the opposite bank. This can occur when

a large estate is split up and sold in parts to third parties. The whole of the river and usually the fishing rights are often kept by the estate.

- Although you may have fishing rights, a riparian owner is still subject to the general laws protecting close seasons for fish. These are set down in the Salmon and Freshwater Fisheries Act 1975. Just because one has the right to access a river, stream, lake or any other water body, one does not automatically have the right to fish in it.
- Byelaws protect fish stocks and fisheries in England and Wales. They apply to all waters whether they are owned by angling clubs, local authorities or private individuals. Owners may impose additional rules, but they cannot dispense with any byelaws that apply to their water. There are national byelaws that apply across the whole of England and Wales and regional byelaws that only apply locally. Both national and regional byelaws need to be adhered to and these are available from the Environment Agency (EA).

Further reading

www.environment-agency.gov.uk (accessed 13/11/2013).
www.landregistryservices.com (accessed 13/11/2013)

9.5 Trees and forestry

Key terms: Forest Authority; felling licence; Tree Preservation Order (TPO); Conservation Area; Town and Country Planning Act 1990; Town and Country Planning (Trees) Regulations 1999

Trees, woodland areas and forests are of particular landscape and conservation importance both in the countryside and in our towns and cities, and accordingly they have a number of special laws and regulations. From a property and/or development perspective, trees can have many benefits. They can slow down wind and therefore reduce heating bills; give shade and reduce summer electricity (e.g. air conditioning) bills; absorb pollution; stabilise sloping sites; and muffle noise. In some cases, trees on site may also provide sustainable crops of timber and other craft materials that can be used in the constriction process. Trees are able to provide shelter and food for wildlife, especially if they form 'green corridors', connecting the site to other trees and woods nearby. Trees can soften the hard edges of property development, giving pleasure to the people who live and work on the site. They can often enhance the development, thereby helping with marketing/sale or letting of the site. In the long term the benefits may include superior property values compared to those with limited landscaping, so long as regular maintenance of the trees is upheld.

However, not all trees are planted, or have been planted by human means. Some existing trees may need to be removed or pruned; as trees grow, you need to think ahead to avoid direct or indirect damage to new buildings. Some trees may be unsafe, and to make the site suitable for building you may need to remove them or have tree surgery done to make them safe. If potentially hazardous trees are left on site, purchasers may be able to claim for damage or for unnecessary tree surgery costs. Building works can easily damage trees, especially the roots. This can be digging with an excavator near the roots or via compaction of the roots by storing building materials close to trees. As a result, and usually after a period of time, the trees may then die back and become unsafe. You need to plan, before even designing the site, which trees to keep and how to protect them, and consider where new planting would be appropriate and would not pose a future risk to people or buildings. Further guidance on trees and development sites can be found in British Standard (BS) 5837.

With regard to felling large numbers of trees and commercial forestry, everyone needs permission from the Forest Authority (a part of the Forestry Commission) to fell growing trees unless the operation is *exempt*. For exemptions to this rule see below. There is a public register of applications for felling licences. All applications have to rest there for 4 weeks and any person or body can comment on them. This facility can be found at the Forestry Commission's Public Register website.

Exemptions – For mainly commercial purposes, landowners are allowed to fell up to 5 cubic metres of timber on their property without a licence, so long as no more than 2 cubic metres are sold. For smaller-scale tree operations, the following exemptions to licences apply:

- lopping and topping, including most tree surgery, pruning and pollarding;
- felling fruit trees, or trees growing in a garden, orchard, churchyard or other designated public open space;
- felling trees that, when measured at a height of 1.3 m from the ground, have a diameter of *8 cm or less*; or
  - if thinnings (trees removed from among other trees to allow their growth) have a diameter of *10 cm or less*; or
  - if coppice or underwood (trees as part of a layer below the main forest canopy) have a diameter of *15 cm or less*;
- felling trees immediately required to implement a planning permission;
- felling trees for work carried out by some service providers (e.g. electricity or water) and that is essential to carry out those services;
- felling necessary to prevent the spread of disease and done in accordance with a notice served by the Forest Authority.

To find out whether your or your neighbour's trees are protected, contact the Local Planning Authority (LPA) for further advice. If your trees are protected, you need written permission to remove them or to do any work to them. If you remove protected trees or do work to them without permission, you could be prosecuted. Trees have protection if they are within a Conservation Area or subject to a Tree Preservation Order (TPO). The owner or the tree contractor can usually apply for work to protected trees on standard forms. You will usually receive a decision within 6 weeks for Conservation Areas and 2 months for TPOs and planning conditions. Trees within a Conservation Area are not strictly protected, but there is a requirement that the owner gives notice to the council of their intention to do works. If you want to remove trees, you may be required to plant replacements of the same species and in the same location. It should be noted that not all mature trees are protected so it is wise to check with the planning authority in the first instance.

TPOs are made under the Town and Country Planning Act 1990 and the Town and Country Planning (Trees) Regulations 1999. A TPO is made by the LPA to protect specific trees or a particular woodland from deliberate damage and destruction. Making a TPO is a 'discretionary' power, which means that the Council doesn't have to – it may choose to make them or it may not. However, once they have decided, they do have a duty to enforce it. Although, it is possible to make TPOs on any trees, in practice they are most commonly used in urban and semi-urban settings, for example, gardens and parklands. A TPO is to protect trees for the public's enjoyment. It is made for the amenity value of the tree and this can include the nature conservation value but more often means its visual amenity. There are presently four types of TPO, although any one order can contain any number of items which can be of one or more types. The types are as follows:

1. *Individual*: can be applied to an individual tree;
2. *Group*: can be applied to a group of individual trees, which together make up a feature of amenity value;
3. *Area*: covers all trees in a defined area at the time the order was made;
4. *Woodland*: covers all trees within a woodland area regardless of how old they are.

Further reading

British Standard BS 5837: 2012 Trees in Relation to Construction.
The Forestry Commission – www.forestry.gov.uk/forestry (accessed 21/11/2013).
Naturenet – www.naturenet.net/trees (accessed 21/11/2013).
The Planning Portal – www.planningportal.gov.uk/planning/appeals/otherappealscasework/treepreservation (accessed 21/11/2013).

9.6 Historic parkland

Key terms: The 1983 Heritage Act; English Heritage; Registered Park; Sites and Monuments Records; Historic Environment Records; National Planning Policy Framework

Most historic parkland today is the product of several phases of design over several centuries. Some originated as, and retain elements of, mediaeval deer parks; others retain remnants of the pre-park agricultural landscape and others contain scheduled monuments and other archaeological features. Historic parks are deemed to be an integral part of the English countryside, making a unique contribution to its character, its biodiversity and cultural heritage. Many historic urban parks and other designed landscapes such as cemeteries have drawn inspiration from the country house park, and many local authority-owned and/or managed country parks were once former country house estates and parklands.

While historic parks may have started from native woodland, they are now considered artificial or man-made landscapes having adapted to different management techniques and different owners over the centuries. The owners of parks were the nobility and gentry, bishops and religious houses. A park usually formed part of a large or medium sized manor. The importance of historic parks can therefore be categorised in many ways in terms of their association with individuals, sites of archaeological importance, architectural buildings but also in more recent times as landscapes to enjoy for leisure and tourism purposes. It is therefore essential that these sites are protected for future generations to enjoy.

The 1983 Heritage Act empowered English Heritage to manage the Register of Parks and Gardens of Special Historic Interest in England. The Register includes a diverse range of designed landscapes including urban parks, allotments, hospital grounds, cemeteries and post-war designs. As with listed buildings, the sites are graded, but the Register has no equivalent statutory powers or consent scheme. However, the Register is a material consideration in the determination of planning applications, which can help to deter inappropriate development, and most development plans now incorporate protection policies. Each Register entry comprises a written description, together with a map showing the boundary of the registered land. The entry includes a summary history and a description of the site as existing, highlighting its special features and characteristics. Register descriptions and maps can be obtained from English Heritage's National Monuments Record Centre.

The Register only records sites of historic importance in a national context. Many more are important locally or regionally, and a number of local authorities and county gardens trusts have compiled or begun to compile local lists of parks and gardens, which may or may not also be subject to development plan policies.

Many county Sites and Monuments Records (SMRs) or Historic Environment Records (HERs) contain information on parks and gardens and their archaeological features.

Parkland is likely to be of historic interest if it contains:

- a listed principal house;
- plantations, avenues, clumps or specimen trees over a hundred years old;
- features such as follies, eye-catchers, lodges, ha-has, old boundary walls, gates or fences.

Similar to listed buildings, historic parks are vulnerable and require ongoing maintenance to protect them from unnecessary neglect and damage. The principle of parkland conservation is to protect and retain the original historic fabric where possible. It is desirable that management should conserve the design and special parkland characteristics as well as accommodate new uses to meet the challenge of sustainability. Examples of changes that can affect the original fabric include lack of maintenance, harm or loss of trees, abandonment of park buildings and monuments, poorly sited or unsympathetic development such as new buildings within or close to the park boundary.

To afford further protection for these parkland sites local authorities can provide HERs. These records are maintained by LPAs; they are used for planning and development control as well as for public benefit and educational use. Access to consistent, up-to-date and high quality information about the historic environment through HERs is an important requirement for the delivery of Heritage Protection Reform (HPR) and the National Planning Policy Framework (NPFF). The NPFF sets out the requirement for LPAs to maintain or have access to a HER and to ensure HERs are used as a matter of course in planning and development matters.

Advice on how to protect historic parklands or relevant grant schemes is currently available from the following organisations: Department for Environment, Food and Rural Affairs (Defra), English Heritage, County Gardens Trusts and the Garden History Society, Local Authority Archaeologists or Conservation Officers.

Further reading

www.algao.org.uk (accessed 13/11/2013).
www.defra.gov.uk (accessed 13/11/2013).
www.english-heritage.org.uk (accessed 13/11/2013).
www.magic.gov.uk (accessed 13/11/2013).
www.wapis.org.uk (accessed 13/11/2013).

9.7 Protected landscapes

Key terms: National Parks; Areas of Outstanding Natural Beauty; Heritage Coasts; National Parks and Access to the Countryside Act (1949); Statutory Designations; Non-statutory Designations; International Union for Conservation of Nature; National Planning Policy Framework

It is important to realise that not only are large areas of our towns and cities, together with their historic buildings and structures, afforded protection from unnecessary change of use and development, large areas of the UK landscape are also protected in a similar way. For example, National Parks and Areas of Outstanding Natural Beauty (AONBs) together cover nearly a quarter of England and represent some of the country's finest landscapes. The rationale for this protection can be traced back to the mid twentieth century. Two defining documents, *The Hobhouse Report* and *The John Dower Report*, were commissioned by the government to respond to the wish of the public to have access to land for recreation purposes. Following these reports, an Act of Parliament was passed, The National Parks and Access to the Countryside Act (1949) in order to conserve and enhance the most sensitive landscapes in the UK. These landscapes are not chosen or designated purely for their natural beauty as their natural elements have generally been shaped by industry, farming and human settlement. It is how they have evolved and been shaped that leads to their 'special qualities' which affords their protection. The main three landscapes that are generally protected either by national law or by local authorities are:

- National Parks – large areas designated by law to protect their special landscape qualities and promote outdoor recreation. National Parks have their own authorities, which among other areas, control planning.

- Areas of Outstanding Natural Beauty – protected by law because of their special landscape qualities, wildlife, geology and geography. They have more protection than other areas under the planning process and, in terms of landscape and scenery, are equal to National Parks.
- Heritage Coasts – these include stretches of outstanding, unspoilt coastline, usually cared for by local authorities and/or charities such as the National Trust.

Landscapes are protected by a range of mechanisms including *statutory* and *non-statutory designations*, national planning policies and European conventions.

Statutory designations – as mentioned above, the National Parks and Access to the Countryside Act (1949) set out to conserve and enhance certain areas for their natural beauty, with areas designated as National Parks or Areas of Outstanding Natural Beauty. Approximately 23 per cent of England is currently protected by one of these two designations, the purposes of which are supported by strategic management plans prepared by the lead local authorities. The importance of the designations has been noted through the planning system, most recently through the National Planning Policy Framework. The legislative system in England and Wales is supported by international guidelines and sharing of experience of designations through the International Union for Conservation of Nature (IUCN) and the European Federation. The England and Wales designation system for designation for National Parks and AONBs is recognised as falling within the parameters of the IUCN's Category V Protected Landscape definitions. The statutory duties/powers for designating new National Parks and AONBs in England and reviewing existing boundaries reside with Natural England.

Non-statutory designations – Heritage Coasts are identified for national purposes of conservation and enhancement of the natural beauty of the coastline, its terrestrial and marine flora and fauna, and heritage features. They also have objectives to:

- facilitate and enhance the public's enjoyment of these resources;
- maintain and improve the environmental health of inshore waters affecting heritage coasts and their beaches.

In delivering these objectives account must be taken of the needs of land management activities and local communities. Heritage Coast designation currently extends across a third of the English coastline, often running in tandem with National Parks and/or AONB designations. An important tool for conserving Heritage Coasts is having supportive policies in LPA development plans. A large proportion of these coastlines have already been incorporated within designation management plans, making it simpler for developers and the local communities to understand the reasons for ongoing protection for future generations.

Further reading

English Heritage, 'Protected Landscapes', available at: www.english-heritage.org.uk/professional/research/landscapes-and-areas/protected-landscapes/ (accessed 19/11/2013).

English Heritage, 'A strategy for English Heritage's Historic Environment Research in Protected Landscapes', available at: www.english-heritage.org.uk/publications/historic-environment-research-in-protected-landscapes/protlandstrat.pdf (accessed 19/11/2013).

Natural England, 'National Parks', available at: www.naturalengland.org.uk/ourwork/conservation/designations/nationalparks/ (accessed 19/11/2013).

9.8 Religious buildings

Key terms: Church of England; listed buildings; listed building consent; conservation area consent; 'ecclesiastical exemption'; total demolition; partial demolition; repair grants for places of worship.

Places of worship of all faiths and religions include many of the most important historic buildings in the UK. They often play a fundamental role within their communities and provide a link to the region's local history. It is estimated that approximately 45 per cent of all Grade I listed buildings are churches. People generally feel strongly about them whether or not they are active members of a worshipping congregation. With this in mind, it is clear to see that these buildings require considerable care in terms of their architectural, archaeological, artistic, historic and cultural interest.

Since the Reformation of the sixteenth century, the Church of England has by far the most of England's oldest and finest churches. Only in the nineteenth century did other denominations such as Anglican and Catholic, and more recently Jewish and Muslim communities, have the financial resources to build on a similar scale. For this reason, the Church of England properties will be discussed in most detail below. The situation in Scotland is also not covered within this text.

The word 'church' means a place of worship of Christian or non-Christian denominations. As previously mentioned, many Church of England churches are Grade I or II* listed and numerous churches are situated in conservation areas. As such, alteration to these buildings would be subject to planning control, especially listed building consent and conservation area consent. These requirements normally mean that:

- listed building cannot be demolished, altered or extended so as to affect its character without consent, according to section 7 of the Planning (Listed Buildings and Conservation Areas) Act 1990 (hereafter PLBA 1990);
- permitted development is often restricted in conservation areas and demolition of unlisted buildings needs consent (s.74 PLBA 1990).

However, because of what is known as the 'ecclesiastical exemption', the requirement for listed building consent does not extend to: 'any ecclesiastical building which is for the time being *used for ecclesiastical purposes* or which would be so used but for the works of demolition, alteration or extension.'

The exemption dates from 1913 when the Church sought independence in its own affairs and was opposed to interference from the State in the form of secular control over alteration to church buildings. Since this time there has been much legislation and case law concerning exemption and listed buildings with the present situation being as follows. Under the Provisions of the Ecclesiastical Exemption (Listed Buildings and Conservation Areas Order) 1994 the exempt denominations are:

- The Church of England
- The Church in Wales
- The Roman Catholic Church
- The Methodist Church
- The United Reformed Church
- The Baptist Union.

Section 60(1) PLBA 1990 gives exemption to: 'any ecclesiastical building which is for the time being used for ecclesiastical purposes'.

Exemption is *from*:

- listed building consent;
- building preservation notices;
- temporary listing (urgent cases);
- compulsory acquisition (of listed buildings in need of repair);
- execution of urgent works for unoccupied listed buildings;
- criminal liability for damaging listed buildings.

You will have noticed that the exemption does *not* cover planning permission. Therefore, any works to an ecclesiastical building that involve 'development' will need planning permission. Neither does the exemption cover listed structures within the churchyard, for example, monuments and gravestones, which may require listed building consent.

With regard to demolition, as far as ecclesiastical buildings were concerned, the law used to distinguish between:

- *total* demolition (outside the exemption); and
- *partial* demolition (within the exemption).

This position has been complicated by the House of Lords decision *Shimizu (UK) Ltd v Westminster CC* (1997).

Total demolition of a church will require:

- listed building consent (if appropriate);
- conservation area consent (if appropriate).

It will not require planning permission following the Town and Country Planning (Demolition – Description of Buildings) (No. 2) Direction 1992.

However, it should be noted that a 'redundant church' does not require listed building consent for demolition works (s.60(7) PLBA 1990). The ecclesiastical exemption is therefore not relevant here. There may be circumstances in which state intervention is needed, for example if a redundant church is in a conservation area and/or there are objections to the demolition by:

- the local planning authority;
- the church commissioners;
- a national amenity group.

Due to their age, a large number of churches will invariably have deteriorated in their condition and state of repair over time. Since the seventeenth century, the Church of England has had a system of regular inspections of its churches, now referred to as 'quinquennial inspections'. The Church Buildings division of the Archbishops Council publishes a useful booklet, *A Guide to Church Inspection and Repair*. With regard to the availability of grants for repair for places of worship, English Heritage provides guidance and a flow chart on current grant schemes.

Further reading

www.english-heritage.org.uk/caring/places-of-worship/ (accessed 13/11/2013).
www.hlf.org.uk/GPOW (accessed 13/11/2013).
www.spabfim.org.uk (accessed 13/11/2013).

9.9 Waste disposal sites

Key terms: landfill; recycling; Environment Agency; Health Protection Agency

In 2011, the UK disposed of 49 per cent of waste to landfill, with 25 per cent recycled, 14 per cent composted and 12 per cent incinerated. Household recycling has risen from just 11 per cent in 2001 to 43 per cent in 2013 and over half of business waste is now recycled. Around 300 million tonnes of waste is produced each year in England and Wales, and all of it must be managed in a way that protects people and the environment. Much of this waste is reused or recycled, but some of it still needs to be disposed of.

The European Environment Agency warned that many countries will fail to meet the European Directive on the Landfill of Waste (1999) target of recycling 50 per cent of waste by 2020. Some countries, such as Germany, Austria and Belgium, already recycle more than half of their waste, but others, in particular those in south-eastern Europe, are far behind: Greece only recycles 18 per cent, up from 9 per cent in 2001, while Romania recycles just 1 per cent. In a few cases, countries have gone backwards, with Norway's rates falling from 44 per cent to 42 per cent, and Finland's dropping from 34 per cent to 33 per cent.

As well as reuse and recycling, other techniques and technologies are being developed to reduce the amount of rubbish sent to landfill each year. These include producing energy from waste by incineration or by anaerobic digestion. One consequence of complying with the European directive on recycling has been a number of serious fires across Europe at recycling plants. In the UK there have been serious fires at recycling plants in Smethwick (July 2013), Wellingborough (September 2013), East Stoke, Dorset (October 2013), Llandow, Wales (November 2013), and many others. Such fires can burn for many days and give out noxious fumes that can endanger the health of nearby residents, who are often advised to stay indoors.

A landfill site is an area of land used to dispose of rubbish, either directly on the ground or by filling a hole in the ground. For example, a disused quarry is often used as a landfill. Up until the 1980s, policies of successive governments had endorsed the 'dilute and disperse' approach, but the UK has since adopted the appropriate European legislation and landfill sites are generally operated as full containment facilities. However, many dilute and disperse sites remain throughout Britain. The local authority decides through the planning process whether to allow the development of a landfill site. It checks the need for the site against waste plans, and sets conditions on size, location and restoration. Local authorities and industry take waste to landfills.

Landfill Tax is a tax on the disposal of waste. It aims to encourage waste producers to produce less waste, recover more value from waste, for example through recycling or composting and to use more environmentally friendly methods of waste disposal. The tax is charged by weight and there are two rates. Inert or inactive waste is subject to the lower rate of £2.50 per tonne in 2013; active waste is subject to the higher rate of £72 per tonne in 2013.

There are three kinds of operational landfill sites categorised by the types of waste they are authorised to take. These are:

- Inert wastes – these sites take only inert wastes such as excavation wastes. They have a low impact on the environment. Inert wastes are unlikely to react with other wastes.
- Non-hazardous wastes – these sites take a mix of inert wastes and biodegradable wastes. They may also be allowed to make special provision for some hazardous waste such as asbestos.
- Hazardous wastes – these sites take hazardous wastes such as contaminated soils and asbestos to landfill. As they do so they produce gas, which is known simply as landfill gas. As rainwater seeps through the landfill a liquid is formed, known as leachate. Landfill gas and leachate must be safely extracted and treated.

There was a landfill disaster in Loscoe, Derbyshire in the 1970s – no one died, but a cottage was blown to pieces due to landfill gas escaping from a nearby capped site. Landfill gas (methane and carbon dioxide) built up in the capped landfill. The air pressure dipped and, as the site had no venting mechanism, the gas followed the least line of resistance through the geology (coal seams) and up into the houses nearby. A spark in one of these ignited the explosive mixture, destroying a bungalow completely and injuring the occupant.

This disaster led to the introduction of key legislation and government guidance and much research into best practice. Over time, landfill sites were designed to vent gas to the atmosphere, then to burn off methane and eventually, in the most productive, to turn the gas into electricity using gas turbines that supply the national grid. Heathfield landfill site in South Devon (Viridor) is a good example. Monitoring of landfill gases must now take place where public housing is 250 m from the site, with reporting of results to the EA – the regulator – and strategies put in place for evacuation should it be necessary.

The EA is responsible for regulating over 2,000 landfills:

- 465 of these are operational sites with a Landfill Directive compliant permit.
- 812 sites have stopped taking waste since July 2001 when the Landfill Directive came into effect. These sites closed in accordance with the Directive requirements.
- 979 sites stopped taking waste before the Landfill Directive came into effect, but continue to have permits from previous regimes.
- about 75 per cent of the EA regulated sites no longer take any waste. However, many of them will continue to pose a risk to the environment for many years to come.

The EA ensures that the operator only takes appropriate waste, that water nearby is not polluted and that nuisance such as odours, waste blowing across the site, noise, and vermin such as gulls do not affect communities in the area. When a site is up and running it is inspected regularly to make sure it is operating within the conditions of the permit and the EA advise on necessary improvements, and may enforce the permit by issuing a warning, serving a notice taking away permission or by prosecuting the operator. When it is time to close the site the operator has to agree with the EA how they are going to do this and agree a plan for its monitoring and management. This aftercare will continue until the EA are satisfied that there is no risk to the environment. Aftercare can last a hundred years or more.

Most waste in the UK has traditionally been disposed of to landfill sites. These can generate considerable public concern about the health effects of emissions and there have been suggested links to a range of health effects including cancer and birth defects. The Health Protection Agency (HPA) recognises that the practice of disposing of waste materials to landfill can present a pollution risk and a potential health risk. Modern landfills are subject to strict regulatory control which requires sites to be designed and operated such that there is no significant impact on the environment or human health. An assessment of the health risks posed by landfill sites and other forms of waste management was published by Defra in 2004, incorporating a review of the assessment by the Royal Society. The HPA carried out a review of more recent research into the suggested links between emissions from landfill sites and effects on health. This review encompassed the results of a number of epidemiological studies, detailed monitoring results from a major project funded by the EA, and advice sought from the Committee on Toxicity of Chemicals in Food, Consumer Products and the Environment. The HPA concluded that there has been no new evidence to change the previous advice that living close to a well-managed landfill site does not pose a significant risk to human health.

All landfill sites must operate without causing pollution of the environment or harm to human health. They are carefully managed to keep risks to a minimum. Anyone who operates a landfill site must have an environmental permit to do so. Landfill operators must now comply with the European Directive on the Landfill of Waste, the Environmental Permitting Regulations (England and Wales 2010) and some additional directives. Other regulations apply in Scotland and Northern Ireland. Regulations set standards for the location, design, monitoring and operation of landfills.

Further reading

www.environment-agency.gov.uk/business/sectors/108918.aspx (accessed 06/12/2013).

www.hpa.org.uk/Publications/Radiation/DocumentsOfTheHPA/RCE18ImpactonHealthofEmissionsfromLandfillSites/ (accessed 06/12/2013).

9.10 UK National Parks

Key terms: moorland; wetlands; lakes; rivers; woodlands; forests; meadows and grasslands

The UK has 15 National Parks (10 in England, two in Scotland and three in Wales), which are areas of protected countryside where visitors are encouraged and where people live and work. They are protected areas because of their beautiful countryside, wildlife and cultural heritage. The first were designated as National Parks in 1951 (Peak District, Lake District, Snowdonia and Dartmoor). A large proportion of land within the National Parks is owned by private landowners. Farmers and organisations such as the National Trust are some of the landowners, along with the thousands of people who live in the villages and towns. National Park authorities sometimes own bits of land, but they work with all landowners in all National Parks to protect the landscape.

All National Parks are funded from central government. In Wales and Scotland the money is allocated by the Welsh and Scottish Governments. The main areas of expenditure are: planning – controlling development (15 per cent); planning – policies and communities (9 per cent); promoting learning and understanding (24 per cent); wardens, estate workers and volunteers (21 per cent); and running the organisation (17 per cent) (Table 9.10.1).

Moorlands are managed and protected by the National Parks, which give advice and grants to landowners to burn heather, graze animals and control the growth of bracken and trees. There have been increases in the numbers of rare birds such as golden plover and merlin on some moorland. Heather is burnt after 12 years so that new growth of young shoots can grow and feed wildlife. Biodiversity action plans list the measures taken to protect rare moorland areas and species.

Wetlands, lakes and rivers are kept clear of mud and unwanted vegetation. Cattle and sheep are fenced in to keep them away from water, fish are encouraged and clean water is monitored. Salmon and water voles are protected in rivers.

Table 9.10.1 National Park expenditure (£)

Authority	*2010–11*	*2011–12*	*2012–13*	*2013–14*
Brecon Beacons	3,302,269	3,215,500	3,402,253	3,387,834
Broads	4,229,502	4,002,149	3,774,799	3,547,447
Cairngorms	4,946,000	4,756,000	4,646,000	TBC
Dartmoor	4,739,642	4,484,867	4,230,095	3,975,321
Exmoor	3,978,580	3,764,715	3,550,853	3,336,989
Lake District	6,921,279	6,549,233	6,177,190	5,805,144
Loch Lomond and The Trossachs	7,122,000	6,768,000	6,648,000	6,499,987
New Forest	4,028,096	3,811,570	3,595,046	3,378,520
North York Moors	5,428,266	5,136,475	4,844,687	4,552,897
Northumberland	3,311,334	3,133,337	2,955,341	2,777,344
Peak District	8,298,814	7,852,720	7,406,630	6,960,536
Pembrokeshire Coast	3,462,640	3,368,100	3,554,853	3,540,433
Snowdonia	4,435,091	4,316,400	4,503,153	4,488,733
South Downs	7,290,000	11,373,133	10,981,271	10,589,405
Yorkshire Dales	5,398,563	5,108,369	4,818,178	4,527,984

Source: UK National Park Authority.

Woodlands and forests are managed in order to encourage biodiversity. Soil conditions are analysed and different trees planted in different places where they will grow best:

- rowan and birch on steep sides of mountain streams;
- ash and alder on flat, wet areas of ground;
- oak and hazel on grassy banks of streams;
- juniper, hawthorn and blackthorn around the edges of woods.

Fences are built around new woods to keep out grazing animals and any dead trees are removed and replanted. Other plants that might inhibit the growth of the trees are removed.

Meadows and grasslands are improved in order to encourage biodiversity. Modern farming uses artificial fertiliser to increase the amount of grass that grows, and the grass is harvested and used over winter to feed sheep and cows. But the grass grows so quickly and so high that other species such as meadow flowers can't compete, so the fields are just full of grass, which is bad for biodiversity. Farmers get grant money to increase biodiversity in their fields. Only using a little organic fertiliser such as cow manure means the grass still grows, but not as quickly, so lots of other plants can also grow. Cutting the plants later in the year gives the plants time to set seed, meaning plants grow again next year. Hay meadows with lots of different flowers in them are great habitats for insects, birds and other animals too. There are some plants that only grow in upland hay meadows in areas such as the Yorkshire Dales, Northumberland and the Peak District.

Further reading

www.nationalparks.gov.uk/home (accessed 06/12/2013).

10 Law

Rachel Williams and Simon Robson

10.1 Contracts

Key terms: agreement; consideration; deed; intention to create legal relations

A contract is an agreement that is legally binding. This means that if there is a failure to perform contractual obligations then there are legal remedies.

There are three key components to a contract:

- agreement;
- consideration or made as a deed; and
- intention to create legal relations.

Agreement is normally defined as offer and acceptance; once an offer is accepted the agreement is formed.

An offer is an indication of willingness by one person to enter into an agreement on specified terms. An offer needs to be differentiated from an invitation to treat which is an invitation to others to make offers. Examples of invitations to treat include goods displayed in a shop window, advertisements and 'For Sale' or 'To Let' boards outside properties. An offer needs to contain all the principal terms of the agreement or else there is insufficient certainty as to what is agreed. For example, if an agent proposed to someone that they act for them on the sale of their house but did not specify the rate of commission this would be characterised as an invitation to treat as the terms of the offer are not established.

The response to an offer may be acceptance, in which case an agreement is formed. For the reply to an offer to be treated as an acceptance the person who receives the offer must accept all the terms of the offer. Frequently the response to an offer is a counter offer, for example the other person might be happy with the other terms suggested but proposes a lower price. The effect of a counter offer is to extinguish the original offer which is no longer capable of acceptance. There may be a series of counter offers made and received before agreement can be reached. For an example of how these principles apply to real estate consider the auction room:

- The auctioneer asking for bids is making invitations to treat.
- The people making bids are making offers.
- The auctioneer saying 'sold to' and hitting the gavel down is indicating the acceptance of the successful bidder's offer, at which point an agreement is formed.

There are agreements that are not legally enforceable because they lack the requisite consideration or the parties did not intend to be legally bound. The concept of a contract includes the notion that some kind of bargain or deal must have been reached. It is a requirement that both parties exchange something but it need not be of equal value. This requirement is easily satisfied in most real estate transactions; for example, in the sale of land, the seller is providing the land and the buyer the purchase price. It is possible to make agreements that are one-sided (i.e. gifts) legally binding by ensuring that the agreement takes effect as a deed. A deed is a more formal legal document that has particular execution requirements.

In some instances there are agreements that are supported by consideration but where the parties do not intend the agreement to have legal force. These types of agreement generally arise between family members and friends. In commercial dealings correspondence is often marked 'subject to contract' to indicate at that stage in the negotiations the parties do not intend to be contractually bound.

Generally under English law contracts can be made orally or in writing. There are a number of exceptions to this rule of which real estate transactions are probably the most important example. Section 2 of the Law of Property (Miscellaneous Provisions) Act 1989 provides that all contracts dealing with interests in land can only be made in writing in an agreement that incorporates all the terms of the transaction and is signed by both parties or their agents. The exceptions to this rule are auction sales and leases for less than 3 years. In *Cobbe* v *Yeoman's Row Management Ltd* (2008) UKHL 55 an oral agreement was reached between a property owner and developer. After the developer secured planning permission for the site the owner reneged on the terms of the deal and the developer sought to enforce the agreement. The developer was unsuccessful: as the agreement did not meet the statutory requirements it was not legally binding.

Further reading

Galbraith, A., Stockdale, M., Wilson, S., Mitchell, R., Hewitson, R., Spurgeon, S. and Woodley, M. (2011) *Galbraith's Building and Land Management Law for Students*, 6th edn, Butterworth-Heinemann, Oxford.

Wood, D., Chynoweth, P., Adshead, J. and Mason, J. (2011) *Law and the Built Environment*, 2nd edn, Wiley-Blackwell, Chichester.

10.2 Legal definition of land

Key terms: land; airspace; mines and minerals; incorporeal hereditaments

It is essential for property professionals to understand how the word 'land' should be construed in legal documents governing the ownership, use and occupation of land.

The definition of land provided by Section 205(1)(ix) of the Law of Property Act 1925 encompasses both a description of physical property (land includes 'land, and mines and minerals … buildings or parts of buildings') and intangible rights in the land ('also a manor, an advowson, and a rent and other incorporeal hereditaments, and an easement, right, privilege, or benefit in, over or derived from land'). This section is concerned with the physical aspects of land.

The definition above does not give a complete explanation of what is meant by the word 'land'. There is a thirteenth-century saying that is still recognised as a good explanation of landownership today: *cujus est solum, ejus est usque coelum et ad inferos*, meaning that the owner of the soil is presumed to own everything 'up to the sky and down to the centre of the earth' (Megarry and Wade, 2012: 53).

Since the advent of air travel the maxim referred to above has had to be adapted and has been the subject of litigation. In *Bernstein of Leigh (Baron) v Skyviews & General Ltd* (1978) Q.B. 479, a claim for trespass was brought against a company that took aerial views of houses to sell the photographs to the homeowners. The claim was unsuccessful as it was held that an owner's rights to airspace above his land extend only to such height as is necessary for the ordinary use and enjoyment of the land. A wide dispensation to aircraft has since been granted under the Civil Aviation Act 1982. In a number of cases involving the development of land it has become an established principle that it is not permissible to swing a crane over another person's land without consent; see for example *Anchor Brewhouse Developments Ltd v Berkley House (Docklands Developments) Ltd* (1987) 2 EGLR 173.

In relation to subterranean areas the starting point is that the owner of the surface of the land is also the owner of the mines and minerals beneath it; however, there are a number of exceptions to this. The Crown is entitled to all gold, silver and petroleum, and all interests in coal are now vested in the Coal Authority (Coal Industry Act 1994). Unlike the rules regarding airspace the physical extent of ownership does not seem to be limited to what is necessary for the reasonable enjoyment of the land. In *Bocardo SA v Star Energy UK Onshore Ltd* (2010) UKSC 35, the defendants were held to have trespassed when oil drilling pipelines were constructed beneath the claimant's property (without its consent) at depths ranging from 800–1,300 ft to around 2,900 ft which had no impact on the claimant's use and enjoyment of the land.

Under the Treasure Act 1996 all treasure is vested in the Crown and anyone finding something they believe to be treasure has a duty to inform the coroner for the district (s. 8). Treasure includes (s. 1, 3) gold or silver coins that are at least 300 years old and other objects that are at least 200 years old and have outstanding historical, archaeological or cultural importance.

Under Section 132(1) of the Land Registration Act 2002 'land' includes land covered by water. The definition of land also includes any plants growing on the land whether or not they are cultivated. During their lifetime wild animals cannot, by definition, be anyone's property but if they are lawfully killed they become the property of the landowner.

The extent to which buildings, and parts of the buildings, form part of the land, is discussed in Concept 10.3.

Further reading

Gray, K. and Gray, S. (2011) *Land Law*, 7th edn, Oxford University Press, Oxford.

Megarry, R. and Wade, W. (2012) *The Law of Real Property*, 8th edn, Sweet & Maxwell, London.

10.3 Fixtures and chattels

Key terms: fixtures; chattels; degree of annexation; purpose of annexation

As seen in Concept 10.2 the legal definition of land includes buildings and other structures attached to it. It can often be challenging to identify what items at a property are included within this definition. It is particularly important to be able to do so when dealing with the transfer of land (identifying what is and is not included in the sale or gift), assessing liability for Stamp Duty Land Tax (SDLT is only payable on land), dealing with mortgages of land and determining the extent of landlords' and tenants' obligations (maintenance responsibilities, status of items at the end of the term etc.).

In *Elitestone Ltd v Morris* (1997) 1 WLR 687 three categories of objects in, on or under the land were identified:

1 things that are 'part and parcel' of the land that are impossible to remove intact, for example a house of conventional construction;
2 fixtures that are capable of being physically removed from the land but are treated in law as part of the land; and
3 chattels that may be removed from the land and are not considered part and parcel of the land.

It is therefore necessary to consider how to differentiate between fixtures and chattels. The two main tests that have evolved to determine whether or not an object should be treated as part of the land or as a chattel are:

(a) the degree of annexation; and
(b) the purpose of annexation.

The first element of the test described above requires us to consider to what extent the object is physically connected to the land or building. If something is fixed to the land in a way that it cannot be removed without causing damage to the property it will normally be treated as a fixture.

The purpose of annexation has become the primary test and it is an objective question to be determined on the 'externally observable circumstances of the case' rather than the subjective purpose of the person who attached the item (Gray and Gray, 2011: 12). An object may be physically attached to a building but this may be for the better enjoyment of the object rather than as a way of enhancing the building; for example, it was held by the House of Lords in *Leigh v Taylor* (1902) AC 157 that although some tapestries had been annexed to a building it was for the purpose of viewing the tapestries rather than improving the building and so they were determined to be chattels. The distinction is well illustrated by Blackburn J. In *Holland v Hodgson* (1872) L.R. 7 C.P. 328 at 335:

> blocks of stone placed one on top of another without any mortar or cement for the purpose of forming a dry stone wall would become part of the land, though the same stones, if deposited in a builder's yard and for convenience sake stacked on top of each other in the form of a wall, would remain chattels.

In the case of a leasehold property the starting point is that all fixtures attached by the tenant belong to the landlord (as part of the land) and therefore must be left at the property at the end of the term of the lease. However, certain fixtures which are known as 'tenant's fixtures' may be removed by the tenant provided that any damage is made good; these consist of: trade fixtures, some ornamental fixtures and some agricultural fixtures. Set out below are a couple of examples of where the designation of an item at a property caused problems for those involved in the transaction *TSB Bank plc v Botham* (1996) EGCS 149. After the home owners defaulted on their mortgage repayments the mortgagee elected to exercise its power of sale but a dispute arose as to whether or not a range of items at the house were fixtures (and therefore subject to the mortgage) or chattels (in which case the mortgagor could remove them).

In the case of *Orsman v HMRC* (2012) UKFTT 227 (TC), a house was sold for £250,000 with an additional £800 to cover the cost of some fitted units in the garage. The SDLT return for the purchase was £2,500 (1 per cent of the price of the land) on the basis that the units were not part of the 'land transaction'. HMRC took a different view and argued that the fitted units were fixtures and therefore part of the land in which case the transaction moved into the next band of SDLT where 3 per cent of the purchase price is payable, which meant that a payment of £7,524 was due. The First Tier Tribunal Tax Chamber applied the two-stage test above and agreed that due to their degree and purpose of annexation they were fixtures and therefore subject to tax.

Further reading

Gray, K. and Gray, S. (2011) *Land Law*, 7th edn, Oxford University Press, Oxford.
TSB Bank plc v Botham (1996) EGCS 149.

10.4 Ownership of land

Key terms: freehold; leasehold; commonhold

Our current system of landownership can date its roots back to the Norman Conquest. Essentially under English law all land is owned by the Crown and landowners own an estate in land rather than the land itself. Gray and Gray (2011: 16) explain that the doctrine of estates means that each landholder owns a 'slice of time in the land'. In medieval times there were a variety of estates, however, in modern times only two remain under section 1(1) Law of Property Act 1925:

1 freehold (also known as a fee simple absolute in possession); and
2 leasehold (a term of years absolute in possession).

The key difference between a freehold estate and a leasehold estate is that a leasehold estate exists for a period of time whereas a freehold is infinite. In *Walsingham's Case* (1573) 2 Plowd. 547, it was held that he 'who has a fee-simple in land has a time in the land without end, or the land for time without end'.

In English law freehold ownership is the closest that anyone (other than the monarch) can get to absolute ownership of land and except on rare occasions the Crown's role is not apparent. There are a few hundred cases of escheat a year (Gray and Gray, 2011: 121), which is where land reverts to the Crown because the freehold is no longer held by anyone; this can arise where someone dies without leaving any successors, or when a company is insolvent or an individual is bankrupt and the insolvency practitioner exercises their statutory powers to disclaim 'onerous property'.

Generally, leasehold estates are regarded as inferior to freeholds; however, as Megarry and Wade (2012: 40) argue, a rent-free lease for 3,000 years may be as valuable as a freehold estate. There are three main types of lease: fixed term, periodic and tenancies at will or sufferance. A fixed term lease can be for any period of time although some conventions have arisen especially within the residential property market where long leases are commonly 99, 125 or 999 years. A periodic lease is one for an initial fixed period but allows the lease to continue until terminated by either landlord or tenant in accordance with the lease, for example a tenancy from month to month could by ended by either party upon giving a month's notice. A tenancy at will is quite rare – either party can end the agreement on demand so it does not satisfy the needs of most property investors or occupiers and is not classed as a leasehold estate in land. If a lease is for an uncertain term then it is invalid, see for example *Lace v Chantler* (1944) KB 368 in which case the term was granted 'until the war ends'.

Commonhold is a special type of freehold that was introduced by the Commonhold and Leasehold Reform Act 2002. It was introduced to deal with multiple unit properties such as apartment blocks and mixed use schemes that would otherwise be set up by granting leases to the purchasers. In commonhold schemes individual unit owners own the freehold of their unit and a share in the commonhold association that owns and manages the common parts of the development. In particular, it was hoped that commonhold would address the following problems:

- tenants finding the value of their property reducing as the residue of the term of the lease diminishes;
- enforceability of positive covenants (see Concept 10.8 on freehold covenants); and
- the enforcement of obligations between unit holders (in a more traditional freehold–leasehold structure only the landlord can enforce tenant covenants).

There have been very few commonhold schemes created so far (Fetherstonhaugh, 2007) It appears that this reform, which aims to create a type of condominium ownership that is popular in other parts of the world such as Singapore, has not been a success. Last decade there was considerable debate among academics and practitioners as to why there had been so little use of commonhold schemes, but in the last few years even the discussion of commonhold appears to have faded away.

Further reading

Fetherstonhaugh, G. (2007) 'Developers need a nudge in the right direction', *Estates Gazette*, October (Online), available at: www.egi.co.uk/Articles/Article.aspx?liArticleID=662085&NavigationID=466, (accessed 20/11/2013).
Gray, K. and Gray, S. (2011) *Land Law*, 7th edn, Oxford University Press, Oxford.
Megarry, R. and Wade, W. (2012) *The Law of Real Property*, 8th edn, Sweet & Maxwell, London.

10.5 Trusts and co-ownership of land

Key terms: joint tenancy; tenancy in common; overreaching

Land may be held on trust by trustees for a variety of reasons:

- For individuals who are entitled to the land in succession. For example, X by her will leaves the family home to her second husband, Y, for life (he has a life interest) and after his death to her children (this is known as an interest in remainder).
- For individuals who hold an interest in land at the same time. For example, where business partners hold their premises jointly or couples own their homes together.
- For an individual where other people are appointed to manage the property on the individual's behalf; this is known as a 'bare trust'. This may arise where the person establishing the trust feels that the individual they would like to have the property would benefit from someone more experienced or cautious having control of it. However, if the beneficiary calls for the land to be transferred to him/her then the trustee must comply and the trust will come to an end (*Saunders v Vautier* (1841) 4 Beav. 1 15).

Under Section 6(1) of the Trusts of Land and Appointment of Trustees Act 1996 (TLATA) trustees 'have all the powers of an absolute owner'. They can therefore sell, lease or mortgage the land. However, in doing so they must take into account the rights of the beneficiaries and are also under a duty to consult beneficiaries aged eighteen or over (ss 6(5) and 11(1) TLATA). Trustees must also ensure that they do not act in breach of trust or their fiduciary duties, for example selling the trust property to themselves would not be lawful (Megarry and Wade, 2012: 470).

Co-ownership is the most common type of trust in modern times. The co-owners hold the property on trust as trustees for themselves as beneficiaries. There are two main types:

1. joint tenancy; and
2. tenancy in common.

The use of the word tenancy here can be misleading – it relates to co-ownership and not leases. In the case of a joint tenancy the co-owners hold the entire property together and upon the death of a joint tenant the surviving joint tenant(s) automatically acquires the deceased's interest in the land. In the case of a tenancy in common the property is held in undivided shares and the owners can leave their interest to whoever they choose. This is illustrated in the examples below.

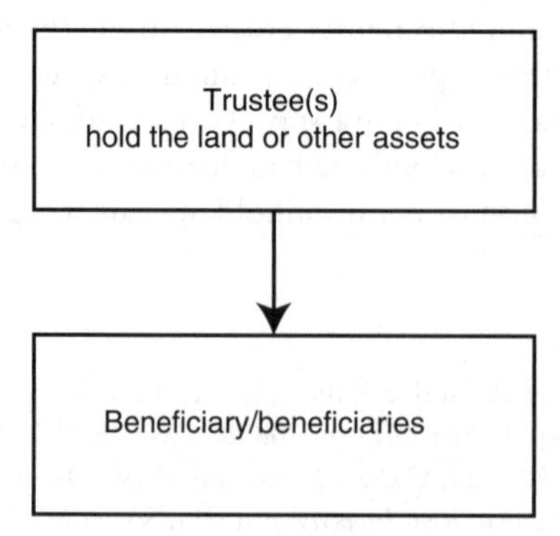

Figure 10.5.1 Trustee holding.

Joint tenancy

Mr and Mrs Smith hold the property together in one holding (Table 10.5.1).

Table 10.5.1 Joint tenancy (1)

Trustee	John Smith and Jane Smith
Beneficiary	John Smith and Jane Smith

John Smith dies and Jane is the sole surviving joint tenant (Table 10.5.2).

Table 10.5.2 Joint tenancy (2)

Trustee	Jane Smith
Beneficiary	Jane Smith

As a consequence the trust comes to an end and Jane is the sole owner of the land.

Tenants in common

Mr and Mrs Smith buy a house together but as Mr Smith contributed a greater amount towards the purchase price it is agreed that they will own the property as tenants in common with John having a 75 per cent share of the land and his wife 25 per cent (Table 10.5.3).

Table 10.5.3 Tenants in common (1)

Trustee	John Smith and Jane Smith	
Beneficiary	John Smith 75%	Jane Smith 25%

John Smith dies (Table 10.5.4).

Table 10.5.4 Tenants in common (2)

Trustee	Jane Smith	
Beneficiary	Person who inherits under John's will or person entitled to inherit under the rules of intestacy (which apply if someone dies without making a will) 75%	Jane Smith 25%

After a property is acquired the co-owners' situation might change, for example a relationship may break down, and the co-owners may no longer want to own the property on a joint tenancy basis. They can at this stage choose to sever the beneficial joint tenancy; this has the effect of transforming a joint tenancy into a tenancy in common. Severance can occur in a number of ways including:

- serving a statutory notice (s36(2) Law of Property Act 1925);
- a joint tenant disposing of their interest in the land during their lifetime;
- by mutual agreement of the joint tenants;
- on the bankruptcy of a joint tenant; and
- if one joint tenant criminally kills another.

Resulting and constructive trusts are a form of implied co-ownership that can arise when the behaviour of the legal owner of the land and another person gives rise to the implication that the owner is holding the property on behalf of him/herself and another. The most common situation where this arises is where a couple is co-habiting but the family home is only in one partner's name. Behaviour that would lead to a presumption that the property is held on trust includes contributions towards the purchase price, acquisition costs or mortgage payments or meeting other household expenses, enabling the legal owner to pay the mortgage (*Dyer v Dyer* (1788) 2 Cox Eq. Cas. 92, *Gissing v Gissing* (1971) AC 886).

If on every occasion co-owned and trust property was sold the buyer had to investigate the details of the trust and the rights of the beneficiaries it would add further delays to the conveyancing process. In order to simplify the investigations that a prospective purchaser of land needs to undertake, the law provides (ss 2 and 27 of the Law of Property Act 1925) that if the proceeds of sale are paid to two or more trustees then the interests of the beneficiaries of the trust are said to have been overreached, which means that the rights of the beneficiaries are detached from the land and attached to the sale proceeds and any action by the beneficiaries would be against the trustees rather than a third party purchaser.

Further reading

Megarry, R. and Wade, W. (2012) *The Law of Real Property*, 8th edn, Sweet & Maxwell, London.
Stroud, A. (2010) *Making Sense of Land Law*, 3rd edn, Palgrave Macmillan, Basingstoke.

10.6 The lease/licence distinction

Key terms: exclusive possession; interest in land; statutory protection

This is a potential trap for the unwary. Documents under which someone is granted the right to use and occupy land for a period of time usually in return for payment are sometimes labelled as licences when in reality they are leases. In the leading cases on this topic, *Street v Mountford* (1985) AC 806, Lord Templeman referred to these types of agreement as 'sham devices' and held that in determining whether or not a document creates a lease or a licence it is necessary to look at the substance rather than the form of the contract. The reasons that it is important to be able to distinguish between leases and licences are:

- A lease creates an interest in land whereas a licence is a personal/contractual right. As a consequence, a lease will bind people beyond those who were parties to the original contract; for example if someone acquires the landlord's interest they will have to observe the terms of the lease, whereas a licence is not an interest in land and in the event of a sale of a property subject to a licence the new owner is not normally bound by the terms of the licence.
- Stamp Duty Land Tax is payable on some leases but not on licences.
- Generally only tenants under leases benefit from statutory protection such as the security of tenure provisions of the Landlord and Tenant Act 1954. Licensees do not normally benefit from such rights.

It is therefore of essential importance to be able to distinguish between a lease and a licence. The key characteristic of a lease is that the tenant (even if they are referred to in the contract as a licensee) will have been granted the right to exclusive possession of the land. Exclusive possession means that they have the right to exclude all others, including the landlord (except for rights under the lease to inspect or repair) from the land.

In order to grant exclusive possession there must be a clearly defined area available for exclusive use. In *Clear Channel UK Ltd v Manchester City Council* (2005) EWCA Civ 1304, an agreement relating to the erection and use of advertisement hoardings was held not to be a lease because there were undefined areas in which the concrete bases and hoardings could be placed.

In *Antoniades v Villiers* (1990) 1 A.C.417, the property owner granted two separate 'licences' to a co-habiting couple. Each licence provided that the small flat was to be shared with anyone designated by the owner. In reality none of the parties intended that anyone other than the couple should live in the flat and it was held that the couple had a joint tenancy as they were together entitled to exclusive possession of the property.

However, according to Megarry and Wade (2012: 757):

> where a contract has been negotiated between parties of equal bargaining power, with the benefit of legal advice, the courts are generally reluctant to go behind express provisions clearly indicating the parties' intentions as to the legal effect of the agreement they have executed.

For example, in *Cameron Ltd v Rolls-Royce Plc* (2007) EWHC 546 Ch, it was held that there were exceptional circumstances in which one party could be enjoying exclusive possession of a property for a rent but a lease was not created. In this case Rolls-Royce had been allowed to occupy the premises under a document described as a licence pending the completion of a formal lease. Mann J determined that despite the fact that the defendant was enjoying exclusive possession the interest granted to them was only a licence.

Pankhania v London Borough of Hackney (2004) EWHC 323 Ch is a case that should serve as a warning of the consequences of not understanding (and acting appropriately) on this distinction. In Pankhania the Council was selling a land at auction which was marketed as a development site and described in the auction catalogue as being sold 'subject to a licence' in favour of NCP. When the successful bidder gave notice to NCP to quit the site they were advised by NCP's solicitors that NCP had a lease and was therefore protected by the Landlord and Tenant Act 1954. After paying NCP £78,931.25 compensation and a delay of about 15 months Pankhania successfully brought a claim for misrepresentation against the seller and was awarded £500,000 in damages. Subsequently, the seller brought a claim against their solicitor and agent who prepared the auction catalogue (Rowe, 2006) but the outcome of these proceedings has not been reported and it is surmised that this matter will have been settled out of court.

Further reading

Megarry, R. and Wade, W. (2012) *The Law of Real Property*, 8th edn, Sweet & Maxwell, London.

Rowe, S. (2006) 'Nelson Bakewell sued for auction misrepresentation', *Estates Gazette*, February, available at: www.egi.co.uk/Articles/Article.aspx?liArticleID=631666&NavigationID=466 (accessed 20/11/2013).

10.7 Land registration

Key terms: registered land; unregistered land; indemnity; sale prices; prescribed clauses

Land registration in England and Wales started in the nineteenth century but our current system of land registration dates back to the Land Registration Act 1925 which has now been replaced with the Land Registration Act 2002.

The Land Registry has registered more than 23 million titles and they estimate that title to around 70 per cent of land in England and Wales is registered. Initially, land registration was voluntary but later certain transactions triggered compulsory registration, e.g. the sale of a freehold or the grant of a long lease. The reason that title to a significant portion of land is unregistered is that there may not have been any trigger events since compulsory registration applied across the whole of England and Wales (1 December 1990, SI 1989/1347). This is particularly likely to be the case with rural land, and trust or corporate properties. There is a campaign by the Land Registry, RICS and the Law Society to promote the advantages of voluntary first registration and there are fee incentives to encourage this.

The system used is a title registration system rather than a land registration system. This means that for one parcel of land there may be more than one title, e.g. a freehold title and a leasehold title. Since 1990 the land register has been open to public inspection. If someone is investigating some land and they do not know the title number they can look this up using the post code on the Land Registry's website or carry out a search of an Index Map.

In the past the Land Registry issued Land Certificates and Charge Certificates (if the property was mortgaged) as evidence of title; however, these are no longer issued as the Land Registry has opted for a paperless system and anyone can order a copy of a title online or through the relevant District Land Registry.

Once title is registered it has a state guarantee which means that if there is a mistake on the register and someone suffers a loss as a consequence they will be indemnified for their loss by the Land Registry. In *Prestige Properties Ltd v Chief Land Registrar* (2002) EWHC 330, Lightman J found that a certificate issued following a search of the Land Registry incorrectly included 5 square metres of land that were not part of the property being sold. It was a term of the sale contract that there would be a retention of part of the purchase price so that if the buyer was unable to register title to the entire property within 6 months of the sale then the seller (Prestige) would reduce the purchase price. When the buyer was unable to register the erroneously included 5 square metres they relied on this contractual provision. The seller then successfully brought a claim against the Land Registry and was awarded £400,000 compensation.

The register of the title for a property consists of three parts:

- Property Register – this section describes the land, defines the extent of the property by reference to the title plan and identifies any rights that the property has the benefit of, for example rights of way and of light.
- Proprietorship Register – this part of the Register identifies the owner of the property and includes any restrictions on the owner's powers to dispose of the property, for example in relation to trust or mortgaged property. For any disposals since 1 April 2000 it may also include the price paid for the property.
- Charges Register – this includes details of any mortgages, easements and covenants or other encumbrances that the property is subject to.

In order to try to standardise the information available to those searching the Land Registry and promote transparency, prescribed clauses for registrable leases were introduced so that key information about the lease such as the term and break options are displayed at the start of the lease.

Further reading

www.landregistry.gov.uk (accessed 20/11/2013).

10.8 Freehold covenants

Key terms: dominant and servient tenements; positive and negative covenants; 'run' with the land; unity of ownership; indemnity insurance

Freehold covenants are defined by Dixon (2012: 323) as 'promises made by deed ("covenant") between freeholders, whereby one party promises to do or not to do certain things on their own land for the benefit of the neighbouring land'.

The difference between covenants and ordinary contractual provisions that seek to impose obligations relating to the use of land is that in some circumstances covenants 'run' with the land and as such they are capable of benefitting and burdening landowners who were not party to the original deed. In order for a covenant to exist there must be two parcels of land in separate freehold ownership (Table 10.8.1).

Table 10.8.1 Dominant and servient land

Plot one	*Plot two*
The person who has made the promise and accepted the burden of the covenant (e.g. their property is subject to a covenant) is known as the covenantor. His/her land is known as the servient land/tenement.	The person who has the benefit of the promise (e.g. their property has the benefit of a covenant – the right to enforce the obligation on the neighbouring land) is known as the covenantee. His/her land is known as the dominant land/tenement.

Covenants can only be created by deed and are negotiated for a number of reasons, including:

- Someone selling part of their land wants to restrict the way the sold land is used so that it does not have a detrimental impact on the use, enjoyment and value of the retained land.
- Someone might want to prevent a property they are selling being used for a business in competition with their own. A covenant of this nature may breach the provisions of the Competition Act 1998 and therefore would not be enforceable.
- Someone might want to impose a covenant restricting development in the hope that in the future the owner of the servient land may be willing to pay the dominant owner additional money in order to allow the covenant to be released or modified.

Covenants can be positive or negative in nature. Examples of some common covenants are shown in Table 10.8.2. Some covenants 'run' with the land and bind and benefit future owners of the land and the benefit of a covenant can also be expressly assigned under s136 Law of Property Act 1925 (Table 10.8.3).

Table 10.8.2 Positive and negative covenants

Positive	*Negative*
To build and maintain a boundary feature	Not to erect any buildings without the consent of the dominant landowner
To contribute towards the maintenance costs of a shared access	To only use the property for residential/office/agricultural purposes

Table 10.8.3 Covenants 'running with the land'

Does it run?	*Dominant land (benefit of the covenant)*	*Servient land (burden of the covenant)*
Positive covenant	In order for the benefit to run with the land the covenant must: 1 'touch and concern'/benefit the land; and 2 the person seeking to enforce the covenant holds a legal estate in the dominant land	The burden does not run
Negative covenant	As above	The burden will run if all the criteria set out in *Tulk v Moxhay* (1848) 2 Ph 774 are satisfied

As can be seen in Table 10.8.3, the burden of a positive covenant does not run with the land. This can pose a problem for landowners. If we take the example of financial contributions towards a shared amenity it is essential that the dominant landowner can enforce payment from the original covenantor's successors in title. A range of solutions are employed in these circumstances which include preventing disposals of the servient land without the consent of the dominant landowner (such consent is only given if the buyer of the servient land enters into a new covenant) and making the right to exercise rights over the shared resources conditional upon making a financial contribution.

If the servient landowner wishes to do something in breach of covenant (for example the covenant restricts any development of the land and they want to build a couple of houses) he/she has a number of options:

1 Ignore the covenant. The dominant landowner could bring a claim against the servient owner for an injunction and/or damages. Even if the dominant owner does nothing it will be very difficult to raise finance for the development or sell the completed houses as the prospective lender and purchasers will not normally be willing to run the risk of the covenant being enforced.
2 Approach the dominant landowner and seek to negotiate a deed of release or modification. This will normally involve a payment to the dominant landowner and may be difficult to secure, especially if the dominant land has been subdivided, in which case there may be multiple dominant landowners.
3 Acquire the dominant land. This may or may not be practical but unity of ownership of the dominant and servient land will extinguish a covenant.
4 Apply to the High Court for a declaration under s84 (2) Law of Property Act 1925 as to whether or not the covenant is still enforceable.

   This can be a useful tool if there are grounds for thinking the covenant is unenforceable but it is quite a lengthy process.
5 Apply to the Upper Tribunal of the Lands Chamber under s84 (1) to modify or discharge the covenant.

   This can be done if it can be shown that the covenant is obsolete, unduly obstructive or of no real benefit. If a claim is acceptable compensation will normally be payable to the dominant landowner.
6 Indemnity Insurance. In some circumstances insurance may be available against the risk of enforcement of the covenant. This course of action will not remove the covenant and cover is not normally available if the servient owner has already attempted to use methods 2–5 above. However, it can be a quick and cost-effective way of dealing with the problem.

Further reading

Dixon, M. (2012) *Modern Land Law*, 8th edn, Routledge, Abingdon.
Stroud, A. (2010) *Making Sense of Land Law*, 3rd edn, Palgrave Macmillan, Basingstoke.

10.9 Easements and *profits à prendre*

Key terms: dominant and servient tenements; implied easements and easements of necessity; prescriptive rights; abatement; abandonment; unity of ownership

An easement is the right to use land in one person's ownership for the benefit of a piece of land owned by another person. Common examples of easements are rights of way, rights to install and keep service apparatus, and rights of light.

It is essential that an easement has a dominant and a servient tenement (two pieces of land) in separate ownership.

In Table 10.9.1 A has granted B an easement over A's land consisting of a right of way over a road to allow B (and B's successors in title) to access property B.

Table 10.9.1 Example of an easement

	Plot A	*Plot B*
The Highway	The private road within A's property This is the servient land; it is subject to the burden of the easement (the right of way).	This is the dominant land; it has the benefit of the easement over the servient land.

If the benefit of the right of way was attached to a person rather than a parcel of land then it would normally exist as a licence or a lease.

- The right must make the dominant land a better and more convenient property and the dominant and servient land must be close but not necessarily adjacent.
- The right must be of a type that can be recognised in law as an easement. Some examples of rights that have been held not capable of existing as easements are the right to an unspoilt view (*William Aldred's Case* (1610) 9 Co. Rep. 57b) and the right to receive television or radio signals (*Hunter v Canary Wharf Ltd* (1997) AC 655).

There are a number of ways in which an easement can be created:

- The most common way is in a deed when someone is disposing of part of their property and needs to either reserve rights over the land that they are selling to benefit the retained land or grant rights to the land being sold so that both parcels of land can be used and enjoyed after they are severed from each other.
- In some circumstances if rights have not been expressly granted or reserved they may be implied.
- Easements of necessity may arise where it is impossible to use the land retained or sold as there is no way of accessing it.
- Acts of Parliament and Compulsory Purchase Orders.
- Easements can also be created by presumed grant or prescription which can arise where a right has been enjoyed for a long time (generally 20 years) without the use of force, secrecy or the permission of the servient landowner.

Disputes frequently arise about the exercise of easements. If someone is prevented from enjoying their easement they can bring a claim against the dominant landowner seeking a declaration of their rights, an injunction or damages or a combination of these remedies. The dominant landowner may also try and resolve the problem through a self-help remedy known as abatement; provided that they only use reasonable force they can remove anything obstructing their right to enjoy the easement, for example by cutting a chain padlocking a gate restricting access to their right of way.

If a servient landowner wishes to change the way they use their land or develop the property they have a number of options:

1. Ignore the easement. The dominant landowner could bring a claim against the servient owner for an injunction and/or damages. Even if the dominant owner does nothing it will be very difficult to raise finance for the development or sell the completed scheme as the prospective lender and purchasers will not normally be willing to run the risk of the easement being enforced.
2. Approach the dominant landowner and seek to negotiate a deed of release or modification. This will normally involve a payment to the dominant landowner and may be difficult to secure, especially if the dominant land has been subdivided, in which case there may be multiple parties to negotiate with.

3 Abandonment. The servient landowner may seek to claim that the dominant landowner has abandoned the easement. This is very difficult to prove, for example, in *Benn v Hardinge* (1992) 66 P & CR 246, 175 years non-use was insufficient evidence that the dominant landowner had abandoned the right.
4 Acquire the dominant land. This may or may not be practical but unity of ownership (the dominant and servient land being in the same ownership) extinguishes an easement.
5 Indemnity Insurance. In some circumstances insurance may be available against the risk of the enforcement of an easement. This course of action will not remove the easement and cover is not normally available if the servient owner has already attempted to use methods 2–4 above; however, it can be a quick and cost-effective way of dealing with the problem.

A *profit à prendre* (profit) is a right to take something from another person's land; for example, timber and peat. A profit can exist in the same way as an easement (the right to take something from the servient land benefits the dominant land) but can also exist 'in gross' which means without a dominant tenement. Profits are not frequently created in modern times as it is more common to use a contractual licence but historic rights, in particular sporting rights (hunting, shooting and fishing) and pasture rights, should not be overlooked when appraising land.

The Law Commission's (2011) report, *Making Land Work: Easements, Covenants and Profits à Prendre*, recommended that the rules regarding the creation of easements by prescription be simplified and that the Upper Tribunal (Lands Chamber) be given additional powers to modify or discharge easements that no longer serve any useful purpose. To date the government has not implemented these proposals.

Further reading

Great Britain Law Commission (2011) *Making Land Work: Easements, Covenants and Profits à Prendre*, Report 327 (Online), available at: http://lawcommission.justice.gov.uk/docs/lc327_easements_summary.pdf (accessed 20/11/2013).
Stroud, A. (2010) *Making Sense of Land Law*, 3rd edn, Palgrave Macmillan, Basingstoke.

10.10 Easements – rights to light

Key terms: right to light; easement; injunction; damages; Law Commission consultation

The ability to access daylight and sunlight from within a building is important for a number of reasons including general amenity, and the health and well-being of occupants. The availability of natural lighting and warmth also supports the sustainability agenda by reducing the energy demands associated with providing artificial light and heating to a building.

The loss of sunlight to a building resulting from a proposed development constitutes a material consideration that planning authorities will take into account when determining a planning application. Planning authorities apply the methodology set out in the Building Research Establishment guide on site layout planning for daylight and sunlight (Littlefair, 2011). The applicant may be required to submit a report carried out in accordance with the guidance, depending on the potential of a proposal to reduce daylight or sunlight levels.

Irrespective of whether planning permission is granted or not a *legal right of light* may exist which planning permission cannot override.

A right to light is a negative *easement*. In England and Wales a right to light is most commonly acquired under the Prescription Act 1832. Under the Act a right to light automatically occurs when light has been enjoyed through defined apertures of a building for a period of 20 years without interruption or consent. A right of light can also be acquired by express grant. Care must be taken to carefully check conveyancing documentation that may grant a right to light; it should be noted however that such documentation may also exclude rights.

To determine whether a right to light has been infringed the test is how much light remains rather than how much has been taken away. In *Colls v Home and Colonial Stores* (1904) Lord Lindley stated:

> An owner of ancient lights is entitled to sufficient light according to the ordinary notions of mankind for the comfortable use and enjoyment of his house as a dwelling-house, if it is a dwelling-house, or for the beneficial use and occupation of the house if it is a warehouse, a shop, or other place of business.

Sufficient light is generally considered to be 1 lumen per square foot at table top height over half of the room in question.

Specialist rights to light surveyors are able to calculate whether an exiting or proposed development infringes or will infringe a right to light. If an interference with a right to light has occurred, or will occur if a development goes ahead, a Court may issue an injunction ordering a developer not to start a development, to stop building or to demolish a building. An alternative to an injunction is for the Court to order the payment of compensation to the party suffering the infringement.

In the majority of cases a negotiated settlement can be arrived at. The owner of the right to light holds the upper hand in such negotiations as the extreme option of an injunction is normally available. The British Property Federation is lobbying for the introduction of a protocol to provide an agreed basis for negotiating loss on the basis of the loss of amenity

Because of increasing concerns that rights to light can cause extra cost and delay to major developments, the Government has referred the matter to the Law Commission who published a consultation paper in February 2013. The provisional proposals are as follows:

- In the future it should no longer be possible to acquire rights to light by long use (known as 'prescription').
- The introduction of a new statutory test to clarify the current law on when courts may order a person to pay damages instead of ordering that person to demolish or stop constructing a building that interferes with a right to light.
- The introduction of a new statutory notice procedure requiring those with the benefit of rights to light to make clear whether they intend to apply to the court for an injunction.
- A process for extinguishing rights to light that are obsolete or have no practical benefit, with payment of compensation in appropriate cases.

Subject to the outcome of the consultation the Government intends to publish a draft Bill in late 2014.

For clarification, access to sunlight and daylight conferred either by the planning system or by virtue of a legal right to light does not give the occupant of a building the *right to a view*. In *Phipps v Pears* (1965), Lord Denning stated that if a neighbour decides to despoil a view that has been enjoyed for many years by building up and blocking it there is no redress in the law. Guidance issued by the Department for Communities and Local Government on determining planning applications makes it clear that the loss of a view is not a material consideration.

Further reading

Law Commission (2013) Consultation Paper 210 'Rights to Light', available at: http://lawcommission.justice.gov.uk/docs/cp210_rights_to_light_version-web.pdf (accessed 21/11/2013).

Littlefair, P. (2011) *Site Layout Planning for Daylight and Sunlight: A guide to good practice*, 2nd edn, BRE Press, Bracknell.

10.11 Manorial land and chancel repair liability

Key terms: overriding interests; sporting rights; mines and minerals; registration

Manorial rights have their roots in feudalism and are the ancient rights of lords of the manor. Examples of manorial rights are sporting rights, rights to mines and minerals and rights to hold fairs or markets. The rights are held separately to the freehold owner of the land and conflicts can arise between the way the freehold owner of the land wishes to use the land and the right holder's requirements. These rights can still be very valuable today, especially mining rights.

Most of the manorial rights that exist today arise where the land was formerly copyhold or the land was the subject of an inclosure award. Copyhold land was a type of property ownership where the tenants held the land from the lord of the manor who normally retained sporting rights and the right to minerals in the land. Copyhold land was abolished in 1926 when all copyhold interests were converted into freehold ownership but the rights of the lord of the manor were preserved. During the eighteenth and nineteenth centuries when land was being inclosed the awards sometimes made provision for the reservation of manorial rights.

Chancel repair liability dates from a time when church rectors were responsible for the repair of the chancel of the church which was maintained using the income from 'glebe' lands. Over time the former glebe lands were sold off but the liability to contribute towards the cost of the chancel repairs remained attached to the land. In modern times these rights were regarded as an archaic anomaly that was no longer significant until the judgment in *Parochial Church of Aston Cantlow and Wilmcote and Billesley, Warwickshire v Wallbank* (2003) UKHL 1 AC 546. In that case the Wallbanks had inherited some former glebe land and then received a bill for almost £200,000 for chancel repairs. The Wallbanks disputed their obligation to pay for the repairs but after a long legal battle were ultimately held liable to pay the full amount. The case was very controversial and was a catalyst for reform.

The position under the Land Registration Act 1925 was that both manorial rights and chancel repair liability took effect as overriding rights. This meant that even if no details of the rights were registered against the title to the property (and therefore the owner could be ignorant of their existence) they would still be enforceable against the freehold owner. The Land Registration Act 2002 has changed the law so that if someone has the benefit of a manorial right or a right to enforce payment of chancel repair liability and has not registered it against the title of the land (or, in the case of unregistered land, registered a caution against first registration) which is subject to the burden of those rights by 13 October 2013, then the rights will be unenforceable against future purchasers of the land. The reform of the land registration rules will, however, not be of assistance to those who already owned the burdened land in October 2013, who will remain bound by these rights regardless of registration.

Further reading

Mocatta, J. (2009) 'Medieval relic causes a modern hangover', *Estates Gazette* (online) available at: www.egi.co.uk/Articles/Article.aspx?liArticleID=705519&NavigationID=466 (accessed 25/11/2013).

10.12 Wayleaves

Key terms: utilities; private land; service apparatus; wayleave fee

Wayleaves are a common tool for granting utility providers rights to install and keep their apparatus in, on or under third party land. There is no statutory definition of a wayleave; they are normally characterised as a contractual licence and are therefore a personal right rather than an interest in land and cannot be registered at the Land Registry.

Northern Powergrid (no date) advise that the standard provisions of one of their wayleaves would include the right to:

- install and keep installed the electric line(s);
- use, inspect, maintain, adjust and remove the line(s);
- cut, lop, trim or fell trees and hedges; and
- enter the land at all reasonable times.

In exchange for granting a wayleave the landowner will normally receive a small fee that could be a one-off payment or annual charge. The amount of the payment is generally based on agricultural land values. Most wayleaves are for a fixed term or include provisions for the termination of the agreement.

If someone buys land that has utility apparatus situated in/on/under it, in accordance with a wayleave agreement the new owner of the land will not be bound by the terms of the wayleave as it is a licence rather than a lease or an easement. However, implied wayleaves are frequently created if the new owner accepts the wayleave payments.

If a statutory undertaker is unable to negotiate a new wayleave with a property owner, or if a new owner requests that apparatus is removed or if a property owner seeks to terminate a wayleave agreement, the utility companies benefit from statutory powers. For example, an electricity undertaker can apply to the Secretary of State for Energy and Climate Change who can grant a wayleave under Schedule Four of the Electricity Act 1989 if it is 'necessary and expedient to do so'. Before granting such a wayleave the Secretary of State must arrange for a 'necessary wayleave hearing' to be held to provide the landowners an opportunity to be heard. If a wayleave is granted the landowner is entitled to compensation under section 5 of the Land Compensation Act 1961.

While wayleaves remain a popular choice for electricity apparatus, other statutory undertakers are making increased use of easements and leases instead as these provide the utility company with greater security of tenure. The utility providers also benefit from compulsory purchase powers under which they can acquire easements to allow them to install their equipment.

Further reading

National Powergrid (no date) 'Frequently Asked Questions', available at: www.northernpowergrid.com/som_download.cfm?t=media:documentmedia&i=1087&p=file (accessed 20/11/2013).

O'Brien, G. and Hamer, C. (2007) *Electricity Wayleaves, Easements and Consents: Litigation, practice and procedure*, EG Books, Abingdon.

10.13 Common land and town and village greens

Key terms: rights of common; commons search; Commons Act 2006

The origins of town and village greens (TVGs) and common land lie in customary law. TVGs and common land are subject to the rights of the public but can often be in private ownership.

According to Defra (2013) around 550,000 hectares of land in England and Wales (4 per cent of the total land area) is registered as common land. Common land is subject to the rights of the public to exercise rights of common over it which includes grazing livestock and cutting turf or wood for fuel. If someone wants to carry out any development of common land they need to secure the consent of the Secretary of State to do so. Failure to obtain consent is a criminal offence.

Most common land was established as such hundreds of years ago and can be protected by registration under the Commons Act 2006. New areas of common land are only created as a replacement for common land that has been lost through compulsory purchase or by the landowner applying for deregistration and exchange of land under section 16 of the Commons Act 2006. Town and village greens are land that has been used by people in the locality for public recreation for a long time. The registration of land as TVGs has caused a certain amount of controversy in recent years. It has been asserted that applications to register land have been used as a weapon against developers. Once land is registered as a TVG it is a criminal offence to enclose it or carry out any building works or disturbance other than works carried out to promote the enjoyment of the land as a TVG. Under Section 15 of the Commons Act 2006 anyone can apply to register land as a TVG and they need to prove that:

- it has been used by local people for at least 20 years;
- the use has been for recreational purposes; and
- the use has been 'as of right' (without force, secrecy or permission).

Due to the concerns about the use of these applications the laws relating to TVGs have been reformed. The Commons Act 2006 was amended in 2013 to restrict people's ability to apply for registration of TVGs. Landowners can now make a statement and register it with their county or unitary authority which has the effect of preventing the use of the land continuing 'as of right' (Sections 15A and 15B of the Commons Act 2006). Section 15C prevents anyone submitting an application for registration when a 'trigger event' has occurred. Trigger events include applications for planning permission or inclusion of the site in a draft local plan.

If someone wants to ascertain whether land is common land or TVG they can carry out a Commons Search with the local authority. This search is routinely carried out when a property is being purchased which includes unenclosed land. The search however, will not reveal whether there is a risk of a TVG application being made so an inspection of the land and enquiries of the owner would need to be undertaken in order to ascertain whether there is any danger of a TVG application being made.

Further reading

Defra (2013) 'Common Land and Village Greens', available at: www.gov.uk/common-land-village-greens (accessed 22/11/2013).

10.14 Highways

Key terms: passage; stopping up; adoption; privately maintained

A highway is the name for any piece of land over which the public has a right to pass and re-pass. They are sometimes referred to as 'public highways' but the word public is superfluous in that expression.

It is obviously very important for anyone dealing with the sale, lease, development or valuation of land to identify whether or not the land can be accessed directly from the highway. This information can be obtained from the highway authority (Highways Agency for motorways and trunk roads or the county or unitary authority in all other cases) and is part of the enquiries included in the Local Search which is routinely carried out in residential and commercial conveyancing.

Some highways pass over private land and are not maintainable at public expense. In most cases highways are maintainable at public expense either because they were created before 1836 or adopted since then. When a developer is constructing a new scheme they will normally enter into a Section 38 Agreement under the Highways Act 1980 under which the new roads are constructed to the highway authority's specification, maintained for a year after completion and then adopted by the highway authority. The interest of the highway authority in a highway maintainable at public expense is in 'the top two spits of the road' (*Tithe Redemption Commission v Runcorn Urban District Council* (1954) Ch 383) the subsoil of the road belongs to the adjoining owners; it is a rebuttable presumption that they own the land up to the centre of the highway.

It is a criminal offence to obstruct the highway in a way that endangers the use of it (section 137 Highways Act 1980). In order to allow property owners and developers the opportunity to maintain and enhance land a number of activities are regulated by licensing; these include skips, scaffolding, tables and chairs licences.

Highways can be created by the highway authority or by the dedication of the route by the landowner and acceptance by the public (this only relates to non-vehicular use). Dedication and acceptance can arise when the public has used a route for 20 years or longer without consent and as if they had the right to be there (section 31(1) Highways Act 1980).

When a new development is going to have a significant impact on the existing highway network a developer can be required to carry out or make a financial contribution towards off-site highway works. 'Once a highway always a highway' is an established legal maxim. However, it is possible to divert or stop up (extinguish) a highway under the rules contained in the Town and Country Planning Act 1990 and Highways Act 1980 where there is an equally convenient route or it is necessary to allow development to proceed.

Further reading

Denyer-Green, B. and Ubhi, N. (2013) *Development and Planning Law*, 4th edn, Routledge, Abingdon.
Hancox, N. and Wald, R. (2002) *Highways Law & Practice*, Butterworths, London.

10.15 Option agreements

Key terms: exclusivity; call options; put options; preconditions; option fee

An option has been defined as:

> [A] unilateral right, created by contract, enabling one of the parties, if he so wishes and the circumstances are covered by the terms of the contract, to exercise a right to do something or require the other party to do (or refrain from doing) something in the future.
>
> (Parsons, 2004)

Options are typically used by a property development company to secure an interest in land that it believes has the potential for development, but currently has no planning permission in place. The option gives the developer a period of exclusivity in which to secure planning permission and then the right to trigger the purchase of the land at a pre agreed price. Options can be described as 'call' options or 'put' options. 'Call' options are the most common and enable the developer to require the landowner to sell (a call for the land). 'Put' options are rare in a property context and enable the seller to require the buyer to go ahead with the purchase on the occurrence of some event.

The main components of an option agreement are as follows:

- The names of the parties involved in the agreement.
- The subject matter – a description of the land involved, normally accompanied by a plan.
- The period of the option.
- The preconditions that will trigger the option. This will normally be the grant of planning permission but could also be the allocation of the subject land in a development plan.
- The option fee or premium – this is the payment made by the developer to secure the option. This may or may not be returnable if a planning application is unsuccessful. The payment reflects the fact that the option agreement has sterilised the land for the length of the option period.
- The agreed purchase price of the land. This may be a fixed amount or could be tied to the value of the land at the date the option is exercised. The purchase price will normally be less than the full market value to reflect the acceptance of risk by the developer. If the price is related to the market value of the land the agreement will identify how this will be determined – normally by an independent valuer acceptable to both parties.

There are various advantages of entering into options. For developers they enable risk to be spread over a number of potential sites. The developer can work on a number of planning applications (or plan promotions) simultaneously in the hope that unsuccessful applications may be offset by successful ones. Options to purchase also help facilitate the purchase of sites in multiple ownership. An example could be a site in four different ownerships, all of which are required to make a viable site. If the developer acquired the freehold in three of the sites the fourth landowner could hold the developer to ransom and seek a price significantly above market value to part with their land. If the developer acquired options in the various parcels of land triggerable when all the necessary options are agreed this will reduce the risk of being held to ransom.

Further reading

Parsons, G. (2004) *The Glossary of Property Terms*, 2nd edn, Jones Lang LaSalle and Estates Gazette, London.

10.16 Conditional contracts

Key terms: conditions; satisfactory search results; waiver; longstop date

A conditional contract is a form of contract used when an agreement has been reached for the sale of land or a lease but the buyer does not want to commit to the purchase unless and until certain issues relating to the property are first resolved. Common conditions you will encounter are that the sale is dependent on the grant of satisfactory planning permission or other consent or permit, securing vacant possession of the land, obtaining landlord's consent and receipt of satisfactory search results/survey/soil tests.

The reason that conditional contracts are used rather than just waiting for the issue the condition relates to being resolved is greater certainty for buyer and seller. A buyer will not want to commit to the cost of securing of planning permission without the land being 'under contract' as that would allow the seller to sell to the highest bidder and the buyer's efforts and expense might all be in vain. Similarly, from the seller's perspective, if they have agreed a sale and ceased marketing the property to allow the buyer time to address the problem they will not want to allow the buyer the freedom to walk away from the deal if the market has fallen or they have changed their mind as this would mean the seller would have tied up the land for months or even years without any financial gain.

Conditional contracts generally contain the following provisions:

- An obligation on the buyer to purchase and for the seller to sell the property if the condition is satisfied or waived.
- Specification of the price or a mechanism for determining the price. If it is anticipated that the condition will be satisfied within a few months of the contract, e.g. satisfactory search results or landlord's consent, then the price will normally be fixed in the contract. On the other hand, if it is likely to take longer to satisfy the condition or the outcome is less certain, for example if it is conditional upon securing planning consent, then the contract will normally provide for the buyer to pay the market value of the land with consent.
- An obligation on the buyer to pay a deposit either on exchange of contracts or when the condition is satisfied.
- An obligation on one or both the parties to try and satisfy the condition. For example, if the contract is conditional on obtaining planning permission then the buyer will normally be obliged to use reasonable endeavours to secure the consent whereas if the sale was conditional upon gaining the landlord's consent the obligation would be on the seller to try and satisfy the condition.
- A provision allowing the buyer to waive the condition so that they can elect to proceed with the purchase even if the condition is not satisfied.

- A longstop date so that if the condition is not satisfied by a specified date the contract will come to an end, otherwise the contract could exist in perpetuity.

One of the drawbacks of conditional contracts over other contracts that could be used instead such as options is that they can be very complicated and have often been the subject of litigation. This has been very much in evidence in recent years when eager developers entered into conditional contracts when the market was at its peak in 2006–7 but by the time the condition had been satisfied property values had fallen and finance for development had become increasingly difficult to secure.

Further reading

Rodell, A. (2013) *Commercial Property*, College of Law Publishing, Guildford.

Vara, B. (2009) 'Killing off enthusiasm'. *Estates Gazette*, April, available at: www.egi.co.uk/Articles/Article.aspx?liArticleID=699265&NavigationID=466 (accessed 20/11/2013).

10.17 Promotion agreements

Key terms: joint venture; risk; development value; profit split

Option agreements and conditional contracts, which were discussed in Concepts 10.15 and 10.16, are the two well-established types of contract that are entered into by developers and landowners for properties that have development potential but for which planning permission has yet to be granted. Both mechanisms have advantages and disadvantages for developers and landowners. Since the economic downturn in 2007 and the problems that have arisen with obtaining development finance, a third way of dealing with the developer–owner relationship has become popular;: promotion agreements.

Promotion agreements are essentially a special form of joint venture. They generally work as shown in Figure 10.17.1.

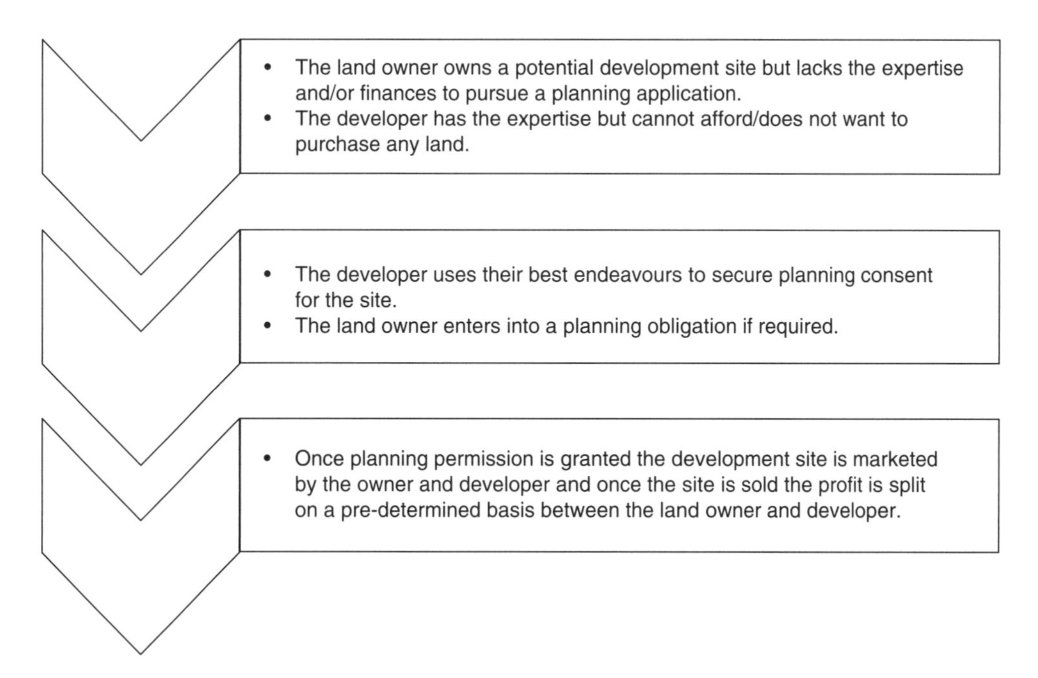

Figure 10.17.1 Main terms of a promotion agreement.

One of the advantages of this method for the landowner is that the developer is incentivised to obtain the planning permission that leads to the highest value of the site. When conditional contracts and options are used, the price paid by the developer is affected by planning consent and they may therefore not pursue the highest value scheme, in order to keep the purchase price as low as possible. From a developers' perspective it means that they do not have to find the finance upfront for an option fee or deposit under a conditional contract.

As the developer is taking on less of the risk of the development the rewards may not be as great as under the more traditional methods. Ultimately, whether or not it is a good option for the developer and owner will depend on the nature of the site, expected timescales for obtaining planning permission and the profit split agreed.

Further reading

Gay, W. (2011) 'A helping hand for developers', *Estates Gazette*, available at: www.egi.co.uk/Articles/Article.aspx?liArticleID=732529&NavigationID=466 (accessed 28/11/2013).

10.18 Overage/clawback

Key terms: development profits; overage period; land banking; restrictive covenant; ransom strips

Overage and clawback refer to the right of a seller of property to receive from the buyer further payment on the occurrence of a certain event following completion of the sale.

The term 'overage' is used where the development of the land following sale is in accordance with the planning permission in existence at the time of sale. Overage refers to an agreement by the developer to pay the original landowner a percentage share in the price obtained on the sale of the development above an agreed threshold level. Overage is offered by a developer to make a land acquisition more attractive to a seller. In essence it represents the prospect of a seller to participate in the future profitability of a development over and above that which was known at the time of the sale.

The term 'clawback' is generally used to represent the agreement by the developer to pay the seller of land a sum of money related to the enhancement in the value of a development site, usually triggered by the granting of planning permission. The payment will generally be a percentage of the difference in value between the site with the planning status at the time of the sale and the value of the site with an enhanced planning permission. Clawback is often used by public authorities (subject to audit) to ensure that land is not sold for an under value. A clawback clause will encourage the sale of development land but also incentivise the buyer to obtain a more valuable planning permission if possible.

Overage and clawback are payable by the buyer and any successors in title. This can be secured contractually via the initial conveyance and registered at the Land Registry. The payment of overage is triggered by the disposal of properties during the overage period. The 'overage period' will be defined in the initial agreement.

Clawback is generally payable on the date a trigger event occurs. The precise details of the trigger are negotiable between the parties to the agreement. A seller will prefer the trigger to be the granting of planning permission whereas the buyer will prefer the trigger to be the implementation of planning permission. This will allow the buyer to procure finance and let construction contracts etc.

The length of overage and clawback obligations will vary. Such obligations will be time limited as buyers will require release after a period of time to avoid titles being burdened indefinitely. Conversely, sellers will want to avoid the period being too short thereby encouraging land banking.

The seller will need to ensure the potential overage payment is secured in some way. This can be done in a variety of ways or combinations of such. The normal approach is secure payments via contractual obligations. This is a basic approach that relies on the covenant strength of the buyer. The main advantage of this method is that it does not muddle the legal title. Other approaches to secure payment include:

- *Parent company/bank guarantee*: although these can be used buyers are generally reluctant to offer either because of the expense involved.
- *Restrictive covenant*: this method can be used if there is adjacent land that can benefit from the covenant. As this is rarely the case its use is limited. This was illustrated in *Cosmichome v Southampton City Council* (2013). Southampton City Council transferred a property to the BBC to use as a regional broadcasting centre. The transfer contained a restrictive covenant requiring the payment of £505 of any increase in value following a change of use. Cosmichome acquired the property in 2004 and sought a declaration that the restrictive covenant was not enforceable against them as successors in title. The Court found that the covenant did not bind Cosmichome because when it was imposed it did not benefit the Council's retained land and had purely been imposed for financial reasons. It is critical that to secure overage obligations they should be registered at the Land Registry.
- *Grant of a long lease*: this method is sometimes used by local authorities and involves the grant of a long lease with freehold being transferred on the payment of the overage.
- *Ransom strips*: these can be used to prevent future development on adjoining land. The seller will agree to release the ransom strip on the payment of a sum representing the market value of the retained land.
- *Option to repurchase*: this method allows the seller to repurchase the land at an agreed price if the buyer defaults on any obligations.

Further reading

Harvard, T. (2008) *Contemporary Property Development*, 2nd edn, RIBA Publishing, pp. 128–9.

10.19 Pre-emption rights

Key terms: first refusal; family/trust land; sales of part; put option

Pre-emption rights are often characterised as a right of first refusal. If a landowner chooses to sell their land they must first offer it to the pre-emption rights holder and give them the opportunity to buy the land before marketing it more widely.

These rights are often created if someone is selling part of their land holding and want the option to buy it back in the future should the buyer choose to sell. As such they most commonly arise in sales of family/trust land and in disposals by local authorities and other government bodies. They can also arise if a developer is seeking to assemble a site but the owner is currently unwilling to sell; it is a way of preventing any competitors acquiring the land if the owner changes their mind about selling in the future.

The main aspects of pre-emption rights are as follows:

- The rights apply for a fixed period of time often 10, 20 or 30 years.
- If the owner decides they want to sell the land they must notify the right holder and give them time to decide whether or not they want to proceed with the purchase, commonly 21 or 28 days.
- If the rights holder does not respond within the time limit or advises the owner they are not interested in purchasing the land the owner is free to dispose of the land.
- If the rights holder responds to the notice in the manner specified in the contract indicating that they intend to buy the land, then a contract is formed for the sale of the land.
- There is a mechanism for determining the purchase price which normally states market value with a dispute resolution procedure if the parties cannot reach agreement on market value.

Further reading

Rodell, A. (2013) *Commercial Property*, College of Law Publishing, Guildford.

10.20 False statements and misleading omissions

Key terms: misrepresentation; Consumer Protection from Unfair Trading Regulations 2008; Business Protection from Misleading Marketing Regulations 2008

This concept is concerned with the law affecting acts and omissions made by sellers/landlords and their agents in the course of marketing and negotiating sales and leases of property.

The starting point under English law has always been *caveat emptor* – let the buyer beware. It was the buyer's responsibility to fully investigate the property and ensure that they were getting a good deal. However, the introduction of the Consumer Protection from Unfair Trading Regulations 2008 (CPR) and Business Protection from Misleading Marketing Regulations 2008 (BPR) has created additional duties for agents and other businesses selling property to disclose important information about the properties they are marketing.

The Property Misdescriptions Act (PMA) 1991 was repealed in October 2013. Under the PMA it was an offence for an agent to make false or misleading statements in the course of a property sale. Under the CPR and BPR this remains an offence but the duties imposed by CPR and BPR are greater as they also make it an offence to make a misleading omission. The duty under CPR on property businesses is to provide the 'material information' that an 'average consumer' needs to make a decision about the transaction (e.g. whether to make an offer, what price to offer etc.). While the agent is not under a duty to investigate matters that will be researched by other parties involved in the transaction (conveyancers and surveyors) once they become aware of a material issue they must disclose it. For example, if they were marketing a property and a buyer pulled out of a purchase when they received the results of their survey and learned that there were major subsidence problems the agent would be under a duty to share this information with other prospective purchasers.

If a business is convicted of an offence under the CPR or BPR they face a fine or imprisonment for up to 2 years. A consumer or business who is aggrieved by a property business's actions could complain to an approved redress scheme or Trading Standards.

If someone has entered into a contract as a consequence of misleading statements made in the course of negotiations by the seller or their agents they may be able to bring a civil claim for breach of contract or misrepresentation. The buyer would only be able to bring a claim for breach of contract if the statement is included as a term of the contract, for example if the area of the land was written in to the contract and it was late discovered to be incorrect.

If a statement has not been incorporated as a term of the contract the buyer may be able to claim misrepresentation if they can prove:

- that a false statement of fact has been made (this does not normally include misleading omissions);
- that the buyer relied upon the false statement; and
- that the false statement induced the buyer to enter into the contract.

In order to protect sellers from claims of misrepresentation many contracts for the sale of land include a statement that the buyer's decision to enter the contract is based on their own research and enquiries and not in reliance on anything said by the seller or their agents. Any clause that seeks to limit or exclude liability for misrepresentation is subject to a test of reasonableness under section 3 of the Misrepresentation Act 1967.

The remedies for misrepresentation are damages and/or rescission (terminating the contract). The availability of remedies depends on whether or not the statement was made fraudulently, negligently or innocently.

One of the most notorious property misrepresentation cases is *Sykes and another v Taylor-Rose and another* (2004) EWCA Civ 299. The house at the heart of the dispute had been the scene of a violent murder. Mr and Mrs Taylor-Rose bought the house without knowing of its history, but after receiving an anonymous letter enlightening them of what had happened they decided to sell it. They took legal advice as to whether or not they were obliged to disclose the house's history when they were selling it and were advised that they need not disclose it but if directly asked they should not lie. The Sykes bought the house from the Taylor-Roses also ignorant of its past. They learned about what happened there through a television documentary and decided that they could no longer live there but they could also not in conscience keep the information from prospective purchasers. Ultimately they sold the house for £75,000 but believed its true market value (disregarding its history) was £100,000. The Sykes sought to recoup their losses by bringing a claim for misrepresentation against the Taylor-Roses. In particular, they argued that in answer to question 13 on the Sellers Property Information Form ('Is there any other information which you think the buyer may have a right to know?'), which had been answered in the negative, was a false statement. However, the Taylor-Roses were able to successfully defend the claim on the grounds that it was an honest answer as they had been professionally advised that the buyer did not have a right to this information.

Further reading

Galbraith, A., Stockdale, M., Wilson, S., Mitchell, R., Hewitson, R., Spurgeon, S. and Woodley, M. (2011) *Galbraith's Building and Land Management Law for Students*, 6th edn, Butterworth-Heinemann, Oxford.

Office of Fair Trading (2012) 'OFT Guidance on Property Sales', available at: www.oft.gov.uk/shared_oft/estate-agents/OFT1364.pdf (accessed 22/11/2013).

11 Planning

Andy Dunhill, Hannah Furness, Paul Greenhalgh, Carol Ludwig, David McGuinness and Rachel Williams

11.1 Legislation and planning policy

Key terms: strategic planning; development management; National Planning Policy Framework; Local Development Plan

The planning system is defined by a range of statutes, regulations, circulars and policy guidance. The planning function within local planning authorities (LPAs) is divided into two key areas: plan making (also known as strategic planning), which involves the preparation of planning policy; and development management, which involves the implementation of that policy when making planning decisions (see Concept 11.4).

Some of the key pieces of legislation relating to planning are set out below:

- Ancient Monuments Protection Act of 1882
- Town and Country Planning Acts of 1932, 1945 and 1947
- Civic Amenities Act 1967
- Town and County Planning Act 1990
- Planning (Listed Buildings and Conservation Areas) Act 1990
- Planning and Compensation Act 1991
- Planning and Compulsory Purchase Act 2004
- Planning Act 2008
- Localism Act 2011.

The planning system is plan led (Cullingworth and Nadin, 2006). As such, planning law requires that applications for planning permission must be determined in accordance with the development plan, unless material considerations indicate otherwise (see Sections 38(1) and (6) of the Planning and Compulsory Purchase Act 2004). In addition to the above key pieces of legislation, the plan-led system incorporates a hierarchy of national policy/guidance and local development plans.

National planning policy guidance

As part of major government reforms designed to make the planning system less complex and more accessible, the Conservative–Liberal Democrat Coalition Government published the National Planning Policy Framework (NPPF) on 27 March 2012. The NPPF is prepared by the Department for Communities and Local Government (DCLG) and is a material consideration in planning decisions (see Section 38(6) of the Planning and Compulsory Purchase Act 2004 and section 70(2) of the Town and Country Planning Act 1990). The NPPF replaces most of the topic-based Planning Policy Statements (PPS) and Guidance Notes (PPG) (44 separate guidance documents) with a single document of under 60 pages. Most PPS/PPGs were superseded by the NPPF except PPS10 on Waste.

While the NPPF shortens national policy and brings it helpfully into one, more useable document, it nevertheless removes some of the detailed guidance that helped to clarify the policy. To try to overcome this limitation, on 28 August 2013, the Government published the National Planning Practice Guidance website which is intended to add detail and flesh out some of the policy contained within the NPPF.

NPPF purpose and role

The NPPF is intended to not only make planning more transparent and accessible, but also to help LPAs plan positively for future growth. As set out in the Foreword of the document: 'The purpose of planning is to help achieve sustainable development', and there must be a 'presumption in favour of sustainable development' (DCLG, 2012: i). As such, the NPPF sets out the Government's planning policies for England and how these are expected to be applied in a positive manner. Moreover, 'it provides a framework within which local people and their accountable LPAs can produce their own distinctive local and neighbourhood plans, which reflect the needs and priorities of their communities' (DCLG, 2012: 1). For instance, it lists the things Local Development Plans must consider, including policies on economic growth, housing, transport, community facilities and climate change. As a material consideration, the NPPF 'must be taken into account in the preparation of local and neighbourhood plans, and ... in planning decisions' (DCLG, 2012: 1).

Other important national policy/guidance:

- Statutory Instruments and National Government Circulars;
- Planning Policy Statement 10 Planning for Sustainable Waste Management;
- Rafts of topic-specific guidance (for example guidance published by other national bodies such as English Heritage or the Environment Agency).

Local policy

Local policy is produced by the LPA in consultation with the community. It becomes the Local Development Plan for the LPA's administrative area. As highlighted above, section 38 of the Planning and Compulsory Purchase Act 2004 provides the Local Development Plan with statutory force. As such, this plan is the most important consideration in deciding planning applications. All Local Development Plans need to conform to national policy (as set out above); otherwise they will not be approved and could be open to legal challenge. They also need to meet particular legal requirements, for example they must conform to the Equality Act 2010 and Human Rights legislation.

The Local Development Plan may be one single document (i.e. a Local Plan) or comprise many documents (i.e. a folder of shorter documents such as the Local Development Framework, or LDF). The LDF suite of documents may include development plan documents (DPDs), Supplementary Planning Documents (SPDs), Area Action Plans (AAPs) and/or Master Plans. All LPAs are also required by law to publish a Statement of Community Involvement (SCI) in advance of preparing their Local Development Plan. This document must set out who the LPA will consult as part of plan preparation as well as how they will be consulted and when. LPAs are also required by law to consider the environmental, economic and social needs of the area, in a Sustainability Appraisal (SA).

What strategic policies should a Local Development Plan include?

Every Local Development Plan will include strategic, high-level policies that set out the long-term vision and strategic priorities for the administrative area. This part of the Local Development Plan is often called a Core Strategy. Some of the strategic elements will include:

- spatial vision and objectives
- spatial strategy
- settlement role and function (see Planning concept, *Settlement Hierarchy*)
- development principles in relation to:
  1. housing, employment, retail, leisure and other commercial development;
  2. infrastructure for transport, telecommunications, waste management, water supply, waste-water, flood risk and coastal change management, and the provision of minerals and energy (including heat);
  3. health, security, community and cultural infrastructure and other local facilities;
  4. climate change mitigation and adaptation, conservation and enhancement of the natural and historic environment, including landscape.

(DCLG, 2012: 38)

What should a Local Development Plan seek to do?

According to the NPPF (DCLG, 2012: 38), Local Development Plans should:

- Plan positively; preferably with a 15-year time horizon and they should be kept up to date.
- Be based on co-operation with neighbours and stakeholders.
- Be informed by a proportionate evidence base.
- Indicate broad locations for strategic development on a key diagram and land-use designations on a proposals map.
- Allocate sites to promote development and flexible use of land, bringing forward new land where necessary, e.g. they must identify a 5 year supply of housing plus a 'buffer' of between 5 per cent and 20 per cent.
- Provide detail on form, scale, access and quantum of development where appropriate.
- Identify areas where it may be necessary to limit freedom to change the uses of buildings (e.g. conservation areas).
- Identify land where development would be inappropriate (e.g. green belt).

The process of local plan making usually takes 3 years and is summarised in Figure 11.1.1.

Complexity of local policy

It is important to note that the local planning policy context can be quite complex and may vary from one LPA to another depending on speed of plan preparation. While the Labour Government introduced Local Development Frameworks (LDF) in 2004 through the Planning and Compulsory Purchase Act, LPAs have been progressing their plans at varying speeds. It has taken a long time for the 'new' system to bed in and indeed at the end of 2013 only 174 of 352 LPAs had an adopted Core Strategy (Planning Inspectorate, 2013).

As such, the local policy context requires careful scrutiny in order to adhere to the correct, valid policies operating in an area. Figure 11.1.2 clarifies visually this complex local policy context.

Saved policies

Many LPAs also have a set of 'saved policies' in operation in their administrative area. The Planning and Compulsory Purchase Act 2004 automatically saved LPAs' local planning policies from their old-style local plans for 3 years from either

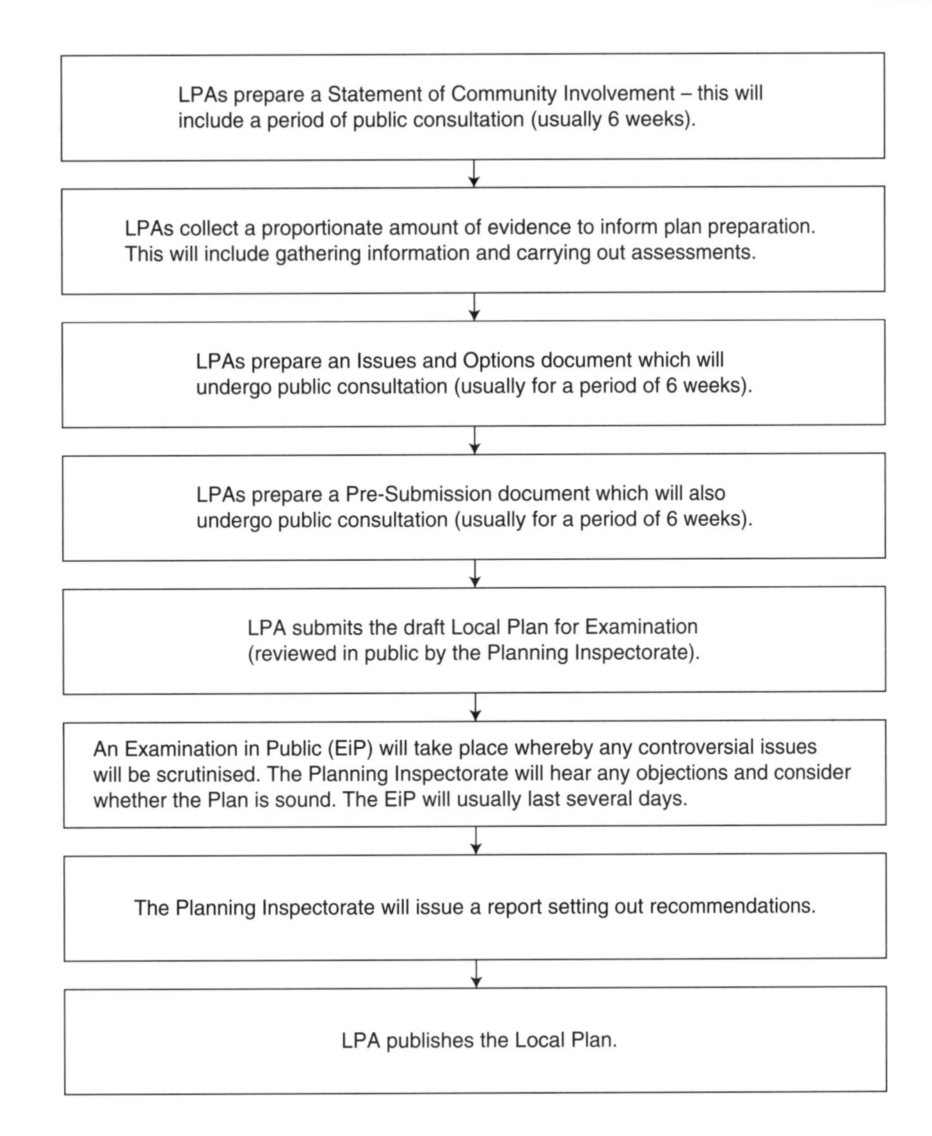

Figure 11.1.1 Local plan-making process.

Source: adapted from Planning Aid (2012).

- the date of commencement of the Act, or
- the date the plan was adopted.

Following this 3-year period, LPAs were then able to apply to the Secretary of State to extend this period of 'saved' policies if they had yet to get their new plans in place. The saved policies could be extended only if the saved policies were deemed necessary and did not repeat national guidance. As such, a number of LPAs currently have 'saved policies' (which form part of the adopted Local Development Plan for the area) and which are used to make planning decisions, while they work on preparation of new policies. Where saved policies are deemed no longer up to date or do not echo the policies of the NPPF, LPAs are compelled to give those policies reduced weight in planning decisions

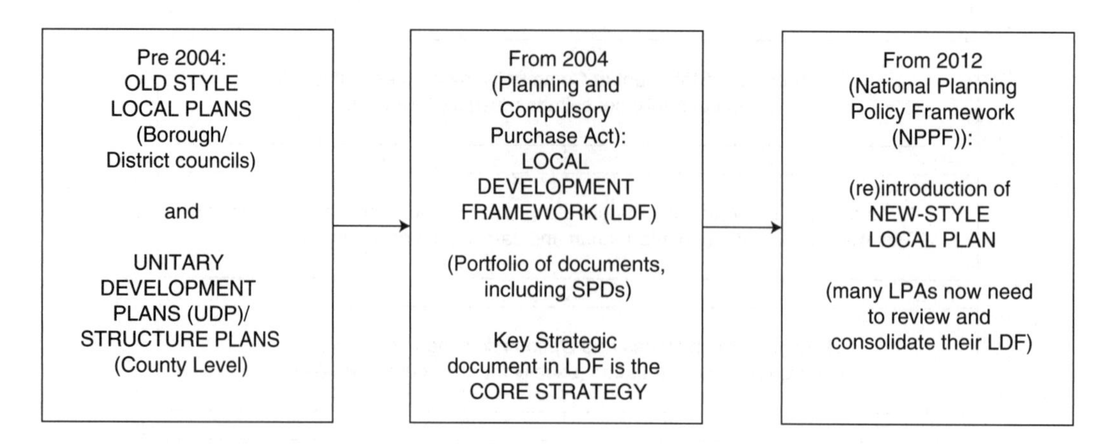

Figure 11.1.2 Demystifying the local policy context.

Source: Ludwig (2013).

(see Concept 11.4). Where this is the case, the NPPF (DCLG, 2012: 4) makes it clear that planning applications should be granted planning permission unless:

1 Any adverse impacts of doing so would significantly and demonstrably outweigh the benefits, when assessed against the policies in the NPPF taken as a whole; or
2 Specific policies in the NPPF indicate development should be restricted.

Regional policy

It is also important to note that the regional layer of planning policy has recently been removed with the abolition of Regional Spatial Strategies (RSS). It has, however, been shown that the RSS evidence base used to justify housing targets, for example, may continue to play an important role when LPAs lack up-to-date Local Development Plans (Geoghegan, 2013).

Further reading

Cullingworth, J. B. and Nadin, V. (2006) *Town and Country Planning in the UK*, 14th edn, Routledge, London.

DCLG (2012) *National Planning Policy Framework*, DCLG, London.

DCLG (2013) 'Plan making manual', available at: www.pas.gov.uk/40-clg-plan-making-manual (accessed 25/11/2013).

Geoghegan, J. (2013) 'Regional housing targets have weight post-abolition', 9 April 2013, available at: www.planningresource.co.uk/article/1177313/regional-housing-targets-have-weight-post-abolition (accessed 25/11/2013).

Planning Aid (2012) *A Handy Guide to the English Planning System*, Urban Forum, London.

Planning Inspectorate (2009) 'Development plan document examination: Procedural advisory notes', available at: www.planningportal.gov.uk/planning/planningsystem/localplans#letter (accessed 25/11/2013).

Planning Inspectorate (2013) 'Preparation and monitoring of local plans', available at: www.planningportal.gov.uk/planning/planningsystem/localplans#dclg (accessed 25/11/2013).

Planning Portal website: www.planningportal.gov.uk

11.2 Strategic planning

Key terms: wider future development; regional planning; Local Enterprise Partnerships

It is argued that effective strategic planning is necessary at UK-wide, national, regional and subregional (local) levels in order to guide development, land use change, and a number of other activities that exist across wider spatial areas that have some clear socio-economic, cultural or ecological identities.

The approach recognises that a number of development schemes and initiatives will have impacts and influences across areas that are wider than a single local authority area. These strategic plans serve to ensure that there is a coordinated approach to planning that encompasses a wide range of sectoral and departmental concerns and policies at all levels of governance. Strategic planning is typically concerned with the development of an understanding or 'vision' for an area, including all of its social, economic and environmental issues that goes beyond just considering land use planning. It is effectively a tool for guiding the wider future development of a spatial area.

In terms of core elements, the Town and Country Planning Association (2003) believe that a strategic plan should be:

- visionary as well as practical;
- aimed at long-term viability in terms of 'sustainable development' criteria;
- prescriptive, not solely informative;
- binding, not just advisory;
- a combination of top-down and bottom-up approaches;
- spatial, relating to specific locations;
- sectoral, relating to specific sectors and areas of activity;
- achievable, in terms of resources and processes of implementation;
- well founded in terms of research and understanding;
- comprehensive in terms of the range of coverage, but not too detailed;
- consistent in content and application;
- open and accessible, with wide consultation, public participation and direct involvement of the stakeholders;
- cyclical, through monitoring and review;
- medium and long term in timescale; and
- accountable to an elected authority, assembly, or *parliament at national/regional level.*

(TCPA, 2003)

Spatial planning was the mechanism used to develop Regional Plans, prior to the removal of the regional tier by the Coalition Government when it came into power in 2010. These underpinned the remit of the Regional Development Agencies' Economic Strategies.

Although no longer relevant at the regional level, the concept of strategic planning is still evident in the strategies that are currently being developed by Local Enterprise Partnerships and Combined Authorities.

Further reading

TCPA (2003) *TCPA Policy Statement Strategic and Regional Planning*, TCPA, London.

11.3 Green belt

Key terms: urban sprawl; conservation; environment; housing; rural planning

Definition and history

The 'green belt movement' was brought about by groups such as the Campaign for the Protection of Rural England (CPRE) lobbying during the 1920s and 1930s to safeguard the countryside against urban sprawl and so-called 'ribbon' development. Green belts were initially introduced around London in the 1940s and then adopted around other major towns and cities after the publication of the Green Belt Circular 42/55 in 1955 (introduced by Duncan Sandys, Conservative MP).

Development pressure was great after 1945, with massive house building and rising car ownership. The green belt was designed to halt urban sprawl by creating protective 'green barriers' around cities to stop towns merging. The Town and Country Planning Act 1947 gave permission to local planning authorities to include proposals for green belts in their development plans.

Green belts are one of the most well-known and supported planning concepts. Though often misapplied to refer to the countryside in general, the green belt refers to specific areas of rural land where development is restricted through adoption of planning policy. The green belt in England is actually much smaller than most people would assume; England has 14 green belts around major cities, making up just 13 per cent of total land in England. Developing in the green belt still causes a massive furore today and there is considerable tension between overwhelming demand for new homes and the significant public support for protection of existing green belt land.

Planning Policy Guidance 2 (PPG2) provided guidance on green belt policy until it was scrapped with the development of the NPPF by the Coalition Government. Much of the NPPF cherry picks policy from previous PPGs and PPSs. In reality little has changed in terms of green belt policy over the last 50 years.

The purpose of the green belt is:

1 To check the unrestricted sprawl of large built-up areas.
2 To prevent neighbouring towns merging into one another.
3 To preserve the setting and special character of historic towns.
4 To assist in urban regeneration, by encouraging the recycling of derelict and other urban land.
5 To assist in safeguarding the countryside from encroachment.

(DCLG, 2012: 19)

The purpose of the green belt has largely remained unchanged over the past 60 years. The final aim of the green belts to assist urban regeneration was added in 1984 to counteract deindustrialisation following the loss of major industries (coal, steel, shipbuilding etc.). The final aim of safeguarding the countryside was added in 1988 as a response to a growing trend of the flight from city centres to more suburban and rural locations.

Case study

Newcastle City Council built on green belt land at the Great Park – part of the reason given was to reduce urban flight. Newcastle was losing residents to neighbouring local authority areas (e.g. North Tyneside) because of the lack of affordable three and four bedroom family homes within the Newcastle City Council boundary. Due to lack of space for development, Newcastle City Council is contemplating releasing a further 10 per cent of its existing green belt in its latest draft of the One Core Strategy. Relaxing green belt boundaries and releasing sections of the green belt is a dilemma that faces many local authorities if they are to meet the inexorable demand for affordable housing, especially within the south of England.

Future issues for green belt policy include:

- the challenge of NIMBYs (Not In My Back Yard) – people opposed to development near their homes;
- building on just 2 per cent of the green belt could provide 8 million homes – desperately needed especially in the South East;
- do we still need green belts in England – there is an argument that green belt has hindered development and stifled growth. It reaffirms the British self-image as a country of rural, pastoral idylls that, in reality, the majority of Britons no longer live in – it is outdated;
- 'New Green Belts should only be established in exceptional circumstances' (NPPF, 2012: 19: 82).

The Policy Exchange, a right-wing think tank has advocated paying resident to accept development on green belts. The Coalition has gone some way to accepting this idea with its initiative to allow a proportion of Community Infrastructure Levy (CIL) payments to be ring fenced and to be kept for the host community to utilise.

Further reading

CPRE and Natural England (2010) *Green Belts: A greener future*, CPRE and Natural England, London.
DCLG (2012) *National Planning Policy Framework*, available at: www.gov.uk/government/uploads/system/uploads/attachment_data/file/6077/2116950.pdf (accessed 30/11/2013).
Rydin, Y. (2011) *The Purpose of Planning*, Policy Press, Bristol, ch. 6.

11.4 Planning decision making

Key terms: principle of development; material considerations; planning process

LPA planning functions are divided into strategic planning (policy makers) and development management (those who implement the policies by using them to make informed decisions about planning applications). Those informed decisions are part of a formal planning decision-making process, governed by many operational rules and regulations.

Establishing the principle of development

When considering planning applications or making strategic planning decisions, section 38(1) and (6) of the Planning and Compulsory Purchase Act 2004 requires that applications for planning permission must be determined in accordance with the development plan, unless material considerations indicate otherwise. The LPA's Local Development Plan is therefore the starting point for considering planning applications and making planning decisions.

Material considerations in decision making

It is important to note that other material considerations are important aspects of decision making; however, the term remains rather ambiguous. Any consideration that relates to the use or development of land is capable of being a material consideration, but other circumstances such as personal hardship and fears of affected residents can be considered in exceptional cases (the House of Lords in *Great Portland Estates v Westminster City Council* (1985)). Government planning policy (the NPPF) is one of the most important material considerations other than the Local Development Plan.

Table 11.4.1 summarises what may/may not be deemed material considerations in planning decision making.

Table 11.4.1 Considerations in planning decision making

Material considerations	*Non-material considerations*
Previous planning decisions (including appeal decisions)	Loss in property value
Proposals/policies in the Local Development Plan	Loss of a private view
National guidance	Private disputes between neighbours, e.g. land ownership
Loss of light/overshadowing	Restrictive covenants
Loss of privacy to a room through distance	Fence lines/boundary positions
Visual amenity	Personal morals or views about the application
Adequacy of parking/loading/turning	Ownership disputes
Highway safety, road layout/access	Applicant's motives
Noise and disturbance resulting from use	Competition
Hazardous materials; Odours	Issues covered by other legislation, e.g. Highways Act
Traffic generation	
Loss of trees/green space	
Landscaping	
Impact upon a listed building or a conservation area	
Design, appearance and materials	
Disabled access	
Nature conservation	
Archaeology	

The planning decision-making process

Is it a planning issue?

It is first important to establish whether a proposal requires planning permission. If proposed activity constitutes 'development' it is likely to require planning permission and therefore a planning decision. Development is defined by section 55 of the Town and Country Planning Act 1990: '"development" means the carrying out of building, engineering, mining or other operations in, on, over or under land, or the making of any material change in the use of any buildings or other land.'

Building operations include the following:

a demolition of buildings;
b rebuilding;
c structural alterations of or additions to buildings; and
d other operations normally undertaken by a person carrying on business as a builder.

Not all development, however, requires planning permission, and thus a planning decision. There are two important pieces of secondary legislation: the Use Classes Order 1987, which sets out the various categories of planning uses and the General Permitted Development Order (GPDO) 1995. Both have been amended many times but are important documents that confirm whether development requires planning permission or not. Development that does not require planning permission is called permitted development. If unsure whether planning permission is required, it is advised to contact the relevant LPA.

Pre-application discussions

Where planning permission is required, the Government encourages a detailed process of negotiation and pre-application discussion. This process is now seen as a key part of a modern development management service. Many LPAs now have formal pre-application advice services that attract fees.

Despite the costs involved, pre-application discussions offer applicant's an early insight into the planning decision-making process and highlight from the outset whether the proposal has any planning issues that need to be overcome. The outcome of such pre-application discussions is usually more appropriate applications which are far more likely to meet planning aspirations for development in the area, are more sustainable, and are ultimately more likely to achieve planning permission.

A great number of LPAs now operate what is termed the Development Team approach. This is an approach whereby a group of key officers likely to be central to considering the development proposal are assembled early on during the pre-application stage to help a proposal through from inception to post-application delivery.

Submitting the application

Validation

Once submitted, planning applications are validated to ensure that all the necessary forms, plans and documents have been submitted. The general process is controlled by the Town and Country Planning (Development Management Procedure) Order 2010, which sets out what needs to be submitted with an application and how the process should be conducted.

Once the planning application has been validated, the LPA will start the determination process (Planning Portal, 2013).

If a planning application is deemed invalid, the LPA will notify the applicant and make clear what documents are required. Additional supporting information will only be requested where it is deemed relevant, necessary and material to the application in question.

Once an application has been validated and registered, the LPA will then publicise and consult on it (Planning Portal, 2013).

Consultation

It is the responsibility of the LPA to notify neighbours and/or put up a notice on or near the site. Applications may also be advertised in a local newspaper. This gives the public the opportunity to make comments. The parish, town or community council will also be directly notified, as well as other bodies such as the Environment Agency for example.

Any comments received on the planning application during the consultation period will be a material consideration in the planning decision-making process. In other words, comments will be collated, assessed, and where relevant will inform decision making. As a result of the comments received, the LPA may suggest minor changes to the application to overcome any issues raised.

The decision

The decision to grant permission or reject a planning application may either be taken by the LPA's planning committee, which is made up of elected councillors, or will be determined by the Chief Planning Officer, through delegated powers. To determine whether the decision will be made by planning committee or will be delegated, the LPA usually have a set of criteria, 'based on size and nature of development' (Planning Aid, 2012: 23). More complex, contentious and/or major developments will usually be determined by the planning committee. Where a decision is to be made by planning committee, the LPA must prepare a report outlining the proposal, and an assessment of the issues raised. The report will culminate in a recommendation for approval or refusal. The report and any background papers (for example, comments of consultees, objectors and supporters) will normally be made available at least three working days before the committee meeting.

The decision-making process up to this point usually takes 8 weeks (minor applications) and 13 weeks for major or complex applications.

Permission granted

The LPA will send a notification letter to applicants informing them of the outcome of the process. The decision will be to either grant permission, grant permission subject to planning conditions (and in some cases subject to the signing up to a section 106 planning obligations agreement) or to refuse the application.

Applicants have three years from the date that planning permission is granted to begin the development. If development has not commenced within the required timescale, applicants will need to reapply for planning permission.

Permission refused or delayed

If the LPA refuses planning permission or if the applicant disagrees with the planning conditions attached, applicants may appeal up until 6 months from the date of the application decision letter (or in the case of non-determination, 6 months from the date the decision should have been made). Appeals, however, are intended as a last resort and may take several months to decide. It may therefore be more appropriate to discuss with the LPA whether there are any changes to your proposal which would make it more acceptable or, in cases of non-determination, when your application might be decided if you choose not to appeal. If your application has been refused, you may be able to submit another application with modified plans free of charge within 12 months of the decision on your first application (Planning Portal, 2013).

The application process is summarised in Figure 11.4.1.

Further reading

General Permitted Development Order (GPDO) (1995), HMSO, London, available at: www.legislation.gov.uk/uksi/1995/418/contents/made (accessed 25/11/2013).

Planning Aid – Royal Town Planning Institute (2012) available at: http://www.rtpi.org.uk/media/6312/Good-Practice-Guide-to-Public-Engagement-in-Development-Scheme-High-Res.pdf (accessed 04/06/2014).

Planning Jungle: http://planningjungle.com/ (accessed 25/11/2013).

Planning Portal (2013) 'How applications are processed', available at: www.planningportal.gov.uk/planning/applications/decisionmaking/process (accessed 25/11/2013).

Town and Country Planning (Development Management Procedure) Order (2010), HMSO, London, available at: www.legislation.gov.uk/uksi/2010/2184/contents/made (accessed 25/11/2013).

Use Classes Order 1987, HMSO, London, available at: http://legislation.data.gov.uk/uksi/1987/764/made/data.htm?wrap=true (accessed 25/11/2013).

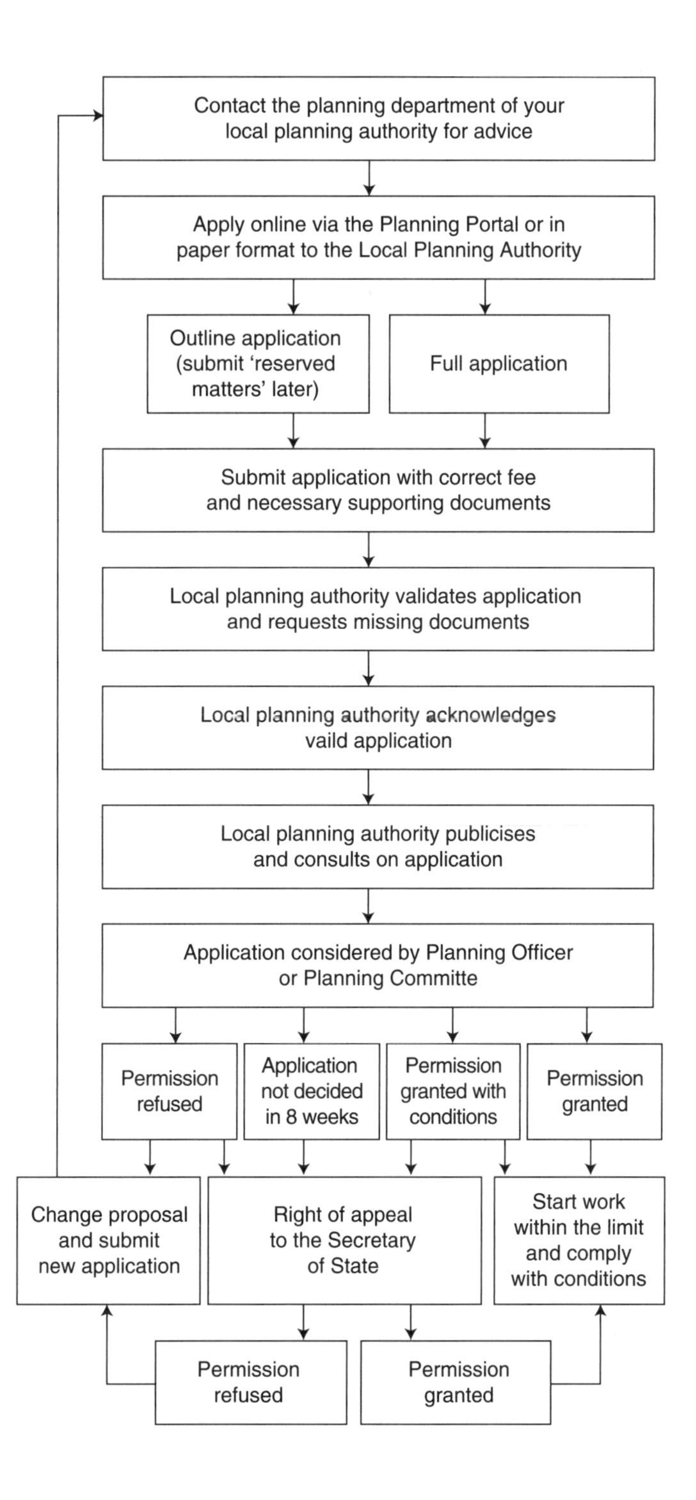

Figure 11.4.1 The planning application process

Source: Planning Portal www.planningportal.gov.uk/uploads/plan_flow_chart_eng.gif.

11.5 Listed buildings and conservation areas

Key terms: listing system; architectural interest; historic interest; conservation areas

Listing buildings

The Statutory Listing system is based around a hierarchy of 'listing' at Grade I (buildings of exceptional importance, around 2.5 per cent of all listed buildings), Grade II* (particularly important buildings of more than special interest) or Grade II (buildings of special interest – see the Planning (Listed Buildings and Conservation Areas) Act 1990.) The 'special architectural or historic character of the building' is the prime determinant of its inclusion in the list (Tait and While, 2009: 722). Given the focus on 'the building', particular emphasis is given to special methods of construction and/or aesthetic elements that lend it its special architectural character (Turnpenny, 2004). Moreover, the ensuing 'art historical' approach to listing decisions means that, 'individual iconic' buildings (and 'iconic' architects) are often prioritised for designation (While, 2007: 658).

Principles of selection

While the criteria used to select buildings for listing have remained relatively stable since the emergence of the Statutory List, they have nevertheless been subtly amended over time. The most recent revisions are set out in the Department for Culture, Media and Sport's (DCMS) *Principles of Selection for Listed Buildings*, published in 2010 Sections 9 and 12–15 of which are summarized below.

Section 9: Statutory criteria

The Secretary of State uses the following criteria when assessing whether a building is of special interest and therefore should be added to the statutory list:

- *Architectural interest.* To be of special architectural interest a building must be of importance in its architectural design, decoration or craftsmanship; special interest may also apply to nationally important examples of particular building types and techniques (e.g. buildings displaying technological innovation or virtuosity) and significant plan forms.
- *Historic interest.* To be of special historic interest a building must illustrate important aspects of the nation's social, economic, cultural, or military history and/or have close historical associations with nationally important people. There should normally be some quality of interest in the physical fabric of the building itself to justify the statutory protection afforded by listing.

Further to these two high-level guiding principles set out above, decisions to list a building or structure are based on several other general principles:

Section 12–15: General principles

Age and rarity. The older a building is, and the fewer the surviving examples of its kind, the more likely it is to have special interest. The relevance of age and rarity will vary according to the particular type of building because for some types, dates other than those outlined below are of significance. However, the general principles used are that:

- before 1700, all buildings that contain a significant proportion of their original fabric are listed;
- from 1700 to 1840, most buildings are listed;
- after 1840, because of the greatly increased number of buildings erected and the much larger numbers that have survived, progressively greater selection is necessary;

- particularly careful selection is required for buildings from the period after 1945;
- buildings of less than 30 years old are normally listed only if they are of outstanding quality and under threat.

Aesthetic merits. The appearance of a building – both its intrinsic architectural merit and any group value – is a key consideration in judging listing proposals, but the special interest of a building will not always be reflected in obvious external visual quality. Buildings that are important for reasons of technological innovation, or as illustrating particular aspects of social or economic history, may have little external visual quality.

Selectivity. A building may be listed primarily because it represents a particular historical type in order to ensure that examples of such a type are preserved. Listing in these circumstances is largely a comparative exercise and needs to be selective where a substantial number of buildings of a similar type and quality survive. In such cases, the Secretary of State's policy is to list only the most representative or most significant examples of the type.

National interest. The emphasis in these criteria is to establish consistency of selection to ensure that not only are all buildings of strong intrinsic architectural interest included on the list, but also the most significant or distinctive regional buildings that together make a major contribution to the national historic stock. For instance, the best examples of local vernacular buildings will normally be listed because together they illustrate the importance of distinctive local and regional traditions (DCMS, 2010).

Evolution of listing principles

The notion of significance in the historic environment is an evolving concept that has broadened over the last decade. Most notable changes include the recognition of industrial and post-war heritage in listings' frame of reference (Stratton, 2000; Orbasli, 2007; Pendlebury, 2009). When national listing was first established during the 1940s, the system was largely restricted to buildings built before 1840. The practical effect of the 1840 threshold was the exclusion from the statutory lists of the building stock associated with the Industrial Revolution (Boland, 1998). Yet in 1987 a government Statutory Instrument opened up the possibility for post-war listing by extending the period of eligibility for listing to any building at least 30 years old. This '30-year rule' enabled constant extension, and additionally included further provision to list buildings over 10 years old if they were deemed, 'outstanding and threatened' (While, 2007: 650). Since then the remit of conservation planning has gradually been further extended (Delafons, 1997).

For instance, prior to the 1970s, 'only the finest examples of Victorian and Edwardian architecture were eligible for inclusion' (While, 2007: 651). Brunskill's (1971) *Illustrated Handbook of Vernacular Architecture*, however, paved the way for an array of vernacular listings (Robertson, 1993). Indeed, even unexpected structures such as a pigeon cree in Sunderland have been listed (Howe, 1998: 9). In the twenty-first century, listing has continued to evolve, evidenced by the recent statutory listing of the Abbey Road zebra crossing in London, England (2010). This is a prime example of further social-philosophical adjustments to the identification and interpretation of significance, as the Grade II listed zebra crossing is deemed important because of its historical association with The Beatles, gained through its international fame on the cover of their 1969 *Abbey Road* album (English Heritage, 2013). Notwithstanding this, such examples are rather atypical, which explains why they receive much media attention.

Objectives of listing

Listed buildings are a finite resource and are irreplaceable assets. As such, their statutory status (as important examples of national history, tradition and culture) is intended to ensure that great care is taken when proposing any changes to them or to buildings/structures in their vicinity. There are two prime objectives to listing. First, listing ensures that any proposed development activity is carried out sensitively and does not cause harm to the listed buildings. This is therefore particularly useful for LPAs carrying out their planning functions. Second, the listing of a building provides a legal basis that prevents its demolition or any alteration that would materially alter it. Listed Building Consent is required from the LPA for any demolition, alterations, (including internal alterations) or extensions that would affect the character of the building. Listed Building Consent is a separate application process to the usual planning consents.

Since 2005, English Heritage have assumed sole responsibility for the administration of listed buildings and while suggestions about what to list are made by English Heritage, proposals for listing can also be made by any member of the public (While, 2007: 648).

Conservation areas

The 1967 Civic Amenities Act was the first to impose a duty on local authorities to designate conservation areas. The duty was to, 'designate as conservation areas any areas of special architectural or historic interest', and this same duty is today imposed in Section 69 of the Planning (Listed Buildings and Conservation Areas) Act 1990. While notions of 'architectural' and 'historic interest' are still organising concepts underpinning conservation area designation (English Heritage, 1997), there are a number of additional aspects to consider that make the remit for conservation area designation wider than that for statutory listing.

Principles of selection

One key difference between listed building and conservation area designation is that LPAs, rather than central government, are responsible for designating conservation areas. As such, conservation areas tend to have a wider remit, focusing more on the notion of local distinctiveness. Local factors, such as a commitment to the preservation of local historic character and/or the industrial heritage, are important factors for conservation area designation. Indeed, PPG15 (now superseded by the NPPF) was the first to make clear that it is reasonable to take account of a wider range of factors when considering conservation area designation than are applicable to listing. For instance, 'special interest' can derive from, 'an area's topography, historical development, archaeological significance and potential, the prevalent building materials of an area, its character and hierarchy of spaces and the quality and relationship of its buildings' (DoE/DNH, 1994: 4.4).

PPG15 was also first to urge a move towards the drawing up of formal character assessments in order to underpin and justify conservation area designations. Again, this represented a positive opportunity to move beyond narrow considerations of artistic or architectural quality and towards an understanding of the evolution of an area and the key interrelationships of all its historic components (Boland, 1998). Moreover, conservation area planning encouraged increased levels of public participation in conservation activity through the creation of conservation area advisory committees (MHLG, 1968: 18–22).

Objectives of conservation areas

When a conservation area has been designated, special attention has to be paid in all planning decisions to the preservation or enhancement of its character and appearance. It is therefore of particular importance to understand the character of the conservation area in order for this to be preserved or enhanced. Many LPAs prepare Conservation Area Character Appraisals and associated management plans which become a material consideration when making planning decisions (see Concept 11.4).

While demolition of all buildings (unlisted as well as listed) is controlled, owners of unlisted buildings usually have permitted development rights. These, however, can be withdrawn by LPAs through the use of an Article 4 direction. Where an Article 4 direction is in place, planning permission must be sought.

Further reading

Boland, P. (1998) 'The role of local lists in the conservation of the historic environment: An assessment of their current status and effectiveness', unpublished thesis (MSc), Oxford Brookes University.

Brunskill, R. W. (1971) *Illustrated Handbook of Vernacular Architecture*, Faber & Faber, London.

DCLG (2012) *National Planning Policy Framework*, DCLG, London.

DCMS (2010) *Principles of Selection for Listed Buildings*, TSO, London.

Delafons, J. (1997) *Politics and Preservation: A policy history of the built heritage 1882–1996*, Spon Press, London.

DoE/DNH (1994) *Planning Policy Guidance Number 15: Planning and the historic environment*, HMSO, London.

English Heritage (1997) *Conservation Area Appraisals: Defining the special architectural or historic interest of conservation areas*, English Heritage, London.

English Heritage (2008) *Conservation Principles, Policies and Guidance*, English Heritage, London.

English Heritage (2013) 'List entry', available at: http://list.english-heritage.org.uk/resultsingle.aspx?uid=1396390 (accessed 10/07/2013).

Howe, J. (1998) 'Not a cree – more a listed building', *Newcastle Journal*, 27 March.

MHLG (1968) *Old Houses into New Homes*, Cmnd 3602, HMSO, London.

Orbasli, A. (2007) *Architectural Conservation: Principles and practice*, Wiley-Blackwell Publishing, London.

Pendlebury, J. (2009) *Conservation in the Age of Consensus*, Routledge, London.

Robertson, M. (1993) 'Listed buildings: The national resurvey of England', *Transactions of the Ancient Monuments Society*, 37: 21–38.

Stratton, M. (2000) *Industrial Buildings: Conservation and regeneration*, Spon Press, London.

Tait, M. and While, A. (2009) 'Ontology and the conservation of built heritage', *Environment and Planning D: Society and Space*, 27: 721–737.

Turnpenny, M. (2004) 'Cultural heritage, an ill-defined concept?: A call for joined-up policy', *International Journal of Heritage Studies*, 10(3): 295–307.

While, A. (2007) 'The state and the controversial demands of cultural built heritage: Modernisation, dirty concrete, and postwar listing in England', *Environment and Planning B: Planning and Design*, 34(4): 645–663.

11.6 Neighbourhood planning

Key terms: localism; neighbourhood; community; planning; grassroots

Neighbourhood planning is part of a wider shift of governance towards the local level that has been by pursued by the Coalition Government. The seeds of this paradigm shift can be found in the Green Paper (Open Source Planning) that was released by the Conservatives in opposition in 2010. The Localism Act 2011 introduced the new statutory level to local plan making called neighbourhood planning. The legislation allows residents and businesses in a neighbourhood to do two things, if there is sufficient demand:

- develop a small-scale plan for their neighbourhood;
- propose that a particular development or type of development should automatically get planning permission in their area (via a Neighbourhood Development Order).

The government described neighbourhood planning as a new way for communities to decide the future of the places where they live and work, choose where they want new homes, shops and offices to be built, allow them to have a say on what those new buildings should look like, grant planning permission for the new buildings they want to see go ahead, influence types of housing and where they are built or lobby for more housing.

Planning at the neighbourhood level is not completely new, as under previous governments local communities could develop parish plans but they were not part of legal framework (statutory). However, in terms of real freedom to make choices for local communities, it has been questioned whether the reality of choice under neighbourhood planning matches the government's rhetoric, as neighbourhood plans must be aligned with wider strategic priorities for an area, such as the local plan and must have regard for the NPPF. In reality this means a neighbourhood plan cannot promote less housing than envisaged in a local plan; although it can comment on issues such as the type of housing, and where it should be built, or say that more housing is required.

A speech in 2010 by the then planning minister Greg Clark summarises the Coalition's aspirations for neighbourhood planning:

> When people know that they will get proper support to cope with the demands of new development; when they have a proper say over what new homes will look like; and when they can influence where those homes go, they have reasons to say 'yes' to growth.

However, it is too early to tell whether this will be the case; it is equally likely that communities may take a NIMBY stance and that the neighbourhood planning process may be dominated/hijacked by people who are opposed to development, especially in rural and suburban green belt areas.

Who can start a neighbourhood planning process?

The Localism Act states that neighbourhood plans and neighbourhood development orders can be initiated by either an existing parish or town council, or (where there is no existing parish or town council) a group that has been designated as a neighbourhood forum. Neighbourhood plans and neighbourhood development orders can be either resident led or led by the local business community.

Currently there are in excess of 8,500 neighbourhoods that have parish or town councils in England. To be eligible to create a plan, a neighbourhood forum must have at least 21 members and membership must be open to all those who live, work or represent the area as councillors, and the forum must have a written constitution. Neighbourhood planning is supposed to be open to all people who live in an area, so it is imperative that forums are open and inclusive.

What can be included in a neighbourhood plan?

Advice from the Department of Communities and Local Government states that there is no fixed format or template for a neighbourhood plan and that they are not intended to be mini local plans. The government envisages that communities should concentrate on a few policies that could have a major impact on their area, e.g. density issues/housing for older people. There is no checklist of evidence or additional reports that a neighbourhood plan must contain but the key is it should be appropriate, proportionate and up to date. The local planning authority must support the process but limited additional central government funding is available. It is estimated that to get national blanket coverage in excess of 10,000+, neighbourhood plans would be needed.

In keeping with the ethos of localism, the focus of neighbourhood plans can vary depending on the context of the local area. They can set out a vision and a set of objectives for the future of the area, or they may be more detailed, setting out planning policies for the development and use of land in that neighbourhood.

The process for developing and agreeing a neighbourhood plan or neighbourhood development order is relatively straightforward:

- First, local authorities need to agree the boundaries of the neighbourhood area, and to check that the boundaries don't overlap with another proposed neighbourhood planning area. A neighbourhood plan can cover more than one town/parish council area but there must be agreement.

- A neighbourhood forum group must apply to the local authority who will check they meet the necessary requirements. Only one neighbourhood forum can exist in any designated neighbourhood area; there cannot be competing neighbourhood forums.
- When agreement is reached the parish/town council or neighbourhood forum will draft the neighbourhood plan or neighbourhood development order. It must be publicised and make available for comments to all people within the neighbourhood for a period of at least 6 weeks. During this consultation period the partnership developing the neighbourhood plan must contact Statutory Consultees.
- When the consultation process has taken place the neighbourhood plan is formally submitted to the LPA for their consideration. The LPA will check that all the relevant supporting information that needs to accompany the draft plan or order has been submitted.
- Finally, the neighbourhood plan and development orders are then submitted to an independent qualified inspector for examination to check the neighbourhood plan and/or neighbourhood development order is in line with local and national planning policy. If the plan is found to be sound by the examiner, the local authority then organises a community referendum, where all people living in the area covered by the neighbourhood plan or neighbourhood development order registered to vote in local elections will be entitled to vote. If it receives the majority of votes *of those voting*, the neighbourhood plan or neighbourhood development order is passed, and incorporated into the local plan.

Alternative viewpoint given in a presentation by a former President of the Royal Town Planning Institute

> For successful neighbourhood planning, communities need consistent and long term engagement. The position we have at the moment is drip feeding of funding and neighbourhoods competing. Perhaps rather than focusing on neighbourhood plans, neighbourhoods should focus on understanding/documenting the issues in their area and trying to influence the LPA and the local plan.

Further reading

CLES (2011) 'Localism – a raw deal for local government', available at: www.cles.org.uk/wp-content/uploads/2011/01/RR-18-Localism.pdf (accessed 30/11/2013).

Conservative Party (2009) 'Open Source Planning', Green Paper, available at: www.conservatives.com/~/media/Files/Green%20Papers/planning-green-paper.ashx (accessed 30/11/2013).

DCLG (2011) 'Plain English guide to the Localism Bill' available at: www.communities.gov.uk/publications/localgovernment/localismplainenglishupdate (accessed 30/11/2013).

DCLG (2012) 'Neighbourhood Planning', available at: www.communities.gov.uk/planningandbuilding/planningsystem/neighbourhoodplanningvanguards/ (accessed 30/11/2013).

Gallant, N. and Robinson, S. (2012) *Neighbourhood Planning: Communities, networks and governance*, Policy Press, Bristol.

11.7 Transport and infrastructure planning

Key terms: transport networks; connectivity; National Planning Policy Framework

Transport is a means of providing access. Transport networks enable us to get from one place to another and there are a range of forms of travel that enable us to do this: walking, cycling, and travelling by car, bus, train, tram, sea or air.

It has been established that transport developments can have positive impacts on populations' welfare, productivity and, ultimately, on GDP. Thus, effective transport systems are fundamental to economic success. For this reason, developments in transport technology have been a key determinant in the spatial distribution of industry and commerce, dictating areas of economic activity and productivity.

Since property and land are static commodities, the influence of transport and industry can be very significant, because it provides the necessary connections to link areas of economic activity, and enable populations to access them. It is this connectivity that is an enabler of economic growth.

Transport planning involves consideration of the relationship between different land uses, and the feasibility of different transport mechanisms and networks within and around these. It is argued that it should consider the relationship between different modes of transport and their relative effectiveness in meeting economic, financial, social and environmental goals.

Increased levels of development in the UK will continue to place strain on the existing road infrastructure. Previously, under PPG13, any developments that carried traffic implications were subject to a Transport Assessment in order to reduce the impact of new developments. The guidance required developments to reduce the growth in and length and number of motorised journeys; encourage alternative means of transport that would have less environmental impact, and therefore reduce reliance on the private car. The Transport Act (2000) introduced further measures including providing half price buses for pensioners and disabled people; and giving powers to Local Authorities to implement road user charging and workplace parking charging, and to reduce the provision of city centre parking.

With the introduction of the NPPF, the UK planning system is undergoing seismic changes, and a number of current policies are in flux. Core principles within the NPPF include 'Promoting sustainable transport', and 'Supporting high quality communications and infrastructure'. In addition, the National Infrastructure Plan (October 2010; updated 2012) sets out a vision for investment in infrastructure in the UK up to 2020, and identifies a pipeline of over 550 planned public and private infrastructure projects worth over £310bn. Infrastructure, as defined in the plan, encompasses: major roads, rail, airports and ports; electricity and gas; communications; water, sewerage and waste; and flood defences.

Further reading

Cullingworth, J. B. and Nadin, V. (2006) *Town and Country Planning in the UK*, 14th edn, Routledge, London, ch. 11.

National Infrastructure Plan (October 2012; updated 2012) available at: www.gov.uk/government/uploads/system/uploads/attachment_data/file/221553/national_infrastructure_plan_051212.pdf (accessed 30/11/2013).

National Planning Policy Framework (2012) available at: www.gov.uk/government/uploads/system/uploads/attachment_data/file/6077/2116950.pdf (accessed 30/11/2013).

11.8 Minerals planning

Key terms: National Planning Policy Framework; aggregates; coal; environmental impact; minerals safeguarding areas

Minerals have been worked in the UK for thousands of years but increased in significance with the industrial revolution in the nineteenth century. Even where surface activity has been on a small scale and worked largely with pick and shovel, the adverse impact on the environment has been considerable, as can be seen to this day in parts of Cornwall, north Wales and the Pennines. In the twentieth century, as a result of technological advance and growing demands, mineral extraction became much greater in scale. However, only after the Second World War through the Town and Country Planning Act 1947, did mineral working become subject to the operation of planning controls as we know them today.

Even after the 1947 Act some mineral sites, particularly in the ironstone industry, were given planning permission by statutory order without detailed consideration of individual sites. Such permissions often covered extensive areas of land, far exceeding the operators' immediate requirements. Only rudimentary provisions were made relating to working methods, waste disposal and site restoration; and environmental controls on such aspects as noise, dust or traffic were virtually unknown.

In response to public pressure, control over new mineral workings was subsequently tightened, especially regarding the imposition of conditions; but mineral operators also showed an increasing awareness of the need to make their operations more environmentally acceptable. At the same time, however, the scale of mineral working increased substantially – in particular in relation to meeting demand for aggregates and this resulted in pressure for working on sites close to urban areas or on high grade agricultural land, increasing the conflict between mineral working and the environment.

The production of planning policy and guidance in the UK has been largely fragmented and piecemeal and evolved gradually from the original Town and Country Planning Act in 1947. Successive governments have published very extensive policy and guidance on minerals, through the publication of a series of Mineral Policy Statements (MPSs), separate from the main series of Planning Policy Statements. The over-arching MPS (Planning and Minerals – MPS1) was published in 2006 and was accompanied by a practice guide offering examples and principles of good practice and background information. Before this, MPS2 (Environmental Effects of Mineral Working) had been published in 2005.

Prior to the publication of MPS1 and MPS2, mineral planning policy and advice was provided in a series of 15 Minerals Planning Guidance Notes, published between 1988 and 2004. These covered general mineral planning matters; guidance on specific minerals including coal (MPG 3), aggregates (MPG 6), peat (MPG 13) and silica sand (MPG 15); and guidance on specific topics such as reclamation (MPG 7) and noise control (MPG 11). This resulted in there being some 1,300 pages of national planning policy by 2009. An aim of the new Government in 2010 was to simplify national planning policy and this was achieved through the publication of the NPPF in March 2012 which has only 59 pages.

The special place of mineral planning is recognised by the NPPF devoting a separate chapter to the subject of 'Facilitating the sustainable use of minerals'. The policy states (paragraph 142):

> [M]inerals are essential to support sustainable economic growth and our quality of life. It is therefore important that there is sufficient supply of material to provide the infrastructure, buildings, energy and goods that the country needs. However, since minerals are a finite natural resource, and can only be worked where they are found, it is important to make the best use of them to secure their long-term conservation.

The NPPF then sets out a list of requirements for LPAs in preparing Local Plans and a similar list of requirements for them in determining planning applications. Mineral planning authorities (MPAs) are required to plan for a steady and adequate supply of aggregates and industrial minerals. An essentially restrictive policy continues to apply in respect of coal extraction (paragraph 149):

> Permission should not be given for the extraction of coal unless the proposal is environmentally acceptable, or can be made so by planning conditions or obligations; or if not, it provides national, local or community benefits which clearly outweigh the likely impacts to justify the grant of planning permission.

The Government also published 'Technical Guidance to the National Planning Policy Framework' in March 2012. This contains additional policy on the proximity of mineral working to communities; dust emissions from mineral workings including the health effects of dust; noise emissions from mineral workings; stability in surface mine workings and tips; the restoration and aftercare of mineral sites; and land banks for industrial minerals.

Aggregates are the most commonly extracted and used construction materials in the UK, comprising about 75 per cent by tonnage of all land-won mineral extraction. In 2009, 119.1 million tonnes were consumed in England and Wales including 10.8 million tonnes from marine landings. Although the mineral planning system in the UK applies to all minerals, the foundations of the system relate to aggregates.

A very different approach applies in relation to coal mining. The NPPF (paragraph 149) states:

> [P]ermission should not be given for the extraction of coal unless the proposal is environmentally acceptable, or can be made so by planning conditions or obligations; or if not, it provides national, local or community benefits which clearly outweigh the likely impacts to justify the grant of planning permission.

This represents a continuation of the essentially restrictive policy that has applied to coal extraction for many years.

Opencast coal mining has been a particularly controversial form of mineral working since it began as a wartime emergency measure in 1942. Opencast coal operators have unsuccessfully lobbied for the presumption against development to be removed in recent years. They considered that many MPAs were refusing opencast applications for non-planning reasons and that because of the 'presumption against' appeal decisions also were being unfairly influenced. This, together with the reduced demand for indigenous coal in the face of competition from overseas coal and indigenous gas, has resulted in significantly reduced opencast production, although there are recent signs of renewed interest by operators as world coal prices increase and the Government has sought a more diversified pattern of energy supply. Three major appeals in Derbyshire, Leicestershire and Northumberland were won by the industry in 2007.

Safeguarding of proven mineral resources against incompatible surface development is a key requirement in the NPPF, and MPAs are required to 'define Minerals Safeguarding Areas and adopt appropriate policies in order that known locations of specific minerals resources of local and national importance are not needlessly sterilised by non-mineral development' (paragraph 143).

Mineral Planning Authorities are County Councils in those parts of England where there continue to be two tiers of local government. Elsewhere Unitary Authorities are the Mineral Planning Authorities for their area. County Councils are required to prepare development plan documents (DPDs) for minerals and waste. They may prepare DPDs dealing solely with minerals (or waste) or DPDs dealing with both minerals and waste. Unitary Authorities may also prepare separate DPDs dealing with minerals and/or waste but would normally include minerals and waste together with other subjects in their DPDs.

The protection of mineral resources from unnecessary sterilisation by other development has been a theme of the planning process since the Town and Country Planning Act 1947. The concept of Mineral Safeguarding Areas (MSAs) is relatively new in minerals planning. It is the purpose of the planning system to address competing demands on land use, but until recently it gave little weight to the protection of mineral resources in comparison with that afforded to environmental assets. As a result, there have been many instances where minerals were needlessly sterilised.

The NPPF (2012) at paragraph 143 states that LPAs in preparing Local Plans should:

> define Mineral Safeguarding Areas and adopt policies in order that known locations of specific minerals resources of local and national importance are not needlessly sterilised by non-mineral development, whilst not creating a presumption that resources defined will be worked: and define Mineral Consultation Areas based on these Mineral Safeguarding Areas.

It continues in paragraph 144 by stating that, when determining planning applications, LPAs should 'not normally permit other development proposals in mineral safeguarding areas where they might constrain potential future use for these purposes'.

The NPPF also makes provision for the safeguarding of essential minerals infrastructure including 'existing, planned and potential rail heads, rail links to quarries, wharfage and associated storage, handling and processing facilities for the bulk transport by rail, sea or inland waterways of minerals, including recycled and secondary aggregate material'. In addition it states that 'existing, planned and potential sites for concrete batching, the manufacture of coated materials, other concrete products and the handling, processing and distribution of substitute, recycled and secondary aggregate material should be safeguarded' (paragraph 143).

The concept of Mineral Safeguarding Areas is relatively new in minerals planning although the protection of mineral resources from unnecessary sterilisation by other development has been a theme of the planning process since the Town and Country Planning Act 1947. MSAs were promoted through MPS 1 and have been given renewed emphasis by their inclusion in the new NPPF. The British Geological Survey (BGS) in collaboration with partner organisations has prepared good practice guidance to assist LPAs in mineral safeguarding and the minerals industry in preparing planning applications.

A number of authorities are now well advanced with plan preparation and examples of their approaches to safeguarding have been referred to in this paper. With the assistance of the BGS broad areas of economic geology are being identified. The policy of pre-extraction, however, presupposes that the minerals within a particular development area are commercially viable during the plan period. This will not always be the case and may lead to the delay or prohibition of surface development to safeguard deposits that may never be capable of commercial extraction, assuming of course that the 'practical' and 'environmentally acceptable' tests can also be met. An effective MSA policy framework should be capable of assessing the merits of safeguarding particular minerals on a site by site basis within a broad area of economic geology.

Environmental impact assessments (EIAs) are frequently required to support mineral planning applications (see Concept 16.8). The purpose of EIA is to ensure that full consideration is given to the environmental, economic and social effects of a proposed development during the determination process. The EIA is compulsory for many types of development, including most new mineral sites. Information detailing the likely effects or impacts of a development is assembled and analysed and collated into an Environmental Statement that supports application. This information is reviewed by the LPA, relevant statutory and non-statutory consultees and the general public prior to the formal determination of a planning application.

Further reading

Communities and Local Government, 'National Planning Policy Framework', March 2012, available at: www.gov.uk/government/publications/national-planning-policy-framework--2 (accessed 24/10/2013).

11.9 Settlement hierarchy

Key terms: Local Planning Authorities; National Planning Policy Framework (NPPF); sustainable development

A settlement hierarchy is the organisation of settlements, usually by size, accessibility and level of service provision, to facilitate a sustainable pattern of development. It enables LPAs to plan positively for new development, ensuring that growth is directed towards the most sustainable locations according to the 'hierarchy' or 'order' of settlements. The NPPF (see Concept 11.1) recommends that a 'presumption in favour of sustainable development' should be the basis for every plan and every planning decision (DCLG, 2012: i). A key aspect of recognising sustainable locations for new development is to identify accessibility and proximity to services and sustainable transport options. As such, the NPPF (DCLG, 2012: 7) recommends undertaking research to understand the role of each settlement to define a network and hierarchy of centres that is informed by evidence. Creating a settlement hierarchy is therefore a fundamental part of local plan making and carrying out this research is a key part of the evidence base required for a new Local Development Plan.

Purpose: planning positively for sustainable development

Plan-led strategies and policies should be underpinned by a robust, yet proportionate evidence base that enables development to be directed towards the most sustainable locations. As Mills (2004: 71) explains, 'at the settlement scale, the measures of sustainability focus on the efficiency of the urban system of which a key aspect is the movement of materials and people'. As such, a key purpose of the settlement hierarchy is to bring housing, employment and other services closer together in and around settlements. This approach contributes to the vitality and sustainability of the settlements by:

- supporting existing and new services and facilities
- helping to create vibrant and lively places, and
- increasing accessibility for all sections of society – thus reducing the need for people to travel by private motor vehicle, bringing multiple environmental and quality of life benefits.

(Guildford Borough Council, 2013: 2: 1.1.6)

Typical methods

To create a settlement hierarchy typically involves collecting data to enable settlements to be grouped by level of services, functions and characteristics. Methodologies may vary from one LPA to another, but a common approach is to score and rank each settlement based on an assessment of the level of accessibility that each settlement has to a range of services and facilities including shops, schools and employment. Sustainability indicators may be created to assist this process and these may relate to, for example:

- the presence of shops, schools and community facilities located in or in the vicinity of the settlement;
- access to public transport;
- access to employment opportunities.

For rural areas, it is also important to assess how well the village functions as a community.

These sustainability indicators will also be supplemented with area profiling work that contains contextual information about each settlement including its size and character, and other economic, social and environmental characteristics (PAS, 2013). Preliminary data relating to each settlement's capacity to accommodate additional development (i.e. from service providers such as water and energy companies) will also be fed into this process.

The score for each settlement will then be used to categorise the settlements into those that are considered the most sustainable for future growth (such as the main towns and services centres) as well as those considered least sustainable for future growth (such as disconnected, remote rural villages and hamlets).

Such groupings may be arranged as follows:

- 'principal towns' (also called 'main towns', with a wide range of services and opportunities for employment, retail and education; high levels of accessibility and public transport provision);
- 'secondary/key service centres' (mid-size settlements with a range of services and opportunities for employment, retail and education with some good public transport links);
- 'local service centres' (smaller settlements with a limited range of services and opportunities for employment, retail and education and a lower level of access to public transport); and
- 'other settlements' (smaller, often rural, inaccessible settlements with no employment or retail opportunities).

The tiers established by the settlement hierarchy are then associated with a set of development principles deemed suitable to that tier. Figure 11.9.1 illustrates this hierarchy.

The settlement hierarchy will not only be used to plan where future development in the area will be located, but it will also be used when assessing speculative planning applications. LPAs will use the hierarchy as a tool to enforce a sequential approach to development. This means that applicants must demonstrate that they have assessed all other deliverable and developable sites and that no other more sustainable locations can be found.

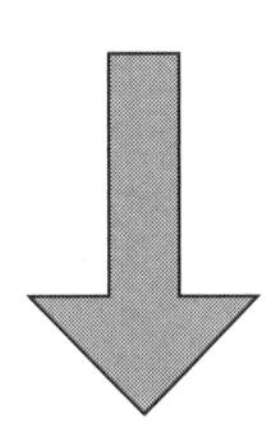

1 Principal towns – the focus for new development and the location for planned housing and employment urban extensions.
2 Secondary Service Centres – development that maintains and strengthens the role of the settlement as a service centre is permitted.
3 Local Service Centres – small-scale development, in-fills, change of use and conversions to meet defined needs and to maintain or enhance local services and facilities is permitted.
4 Other Settlements – the priority in these settlements is the re-use of existing buildings or conversions only.

Figure 11.9.1 Settlement hierarchies in action.

Further reading

DCLG (2012) *National Planning Policy Framework*, DCLG, London.

Guildford Borough Council (2013) *Settlement Hierarchy*, Guildford Borough Council, Guildford.

Mills, G. (2004) 'Progress toward sustainable settlements: A role for urban climatology', *Theoretical and Applied Climatology*, 84: 69–76.

PAS (2013) 'Area profiling', available at: www.pas.gov.uk/ (accessed 25/11/2013).

11.10 Planning obligations

Key terms: planning gain; Section 106; mitigation; Community Infrastructure Levy

As we have seen in Concept 11.4, planning permission is usually granted on a conditional basis. Generally, when considering how to regulate a development the preference is to use conditions, but there are some circumstances where it is not possible to use a condition; for example, you cannot impose a planning condition requiring someone to make a payment. Planning authorities use agreements with developers to further regulate the development and to address any negative impacts arising from it, for example, pressure on school accommodation, local traffic, and availability of play spaces. These agreements are made under Section 106 of the Town and Country Planning Act 1990 and are generally referred to as section 106 agreements or planning obligations. Examples of typical planning obligations are the provision of affordable housing, the transfer of land for play areas or other public uses and financial contributions towards public open spaces/public art/security etc.

Regulation 122 of the Community Infrastructure Levy Regulations 2010 (as amended) and paragraph 204 of the NPPF provide that planning obligations must:

- be necessary to make the proposed development acceptable in planning terms;
- directly relate to the development; and
- fairly and reasonably relate in scale and kind to the development.

As part of the formulation of their development plans authorities have and continue to specify what types of obligation they would seek (e.g. what proportion of affordable housing would be sought) and how any financial contributions would be calculated. These policies are the starting point of negotiations with the developers who will seek to minimise the level of obligations. There will frequently be arguments as to viability. Sometimes developers will offer or agree to accept obligations that do not directly relate to their development. This gives rise to implications of bribery and to suggestions that planning permissions can be bought and any permission granted in connection to such an obligation could be the subject of a judicial challenge.

Section 106 agreements are normally negotiated and completed before planning permission is issued but are conditional upon the granting of permission. These agreements 'run with the land' and the planning authority will therefore ensure that everyone with an interest in the land is a party to the agreement to ensure that it can be enforced. In some cases the developer will sign a unilateral undertaking – this usually arises in planning appeals where the authority is opposed to the principle of development or cannot reach agreement as to the extent of the obligations.

Under Section 106A of the TCPA 1990 a person can apply to the LPA to modify or discharge a planning obligation provided that it has been at least 5 years since the obligation was entered into. If the LPA does not determine the application within the appropriate time period (normally 8 weeks) or refuses to modify or discharge the agreement or undertaking, then there is a right to appeal to the Secretary of State. The Growth and Infrastructure Act 2013 introduced a new application and appeal procedure for the review of planning obligations on planning permissions that relate to the provision of affordable housing.

There have been a number of proposals to reform the planning obligation system, notably the Barker Review, Planning Gain Supplement and the Planning Charge. Some of the main complaints levelled at the section 106 system is that it is a slow, uncertain and potentially costly (in terms of professional fees) process. In 2007 the Government decided to introduce CIL which is discussed in Concept 11.11. The introduction of CIL does not mean that section 106 agreements will cease to exist. CIL is to be used to fund infrastructure with a wide definition that could be used for a number of purposes that section 106 agreements are currently used for, including school extensions and play areas. However, it cannot be used for the provision of affordable housing, and planning obligations will still be required where the developer is providing land. CIL is in its infancy so we will have to wait and see how CIL and planning obligations work together.

Further reading

Moore V. and Purdue M. (2012) *A Practical Approach to Planning Law*, Oxford University Press, Oxford.

Planning Portal (no date) 'Conditions and obligations', available at www.planningportal.gov.uk/planning/applications/decisionmaking/conditionsandobligations (accessed 27/11/2013).

11.11 Community infrastructure levy

Key terms: tax; planning obligations; infrastructure

Community Infrastructure Levy (CIL) is a levy that local authorities in England and Wales can choose to charge on new development in their area (Communities and Local Government, 2013). It was introduced by the 2008 Planning Act to address a long-held recognition that planning obligations in the UK have often struggled to contribute to large-scale infrastructure requirements or the cumulative infrastructure needs resulting from incremental development.

The statutory purpose of CIL, as set out in Part 11 of the 2008 Planning Act, is 'to ensure that costs incurred in providing infrastructure to support the development of an area can be funded, wholly or in part, by owners or developers of land'. Subsequent regulations (HoC 2010, 2011 and 2012) make it clear that, while (Section 106) planning obligations should aim to secure necessary requirements that facilitate the granting of planning permission for a particular development, CIL contributions are for general infrastructure needs. Thus CIL sits beside and is in addition to Section 106 obligations (see Concept 11.10) which are contributions to specific infrastructure or other works to mitigate the impact of a particular consented development. CIL excludes affordable housing, which is often included in S106 agreements, consistent with the general principle that developers and landowners should not be charged twice for the same development. It is not a tax on land value uplift.

CIL operates by local authorities opting to charge a levy on new development in their area by introducing a charging schedule that identifies taxable development. This is achieved by setting out the types of development on which a charge is to be levied e.g. residential, office, retail etc., the areas in which a charge is to be levied and the rate at which each type is charged. It is only charged on new development that constitutes construction of new buildings, and thus excludes telecommunication and utility based construction. Some developments are excluded because of their type (e.g. social housing or charitable activity) or location or, indeed, because the local authority has opted not to levy CIL. The rate should not put at risk overall development and must strike a balance between funding infrastructure from the levy and the potential effects of the imposition of CIL on the viability of development in an area.

The proceeds raised by CIL are to be spent on local infrastructure and charging authorities must identify an infrastructure gap that may be filled by a selection of indicative infrastructure projects that underpin the development plan. The charging authority must spend the money on funding infrastructure to support development of its area or may pool funds with other LAs to fund conurbation or region-wide projects.

In practice, a landowner or lessee must determine whether they are liable for CIL and serve an assumption of liability notice on the charging authority, which responds with a liability notice. The landowner or lessee would then serve a commencement notice on the charging authority after which they would receive a demand notice from the charging authority. The CIL rate is determined at the point at which planning permission first permits chargeable development to take place.

In theory, according to the Department of Communities and Local Government (2013), CIL should deliver the benefits for the following stakeholders:

1 for local authorities – freedom to set their own priorities regarding what the money should be spent on and a predictable funding stream to allow them to plan ahead more effectively;
2 for developers – greater certainty from the outset about how much money they should expect to contribute;
3 for local communities – greater transparency, as local authorities must report what they have spent the levy on each year, and potential to reward the communities that receive new development with a share of the proceeds of the CIL levy collected in their area (15 per cent to 25 per cent dependent on whether they have a Neighbourhood Plan).

The actual impact of CIL is uncertain for two main reasons: first, the voluntary nature of the levy, such that some local authorities will choose not to charge the levy on new development in their area, which may have consequences for neighbouring authorities that do opt to charge the levy, and second, variable rates for certain types of development, both of which potentially will have spatial implications and consequences.

Further reading

DCLG (2010) *The Community Infrastructure Levy: An overview*, DCLG, London.

DCLG (2013) *Community Infrastructure Levy Guidance*, DCLG, London.

Planning Portal (2013) 'Community Infrastructure Levy: about the Community Infrastructure Levy', available at: www.planningportal.gov.uk/planning/applications/howtoapply/whattosubmit/cil (accessed 14/10/2013).

11.12 Planning appeals

Key terms: planning appeals; written representations; informal hearings; public inquiries

Introduction: rationale for planning appeals

Planning appeals provide the opportunity for planning decisions to be challenged in an independent and impartial environment. The planning appeal system is underpinned by the notion of natural justice, openness and fairness. As such, appealing against a decision not only provides an opportunity to challenge planning decisions, but also enables particular issues of contention and vague and/or contradictory policy to be fine-tuned and clarified, setting a precedent for future decision making.

Criteria for appealing

Appeals to the Secretary of State can be made by an unsuccessful planning applicant under Section 78 of the 1990 Town and Country Planning Act.

The main reasons to make an appeal are as follows:

1 refusal of planning permission;
2 appeal against (unacceptable) planning conditions attached to permission (e.g. hours of opening);
3 appeal against non-determination (where a LPA has failed to give a decision within the prescribed time period);
4 appeal against enforcement notices (such as to remove a building structure or cease a use).

The last resort

The Planning Inspectorate is an independent group of qualified planning inspectors, appointed nationally by the Government, and these are responsible for planning appeals. The Planning Inspectorate, also known by the acronym 'PINS', do not encourage appeals. Instead, they advise parties initially to attempt to resolve issues through dialogue and negotiation. If such mediation proves unsuccessful, only at this stage should the appeal system be considered. Even once an appeal process has been commenced, PINS encourage that open communication channels are maintained to attempt to narrow down areas of dispute where possible. Such communication can save vast amounts of time and money.

Appeal process

The Planning Inspectorate provides detailed advice and guidance for making an appeal (PINS, 2010). This publication replaces PINS 01/2009, first introduced on 6 April 2009 and the suite of Good Practice Advice Notes which were published in 2009. More information can also be found on the Planning Portal website: (www.planningportal.gov.uk/planning/appeals/online/makeanappeal).

Appeal forms can be completed online (see the link to the Planning Portal website above). The appellant must outline the grounds for appeal and respond directly to the reasons for refusal given in the LPA's decision notice/letter. The main areas of dispute should be highlighted and explained in clear planning terms. Note that an appeal against a successful application cannot be lodged by a third party objector to the scheme.

There are clear deadlines for making an appeal which must be adhered to. Applicants have:

- 12 weeks to appeal against a householder application;
- 6 months for other cases from the date on the decision notice; or
- 6 months from the expiry of the period which the LPA had to determine the application;
- 28 days in respect of an enforcement notice.

Once the Planning Inspectorate has received the appeal form and confirmed validity, they will decide how the appeal will be heard. Section 319A of the Town and Country Planning Act 1990 gives the Planning Inspectorate/Secretary of State the statutory power to determine the procedure.

Modes of planning appeal

There are three different ways in which an appeal can be assessed (see Figure 11.12.1).

The following summarises the key aspects of each method, together with the main advantage/disadvantage of each.

1 *Written representations*

- Suitable for minor cases – straightforward planning issues such as most household appeals, cases of low public interest and little complexity.
- Appellant and LPA submit a written report to the Planning Inspector to consider. The Inspector will conduct a site visit.
- Approximately 80–85 per cent of all appeals are conducted using this method.
- Advantage: allows for a quick and relatively low cost decision (in terms of professional fees).
- Disadvantage: little chance to argue your case and challenge the case of the LPA.

2 *Hearing*

- *Inquisitorial* – less formal than a Public Inquiry but enables Planning Inspector to ask questions. Highly complex legal, technical issues are unlikely to arise.
- A 'half way house' between written representations and public inquiries (1-day hearing suitable for small-scale development with little or no third party interest).
- Site visit, exchange of written evidence.
- No formal cross examination. Instead Inspector facilitates roundtable debate.
- Approximately 15–20 per cent of all appeals are conducted using this method.
- Advantage: allows for debate and allows appellants to challenge/respond to the LPA's decision making verbally.
- Disadvantage: more costly and requires more time than written representations.

3 *Public inquiry*

- *Adversarial* – suitable for major, controversial and complex cases.
- Format involves the questioning and cross examination of evidence, with expert witnesses called (third party input, i.e. Highways, Environment Agency, English Heritage).
- Proceedings are managed by an independent Planning Inspector, but barristers take a lead role in cross-examination of witnesses.
- The LPA must publicise the inquiry through local newspapers and post notices close to the site, etc.

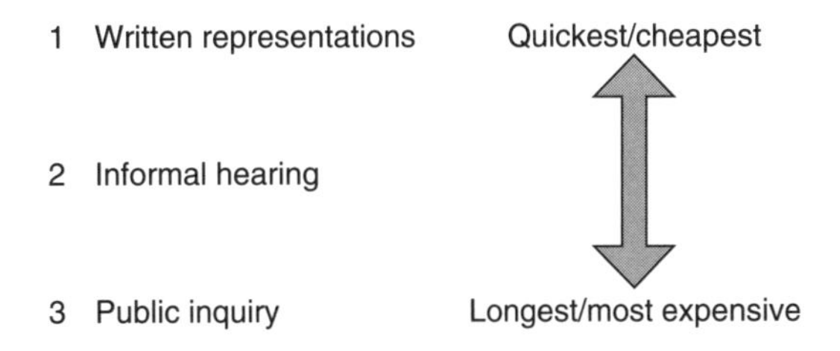

Figure 11.12.1 Assessment of appeals.

- Approximately 5 per cent of all appeals are conducted using this method.
- Advantage: thorough, open, impartial and fair consideration of issues.
- Disadvantage: most expensive in terms of professional/legal fees and time consuming (average appeal lasts 29 weeks). Parties must be sure of the strength of the case because costs will need to be covered.

The decision

The Planning Inspector will make one of the following decisions:

1 Allow the appeal (with conditions; overturns the LPA's decision) or
2 Dismiss the appeal (confirms the LPA's decision).

Recent changes to planning appeal procedures

On 1 October 2013, the 'Review of Planning Appeal Procedures', undertaken by the Department for Communities and Local Government (DCLG) came into effect. The review includes a number of measures intended to make the planning appeals process faster and more transparent. Key changes include:

- frontloading the procedures: appellants submit full case statement at the very start of the process and an agreed 'Statement of Common Ground' (SCG) must be submitted to PINS within 5 weeks (relevant for both hearings and inquiries);
- faster start dates: *new* target dates for commencement of appeal;
- faster decision times: 80 per cent of written representations and hearing appeals to be decided within 14 weeks, and 80 per cent of inquiries to be decided within 22 weeks.

Further reading

HOC (2012) 'Planning Appeals Policy', available at: www.parliament.uk/briefing-papers/SN01031 (accessed 4/12/2013).

Moore, V. and Purdue, M. (2012) *A Practical Approach to Planning Law*, Oxford University Press, Oxford.

PINS (2010) 'Procedural guidance – Planning appeals and called-in planning applications', available at: www.planningportal.gov.uk/uploads/pins/procedural_guidance_planning_appeals.pdf (accessed 4/12/2013).

PINS (2012) 'Making your appeal', available at: www.planningportal.gov.uk/uploads/pins/enforcement_making_your_appeal.pdf (accessed 4/12/2013).

Planning Portal (2013) 'Technical review of planning appeal procedures', available at: www.planningportal.gov.uk/planning/appeals/news (accessed 4/12/2013).

Ratcliffe, J., Stubbs, M. and Keeping, M. (2009) 'Urban Planning and Real Estate Development', Routledge, Abingdon.

Tait, M. (2012) 'Building trust in planning professionals: Understanding the contested legitimacy of a planning decision', *Town Planning Review*, 83(5): 597–618.

12 Property asset management

Cheryl Williamson, Dom Fearon and Kenneth Kelly

12.1 Property asset management

Key terms: real estate; adding value; physical changes; re-gearing; additional revenue streams

The role of the real estate asset manager in connection with real estate portfolios is to ensure that the value is retained and to increase value where possible. If a client purchases a portfolio worth £5 million today, then in a year's time they would want it to be worth more than that to cover inflation and to make a capital profit. This is a relatively easy job when the markets are hot but during economic downturns and for poorer quality portfolios, the adding value part can be much more difficult. On a day-to-day basis the surveyor needs to ensure that the portfolios are managed prudently – the leases need to be 'policed'. This means that the tenants are paying their rents promptly, maintaining the buildings and complying with their covenants in the lease. Hard decisions can need to be made on whether it is better to keep a tenant even if they are struggling to pay their rent and avoid costs such as empty business rates, or whether a new tenant of a better covenant strength can be found as a replacement.

A further role for the real estate asset manager is to add value to the portfolio. This can be achieved by moving funds between different investment groups/opportunities. The returns from different investments will be compared and then decisions will be taken on buying or selling the portfolios. The aim here is often to purchase at the bottom of the market and sell at the top to maximise returns. However, actually being able to spot these opportunities is not always easy. On a smaller scale, investors can realign their portfolios by selling off the poorer stock and purchasing better quality properties. These may have better quality tenants with stronger covenant strengths, better leases, and improved quality of buildings or better locations. Other ways of adding value involve physical changes to the buildings. Existing properties can be demolished and rebuilt to modern standards, achieving higher rental levels and better tenant covenants, although the decision will depend upon the balance between costs and value. If the buildings cannot be demolished, then existing buildings can be merged or split to provide more valuable space that the market requires. Finding more valuable uses for the portfolio is another way to add value. The surveyor needs to look at the planning situation and whether there is a possibility of obtaining a change of use to a more valuable use. Examples of this are the current developments of old 1960s office blocks into residential units or the proposed changes from obsolete retail units into small-scale residential units. In many areas residential use is more valuable than office or retail particularly if the other uses are obsolete in that area. Changes like this can add considerable value to a portfolio.

Refurbishment of existing buildings can also add value to a portfolio. Discussions need to be held on costs versus any uplift in price as a landlord will not want to spend money if there is no return; however, refurbishment can make a building more marketable which is an advantage in a recessionary market. This means that it can reduce the amount of time that a landlord has no income from the property while having to pay the outgoings.

As can be seen from the above the real estate asset manager's role is to review the portfolios and find a variety of ways to add value from physical changes and improvements. In addition, another way of adding value to a portfolio is to look at the leases and the tenants to see ways of making changes that can add value. This is known as lease re-gearing. Negotiations between landlords and tenants can result in both parties obtaining something of value; an example of this would be a tenant giving up a break clause in a lease in return for a reduction in their rent. The landlord would be given more certainty of income for a longer period and the tenant would reduce its outgoings, thus both parties would benefit from the new arrangement. Landlords can also look for additional revenue streams and the asset managers need to 'think outside the box'. This income stream can come from a variety of sources such as advertising space on gable walls, allowing concessions to have space in a shopping centre or renting space on high buildings for telecommunication equipment.

An asset manager needs to have a good knowledge of property management principles together with the added ability to seek out value in a portfolio in a variety of ways.

Further reading

Edwards, V. and Ellison, L. (2004) *Corporate Property Management*, Blackwell, Oxford.
Haynes, B. P. and Nunnington, N. (2010) *Corporate Real Estate Asset Management*, EG Books, Oxford.
Scarrett, D. (2011) *Property Asset Management*, Routledge, Oxford.
www.isurv.com (accessed 14/11/2013).
www.rics.org/uk (accessed 14/11/2013).

12.2 Leases in commercial property

Key terms: legally binding contract; exclusive possession of the property; leasehold interest; full repairing and insuring lease; building insurance

In commercial property, i.e. retail, offices, industrial and leisure, a commercial property lease or business lease is a legally binding contract between the legal owner of the property (landlord) and the occupier (tenant) of the property. As the lease is a legally binding contract, failure by either party to comply with the terms of the agreement could result in court action. In some cases, the landlord is also the tenant of another owner which may restrict the flexibility of the terms the landlord can offer. If this is the case, the landlord should always state in advance and provide a copy of the current lease between themselves and the superior landlord. It should be remembered that most commercial leases are generally drafted on behalf of, and very much to the advantage of, landlords, therefore a tenant should take great care before entering into such a lease agreement.

A key distinguishing element of a commercial property lease is that it grants exclusive possession of the property upon the tenant. This means that the landlord cannot enter the property except in certain prescribed circumstances as set out within the lease. A lease is a contractual obligation but it also creates an interest in property. This is known as a *leasehold interest* as opposed to a freehold interest in property. It must therefore be granted for a distinct period of time but can carry on beyond the fixed term as set out in the contract (lease). If this happens the tenant is referred to as 'holding over' on the lease and the lease becomes a *Tenancy of Will* which is then terminated by notice.

Most commercial property leases are of a particular type, being either FRI (full repairing and insuring) leases or IRI (internal repairing and insuring) leases. It is important for a tenant at the outset to understand the difference between these leases and the financial implications each one offers. Under an FRI lease, tenants will be responsible for the repair and maintenance of the whole of the property. This becomes especially important when the lease relates to an older/listed or unusually designed or specialist property where costs can be higher. Under an IRI lease tenants will only be responsible for internal areas of their demise ('demise': the area occupied under the terms of the lease) with the landlord being responsible for the main parts of the property being the foundations, structural walls and roof of the building. Obviously, most tenants would prefer the less risky obligations of an IRI lease, while the landlord would prefer the tenant take full repairing liability under a FRI lease. Often the type of lease offered is negotiated between both parties and has a direct link to the rental sum offered. Generally speaking, because of the repairing obligations, a tenant would expect to pay less rent per annum for a FRI lease and slightly more rent per annum for an IRI lease. In large, multi-occupied buildings such as office blocks and shopping centres, landlords will request from tenants, an IRI lease with the addition of a service charge to cover additional costs towards common areas of the property.

Insurance provisions under an FRI/IRI lease vary, depending on the type of premises being leased. In the majority of cases, the landlord insures the property with the tenants repaying the premium to the landlords. The landlord will usually insist on insuring the property as they are then in control and can ensure that the property is properly insured at all times. If, unusually, the tenants are to insure the premises, the lease will contain provisions to ensure that the tenants insure the property to the level required by the landlord.

In both situations the lease should say what risks will be insured against. The insured risks are important as they relate directly to the repairing obligation. If any damage is caused that is not an insured risk the tenants will have to repair or reinstate the property. Likewise, if the tenants breach the insurance policy, they will be required to reinstate the property. The insurance provision discussed above refers to *building insurance* against risks of fire, flooding and damage to the building which is separate to contents insurance which is usually the responsibility of the tenant for their own stock, fittings etc.

The other *key terms* typically found in a commercial property lease which, as above, are generally open to negotiation between the parties, are as follows:

- length of term
- rent payable
- payment dates
- VAT
- lease costs
- business rates
- service charge
- rent review provision
- break clause
- transfer (by way of assignment or subletting)
- fitting out/alterations
- dilapidations
- use of the premises.

The Code for Leasing Business Premises in England and Wales 2007 provides a framework within which a prospective tenant can reasonably expect a landlord to operate. As a prospective tenant, you should not expect a landlord or landlord's agent to always comply with this code as it provides guidance or best practice rather than full legislation. The government, however, takes a keen interest in ensuring the property industry complies with this voluntary code.

Further reading

The Code for Leasing Business Premises in England and Wales 2007, available at: www.leasingbusinesspremises.co.uk (accessed 20/11/2013).

12.3 Breach of covenant

Key terms: breach of covenant; remedies; court action; forfeiture; specific performance; injunction

It is one of the main responsibilities of the asset manager to ensure that the portfolio is protected by 'policing' the leases. This means ensuring that the tenants all abide by the covenants in their leases; if they don't then they are in breach of their lease. At this point the landlord or their agent can take action against the tenants in a variety of ways. There is a process to go through that needs to be thought through at the beginning. Once a landlord is aware of a breach then he must be careful not to 'waive' it, which means accepting the breach. One of the ways that a landlord could waive a breach is to accept rental payments and this will be discussed later in this section.

Once the breach is known, then the landlord must ask a series of questions: Is the breach important or significant? Will it impact on the value of my portfolio either today or in the future? Does it affect the property itself? Will it have an effect on neighbouring tenants? Is it important that something is done about the breach? If the answer to any of the above is yes then the next question needs to be asked and that is can the breach be remedied? This means can the situation be put back to what it was originally? Breaches can be once and for all or continuing, capable of being remedied or not. An illegal assignment or subletting is a once and for all breach that cannot be remedied, however disrepair of a property can be continuing and capable of being put right. So, if a breach of covenant is significant then further action may be required. It is also important to understand that landlords can also be in breach of covenant and tenants can also take action against them.

Breaches of covenant can be breach of the repairing clause, unlawful assignment or subletting, unauthorised alterations or an unauthorised change of use. In fact any covenant or clause in the lease if not complied with is technically a breach; however, some may not have much of an effect on the property and could be ignored and documented.

The main remedies available to either a landlord or a tenant where there is a breach of covenant are as follows:

1. *Action for damages*
   This is the most common remedy for non-performance of a covenant. The amount of damages awarded to the plaintiff will be on the basis that he will be in a similar position after receipt of the damages to where he would be had the breach not occurred.
2. *Distress*
   This is a remedy available to landlords when the tenant has not paid the rent. It allows the landlord to enter the premises and take goods to the value of unpaid rent and hold them until the tenant pays the due sums. Failure to pay the due sums means the landlord can sell the goods and keep the proceeds up to the value of outstanding rent. This will now be regulated by Commercial Rent Arrears Recovery (CRAR) with effect from April 2014.
3. *Forfeiture*
   Provided the lease contains a forfeiture clause for a tenant's breach of covenant under certain circumstances, the landlord can seek to forfeit the lease. This means that if the landlord is successful he will generally be left with vacant premises and no one to pay the rent. This can be a Pyrrhic victory for the landlord and should only be used after a great deal of consideration. Prior to forfeiting the lease he will have to serve a S.146 notice of the Law of Property Act 1925. The notice must specify the breach, and require it to be remedied within a reasonable time. The breach must be capable of being put right. The landlord can ask for damages.
4. *Specific performance*
   This is not a usual remedy but may be available where damages would not achieve the objective. For example, a covenant to provide services could be remedied by specific performance as it is unjust

to expect the tenant to continue working in unheated premises because the landlord is failing to provide heating to the building. Specific performance tries to make a party actually do something.

5 *Injunction*
This is a discretionary remedy and not often used in the field of landlord and tenant as the majority of breaches of covenant can be compensated for by the payment of damages. It may, very unusually, be used to force a tenant to stay open for business at specific times (for example if the tenant was trying to assign the lease, he should continue trading). An injunction tries to stop a party doing something.

Breach of covenant can be a difficult area of property law to understand and deal with, particularly as other legislation such as the Leasehold Property (Repairs) Act 1938 interacts with the law concerning breach of covenant. In addition it is important to look at the remedies in light of the economic climate, the local market conditions, the type of tenant, the effect of the breach on the property and other tenants. The decision about what action to take is individual to each set of circumstances.

Further reading

Rodell, A. (2013) *Commercial Property*, College of Law Publishing, Guildford.
www.isurv.com (accessed 20/11/2013).
www.rics.org/uk (accessed 20/11/2013).
www.insolvencyhelpline.co.uk.

12.4 Commercial service charges

Key terms: multiple occupation; budget; RICS Code of Service Charges in Commercial Property; apportionment methods; sinking fund; reserve fund

Service charges are defined as charges and associated administrative costs properly incurred by an owner in the everyday running of the property, including maintenance, repair, replacement (where beyond economic repair) of the building, plant, equipment and other materials. A service charge for a building will vary depending on the size and type of building. Obviously a service charge for a large shopping centre will have more headings within it and a higher overall cost than a service charge for a small industrial estate. They are used in properties where there is multiple occupation and the separate units are normally let on an IR (internal repairing) basis with the landlord recovering costs for the external repair, common areas and services provided by way of the service charge. Details of what is included in the service charges and how they will be recovered will be included in the lease.

The landlord will set a budget for a service charge year usually taking into account the costs for the previous year and adding any additional costs that may occur in the following year, for example, legislation changes affecting health and safety requirements. The budget should where possible use industry standard cost headings which make it easier for service charges to be compared. This budget figure will then be charged to each tenant one quarter in advance for the next service charge year. Where there are any vacant units within the building the landlord will have to pay the service charge for those units as they are not allowed to pass the costs on to the other tenants. There are various ways of apportioning out the service charge to the tenants but whichever method is used the landlord must be fair and reasonable, as demonstrated in the RICS Code of Service Charges in Commercial Property third edition which came into effect from February 2014. One way is to base the apportionment on the amount of floor space occupied by the tenant as a percentage of the total floor space. So if the tenant's unit occupies 10 per cent of the total floor space of the building then they would pay 10 per cent of the service charge. This appears to be a fair system, but tenants who occupy large amounts of space feel that this is an unfair system and they prefer a weighted floor area basis. This zones the space into areas, for example the first 1,000 sq m would pay the full price per sq m, the next 1,000 sq m would pay half the rate, the next 1,000 sq m would be at a quarter of the rate and anything over that would pay at an even lower rate. The problem with this is that the tenants occupying smaller units pay proportionately more for their units. A final method that is sometimes used is to base the apportionment on RVs.

At the end of the service charge year, the accounts will be audited and it is good practice for an external auditor to be used, although the person responsible for managing the service charge also has a duty of care to certify and sign off the audit under the new code. This is called the service charge reconciliation. The actual expenditure will be ascertained and apportioned on the same basis as before and if the budget has overspent the tenant will be asked for more money and if it is underspent then landlords will credit the tenant's account. A full explanation of the expenditure should be given to the tenants. According to the code this should be achieved within 4 months of the year end.

Service charges in commercial properties can be contentious between landlords and tenants over the amounts of money spent and services provided. As part of this, tenants prefer service charge rates to remain stable and do not like wide variances in the rates and this is sometimes difficult for landlords if unforeseen expenditure arises. One way to try to smooth out wide variances in the level of the service charge is to use a sinking fund or a reserve fund. A sinking fund collects monies over a period of time for specific items of repair or replacement, for example, if a building needs a new roof in 5 years' time at a cost of £50,000 then a sum of £10,000 would be put into the sinking fund every year. Then, after 5 years when the roof needs to be renewed the monies are already there to pay the bill rather than adding £50,000 to that particular service charge year. A reserve fund works in the same way but is usually used for works of a recurring nature – an example would be repainting external paintwork. While both funds are useful, they have problems with VAT recovery for tenants and other problems.

Further reading

Forrester, P. (2004) *Service Charges in Commercial Property*, 2nd edn, RICS Books, Coventry.
Rodell, A. (2013) *Commercial Property*, College of Law Publishing, Guildford.
www.isurv.com
www.rics.org/uk/servicechargecode.

12.5 Rent

Key terms: rent; negotiated agreement; market rent; concessionary rents; headline rents; stepped rents; turnover rents

Rent can be described as:

- The amount of money that a landlord and tenant agree reflects the terms of the agreement which they have reached for the use of the premises. It reflects who does the repairs, who pays the insurance and who pays the rates etc.
- The amount of money that gives the landlord a return on the capital he has invested in the property concerned but enables the tenant to trade from the premises and generate enough profit so that he can stay in business.
- Freely negotiated between the parties. There is no compulsion on either party to agree to a rent that they are not happy with and there is no legislation to decide what the level of rent should be (except sometimes in the case of residential property).

So, rent is a freely negotiated sum of money paid by the tenant (occupier) to the landlord (investor) so that the tenant can occupy premises and run a business from them. If the landlord and proposed tenant cannot come to an agreement about rental terms, they can agree to disagree and no letting will take place.

However, once an agreement is reached and the amount of rent is stated in the lease document, the tenant is legally bound to pay it in the agreed way until the lease expires, the rent is reviewed or some other mutually acceptable action happens.

Generally the rent agreed between the landlord and the tenant is based on market value as defined by the International Valuation Standards Council VS.3.2 and comparable evidence is used to ascertain the value of the property. This is not the only way that rental levels can be set, there are other methods that will be outlined below.

A concessionary rent is lower than market value and can be used by the landlord to encourage an anchor tenant to take a lease in a new shopping area or to get lettings in an office block or on an industrial estate. Once tenants take space in buildings it can be easier to attract further tenants. Sometimes concessionary rents can be used if an existing tenant is struggling to pay a market rent, so it can be a useful tool for a landlord. A premium rent is really the opposite to a concessionary rent as it is higher than market value. It is an artificial rent that sometimes occurs when a tenant has to overbid for a particular property or it might reflect an unusual rent review pattern.

During difficult markets landlords may want to protect their investments and will not want the market rental levels to reduce as this will have an effect on their investments. At the same time it is difficult for them to attract tenants, and tenants understanding their strong bargaining position often negotiate lower rents, longer rent-free periods and other lease advantages or inducements. It is not uncommon for a headline rent, i.e. the asking rent, to be declared but details of inducements to be kept confidential. The inducements will have an effect on the rental level, normally reducing it, so the landlord will want the headline rent to be used to keep the value of the investment higher, particularly if they have more tenancies either in the building or in the surrounding area.

Sometimes landlords have poorer quality properties within their portfolios that are difficult to let and may only attract poorer covenants or tenants who have no business history. These types of tenants may find it difficult to immediately pay a market rent. A landlord may then use a stepped rent as shown in Table 12.5.1. The tenant pays a lower rent in the early years of the lease until their business is more settled and able to pay a market rent. This can be a useful tool for both the landlord and tenant.

Table 12.5.1 Stepped rent

Year 1	*Year 2*	*Year 3*	*Year 4*	*Year 5*
£8,000	£8,500	£9,000	£10,000	£10,000

The final type of rent that can be used by landlords is a turnover rent. It is commonly used in shopping centres, airports and petrol stations. It is where the tenant pays a percentage of their takings as rent. It can be either based on a pure turnover basis or on the sum of a base rent plus a percentage of the turnover.

Further reading

Rodell, A. (2013) *Commercial Property*, College of Law Publishing, Guildford.
www.bpf.co.uk (accessed 20/11/2013).
www.isurv.com (accessed 20/11/2013).
www.rics.org/uk (accessed 20/11/2013).

12.6 Rent reviews

Key terms: upward only; time of the essence; arbitrator; independent expert; assumptions; disregards

Historically, if a landlord let a building to a tenant, they agreed on the rent to be paid and it was only reviewed if or when the lease was renewed. However, once the economy was affected by inflation then the initial rental levels became less valuable to the landlord as inflation ate away at the income. To stop this happening landlords began to insert rent review clauses in their leases, which gave them the right to review the rent at certain points during the lease. The original rent review pattern was multiples of 7 years but this reduced to a 5-year pattern, although some landlords have 3-yearly review patterns. Landlords used what was called an upward rent review, this meant that the rent either stayed at the same level as the existing rent or went up to a higher level. The rent could never reduce and it meant in times of recession when rental levels can drop that tenants were stuck paying rents higher than market rents. The Code of Leasing Business Premises talks about using upward and downward rent reviews as a fairer system for both parties and this view is backed by many property organisations.

Modern forms of lease usually include a rent review schedule and this gives details of when the rent is to be reviewed, how the procedure starts and what happens if the landlord and tenant cannot agree on the new rent. Occasionally a rent review may be 'time of the essence' and this means that if there is a timetable laid down in the schedule that it must be abided by, otherwise the landlord loses their right to review the rent (*United Scientific Holdings v Burnley Borough Council* 1978). This can be a complex area so it is usually a good idea to take further advice on this matter. If the rent cannot be agreed, then the lease will normally specify that it will be determined by either an arbitrator or an independent expert. Their roles are slightly different, although both will decide what the new rent will be. The arbitrator is governed by the Arbitration Act 1996 and has a quasi-judicial role. They can ask for disclosure of documents and subpoena witnesses just like a judge. Evidence to support the rental figure is provided by both parties and the arbitrator bases their decision on this information; they are not allowed to make a decision based on anything not raised by either party. They cannot be sued for negligence and there are very limited grounds on which to appeal the decision. The arbitrator can award costs which means that if they feel one party was not being professional then they can make them pay more of the costs. The independent expert does not have quasi-judicial powers. They decide the rental level by looking at the evidence put before them but they are allowed to use their own knowledge when making the decision. They can be sued for negligence (none have been successfully sued) and they cannot award costs unless the lease allows them to. Both the landlord and tenant can agree on who to appoint, but if there is a disagreement it is usual for the president of the RICS to make the appointment. Once the rent is agreed either between the landlord and tenant or by a third party (arbitrator or independent expert) it is documented in a rent review memorandum which is then attached to both copies of the lease.

In the majority of cases the intention at rent review is to get the rent to a level that reflects the rent the landlord would be able to achieve and the tenant would be willing and able to pay if the property was on the open market. The rent review schedule will try to reflect this situation and give details of what is to be valued. Generally, there will be a list of assumptions to be made by the surveyor. For example:

- the length of the lease;
- that the property is vacant;
- that it is fit for immediate use and occupation;
- obligations under the lease;
- any consents required.

All of these may have an effect on the market value of the property and can be the subject of debate between the two parties. Leases can also include matters that the surveyors must not take into account when valuing the property, these are commonly known as the S.34 disregards and include things such as the occupation of the tenant or any improvements that they have made to the property. This method is the most common way of reviewing rents; however, there are other ways that can be used. Fixed rent reviews can be agreed when the lease negotiations are taking place which gives both the landlord and the tenant certainty of future income and expenditure. Reviews can be linked to a particular index such as the Retail Price Index or can be geared to another type of property use.

Further reading

Code for Leasing Business Premises in England and Wales 2007, available at: www.leasingbusinesspremises.co.uk (accessed 20/11/2013).

Rodell, A. (2013) *Commercial Property*, College of Law Publishing, Guildford.

www.isurv.com (accessed 20/11/2013).

www.rics.org/uk (accessed 20/11/2013).

12.7 Proactive management to recover rent

Key terms: income stream; breach of covenant; proactive management; CRAR; enforcement agents; court action; forfeiture

It is one of the main responsibilities of the asset manager to protect the income stream and to ensure that the income is collected on the due dates. The better the quality of the portfolio, the easier this task is, as blue chip tenants usually make their payments on time, but the poorer the portfolio, the more difficult this can be and the more time intensive. On large portfolios for pension funds or property companies the collection time for obtaining rents is often used as a key performance indicator (KPI) to judge the abilities of the asset managers and their companies. Therefore, it is important that the asset manager understands all the processes/methods available in dealing with this area of debt recovery. In multi-let buildings the asset manager will also be responsible for the collection of the service charges and again this is an important area for the asset manager to ensure that the monies are collected on time.

This is the most common breach of covenant and it can be dealt with in a slightly different way from most other breaches. There are a variety of methods of dealing with tenants with rent arrears; some are very 'light' touch while others have serious implications for tenants. First, check the lease. If the tenant has provided a guarantor then ask them to make the payment.

Are there previous tenants? If yes, check the lease; is it an old or new lease? Is there an Authorised Guarantee Agreement (AGA)? If the lease has been assigned, then serve a S.17 notice under Landlord and Tenant (Covenants) Act 1995, within 6 months of the debt arising. If the property has been sublet and the head tenant is withholding the rental payments the landlord can serve a notice on the subtenant requesting that they pay the rent directly to the landlord. This is covered by S.6 of the Law of Distress Amendment Act 1908.

Commercial Rent Arrears Recovery

Changes to the way that landlords could recover their arrears of rent were first decided under the Tribunal, Courts and Enforcement Act 2007; however, the detail of these changes was not completed until the Taking Control of Goods Regulations became a Statutory Instrument on 30 July 2013. The changes to the procedures will take effect from 6 April 2014. From this date the remedy of distress will be abolished and CRAR will take its place. This applies only to commercial property where the regulations provide for landlords being still able to seize tenants' goods where tenants have not paid their rent. Landlords are now only able to recover any rent arrears but cannot use this process to recover arrears of service charge or non-payment of an insurance premium.

Under S.76 of the Tribunal, Courts and Enforcement Act 2007, a landlord can only use CRAR if there is a legal or equitable lease in place, which is in a written format. Seven days net rent including VAT and interest must be in arrears for CRAR to also apply. The processes have been standardised and enforcement agents (previously known as bailiffs) can enter commercial premises between the hours of 6 a.m. and 9 p.m. and take enforcement action on Sundays. Only enforcement agents can undertake this work. Before they can take action a notice has to be sent to the tenant giving 7 clear days' notice that they will be calling. On arrival at the property if the tenant is still unable to make the payment, the enforcement agent will take 'possession' of goods under a controlled goods agreement. A further 2-day written notice must be given before the goods can be collected and removed. Then a further 7 days must elapse before the goods can be sold at auction.

The new regulations will make it more difficult for landlords to recover rent arrears from tenants and may lead to the Courts being used for the recovery of other debts such as service charges which can no longer be recovered even if reserved as rent in the lease.

Court action

This is probably more effective as a threat and it is worth writing to the tenant explaining that if the matter proceeds to court that there will be additional costs and they risk having a court judgement awarded which will affect their credit worthiness. If this does not produce a payment then it is probably better to use the small claims court rather than involve solicitors if the debt does not exceed £5,000. Debts over that amount can still go through the small claims court under the Fast Track or Multi-Track systems, but may require legal input. The factors to take into account when deciding whether to pursue this course of action are:

1 the odds of receiving the monies owed;
2 the slowness of the court system;
3 the possibility that the court will be sympathetic to the tenant;
4 whether it is more cost effective to write off the debt;
5 what message is sent to other tenants if the debt is not pursued.

As with many real estate issues this is not a simple solution and requires a great deal of thought before action is taken.

Forfeiture

At the present time landlords can still peaceably re-enter commercial premises for non-payment of rent, provided that there is an express clause within the lease. Forfeiture brings the lease to an end but does not obtain the rent arrears, so it should only be used as a final solution to poor tenant behaviour.

Further reading

Rodell, A. (2013) *Commercial Property*, College of Law Publishing, Guildford.
www.hmcourts-service.gov.uk (accessed 20/11/2013).
www.insolvencyhelpline.co.uk (accessed 20/11/2013).
www.isurv.com (accessed 20/11/2013).

12.8 Landlord and Tenant Act 1954 part 2

Key terms: service of S.25 and S.26 notices; time limits; interim rents; contracting out; security of tenure

The termination of business leases is significantly affected by statute. A lease of business premises will not normally come to an end through effluxion of time as a positive action has to be undertaken to bring it to an end. The main legislation dealing with this is the Landlord and Tenant Act 1954 part 2, as amended by the Regulatory Reform (Business Tenancies) (England and Wales) Order 2003.

The Act describes itself as: 'An Act ... To enable tenants occupying property for business, professional or certain other purposes to obtain new tenancies in certain cases ... and for purposes connected with matters aforesaid' (Landlord and Tenant Act, 1954). The original legislation has remained very much untouched since 1954, although changes have been enacted in the Regulatory Reform (Business Tenancies) (England and Wales) Order 2003. These changes were aimed at making lease renewals quicker, easier and cheaper for all parties. The basis of the act did not change, only certain processes and procedures.

The scope and purpose of the act is that:

1 It ensures the continuation of the tenancy/lease until proper steps are taken to terminate them.
2 It grants the tenant a right to apply for a new tenancy/lease.
3 It specifies the limited grounds upon which the landlord can oppose the grant of a new tenancy/lease.
4 It covers the terms of a new tenancy/lease if not agreed between parties.
5 It lays down a basis for compensation.

Not all commercial leases/tenancies come under the umbrella of this act. In order for the act to apply there are three main areas that have to be satisfied:

1 There must be a tenancy – it must not be an excluded tenancy, examples of which are agricultural leases, fixed term tenancies, tenancies at will or licences. If the tenancy is not excluded from the act then it may fall within the protection of the act.
2 There must be occupation – the tenant must be in occupation, although this can be through an agent or manager. If premises are sublet then the tenant is not in occupation and cannot claim the security of the act.
3 There must be a trade, profession or business use in the premises. The business does not have to be a profit-making organisation. If the premises are of a mixed use then the business use has to be significant.

If the lease or tenancy is not on the excluded list, the tenant is in occupation and uses the premises for a business, then their lease is protected by this legislation. This means that the tenant has a right to a new tenancy and that the landlord has limited grounds to prevent this happening. This makes it difficult for landlords to asset manage portfolios in terms of moving tenants.

The timing of the process of lease renewal can be started by either the landlord or the tenant. The landlord serves either a S.25 LT1 or LT2 notice depending upon whether the landlord is offering new lease terms or is opposing the grant of a new lease. The notices are in a prescribed format and if a S.25 LT1 notice is served then it must include some of the new terms being offered by the landlord. If a S.25 LT2 notice is being served it also gives full details of the seven grounds (S.30) that a landlord can use to try to obtain possession of their premises at the end of a lease, together with information on lease renewals and compensation issues.

Rather than wait for a landlord to serve a notice a tenant can serve a S.26 notice requesting a new lease. If the landlord is unwilling to grant a new lease then they have 2 months in which to counter notice stating which of the seven grounds they are relying upon.

In order to bring the lease to an end on the lease expiry date, using any of the above notices, it must be served no more than 12 months and no less than 6 months from the expiry date. Once either the S.25 or 26 notices have been served, either party can make an application to the courts. The tenant must make an application to court before the date in the notice in order to gain security of tenure. This means that the landlord cannot make them leave without the courts approval and they have the legal right to a new lease.

Interim rents (S.29A) deal with the situation where the tenant has applied to court and has security of tenure. A situation could arise where rental levels have risen or fallen since the rent was last reviewed under the lease and the tenant would still pay the existing rent until the matter was settled. Depending on the rise or fall in rental levels, one of the parties could be disadvantaged so this process allows landlord or tenant to apply for an interim rent which is normally open market rent.

Contracting out is possible under S.38 of the legislation. This means that the tenant will have no rights to renew their lease when it expires. The important point to remember here is that the contracting out process must happen before the lease is completed.

Further reading

Rodell, A. (2013) *Commercial Property*, College of Law Publishing, Guildford.
www.isurv.com (accessed 20/11/2013).
www.rics.org/uk (accessed 20/11/2013).

12.9 Squatters and adverse possession

Key terms: squatters; civil law; IPO or court action; Land Registration Act 2002; possessory title; non-possessory title; rights of light

Owners of property portfolios need to protect their properties from a variety of dangers both physical and financial as these can have a serious effect on the value of the portfolio and in certain circumstances the landowner can lose possession of their property. The first threat to a landlord's portfolio can come from squatting, which means that people occupy property without the permission of the owner. Recent changes in the law have made squatting in residential properties illegal (S.144, Legal Aid, Sentencing and Punishment of Offenders Act 2012). It is now a criminal offence with fines and potentially a prison sentence. Squatting, however, in commercial properties is not included within the legislation and it remains a civil problem. If squatters enter a property either as fly traders or to live there then it is difficult for landlords to regain possession. Trying to regain possession physically, and using violence or threatening to, is a criminal offence, so a landlord can only enter if the property is temporarily absent. Professional squatters are usually aware of their rights and will ensure that landlords are not given an opportunity to regain possession.

The landlord has two avenues to pursue – either through possession proceedings or an Interim Possession Order (IPO). If a trespass is discovered within 28 days of it starting the landlord can apply to court for an IPO, and the squatters have to vacate within 24 hours until the court hearing. Alternatively, the landlord can apply for a possession order through the normal court process which can take a length of time and allows the squatter/trespasser to remain in possession of the property until the court makes a decision. This applies to both land and buildings.

Prior to the Land Registration Act 2002 coming into force, a person using land or property without the permission or knowledge of the landlord could after 12 years apply for the legal title to it. This is known as possessory title. Since the act came into force there have been procedural changes that partly protect landlords (the land must be registered at the Land Registry). If land or buildings are being occupied, the 'illegal' occupier must after 10 years apply to the Land Registry who then advise the landowner. They can then accept the claim or take no action at all. In either of these situations the title transfers to the squatter. If the landowner objects then the squatter has to make a case under certain circumstances to prove that the title should transfer to them. It should also be noted that this can affect air space around the land, as well as any foundations from adjoining buildings. So it is important to check building plans for adjoining buildings to see whether parts of the building will overhang or whether the foundations will encroach onto the land, as legal steps need to be taken at an early stage. Another way in which rights can be acquired is through prescriptive rights. This is the acquisition of non-possessory interests in land through long continuous use of the land. The use needs to be open, and continuous for more than 20 years and can be rights of way, access rights or drainage rights. To prevent these being acquired it is important that notices are erected giving people permission to use the land. It is common to see 'permission' notices on large landed estates where there are bridle paths and walks through the land giving the public permission to use the right of way or bridle path without allowing them to obtain legal rights.

Rights of light are another form of rights that can be acquired to the disadvantage of the landowner through the Rights to Light Act 1959. These are acquired once light has been enjoyed through defined apertures of a building for an uninterrupted period of 20 years. The amount of light is worked out by way of a formula and if these rights are infringed upon then neighbours can claim compensation or, in one recent case, the developer was made to remove the top two floors of the recently built office block – an expensive infringement for the developer.

Generally it is more efficient for the landowner or landlord to ensure that rights are not acquired by other parties and a variety of strategies need to be put in place to prevent this happening.

Further reading

Rodell, A. (2013) *Commercial Property*, College of Law Publishing, Guildford.
www.hncourts-service.gov.uk/infoabout/housing/occupants/squatters (accessed 20/11/2013).
www.squatter.org.uk (accessed 20/11/2013).

12.10 Alienation

Key terms: privity of contract; privity of estate; assignment; subletting; pre/post 1 January 1996; S.19 (1)a and S.19 (1)A

There are instances when tenants need to be able to rid themselves of their lease liabilities. Obviously this is possible if break clauses are contained within the lease or if negotiations to surrender the lease are successful. However if these options are not available to the tenant then their only option is to either assign, sublet or part with or share possession. There must be an express clause in the lease allowing this. It is known as alienation and gives a tenant flexibility to leave their premises.

However, from a landlord's point of view (investor), the important thing is to maintain a quality income flow and maintain the tenant mix quality or quality of the occupier generally. The way in which the legislation is applied by the landlord must always bear these considerations in mind. Sometimes the legal outlook and the estate management outlook are at loggerheads and a landlord is not allowed to prevent an assignment on estate management grounds alone.

The legal position regarding assignments has been radically altered with effect from 1 January 1996 with the introduction of the Landlord and Tenant (Covenants) Act 1995. This Act removes the privity of contract between the landlord and the original tenant, for all leases granted on or after 1 January 1996 in return for giving the landlord more control over the assignment process.

Before 1 January 1996

For 'old' leases, the main legislation governing the grant of assignments is the Landlord and Tenant Acts 1927 and 1988. These Acts put the onus on the landlords to prove that they are being reasonable in refusing consent to the assignment (S.19 (1)(a). The 1988 Act also requires the landlord to make a decision in 'reasonable time' when he has been approached for the consent to an assignment.

Various grounds have been held by the courts to be valid reasons for refusal:

- serious breach of repairing covenant by the tenant;
- objection to the proposed user;
- status of the assignee;
- lack of information from the assignee;
- reduction in value of the landlord's reversion;
- no suitable surety available.

After 1 January 1996

On granting the lease, the landlord is entitled to include circumstances in which the landlord may withhold his consent and conditions subject to which an assignment would be acceptable (S.19 (1)(A)). These circumstances and conditions must be fact and not opinion. If they must be determinable by the landlord or any other party, then the tenant must have the right to challenge this determination by asking a third party to determine absolutely, or else the landlord or other specified party must be required to act reasonably. These will normally be included within the AGA.

Examples of circumstances might be:

- The proposed assignee must, in the landlord's reasonable opinion, be of equal tenant strength.
- The proposed assignee must have a trading profit equal to 2 years rent.
- The proposed assignee must be a public limited company.
- No assignment may take place within the first or penultimate years of the term.

An example of conditions might be:

- The proposed assignee will pay a rent deposit equal to 6 month's rent at the then current rate.

It is important that the landlord, on granting the lease, thinks very seriously about the circumstances or conditions to be included in the lease, as these will have an impact on the rent at review and the initial rent offered by the tenant. Issues of tenant mix may be crucial if the letting is of retail premises in a shopping centre.

It is still possible to have an absolute ban on assignment but this will obviously depress the amount of rent offered by the tenant.

The 95 Act also affected the position of the landlord on the transfer of their lease to another landlord. As from 1 January 1996 the landlord has to serve a S.8 notice on the tenant informing them of the assignment and requesting a release from the landlord's covenants within the lease.

Subletting

On occasions a landlord may not be happy with a proposed assignment but may be happy to agree to a subletting of the lease. This allows the existing tenant to move to new premises but keeps them responsible for the property via the existing lease; the subtenant occupies the property and pays the rent to the tenant. Additionally, a tenant may decide to sublet the property with consent from the landlord to suit their own property and business strategy. A landlord will usually seek to exercise very strict control over the subletting and the terms of the sublease. This is due to the fact that the subtenant may become a direct tenant at a later date and any change to the rental levels agreed could affect values as this provides evidence. Landlords try to ensure that the rent for the subletting is the same as the rent being paid by the tenant and this can produce problems for tenants who are in over-rented properties. The British Property Federation (BPF) is campaigning to stop landlords insisting on this clause in leases. See *Allied Dunbar Assurance plc v Homebase Ltd* (2002) and other case law for more details on this subject.

Parting with or sharing possession of premises by tenants are usually prohibited within the terms of the lease. Parting with possession means that the tenant is no longer in physical possession of the premises while sharing possession means that another person or company is also using the premises.

Further reading

Rodell, A. (2013) *Commercial Property*, College of Law Publishing, Guildford.
Scarrett, D. (2011) *Property Asset Management*, 3rd edn, Routledge, Abingdon.
www.isurv.com (accessed 20/11/2013).
www.rics.org/uk (accessed 20/11/2013).

12.11 Exit strategies

Key terms: exit strategies and methods of terminating leases; forfeiture; surrender; break clauses; efflux-ion of time

The role of the real estate asset manager in connection with real estate portfolios is to ensure that value is retained, value is added and flexibility is built in, depending on whether they are acting for a landlord or tenant. Landlords may want to move or remove tenants and tenants may want to leave before their leases expire however leases are legally binding contracts so there are limited ways to terminate them. The various methods are considered below.

Forfeiture

This is a way of terminating leases for landlords. In order to forfeit a commercial lease there needs to be:

(a) a breach of covenant;
(b) an express clause in the lease allowing forfeiture.

For the majority of breaches of covenant a S.146 notice needs to be served and leave of the court is required before the landlord can re-enter the premises and terminate the lease. The exception to this is breach of covenant for non-payment of rent. At the moment a landlord can still serve a notice of forfeiture on a tenant, enter the premises and change the locks, thus bringing the lease to an end. Care must be taken if the tenant is still in possession as it is an offence to use force to evict the tenant. In these circumstances a court order should be sought.

Landlords have to be careful not to waive the right to forfeit by performing some act that is deemed to be an acknowledgement of the continuance of the lease. An example would be accepting rent where it was known that there had been an illegal subletting.

Under S.146(2) Law of Property Act 1925 a tenant has a statutory right to request the reinstatement of their lease through the courts upon full payment of arrears up to 6 months after the forfeiture.

This method of terminating leases is obviously adversarial and with the tenant's right of relief gives landlords an element of uncertainty for a period of time after the forfeiture. The question to ask is, is this the correct course of action to take in these circumstances? Or is there another way of terminating the lease that is more effective? The market, the property and the economic climate will all be factors to be taken into account in the decision making. Also it must be remembered that once the lease is forfeited, it comes to an end and the landlord then becomes responsible for the outgoings including the empty rates. This needs to be thought through and discussed with the landlord before action is taken.

Surrender

This can be used by landlords and tenants. A lease can only be surrendered by agreement, so this is a non-adversarial method of bringing leases to an end. It is a useful tool for both parties. A payment can be made by a tenant to a landlord to surrender their lease if the lease is now surplus to their requirements or if there are problems paying the rent. Equally, a landlord can make a payment to a tenant in return for the tenant surrendering their lease. This may allow the landlord to re-let the premises to a better covenant, on better lease terms or at a higher rent, or to proceed with a redevelopment scheme.

There are two ways of surrendering a lease:

(a) express surrender
(b) operation of law.

Express surrender is by deed. The lease is surrendered by a conveyance of the leasehold interest to the landlord. Once the lease is surrendered all obligations come to an end. It is important that any conditions or payments to be made are included within the agreement. The asset manager needs to check that the conditions have been complied with and needs to liaise with the solicitor throughout the process.

Surrender by operation of law occurs when the landlord and tenant act in a way wholly inconsistent with the continuation of a tenancy. An example would be where a tenant gives up possession of the property and a landlord re-lets it. A tenant returning keys does not constitute a surrender of the lease! A surrender of a head lease does not bring a subletting to an end.

Break clauses

Again, both landlords and tenants can have break clauses inserted into the lease. These are specific clauses allowing landlords or tenants to bring leases to an end at specific times during the lease. This often gives both parties flexibility in certain circumstances.

Often if a landlord has allowed a break clause in favour of the tenant it will be a conditional break clause. This means that the break will be subject to the tenant having to do something, i.e. comply with repairing covenants, making an additional payment and often adhering to a strict timetable. If there are conditions, then tenants must comply with them and if they fail to do this then the right to break can be lost. See *Avocet Industrial Estates LLP v Merol Limited* (2011), a recent case that demonstrates how critical it is for tenants to ensure that all the conditions are met. It is quite normal for the tenants' break clause to be tied in with their rent reviews, on the basis that if the rent is too high at review they can leave the premises. This gives the tenant flexibility in their business. A tenant may also be prepared to pay more for the lease initially if a break clause can be negotiated. A landlord may want a break clause inserted into a lease if there is potential for a development at a later date. This could be a fixed term break or a floating break on giving a certain notice period. If a landlord negotiates a break clause during the lease negotiations then it is more likely that the tenant will reduce their rental bid as they have lost certainty in their lease term.

Effluxion of time

Under common law a lease granted for a fixed period of time should expire on the last day of the lease; however, this position has been affected by statute, particularly for business premises. Leases that are 'contracted out' under section 38 (4) of the Landlord and Tenant Act 1954 pt 2 do come to an end on the last day of the lease.

Further reading

Rodell, A. (2013) *Commercial Property*, College of Law Publishing, Guildford.
www.bpf.org/uk (accessed 20/11/2013).
www.isurv.com (accessed 20/11/2013).
www.rics.org/uk (accessed 20/11/2013).

12.12 Health and safety

Key terms: Health and Safety Executive; duty of care; reasonably practicable; Sections 2, 3 and 4 of HSAW Act 1974; risk assessments

All landlords will have some responsibility for the health and safety of their tenants, visitors, contractors and even trespassers. They have to ensure that the premises are safe to use by all of the different groups and this is further complicated by the leases and the different uses within the building. The main legislation controlling health and safety is the Health and Safety at Work Act 1974 which is the primary legislation in the UK and is enforced by the Health and Safety Executive (HSE), together with other legislation and Statutory Instruments. The HSAW Act, gives employers and landlords a duty of care towards a variety of groups from employees, visitors/contractors through to the general public. Section 2 of the Act gives a general duty of care to employees and the employer must where reasonably practicable ensure the health, safety and welfare of the employees. This means that employers must have a health and safety policy which is reviewed and revised when necessary and that risk assessments must be carried out and adequate controls put into place. Employers must be aware of other legislation that affects this particular section, for example, the Personal Protective Equipment Regulations 2002 whereby employers must ensure that all staff carrying out work on a construction site have the appropriate safety equipment to wear. In addition they must ensure where possible that the equipment is used properly.

Section 3 of the act gives an employer a duty of care to other people who are not direct employees, which includes contractors. The employer must ensure where reasonably practicable that they can carry out any works in the building without risk to their health and safety. For example it is a legal requirement under the Control of Asbestos Regulations 2006 for an asbestos register to be kept showing all the known locations of asbestos within a building. A contractor before starting work on the building must be advised of these locations. An employer would be considered negligent if they allowed window cleaners to work at heights without using the appropriate safety harnesses.

Section 4 of the act covers the general duty of persons concerned with premises to other people. Again, where reasonably practicable, that person has a duty of care to ensure that the premises including plant and machinery, are safe for people to use. Responsibilities include ensuring that the water quality is regularly tested to prevent Legionnaires' disease and that regular cleaning of shopping malls is carried out to help prevent people slipping and falling. Risk assessments need to be carried out to identify the possible risks and then an action plan needs to be put in place that addresses the risks and the actions required to either eliminate the problem or to minimise the chance of it happening. For example if a member of staff worked in an area of loud noise then ear protectors would be issued to protect their hearing, but in a nightclub how would bar staff be able to hear customers' orders if they wore ear protectors? A reasonable step would be to ensure that the noise level was kept within any legal standards or that staff were moved round into areas where the noise levels were lower.

In addition to the above act the following legislation can also impact on the health and safety of people:

- Corporate Manslaughter and Corporate Homicide Act 2007
- Management of Health and Safety at Work Regulations 1999
- Regulatory Reform (Fire Safety) Order 2005
- Control of Substances Hazardous to Health Regulations 2002
- Electrical Equipment (Safety) Regulations 1994
- Gas Safety (Installation and Use) Regulations 1998
- Construction (Design and Management) Regulations 2007
- Work at Height Regulations 2005
- Occupiers Liability Act 1957 and 1984.

This is not an exhaustive list of legislation affecting health and safety in the UK – a full list can be found on the HSE website at www.hse.gov.uk.

Further reading

www.hse.gov.uk (accessed 20/11/2013).
www.isurv.com (accessed 20/11/2013).

12.13 Dilapidations

Key terms: Full Repairing and Insuring (FRI); Internal Repairing and Insuring (IRI); repair clauses; yielding up clause; schedule of condition; interim and terminal schedules; Civil Procedure Rules and the Dilapidations Protocol; RICS Dilapidations Guidance Note 6th edition; Law of Property Act 1925; Landlord and Tenant Act 1925; Leasehold Property (Repairs) Act 1938

Dilapidations are the disrepair items that a tenant is liable for on a property when they have agreed lease obligations within their lease that require them to keep the property in good repair and condition. This liability for repair comes from the lease and there will be differing levels of repair arising from the wording in the lease. FRI leases mean that the tenant is responsible for all the repairs, both internal and external to the property, IRI + S/C means that the tenant is responsible for all the internal repairs but that the landlord recovers the costs of the external repairs through the service charge. Effectively the tenant pays for the complete repair of the property. IRI means that the tenant is only responsible for the internal repairs of the property. The wording of the repairing clause and the yielding up clause has an effect on the responsibilities of the tenant and what a landlord can claim as does any alteration clause. Dilapidations claims can be made at the end of the lease, known as a terminal schedule of dilapidations, or during a lease if a tenant is in breach of their repairing liability. This is known as an interim schedule of dilapidations. Dilapidation claims can also be brought by landlords when leases are terminated by way of forfeiture or break clauses. Tenants can also start the procedure if a landlord has a repairing liability to them and they are in breach of their responsibility. A schedule of dilapidations is a document that identifies the lease covenants and obligations of the tenant/landlord, alleged breaches of covenant, remedial works required and costs to rectify the breach.

Failure to comply with repairing liabilities is a breach of covenant and this gives a landlord and tenant certain ways to put the situation right. The repair works could be carried out to put the property back into repair or the repair works could be quantified and paid as damages. In the past this area has been contentious between landlords and tenants and has resulted in expensive court cases. In a bid to change this, the Civil Procedure Rules and the Dilapidations Protocol came into effect from 1 January 2012 with its aim to effectively resolve disputes regarding dilapidations. Failure to comply with the Protocol may result in the party deemed to be 'guilty' having to pay additional costs. The Protocol aims to give clarity to the process as it lays down how the dilapidations disputes are to be run prior to a claim and to ensure that all parties now know exactly what their obligations are. The majority of the content of the above Protocol is now incorporated into the RICS Dilapidations Guidance Note 6th edition. The main points included are as follows:

- Reasonable conduct – both parties must have a genuine intention to reach an agreement and surveyors must act in an objective, honest and professional manner.
- Endorsements – both landlords and tenants have to provide endorsements on future intentions for the property and a response to the quantified demand.
- Quantified demand – a complete statement of all the costs being claimed including all ancillary and consequential losses.
- Glossary of terms – to clarify certain terms and ensure that all parties are using the same terms in the correct way.
- ADR – alternative dispute resolution should be used and court action should only be taken as a last resort.

The main legislation affecting this area is the Law of Property Act 1925, Landlord and Tenant Act 1927 and the Leasehold Property (Repairs) Act 1938.

Further reading

Williams, D., Shapiro E. and Thom, J. (2005) *Handbook of Dilapidations*, Sweet & Maxwell, Andover.
www.isurv.com (accessed 20/11/2013).
www.rics.org/uk (accessed 20/11/2013).

12.14 Insolvency

Key terms: debtors; creditors; bankruptcy; IVA; liquidation; administration; receiverships; Insolvency Act 1986; Enterprise Act 2002

An individual or organisation can be insolvent but it doesn't necessarily always end up in bankruptcy or liquidation as sometimes steps can be taken to stop this from happening. It is a very complex area, often requiring properly qualified practitioners, but there are certain basics that can be understood and some of these are outlines in this concept. The legislation that affects insolvency is mainly the Insolvency Act 1986, as amended, and the Enterprise Act 2002.

A debtor is the person or organisation that owes money to other people and organisations, which are called creditors. There are secured and non-secured creditors, the difference being that if a person or organisation becomes bankrupt or goes into liquidation and their assets do not cover the amount owed, then only secured creditors get a share of any money. For example, if a person owed £25,000 to their creditors, that person would need to sell everything they owned (their assets). If the money raised was only £5,000, then only secured creditors would get some or all of their money back. If their assets were worth £25,000 then they could repay all of their creditors. An example of a secured creditor would be a bank where a secured loan had been taken out.

An individual can enter into bankruptcy or an IVA, which stands for individual voluntary arrangement. For bankruptcy, there has to be a personal minimum debt of £750. The Official Receiver (a civil servant) deals with the case until a Trustee in Bankruptcy is appointed at the creditors' meeting – this can be either the Official Receiver or a licensed insolvency practitioner. The debtor can be released from their debt after 1 year but being made bankrupt means that they lose their assets and it will affect their credit rating in the future. An IVA is an alternative to bankruptcy and it is a formal debt repayment scheme. A licensed insolvency practitioner helps to draw up a set of proposals on how to repay the creditors and how long it will take to do this. A creditors meeting is then held and if 75 per cent of the creditors agree to the proposals then all creditors are bound. Again the credit rating will be affected.

If the debtor is an organisation then there are various ways in which creditors can begin insolvency proceedings to recover their debts. A company can enter into a compulsory liquidation which is ordered by the court, usually when a creditor has petitioned the court to do this. The liquidator is appointed and their job is to 'realise' the assets of the company, agree what the total owed is and distribute the liquidised assets, either in total or by way of a dividend. A creditors' voluntary liquidation is commenced by the company to deal with its own insolvency and it has a duty to realise the assets, agree the claims and distribute the liquidised assets as before. A members' voluntary liquidation is a 'solvent' liquidation used when a company is being wound up, as all the creditors are paid in full and the company has sufficient funds to meet all their liabilities for the following 12 months. Instead of liquidation, a company can elect to enter into a company voluntary arrangement (CVA) which allows the company to reach an agreement with their creditors to pay back their debts over an agreed period of time. Again, as in an IVA, 75 per cent of the creditors have to agree to the proposals and this then binds all the creditors.

A company can also enter into administration, which was set up to help ailing companies and give them some time to sort the problems out without fear of the creditors forcing liquidation or taking other routes to collect their debts. Administration is a short-term intensive care plan. An application is made to court and if this is agreed the court appoint an administrator whose job is to manage the company and help it trade out of its difficulties. If the company can be saved it will leave the administration or if it cannot be saved it will go into liquidation. Recently there have been 'pre-pacs' where a company goes into administration and is sold before news of the administration is public. The reasons for 'pre-pacs' are that it is a relatively smooth transfer, it minimises a loss of suppliers and customers and keeps jobs. The arguments against are that the process is not transparent and negotiations take place without being tested in the open market.

Where there are secured creditors either by way of a fixed charge asset or a floating charge, then in the event of the loan not being repaid a receiver is appointed to seize and sell the assets. If the loan is guaranteed by a specific asset or a fixed charge, then a LPA Receiver is appointed under the Law of Property Act 1925 to do this. Where a floating charge has been taken before 15 September 2003, then an Administrative Receiver is appointed to collect the debt. These last cases will become less common due to time passing.

Further reading

Rodell, A. (2013) *Commercial Property*, College of Law Publishing, Guildford.
www.isurv.com (accessed 20/11/2013).
www.insolvency.gov.uk (accessed 20/11/2013).
www.insolvencyhelpline.co.uk (accessed 20/11/2013).

12.15 Facilities management

Key terms: facilities management, core function, Private Finance Initiative, efficiencies, hard and soft services

The Facilities Manager is now a fully recognised professional individual within the construction industry. A short time ago the same role, although probably less technical, would have been called an Estate Manager. The new role is seen as more proactive and offers a cohesive approach to work–space management and the operational process within an enclosed space, as well as managing the various traditional aspects of keeping a building or estate viable. Below are two definitions seeking to condense the role into a manageable statement:

> Facilities Management is the active management and co-ordination of an organisation's non core business services, together with the associated human resources and its buildings (including their

systems, plant, IT equipment, fittings and furniture) necessary to assist that organisation to achieve its strategic objectives.

(RICS, 2013)

Facilities management is the integration of processes within an organisation to maintain and develop the agreed services which support and improve the effectiveness of its primary activities.

(British Institute of Facilities Management, 2013)

Facilities management (FM) includes all of the services orientated towards supporting the core business functions, i.e. their purpose for being, what they are there to do, be it manufacturing, health care, education etc.

It can sometimes be difficult to decipher the core from the non-core functions as they can be intrinsically linked; for instance, could Nissan be a car manufacturer without buildings in which to facilitate this core function?

Drivers for change

Cost-cutting initiatives emerged in the 1970s and 1980s which saw the beginning of companies contracting out or outsourcing non-core services to allow managers to concentrate on the core function of the organisation. Further professionalising of the emerging FM sector saw dedicated FM organisations coming to the market with the introduction of private finance initiatives (PFIs) and public–private partnerships (PPPs). These were seen as a new way for private finance to manage, replace and upgrade outdated public sector facilities and infrastructure without the need for public financing the large capital sums required for the construction of such projects; however, there are significant revenue costs to the public purse (see Concept 3.3).

Statistics show that companies'/organisations' overheads for the property they occupy could be as much as 50 per cent of their total overhead costs. A good facilities manager is thought to be able to reduce this by up to 10 per cent. This is done through effective management of an organisation's assets, be it in house or outsourced to an external organisation (therefore transferring the risk).

Many organisations view a building as a vehicle that allows them to conduct their business. Therefore, efficient use of the buildings and in particular space can improve organisations' efficiencies and enable them to achieve their business plans, strategies and ultimately reduce building-related costs. This is a key driver within facilities management.

Organisations are keen to enable new working styles and processes in a bid to create a more flexible and content workforce that can invariably lead to better productivity. Modern methods of working are becoming more of the normal within many of today's businesses and can include the likes of 'hot-desking' and working remotely or at home, thus freeing up space within the workplace. FM has a major role to play in facilitating these practices by providing a more flexible workplace with the knowledge gained in the post occupation management of buildings.

Facilities managers are using this expertise along with the knowledge gained and are becoming increasingly involved within the design phase of new construction or refurbishments. The areas of beneficial input are:

- space planning;
- asset tracking;
- future maintenance;
- life cycle costing;
- energy management.

Clearly, although important, cost saving alone will not result in a smooth-running organisation. There are other elements of the built environment and operational procedures and services that will come under the scrutiny of the facilities manager to support the organisation's core function/s. This will include all hard, i.e. building maintenance, PPM etc. and soft services, i.e. cleaning, catering, security, logistics etc.

As construction related or employment focused legislation is enacted, new responsibilities are coming under the remit of the facilities manager who must assume responsibility for compliance. In recent years the areas this has encompassed are:

- The Regulatory Reform (Fire Safety) Order 2005;
- Control of Asbestos Regulations 2012;
- Water Hygiene including Legionella Control;
- Waste;
- Display of Energy Certificates (DECs);
- Carbon Reduction Scheme (CRC).

This is not an exhaustive list and is subject to frequent change.

In summary: FM is there to seamlessly support the core business function and to have a dedicated person, department or external organisation to manage buildings related services, allowing managers to concentrate on the core functions of the business.

The main areas where these skills can be beneficially employed are recognised to be:

- health and safety monitoring – legislative compliance;
- component specifications;
- systems and software;
- services – both hard and soft.

Further reading

Atkins, B. and Brooks, A. (2009) *Total Facilities Management*, 3rd edn, Blackwell, Oxford.

British Institute of Facilities Management (2013) *Facilities Management: an Introduction*, available at: http://www.bifm.org.uk/bifm/about/facilities, (accessed 20/10/2013).

RICS (2009) *The Strategic Role of Facilities Management in Business*, RICS Books, Coventry.

RICS (2013) *Strategic Facilities Management*, 1st edition, RICS Books, Coventry.

13 Quantity surveying

Glenn Steel

13.1 Measurement and quantification

Key terms: drawings; specification; bill of quantities; standard methods of measurement

The ability to measure, and thereby quantify, the work a contractor must provide in the construction of a building is regarded as a core skill of the quantity surveying profession. In recent times, advances in computer technology have seen the development of automated measurement systems producing quantities without the traditional manual input of the quantity surveyor (QS).

Despite the potential demise of the measurement aspect of the QS role, measurement skills are still in demand by the construction industry. The establishment and promotion of collaborative strategies through methodologies such as building information modelling (BIM), which rely on technological solutions to measurement, have changed the role of the QS from the manual/technical function it once was to a more interpretive and analytical task.

To understand the importance and necessity of measurement and quantification, it is necessary to go back in history where 'traditional' methods of construction procurement were used – the word traditional implying that a building design would be completed in totality *before* contract documents were produced and the construction process commenced.

For any construction project a design team designs the building and the production team (i.e. the contractor) construct it. Traditionally, the design team consists of an architect, structural engineer, building services engineer and quantity surveyor, each responsible for producing different types of design information. The design team communicate design information to the production team that explains what has to be built.

Conventional design information represents recognisable design features of the building – the components and materials that the contractor must bring to the construction site and assemble. The materials and components are provided and assembled by the contractor using appropriate labour and plant resources, but the design team tell the production team neither what labour and plant resources to employ nor how to deploy the resources. This decision is left to the contractor.

Design information is usually communicated to the production team via one or more of the following project documents (Figure 13.1.1):

- Drawings – a drawing is a pictorial representation or diagrammatic schedule of the required materials and components.
- Specification – a specification is a statement of the required quality of the materials and components and of the workmanship associated with them.
- A bill of quantities (BoQ) is an ordered list of items consisting of worded descriptions of the design features on the drawings. Each description is accompanied by a quantity measured from the design features on the drawings. BoQ production is invariably the responsibility of the quantity surveyor.

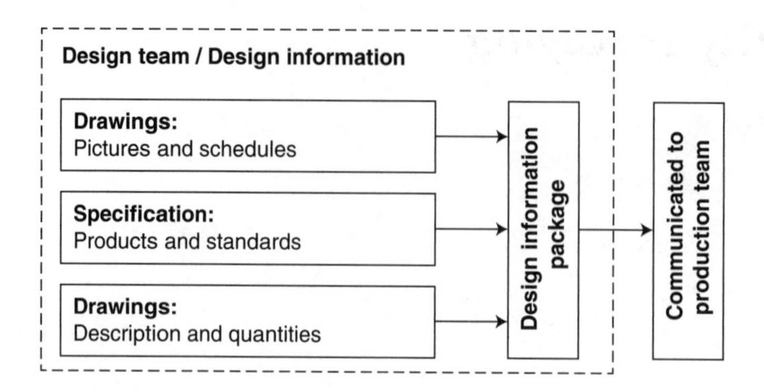

Figure 13.1.1 Design information.

Source: Dr Alan Davies, Northumbria University.

The BoQ does not always possess legal status and is of debatable utility to anybody but the QS and the contractor. A distinction must be drawn between a project document and a contract document. A contract document has legal status enforceable in a law court. It legally defines what the parties to the construction contract must provide. The BoQ does not legally define these things. If there is doubt, the parties invariably refer to the drawings and specification, which invariably have legal status. The role of the BoQ is merely a project document. Its status is 'quasi-legal' in that the building contract frequently defines the administrative functions for which it should be used.

In the BoQ, the QS measures the finished products that the contractor has to furnish. This is an important distinction. To achieve this, the QS must apply the rules contained in a standard method of measurement (SMM).

SMMs provide:

- rules for what to measure and what not to measure;
- rules for what units of measurement to express;
- accepted technical phraseology in which to write descriptions of what has been measured;
- definitions of the legal meanings of these item descriptions and of what work the contractor should allow for when pricing them.

There are several types of SMM, drafted to cater for different specialist types of construction work. For example SMM7 is for building work whereas CESMM4 (*Civil Engineering Method of Measurement*, 4th edition) relates to civil engineering work. The parties to the contract are deemed to understand the technical phraseology and rules contained in the SMM that is being used. The QS is deemed to have obeyed the rules contained therein when measuring the items and the contractor is assumed to have obeyed them when pricing them.

The SMMs have been revised and reviewed over the years since the first edition (known as SMM1) which was published in 1922, up until SMM7 which was introduced in 1988 and reissued in 2000. The most recent version (issued for use in 2013) represents a radical shift in approach where the standard method has been issued in three distinct volumes. The documents have been renamed 'New Rules of Measurement' (see Concept 13.2).

Further reading

Ashworth, A. and Hogg, K. (2007) *Willis's Practice and Procedure for the Quantity Surveyor*, 12th edn, Blackwell, Oxford.
Cartlidge, D. (2013) *Quantity Surveyor's Pocket Book*, 2nd edn, Taylor & Francis, London.
Ramus, J. W., Birchall, S. and Griffiths, P. (2006) *Contract Practice for Surveyors*, 4th edn, Butterworth Heinemann, London.

13.2 New Rules of Measurement

Key terms: measurement rules; bill of quantities; NRM1, 2 and 3.

The New Rules of Measurement (NRM) is a suite of documents issued by the RICS to provide a globally acceptable set of measurement rules that are useful to and applied by any party involved in a construction project. Despite the principles being based on UK practice, the need for a coordinated set of rules that can be applied in an international situation is seen as being beneficial to the industry. The previous set of measurement rules, SMM7, had been found by the RICS to be used by different parties in the industry in different ways. The key output from the application of SMM7 was the BoQ and the industry has been found to have moved away from this 'traditional' approach. Although the use of them still exists, it is considered that due to the wide variety of procurement routes being adopted (for example, the increase in the use of design and build procurement) other methods were being used to obtain tender/quotation prices in the industry (e.g. the use of drawings and specifications).

Paradoxically, the move away from 'general' contracting to subcontracting has resulted in contractors using a BoQ approach much more than clients in respect of main contracts. However, the format of these documents was not 'pure' SMM7 and was rather more a 'builder's quantity' approach, this emphasising the need to reflect reality in the industry and to move towards measurement standards that could be used by all involved. Indeed, the focus on the increase in the subcontracting element of works is reflected in the fact that subcontracting organisations were involved in the design of NRM and not main contractors.

The other perceived weaknesses were the lack of a standardised measurement approach to the estimating stage of the construction process. This was remedied by the introduction of NRM1, which has been generally accepted by the industry as a useful development. There will be a clear link from NRM1 to NRM3 (which is to be published in 2014) to allow recognition of issues related to sustainability and life cycle costing (see Concept 13.4). The coordination of the NRM volumes is seen as a positive contribution to the successful delivery of value to clients.

NRM2 is more directly developed from the old SSM7 document and is concerned with the measurement of construction work in detail. Initial consultation on NRM2 suggested that this would be a radical departure from the principles of SMM7, but it can be said that NRM2 is a significant development of SMM7 instead – many of the basic principles created in SMM7 have been retained in the new rules. One immediate observation is the length of the document, which exceeds 300 pages – a significant increase in volume. This is because the number of trades included has increased to reflect the reality of the industry.

In summary, the NRM suite comprises three volumes. The three guides together will provide industry professionals with reliable guidelines in terms of quantifying and measuring capital, maintenance and renewal works on existing built assets.

NRM1: Order of cost estimating and cost planning for capital building works

NRM1 provides guidance on the quantification of building works for the purpose of preparing cost estimates and cost plans. Guidance is given on quantifying other construction project costs which are not reflected in the measurable building work items (for example, preliminaries, overheads and profit, project team and design team fees, risk allowances, inflation, and other development and project costs). This is seen as a very positive development. The intention is to enable better cost advice to be given to clients and other project team members and to facilitate better cost control. The second edition became operative on 1 January 2013.

NRM2: Detailed measurement for building works

NRM2 provides guidance on the detailed measurement and description of building works for the purpose of obtaining a tender price. The rules address all aspects of BoQ production, including setting out the information required from the employer and other construction consultants to enable a BoQ to be prepared, as well as dealing with the quantification of non-measurable work items, contractor designed works and risks. The first edition of NRM2 became operative on 1 January 2013 and SMM7 will run alongside it until the latter is removed from publication in due course.

NRM3: Order of cost estimating and cost planning for building maintenance works

The alignment of NRM1 and NRM3 will allow the cost estimating, cost planning, cost reporting and benchmarking of the total cost of capital building and maintenance works. It makes use of a common format aligned with capital building costs and is regarded as a groundbreaking development. In particular, agreement has been struck with the Chartered Institution of Building Services Engineers (CIBSE) and the Building and Engineering Services Association (B&ES) to adopt the new NRM3 cost structure. This implies that NRM3 may realise significant benefits for the construction industry (and for related maintenance industries). It will enable comparison of costs on a like-for-like basis. Together with NRM1 and 2, NRM3 will provide a basis for life cycle cost management.

Further reading

Cartlidge, D. (2013) *Quantity Surveyor's Pocket Book*, 2nd edn, Taylor & Francis, London.

13.3 Cost planning and cost control

Key terms: early stage estimate; outline/detailed cost plan; cost checks

Cost planning, in the context of a construction project, is concerned with planning the cost of a proposed project in advance of it being built. It is a widely used term, meaning different things to different people. The term can be used in a general way, covering all aspects of costing a project from beginning to end, or in a narrower way, relating to a specific part of the design and construction process. For example, a cost plan could be produced for one or more of the following reasons:

- a detailed estimate for a client, as part of the design process;
- a detailed breakdown of anticipated costs, as part of a tender submission;
- as a basis for cost control of a project on site;
- as a basis for planning 'whole life' costs over the anticipated lifetime of a building once it is in use.

Therefore, to avoid confusion it is important to be clear about the type of cost planning under examination. This section focuses on cost planning during the design stage of a project. Using the design stages of the RIBA Outline Plan of Work as a framework, a simple outline of the cost planning procedure is shown in Table 13.3.1.

Table 13.3.1 Cost planning procedure

Design stage	*QS tasks*	*Deliverable*
Feasibility	Prepare feasibility study/early stage estimate	Budget/estimate
Outline proposals	Prepare cost plan (perhaps of different possible solutions)	Outline cost plan
Scheme design	Detailed estimates of project elements	Detailed cost plan
Detail design	Cost checks	Reconciled cost plan

In practice the various tasks and deliverables shown in Table 13.3.1 are unlikely to be one-off activities, but are more likely to be iterative, the tasks being revisited as often as necessary while the design is being developed. As such, the activities are not always as clearly delineated as the table might indicate. However, due to the need for certainty in a project, there will usually be definite points or milestones in the design process, where a firm estimate and a detailed cost plan are required.

The following introduces the principles of estimates, cost plans and cost checks in turn. However, it is important to remember that they can overlap in practice.

Early stage estimate

The minimum information required to produce an estimate for a proposed construction project in the early stages of the design process is as follows:

- Knowledge of the proposed project, including:
  - the type of building (e.g. a hotel or a warehouse or an office development);
  - the anticipated quality of specification (often implicit in the type of building, but still highly variable and a big influence on cost); and
  - an indication of size or quantity (e.g. m^2 of office floor space, or number of bedrooms in the hotel).
- Cost information, relevant to the proposed project. This could be available from a number of sources, including information from similar past projects, published information from reliable sources and quotations upon request from specialist sources (although the latter is more likely to occur during detailed design later on in the process).
- Adjustment factors, to update and adjust the cost information so that it can be applied to the proposed project with as much confidence as possible.

Cost information

Surveyors generally prefer to use cost information from their own 'in-house' sources, usually based on past projects familiar to them. Sometimes this is not sufficient and other sources in the public domain are used.

Adjustment for time and location

The effect of inflation on costs and prices over time is well known in general terms. Inflation in construction costs and prices is no different. Several indices are published that track the movement of construction costs and prices over time. An example is an all-in TPI (tender price index) published by the BCIS (Building Cost Information Service).

In practice, a surveyor may prefer to use information about adjustments for time that are derived from the organisation's own sources about the movement of tender prices, rather than using published information that may be considered too general, but the principle is the same.

Construction costs and prices for similar work also vary between locations. In a similar way to the TPI, an index can be used to compare the construction costs and prices between different locations. In the UK, BCIS also publish location factors for this purpose. These factors can change over time but are generally stable.

Other adjustments

The adjustments for time and location are mechanical and fairly easy to calculate, providing the surveyor is confident in the indices being used. However, there will be other adjustments to make that are more subtle and require the surveyor to exercise professional judgement in their application. Factors influencing this judgement may include (this list is not exhaustive):

- size of site
- existing buildings on the site
- existing ground conditions
- progress of the design.

Other potential factors that may influence the estimate include planning constraints, access to the site (easy or difficult), number of storeys and storey height requirements. Again, this list is not exhaustive, but does demonstrate that each project is unique and professional judgement is required when considering the impact of all potential factors on the estimate.

Price and design risk

Because of these uncertainties, a percentage allowance called 'price and design risk' is usually applied to estimates produced at the early stages in the design period. The less certain the brief and available information about the proposed project, the higher is the likely percentage addition for price and design risk. A common addition for price and design risk is 2.5 per cent , but this will obviously vary, as described above. This figure acts as a kind of contingency allowance for 'extras' that may emerge as the design develops.

Given the level of uncertainty that may surround an estimate, the surveyor should not 'gloss over' this fact when communicating with the client. All assumptions made in pricing an incomplete design should be stated, with any exclusion clear and explicit. Again, given the uncertainty, the surveyor would be quite justified in stating a 'confidence range' within which it is felt the project cost will fall. However, this rarely occurs in practice, with the estimate presented to the client usually becoming the target figure for development.

Outline/detailed cost plan

Whereas an estimate is a statement that predicts how much money will be needed for a project, a cost plan, by comparison, is a statement that provides a basis for managing how this money will be spent.

As stated above, 'cost planning' is often used as a catch-all term covering all aspects of costing a project from beginning to end. However, in the strict sense of the term, a cost plan is a management tool, with the following objectives:

- to decide upon the distribution of the budget that best reflects the client's needs;
- to provide the design team with cost, quality and quantity parameters;

- to set cost targets for the elements of the project – this will provide a basis for checking and adjustments as necessary as the design proceeds.

While any suitably detailed breakdown of an estimate may be called a 'cost plan', the industry has found it useful to be able to refer to a standard set of elements when preparing cost plans.

Cost checks

Cost checking could be defined as the process of calculating the costs of specific design proposals and comparing them with the cost plan during the whole design process. The objectives of cost checking are:

- to confirm that, as the design develops, the cost remains within the budget;
- to manage changes to the design if the cost checks indicate that the budget is likely to be exceeded.

Cost checking may be carried out on one or more elements of a project, or perhaps the entire project depending on the scope of the project and the information available. Cost checks can be produced using any suitable method, depending on the information available:

- 'Approximate quantities' may be measured, which together with suitable prices will allow a more detailed estimate for an element to be produced.
- Element unit quantities and rates may be used. BCIS provides data in this format.
- For specialist work, quotations may be obtained from suitable sub-contractors/suppliers.

Approximate quantities

The techniques for producing approximate quantities are not based on an agreed method of measurement. The level of detail will depend on the information available. The quantities for these items would be measured from drawings, while the prices would be gathered from similar work elsewhere or possibly from published price books (e.g. Spon's) or specialist quotations.

Element unit quantities and rates

If the quantity of an element can be expressed in a single figure (e.g. m^2 of wall) and a suitable unit price for that element can be sourced (e.g. £/m^2 for that specification of wall), then a cost check for an element could be produced. The BCIS database contains this type of information, with some past projects providing actual element unit quantities (EUQs) and rates (EURs), as shown in the example below (the services elements, missing here, rarely have element unit quantities).

The element unit quantity allows the cost of the element to be represented as an element unit rate, which directly relates to the quantity of the element, rather than the cost per square metre, which relates to the gross internal floor area. This information (element unit rate) may then be applied to elements in proposed projects that have a similar specification (although much care and judgement is needed in using cost information in this way).

BCIS also provides average prices for element unit rates. The rates are derived from all of the similar projects in the BCIS database that actually provide element unit information. Rates are presented in a statistical format, with the spread of prices indicating the high level of variation between projects, depending on the specification. Therefore, these prices can only provide a guide. A more specific specification and associated rate, perhaps obtained from an individual past project analysis as above, are more likely to be of use in practice.

Reconciliation with cost plan

Whatever method is used for the cost checking of one or more elements, the result must be compared with the targets in the cost plan and suitably reconciled. If the cost check suggests that an element is likely to cost more than the allowance in the cost plan, then either the specification must be examined to see whether there is scope for a saving, or any savings or contingencies elsewhere in the cost plan may be used to make up the difference. Sometimes a situation of 'swings and roundabouts' will occur where increases in one element can be offset by savings from another as the design progresses, but clearly this needs careful monitoring and management. The surveyor will not usually be in a position to suggest that the budget should be increased to cover anticipated increases in the overall estimate.

Further reading

Ashworth, A. (2010) *Cost Studies of Buildings*, 5th edn, Pearson, Harlow.

Kirkham, R. J., Ferry, D. J. and Brandon, P. S. (2007) *Ferry and Brandon's Cost Planning of Buildings*, 8th edn, Blackwell, Oxford.

13.4 Life cycle costing

Key terms: whole life costing; BREEAM; embodied carbon; sustainability

Life cycle costing (LCC), also referred to as whole life costing, could be defined as the systematic consideration of all relevant costs and revenues associated with the acquisition, use, maintenance and disposal of an asset. It is concerned with predicting the costs of a building over its entire lifetime, not just its initial construction.

In recent times, LCC has become vital with the advent of long-term methods of procurement, such as PFI projects, and the international focus on global warming and the efforts to reduce the 'carbon footprint' of buildings. When it is considered that 40 per cent of total energy consumption and greenhouse gas emissions are attributable to constructing and operating buildings, the importance of whole life costing of buildings becomes apparent. Furthermore, it is estimated that for a typical office block with a lifespan of 30 years the ratio of construction to maintenance to running costs is approximately 1:5:200. In other words, the capital cost of construction (where most attention is focused!) is relatively insignificant.

There is now a significant focus on issues related to sustainable construction. This term is essentially very vague and a practitioner needs to be clear about the area of sustainability that they are referring to. For example, clients and designers currently focus on achieving a high score on the BREEAM (Building Research Establishment Environmental Assessment Method) which is intended to increase the environmental performance of a building. Whereas the objectives are sound and have been readily adopted by the industry, the resulting building may not necessarily be energy efficient. This is because BREEAM 'points' are gained for other 'sustainability' related practices such as provision of bike racks and locating the building near public transport routes (thus reducing carbon).

A developing area of research is into the issue of embodied carbon. This refers to the fact that the construction phase of the building will use significant amounts of carbon in both manufacturing the building materials and in the construction process. This includes the amounts of carbon used to transport to site – for example, the use of Italian marble as a floor finish is very durable and therefore has a positive impact on LCC, but transporting from Italy incurs a significant 'carbon cost' in the construction process. Therefore, embodied and operating carbon is a significant sustainability issue where a difficult 'balance' needs to be struck.

However, LCC does carry some risks. It requires different skills that may be lacking and clients themselves are often unable to adequately describe how they expect the building or asset to be used.

Furthermore, LCC is not an exact science as practitioners are attempting to model costs over a protracted period into the future. The future can be predicted to some extent but is in effect 'unknowable'. It cannot be predicted how (and more importantly when) significant change will occur that will affect the building. For example, legal changes related to sustainability and energy efficiency are predicted to alter in the coming years. Governments will inevitably change in the coming years. Technological change is accelerating. All of these issues make LCC more difficult.

LCC techniques require 'estimates' of the likely life expectation of a variety of components, materials and equipment. Furthermore, the calculations require predictions of future interest rates and cash flow requirements, which will be educated guesswork at best. However, weaknesses in the techniques can be overcome by calculating a range of possible outcomes using sensitivity analysis.

Overall, there is no doubt that LCC is of fundamental importance to clients and the future sustainability and carbon reduction in buildings. The key attributes of successful application of LCC can be seen as:

- Effective risk assessment – particularly if an alternative form of construction is proposed.
- Applying LCC as early as possible in the construction process. The development of the New Rules of Measurement by the RICS is seen as a significant benefit in forcing clients to consider LCC from the inception of the project. NRM1 ensure consideration of LCC at the concept/budget estimating phase and this flows through into NRM3 which considers maintenance and operation of the building upon completion (see Concept 13.2).
- Knowing the long-term use and predicted year of disposal of the building, which is central to LCC, so a strategy in this area is crucial.

The developing importance of sustainability in the built environment implies that LCC is crucial in the investment process. Skills in this field are developing quickly through the application of technology and the advent of building information modelling (BIM) techniques (see Concept 13.19).

Further reading

Cartlidge, D. (2011) *New Aspects of Quantity Surveying Practice*, 3rd edn, Spon, London.
Potts, K. (2008) *Construction Cost Management: Learning from case studies*, Spon, London.

13.5 Construction law

Key terms: subcontracting; bespoke contracts; Technology and Construction Court; adjudication

The term construction law refers to a body of law that has developed rapidly over the years and covers building, engineering and the oil and gas industries. It has a far reaching involvement in common, statutory and regulatory law and its complexity stems from the range of activities it covers. The word 'construction' itself hides the range of types of project that can be subject to legal proceedings. Projects can include the erection, repair and demolition of a diverse range of building types such as house, shops, offices, hospitals as well as engineering projects such as dams, bridges and motorways. Furthermore, the scale of projects can vary enormously, from small domestic extensions to 'mega-projects' such as the Olympic Games.

Further complexity is introduced when one considers the fragmented nature of the industry. In construction, there are approximately 250,000 companies registered in the UK, of which over 93 per cent can be regarded as small companies (employing 13 people or fewer). This makes the industry very complex. Contracting organisations tend to subcontract 75–90 per cent of the work on projects, thus creating a long and complex supply chain to serve construction clients. It is this chain of contracts and associated liabilities that is at the core of the expansion in construction law. In addition to the 'supply' side of the industry, fragmentation prevails on the 'demand' side as there is an increasing range of professional disciplines involved in projects caused by advances in technology and associated specialists.

Despite the majority of the multitude of contracts administered in the industry being based on standard forms (see Concept 13.7), still around 10 per cent of contracts are 'bespoke' in nature (see Concept 13.8) and by implication this will lead to an increase in disputes and therefore recourse to the law in some form. The use of bespoke construction contracts is particularly prevalent in the contractor/subcontractor relationship (and for subcontractor/sub-subcontractor contracts) and also in the professional service side of the industry where almost half of appointment contracts are bespoke.

The quantity of cases reaching court developed into something of a crisis in the 1970s and 80s. This resulted in the establishment of the Technology and Construction Court (TCC) in October 1998. Until this time the court was known as the Official Referees' Court but had no jurisdiction and simply reported to The Queen's Bench Division. Upon the implementation of new Civil Procedure Rules by Lord Woolf in 1999 the industry developed more of a focus on alternative dispute resolution (see Concept 13.6).

In addition, the Latham report of 1994 (see Concept 13.11) led directly to The Housing Grants, Construction and Regeneration Act 1996 (known as the Construction Act 1996) and this had a significant impact on dispute resolution in the industry and a reduction in cases reaching the TCC. The proliferation of adjudication following its introduction did give the court an additional role in enforcing adjudication decisions. The Construction Act gives parties to a 'construction contract' a right to refer matters to adjudicators, with the aim of aiding cash flow in the construction sector by allowing disputes to be settled without the need for lengthy and costly court proceedings. Changes to the Construction Act 1996 brought in by the Local Democracy, Economic Development and Construction Act 2009, are likely to see even more disputes referred to adjudication before reaching the TCC.

The development of legislation with a specific focus on the construction industry brings into focus another developing aspect of construction law. There is an increasing rate of development in this field. For example, the Construction Design and Management Regulations (CDM) were introduced in 1994 and these were revised substantially in 2007. There are proposals to revise them again in 2014. In both cases it would seem that the industry fails to implement change to the level that satisfies the legislature. The changes to the CDM Regulations coupled with the revisions to the Construction Act reflect an increasing preparedness for Government to influence the construction industry, thereby affecting the developing field of construction law.

Further reading

Adriaanse, J. (2010) *Construction Contract Law*, 3rd edn, Palgrave McMillan, Basingstoke.

Murdoch, J. R. and Hughes, W. (2008) *Construction Contracts: Law and management*, 4th edn, Taylor & Francis, London.

13.6 Alternative methods of dispute resolution

Key terms: disputes; negotiation; alternative dispute resolution; Dispute Resolution Boards; adjudication

Construction contracts in the distant past were much more simply constructed than they are today. They were generally concluded with a handshake, but underlying such agreements was an essential set of values of competence, fairness and honesty. Unfortunately, the industry has developed a complex and onerous set of conditions that attempt to cover every eventuality and in so doing create loopholes from which the legal professions earn a substantial living. Construction contracts need to become less adversarial and more simply constructed.

Disputes can usefully be divided into two distinct types:

1 those that rely on the involvement of the parties alone or with a 'non decision making' third party facilitating the settlement of the dispute – i.e. alternative dispute resolution (ADR);

2 those that involve a third party in making a decision that is, to greater or lesser extent, binding on the parties, i.e. the involvement of an 'umpire' who decides on the issue.

Resolution of disputes

Accordingly, it can be said that there are three separate genres of dispute resolution methods – those that are resolved:

- by the parties themselves;
- with help from a third party who plays a neutral facilitative role, but does *not* make a decision about the dispute;
- where a third party acts as an 'umpire' and makes a decision based on evidence presented by the parties in relation to the dispute, in a similar way that an umpire in tennis or cricket decides the outcome.

These alternatives are represented in Figure 13.6.1, the key point being whether or not the dispute resolution process remains under the control of the parties.

Negotiation

This is often not regarded as a dispute resolution technique – however, this method of dispute resolution is undertaken often without realising it or without a conscious decision to do so being made by either party.

It is a form of joint decision making that progresses from antagonism in the early stages to coordination in the later stages. Outside factors, such as the threat of long and expensive dispute resolution methods such as litigation and arbitration can be used as a 'lever' to persuade an opponent to continue negotiation.

Alternative dispute resolution

Alternative dispute resolution is the term used to describe a number of formal, non-adversarial methods of resolving disputes without resorting to the courts or arbitration. It is claimed to be a less expensive, fast, effective and more palatable alternative to litigation and arbitration. It is also less threatening and stressful. ADR offers the parties who are in dispute the opportunity to participate in a process that encourages them to solve their differences in the most amicable way that is possible.

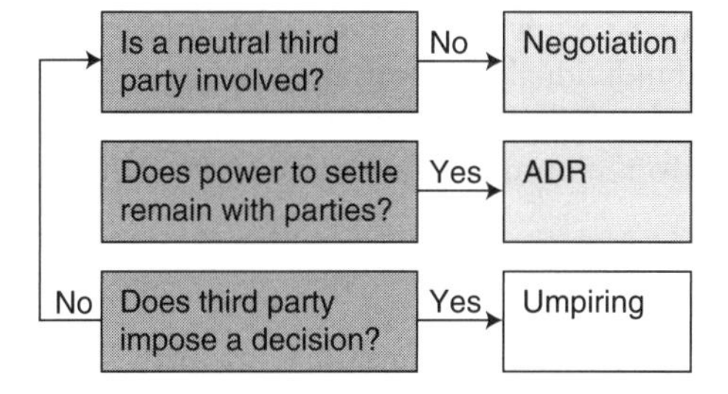

Figure 13.6.1 Resolution of disputes.

Source: Adapted from Gould et al (1999: 44).

As with arbitration and litigation, the processes involve a third party. However, unlike arbitration and litigation, the role of the third party is not to sit in judgement, but to act as a neutral facilitator to a negotiated settlement between the parties.

Before the commencement of an ADR negotiation the parties who are in dispute should have a genuine desire to settle their differences without recourse to either litigation or arbitration. They must, therefore, be prepared to compromise some of their rights and move from their stated position in order to achieve a settlement. Proceedings are non-binding until a mutually agreed settlement is achieved. Either party can resort to arbitration or litigation if the ADR process fails.

The advantages of ADR can be termed the four Cs. These are:

1 Consensus – this approach requires the agreement of all parties to a resolution of the dispute. The emphasis is on finding a business, rather than a legal or adversarial, solution to the dispute.
2 Continuity of business relations – the processes are concerned with resolving disputes within the context of, and without permanently damaging, ongoing business relations.
3 Control – resolution of the dispute remains in the control of the parties to it. The parties can concentrate on forging a settlement that focuses on commercial issues rather than the letter of the law and may thus be less damaging for all parties. Once a dispute is referred to the courts or to arbitration, the parties effectively lose control of the process.
4 Confidentiality – the proceedings are not published, and therefore the damage resulting from adverse publicity is avoided.

One of the requirements of ADR is that it must be non-binding. If the process is not working, recourse to litigation or arbitration, as appropriate, must be available. If agreement is not reached, the process will seldom have been a waste of time, effort and expense. It will probably have clarified or narrowed the scope of a dispute. Where an agreement is reached using one of the alternative methods of dispute resolution, it should be formalised into a written agreement between the parties. Some of the methods of ADR are described briefly below (this is not exhaustive!).

Conciliation

Conciliation and mediation are confused as methods of dispute resolution. Conciliation is a process where a neutral adviser listens to the disputed points of each party and then explains the views of one party to the other. An agreed solution may be found by encouraging each party to see the other's point of view. With this approach the neutral adviser plays the passive role of a facilitator. Recommendations are not made by the adviser. Any agreement to settle their differences is reached by the parties themselves. Where an agreement is achieved, then the neutral adviser will put this in writing for each of the parties to sign.

Conciliation is a similar process to mediation. The distinction is that a conciliator will actively participate in the discussions between the parties, offering views on the cases put forward. There are no private meetings between the conciliator and individual parties to the dispute. It is more informal than mediation, perhaps aimed at getting the parties to discuss differences. Should the parties fail to reach agreement, it is common for the conciliator to recommend how the dispute should be settled.

Mediation

Mediation has come to great prominence in the construction industry, particularly since the introduction of mediation into Joint Contracts Tribunal (JCT) contracts. Within these contracts, an express (written) provision has been introduced which makes mediation one of a range of three possible dispute resolution methodologies available to parties to the contract. The other two are arbitration and adjudication.

In mediation, the neutral adviser listens to the representations from both parties and then helps them to agree upon an overall solution. An active role is played by the adviser by putting forward suggestions, encouraging discussions and persuading the parties to focus upon the key issues. Within the generic term 'mediation' there is a range of possible options. The most well used are:

- facilitative mediation (mediator does not make recommendations);
- evaluative mediation (mediator makes recommendations as to outcome).

Under mediation, the parties in dispute select an independent third party to assist them in reaching an acceptable settlement to their dispute. This mediator should be skilled in problem solving and preferably have expertise relative to the dispute in question. The role of the mediator is not to make a judgement of the dispute, but to facilitate a settlement between the parties. The normal process involves the mediator meeting with the parties to agree the format and programme. The sessions usually begin jointly with each of the parties presenting their case, informally, to the mediator. This is followed by private sessions, known as 'caucuses', between the mediator and each of the parties. The mediator's role will be to get each of the parties to focus on their main interests and to get them to move to a common position. The mediator will move between caucuses, often passing offers from one party to the other. As he or she will be in a position of knowledge, confidentiality and impartiality are essential.

Dispute resolution boards

A Dispute Resolution Board normally comprises a panel of three independent technical experts jointly selected by the employer and the contractor. Where an issue is not resolved by negotiation between the contracting parties, it is formally referred to the Board which is required to present a recommendation within a short period of time. The recommendation is usually non-binding.

Dispute boards are a relatively new form of dispute resolution and generally exist in one of two forms: a dispute review board (DRB) or a dispute adjudication board (DAB). They differ only in the fact that those recommendations made by a DAB are binding, and must be immediately implemented unless overturned by arbitration or litigation. Both are financed at contract level, and made up of a panel of usually three experienced and impartial reviewers who, in an ideal scenario, are appointed before a contract begins, to reign in and where possible, prevent problems before they arise!

Expert determination

In expert determination, an independent third party considers the claims made by each side and issues a binding decision. The third party is usually an expert in the subject of the dispute and is chosen by the parties, who agree at the outset to be bound by the expert's decision. Therefore, this approach is suitable for the construction industry as the 'expert' can appreciate and understand the technical aspects of a complex dispute.

Adjudication

Adjudication has risen to prominence in the construction industry owing to the implementation of adjudication as a statutory right due to the passing of the Housing Grants Construction and Regeneration Act 1996. In effect, adjudication, if not specifically included in a construction contract, will be implied into that contract with standard clauses provided in an instrument known as the Scheme for Construction Contracts (England and Wales) Regulations 1998.

Whereas 'the Scheme' is applied across all projects, commercial adjudication is much more narrow.

- Adjudication is a statutory procedure by which any party to a construction contract has a right to have a dispute decided by an adjudicator.
- It is intended to be quicker and more cost effective than litigation or arbitration.
- It is normally used to ensure payment (although most types of dispute can be adjudicated).
- The adjudicator must generally decide the dispute in less than 42 days.

The decision is temporarily binding until a party decides to proceed to arbitration or litigation. However, adjudicator's decisions are usually upheld by the Courts.

Further reading

Ashworth, A. and Hogg, K. (2007) *Willis's Practice and Procedure for the Quantity Surveyor*, 12th edn, Blackwell, Oxford.
Gould, N., Capper, P., Dixon, G. and Cohen, M. (1999) *Dispute Resolution in the Construction Industry*, Thomas Telford, London.
Uff J. (2009) *Construction Law: Law and practice relating to the construction industry*, 10th edn, Sweet & Maxwell, London.

13.7 Standard forms of contract

Key terms: Joint Contracts Tribunal; New Engineering Contract; amendments

There has been a steep increase in the number and range of standard forms of contract available in recent years. The advance of collaborative construction strategies has resulted in different types of contract coming to market. The advent of technology and the ease with which documents can be published have resulted in nationally recognised bodies such as the CIOB and Constructing Excellence producing their own 'offerings' to the range of contracts available. The predominant forms of contract used in the industry over the years, certainly since the 1960s have been those published by the JCT. The current list of standard contracts published by the JCT now stands at over 50 as they produce many different versions for different purposes – for example, the main contract, a subcontract and a sub-subcontract form.

In recent years, JCT has become less popular in favour of NEC3 (New Engineering Contract) which was promoted by being mentioned by Sir Michael Latham in his seminal report *Constructing the Team* (see Concept 13.11). Parties are generally free to choose their own terms of contract. In practice, however, it is a lengthy and expert process to write a contract (see Concept 13.8) and it is rare for either the employer or the contractor to have the desire or ability to work out all the detailed terms and conditions that are required to govern their contract. Therefore a number of standard forms (or 'conditions of contract') are published and available for use in construction projects. Independent groups who represent the interests of the various disciplines and groups within the industry are usually responsible for publishing them.

The general layout of standards forms does not vary greatly, but the detail in the clause and their interpretation vary immensely.

Advantages of standard forms of contract

- Economy of production and use: it is a lengthy and expert process to write a contract for even an averagely complex construction project. A standard form can be used 'off the shelf' rather than starting from scratch with every new project.
- Should a party to a contract draft their own terms and these are considered to be harsh or badly written, courts are likely to find against them under the principle of '*contra proferentem*'.

- Similar projects demand similar contract conditions.
- Familiarity: with repeated use on different contracts, the parties become familiar with the forms and therefore more confident in using the forms over time.
- Common understanding of risk allocation: this should be one of the benefits of familiarity, i.e. a common knowledge of how the contract apportions risk between the parties. However because this is implicit in most forms, it is not very easy to assess whether or not the risk is apportioned in a manner that is suitable for the particular proposed project. An employer (or a contractor) may be taking on a higher level of risk than intended without even knowing about it.
- Tender comparisons are made easier since the risk allocation is the same for each tenderer. Parties are assumed to understand this and therefore their prices can be accurately compared.
- Industry-wide consensus and recognition: this is through frequency of use rather than through a voted mandate. Standard forms are often a compromise solution representing a balance of interests across the industry.

Problems in using standard forms of contract

- A universal standard form of contract is unrealistic: different standard forms are needed for different types of project. This has led to a plethora of forms and it is not always obvious which is the most suitable one to use.
- The forms are often cumbersome, written in legal jargon and complex, leading to lack of ability to understand them. However, some (are written in simpler language e.g. NEC3) and this leads to doubts whether clauses would bear the scrutiny of the courts in the event of legal disputes.
- Narrow understanding of the law: use of one standard form can lead to a narrow understanding of the contractual issues involved and perhaps a failure to appreciate the wider legal issues. The answer to this potential problem is to understand the principles of contract law first and then consider the allocation of risk within the standard forms.
- Amendments: it is common practice to amend the published version of a standard form of contract (by striking out some clauses and adding others). Such amendments are often minor, but occasionally they are onerous and result in a disadvantage to one or other of the parties. This practice can be unfair and deceptive; it is therefore not normally advisable. However, some employers wish to change the balance of risk contained in a standard form for genuine reasons and there is nothing to prevent the parties from altering the terms and conditions to suit themselves if they so wish.

Which form of contract?

There are many influences on which form of contract to choose for use on a particular construction project, including:

- client objectives;
- type of client (public or private sector);
- procurement method;
- the extent to which the design is complete before tender, and the extent to which bills of quantities are used to describe and quantify the work for the purposes of tendering;
- the extent of design by the contractor (rather than the client's consultant);
- method of pricing and payment;
- The size and complexity of the project.

Further reading

Adriaanse, J. (2010) 'Construction Contract Law', 3rd edn, Palgrave McMillan, Basingstoke.
Murdoch, J. R. and Hughes, W. (2008) *Construction Contracts: Law and management*, 4th edn, Taylor & Francis, London.

13.8 Bespoke contracts

Key terms: precedent; contract; rules of interpretation

The literal meaning of bespoke is 'made to order', the most obvious example being that of a tailor making a made-to-measure suit. In the construction and property industries, making a contract fit the clients' requirements is a key function of the chartered surveyor. A bespoke contract establishes a contract where no precedent exists. The principle of precedence is central to the English legal system which can be defined as a 'previous example or occurrence taken as rule'.

It could be argued that *all* contracts are bespoke to some extent in that no project is the same as another. Indeed, all standard contracts need to be altered in some fashion even if the amendments are required for changes in name, choices of alternative optional clauses etc. It is rare that a contract will need no amendment at all.

In drafting conditions of contract, there are effectively two starting points – take an existing contract (precedent) as a 'model' or start from scratch. Companies will adopt different solutions depending on their in-house expertise and time pressures. Careful scrutiny of model contracts as precedents is required to match the aims and objectives of the drafting organisations (client or contractor). Major areas that can cause problems are payment, termination and default, dispute resolution and variations clauses. These need to be carefully considered to ensure comprehensive coverage of key items is provided as well as being drafted with clarity and certainty in mind.

Quantity surveyors have been involved in drafting contracts 'from scratch' for a diverse range of projects. In particular, in the 1990s, contracts were required in local government for service areas ranging from refuse collection to legal services and in some authorities the 'in-house' QS professional played a key role in exposing such service to competition with the private sector under Compulsory Competitive Tendering legislation. Furthermore, the public sector required bespoke contracts for the maintenance of public facilities, although standard forms of contract (see Concept 13.7) now exist for these (e.g. measured term contracts).

The construction industry, in particular, has faced significant problems in the contractor/subcontractor relationship due to harsh and draconian terms of contract imposed in bespoke contracts – i.e. contracts written (often from scratch) where the terms favour one party over another. This resulted in the introduction of the Housing Grants, Construction and Regeneration Act 1996 and the Local Democracy, Economic Development and Construction Act 2009 (known as the Construction Acts 1996 and 2009) both of which outlawed harsh contractual terms related to payments to subcontractors – which had resulted in small companies being starved of cash.

Attributes of good contract conditions

A good contract clause should be understandable to anyone who is looking at the contract and it should be clear what the contract requires. Contracts in the construction industry have become very 'legalistic' in approach (e.g. the JCT suite of contracts) and it is arguably difficult to understand the meaning behind the words. A contract that has attempted to alter this trend is the NEC3 suite of contracts which has been drafted in plain English. It remains to be seen whether the perception that using simple language will lead to more disputes due to 'loose' wording.

A good contract will be organised in a logical fashion and will introduce certainty and clarity into the administration of the contract. If any vagueness is introduced into the contract, it is done so deliberately – for example the use of the word 'reasonable'. The text should be broken up by using paragraphs and sub-paragraphs and each clause should relate to a separate issue. Clauses should be arranged in a logical fashion and be numbered for ease of use. For example, the versions of JCT contracts since the 1998 edition have undergone a significant review of clause numbering to make the use of the contract easier and more logical.

Rules of interpretation

When drafting contract clauses, whether it be a contract from scratch or individual amendments to clauses from a standard form, a number of legal issues need to be taken into account:

- The Unfair Contract Terms Act, 1977 which states that contracts should not impose restrictive or onerous terms upon a party, or exclude right and remedies for that party.
- *Contra proferentem* – this is a legal principle that provides that a court will regard the 'drafter' of a contract less favourably than the other party. The view of the courts is that the drafter had the opportunity to be fair/accurate/unambiguous through their choice of words. Therefore, any issues with a contact may be resolved by the court in favour of the other party.
- Implied terms – if the contract is silent in respect of some issues that are in dispute the court can 'imply' clauses into the contract as if they were written therein. Implied terms can come from law (a good example is the adjudication provisions of the Construction Act – see Concept 13.6), custom (i.e. usual business practice) or simply by the court itself.

Further reading

Isurv (2013) 'Non-standard bespoke contracts', available at: www.isurv.com/site/scripts/documents.aspx?categoryID=70 (accessed 13/12/2013).

13.9 Contractual claims

Key terms: claim; contractual quantum; disruption; prolongation; acceleration

In the construction industry the word 'claim' is commonly used to describe any application by the contractor for payment that arises other than under the ordinary contract payment provisions. All standard forms of contract contain terms and conditions that define the ways in which contractors are paid by their clients – the most common are clauses that state how work in progress is paid (interim valuations) and how variations to the contract (e.g. changes in specification) are valued and paid. A claim is made by a contractor for reimbursement of expenditure and loss caused by the client for which the contractor would not normally be paid via these clauses.

The word 'claim' is defined as a demand for something as due (*Oxford English Dictionary*). This basic definition provides a clear perspective concerning one of the key attributes of contractual claims – a contractor must *demand* or claim their rights under the terms and conditions of a construction contract. These rights occur where they have suffered disruption, low productivity, loss and/or expense for something that the client has done and must take responsibility for under the terms of that contract. A claim can involve a demand for reimbursement of such costs or allowance for more time or both. In a legal sense, by doing something that results in the contractor losing money, or spending more than they planned (or both), the client has breached the contract with the contractor. The calculation of the amount of the claim that becomes payable is therefore equivalent (no more and no less) to the amount the contractor *would* have received in court under the process of common law. Indeed, should the contractor be unable to persuade the client of their entitlement to reimbursement under the terms of the contract, the contractor will reserve the right to proceed to court if necessary.

A claim is often seen as a 'dirty' word by a client as it is likely to lead to their budget being exceeded. Whereas a client may see a claim as a way in which a contractor can exploit the weak position a client may find themselves in, it is clear that a claim is an entitlement under the form of contract. The terms and conditions of contract in fact define those items for which the client takes the risk – in other words if the client follows a particular course of action that costs the contractor time and/or money and those actions are defined in the contract then it is inevitable that the contractor will seek to recover the costs (related to the disruptive impact of the client's actions) and the time lost (to prevent the client deducting liquidated and ascertained damages for a late finish).

Legal criteria of claims

All standard contracts lay down in express terms the procedure to be adopted by the contractor in pursuing a claim. This normally commences with giving 'notice' to the client that an event has taken place that is likely to result in the contractor taking more time and/or incurring loss and expense. Such notices are vital in contracts and this issue has been tested in courts many times. It is normally the case that unless a contractor satisfies the condition to submit a notice first, then they lose their rights/entitlement to a claim. This is a legal principle called 'condition precedent' – in others words a written notice *must* be given by the contractor and then a lack of notice will be *fatal* to their rights under the contract. If a contractor does not comply they may be required to carry out the actions instructed by the client (e.g. carry out different work to that defined in the contract documents) but would lose the right to additional time and additional money. Therefore, payment to the contractor by the client will not take place unless the contractor follows the requirements of that contract.

Courts have established the types of loss and expense a contractor is allowed to claim – many of these items have changed over time due to courts reconsidering each legal position establishing a new 'precedence' in court. Courts have established the following:

- Contractors *must* prove that loss and expense has been suffered.
- A claim *must* show that the loss arises from the items listed in the contract that the client takes responsibility (risk) for. This list is an express term of the contract.
- A claim *must* be 'Direct' – in others words, the loss and expense (costs) and time delays giving rise to the claim must flow directly from the actions of the client.
- The act of the client (as listed in the contract) must be shown to be the cause of the costs and delays, not simply a contributory cause.
- The contractor is under a contractual 'burden of proof' to show a link between the event (cause) and delay to completion (effect).

Contractual quantum

Once the contractor has established his right to a claim (time and/or money) the contractual quantum (how much?) of the claim must be established. As discussed above, a fundamental rule of common law is where a party suffers loss (by breach of contract) – they are placed in the same situation, with respect to damages, as though the contract had not been performed. The damages in this case of a claim equate to the damages the contractor would have obtained in court. It should be noted that a contractor is not entitled to earn additional profit on a claim.

It is almost certain that the event for which the client takes responsibility will cause the contractor *disruption*. A dictionary definition of disruption is 'to interrupt or impede the progress, movement, or procedure of'. This is a reasonably accurate representation of disruption in construction contracts. In effect, this means that the actions of the client have resulted in the contractor's planned works being rendered impossible (wholly or in part) in terms of their intended sequence, duration of activities and/or loss of productivity of labour, mechanical plant, supervision and any or all of them.

Types of claim

There are three types of claim a contractor can make:

1. *Disruption*
   Whereas all claims have disruption at their heart, certain actions by the client will result in an activity or series of activities being disrupted without them taking any longer in terms of time. For example, if the client changes specification of external walls from brick/block to concrete cladding panels, the time may be the same (or even less!). However, the disruptive impact will be related to the fact that the contractor's method of working will be totally changed, different subcontractors will be required, work will be carried out in a different sequence, alternative items of mechanical plant and equipment will be required. All of this leads to loss of productivity, loss of money already spent (loss) and additional expenditure (expense).
2. *Prolongation*
   This type of claim occurs when the event that has led to the claim has caused a delay that prolongs the overall contract period. The quantum of the claim will thus have two elements. First, the costs of disruption (similar to 1 above) and second, the costs of the contractor staying on site for a greater period of time (i.e. site set-up costs). Figure 13.9.1 represents the logic.
3. *Acceleration*
   This type of claim occurs when the event that has lead to the claim has caused a delay, but the programme is accelerated to 'catch up' to the original contract programme. Therefore, the quantum of the claim will comprise additional resources and expenses such as weekend and overtime working. However, it is unwise and commercially risky for a contractor to assume such costs will be automatically payable. It is a principle long established in common law that a contractor must 'take all reasonable steps to mitigate the loss consequent to the breach'. Therefore, a wise contractor will seek written instruction form their client to proceed with such acceleration as the client might assume that the contractor (out of the goodness of their heart!) has caught up the lost time at their own expense (Figure 13.9.2).

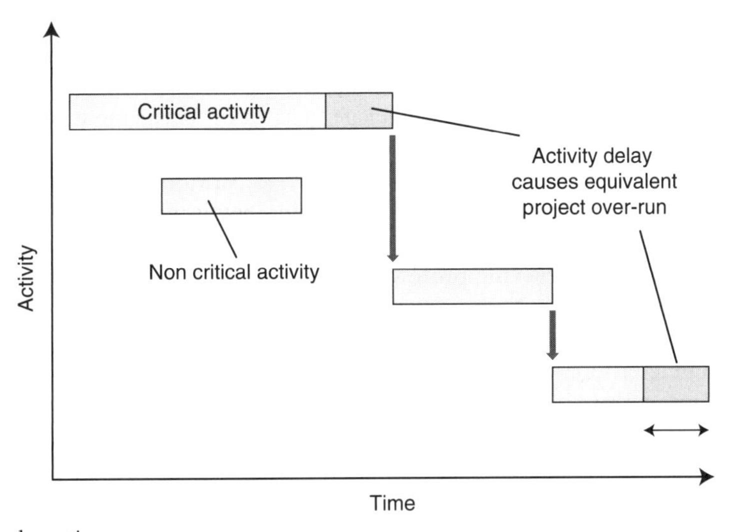

Figure 13.9.1 Prolongation.

Source: Glenn Steel (2013).

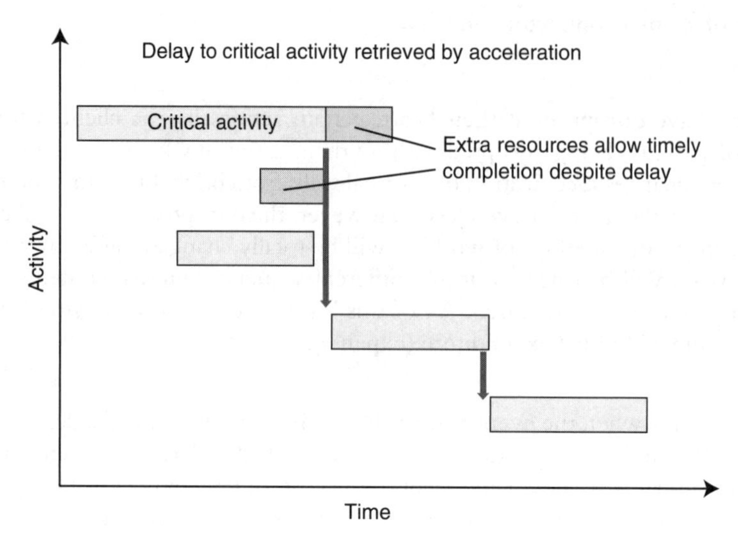

Figure 13.9.2 Acceleration.

Source: Glenn Steel (2013).

Further reading

Adriaanse, J. (2010) *Construction Contract Law*, 3rd edn, Palgrave McMillan, Basingstoke.
Murdoch, J. R. and Hughes, W. (2008) *Construction Contracts: Law and management*, 4th edn, Taylor & Francis, London.

13.10 Project management

Key terms: generic project management; project plan; critical path methodology

The quantity surveying profession provides a significant percentage of the membership of the Project Management Faculty of the RICS. The skills learned in the profession (both during undergraduate degree programmes and experience in practice) hold them in good stead to be effective project managers. Transferable skills in subjects such as forecasting, planning, analysis, attention to detail through document production, budgeting, control, cost management, risk management and value management are subjects that support the extension of the role into project management.

Figure 13.10.1 indicates the terminology associated with 'generic' project management, i.e. not associated directly with the construction industry. The terminology is not often observed in the construction industry, but appears to be coming more and more into the delivery of construction projects. New approaches by the quantity surveying profession (e.g. the links through the new suite of New Rules of Measurement documentation) are implementing good practice principles from the field of project management.

In particular, the contractor's QS (quite often their job title is that of 'Commercial Manager') needs to implement such plans as that shown above to successfully deliver a project, although, as stated above, the terminology is quite different.

The question 'why' is perhaps not as relevant to a contractor as other considerations shown in the above diagram. A contractor that has been successful in securing work needs to deliver a project to the satisfaction of the client and achieve an acceptable return. In project management theory, a company that cannot deliver the benefits of a proposed project should not be starting the project – this is quite

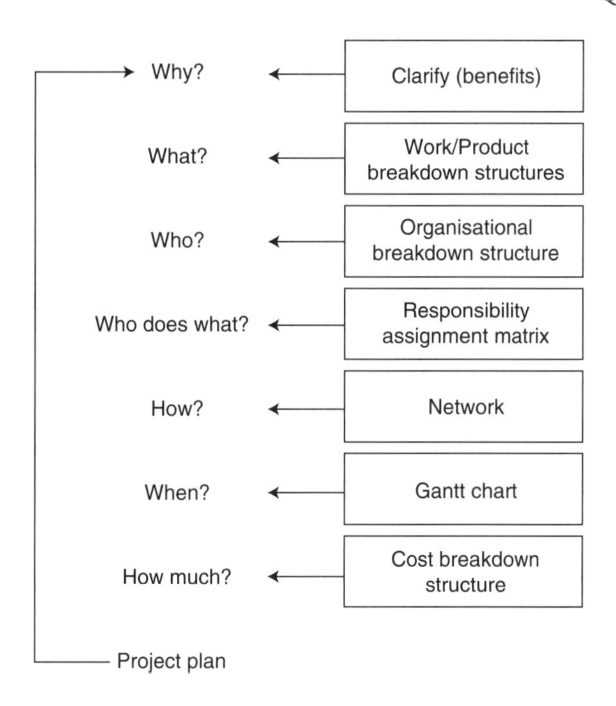

Figure 13.10.1 Project plan.
Source: Glenn Steel (2013).

different in construction for a contractor. Nevertheless, consideration of what the objectives are for the contractor in completing the project is good practice, even though the benefits are likely to be client satisfaction and financially based.

The question of 'what' is a significant issue for the contractor. The subdivision of the work into packages for subletting to subcontractors is a very important stage in the project management process and this is managed by the commercial QS. Quite often poor procurement strategy results in the contractor losing money or being delayed by poorly selected subcontractors which has a detrimental impact on project delivery. Choosing the correct subcontractor that is financially sound, technically capable with good health and safety practice is crucial to the success of the project. Well-managed demarcation between packages with no overlap results in better working arrangements and can also avoid 'scope creep' when on site. Defining the scope of a project is one of the key phases of project planning.

The organisational breakdown structure shown in Figure 13.10.1 refers to the process of assigning work to the teams or personnel responsible for undertaking the work. However, allocating the contractor's team to the project is not a complex process, but the allocation of the right people for the right project is crucial to success. Similarly, the identification of 'who does what' is also a relatively simple process and a not insignificant issue in construction.

The next two stages in Figure 13.10.1 are central to the commercial success of the project for the contractor. Time planning, method statement and sequencing of work are crucial factors in controlling the work and ensuring timely completion, thus avoiding liquidated and ascertained damages at the end of the project. Critical path methodology is a key contributor to managing time and recording and reviewing progress is a critical tool in the project life cycle. On a more negative aspect, keeping records is of fundamental importance in compiling contractual claims (see Concept 13.9) as any delays will need to be proven to the client.

The final stage in the planning process is the cost breakdown structure. Once again, this is not normal terminology but is analogous to the financial control systems a contractor would adopt. In particular, costing systems will need to have sufficient detail to allow effective financial control to take place. The cost value reconciliation process (CVR – see Concept 13.14) is the vehicle by which the contractor will identify problems and take corrective action in order to meet their project objectives.

Planning is crucial in projects to ensure objectives are achieved. Objectives in construction projects can be structured around the 'iron triangle' – that is time/cost/quality. In recent years, a fourth dimension has been added, namely health and safety. One issue that has not been mentioned in the above plan is the contractor's need to be aware of the project stakeholders. Obviously, a key stakeholder is the client, as well as all of the subcontractors and their own employees. Another issue that has become more apparent over recent years is the need to have regard to the sustainability. This has become crucial and contractors need to have appropriate policies in place and publicise them to the general public, particularly in the locality where the site is situated. This attention to what may be termed 'corporate social responsibility' needs to be visible to be implemented effectively. Certainly, under initiatives such as the Considerate Constructors Scheme and the implementation of waste management plans, construction sites have become a much improved environment and cause much less disturbance to the local environment and infrastructure.

In summary, whereas the terminology associated with the construction industry is different and methods vary, the overriding need for effective planning remains the key issue in projects. Compliance with time, cost, quality and health and safety issues has a significant impact on project success and company reputation.

Further reading

Harris, F., McCaffer, R. and Edum-Fotwe, F. (2013) *Modern Construction Management*, 7th edn, Wiley-Blackwell, Chichester.

Maylor, H. (2003) *Project Management*, 3rd edn, Prentice Hall, Harlow.

13.11 Partnering

Key terms: competitive tendering; Latham Report; Construction Industry Board; Egan Report; strategic partnering

The construction industry traditionally for many years adopted an ethos of doing business on a competitive footing based on a procurement process that was based solely on completing a building design (via a range of consultants) before submitting the design (recorded in detail in a BoQ) to a range of contractors for tender. The competitive nature of the industry over many years has led to a range of problems which have been impossible to resolve. The reliance on pure competitive tendering has driven prices and therefore profit levels so low that return on investment is poor in construction companies. Furthermore, the lack of retained profits from companies has meant that there has been a lack of innovation in the industry resulting in a state of inertia.

The industry has a strong reputation for confrontational and adversarial relationships. The competitive nature of the industry was seen as one of the major causes of this as the need to win work in competition drove prices so low that the winning tender would attempt to exploit any 'gaps' in the client's design by making claims for extra money and time. The very nature of the competitive environment created a 'race to the bottom' and resulted in poor quality projects as well as poor quality and short-term relationships. High numbers of disputes led to the court system being over extended and resulted in long and protracted legal proceedings. This further exacerbated poor relationships and costs of legal proceedings were spiralling out of control.

Starting in the 1960s, there have been several attempts by the government to influence change in the industry which have generally failed. Competitive tendering was one of the key issues tackled by the Latham report of 1994 (called *Constructing the Team*). Sir Michael Latham led a joint industry – government review – and saw traditional competitive tendering as the main reason for adversarial relationships in the industry. In an earlier report entitled 'Trust and Money' he criticised the ritual tendering process prevailing in the industry and particularly the level of waste incurred by contractors (particularly subcontractors) in abortive tendering costs due to unreasonably long tender lists. In addition, Latham referenced some unethical bidding and tendering procedures at subcontractor level which had become endemic in the industry.

Latham also discussed the pressure faced by public sector bodies that were forced to accept the lowest bids and that clients should take a balanced view by considering 'non price' or quality criteria in their bid evaluation. The government established the Construction Industry Board to implement the findings of the Latham report which made 30 recommendations. This led to the implementation of several codes of practice for tendering which implemented some of Latham's recommendations. However, the most visible and well documented impact of the Latham report was the legislation introduced (Housing Grants, Construction and Re-generation Act, 1996) which implemented new payment laws that intended to improve cash flow from contractor to subcontractor and the statutory right to adjudication to introduce rapid resolution of disputes (within a 42-day period).

In 1998 a more strategic and radical approach to tendering was shown in another industry-wide report, *Rethinking Construction* (the Egan report, named after the chair of the commission Sir John Egan), which stated a deep concern that the industry was failing its clients. The report implemented stiff and ambitious targets, one of which was to ensure that 20 per cent of construction projects (rising to 50 per cent by the end of 2007) would be delivered using integrated teams and supply chains. This was one of many 'benchmarks' established by Egan, and measurement of continuous improvement is one of the positive and lasting developments resulting from the Egan report. The report led directly to the development of partnering as a method of procurement, relying on the principle that cooperation is a more efficient method of working than traditional competitive contracting. There are three main components to a partnering arrangement:

- mutual objectives;
- an agreed method of problem resolution;
- an active search for continuous and measurable improvements.

Partnering can take many different forms. In its purest sense, a 'partner' has absolute trust in the other and will work together without any need for a contract. This is unrealistic in a legal and risk based sense, but spirit of working together and openness in business relationships (even going as far as sharing detailed financial information in an 'open book' accounting arrangement) is seen as the ideal. The birth of partnering saw the introduction of standard forms of contract for partnering arrangements (e.g. PPC2000) and the introduction of appropriate clauses and working arrangements intended to encourage partnership approaches. For example, clause 10.1 of NEC3 states that the contract should be performed in a spirit of mutual trust and cooperation. It is doubtful whether such a clause will be enforceable at law despite it being an express term of the contract.

There are two types of partnering arrangement – project specific partnering and strategic partnering. Project specific partnering, as the name suggests, relates to specific 'one-off' projects. This may be appropriate to clients that build infrequently, but measuring 'continuous improvement' will not be possible over time. Strategic partnering appears to be the basic intention of the Egan report as the potential for long-term savings and procedural improvements is created. By establishing long-term relationships, the client and contractor (and subcontractors?) can create common facilities and learning from projects can be implemented in repeat projects. There is the capacity to make measurable improvement and to meet the benchmarking targets established in the Egan report (see Concept 13.16).

The application of strategic partnering can lead to significant advantages. These may include a reduction in disputes, reduction in costs (due to development of more efficient working systems), improved quality and safety, more stable and regular workloads for contractors and subcontractors, and improved working environments.

Large clients with significant capital programmes introduced framework agreements, where a small number of large contractors 'competed' to be included on a framework, which was in effect a partnership between the client and multiple contractors. The work would be shared between the partner organisations based on ability, financial efficiency and availability. In this case, the client would receive the benefit of the experiences of several contractors and thereby improve performance and increase organisational learning.

The weakness of partnering is that the means of agreeing prices is by negotiation. Whereas the improvements in efficiency should outweigh any extra costs, the client 'feels' that they have lost the feel for the market price. This is particularly true of public sector clients who will need to demonstrate they have achieved 'Best Value' for the public purse. This will involve quite complex and therefore expensive analyses of costs. It is easier, in some respects, to go back to the 'old days' of competitive tendering.

This is a key point – in times of recession, the client is tempted to revert back to 'type' and insist on competitive bids to ensure the market price is achieved. After perhaps years of negotiation and due to tight financial constraints the client may feel that the costs of negotiation outweigh the potential savings of competition. This approach is evidenced by some framework agreements resorting to 'mini competitions' rather than maintaining the true philosophy of a strategic partnering approach.

Further reading

Cartlidge, D. (2011) *New Aspects of Quantity Surveying Practice*, 3rd edn, Spon, London.
Egan, J. (1998) *Rethinking Construction*, HMSO, London.
Latham, M. (1994) *Constructing the Team*, HMSO, London.
Potts, K. (2008) *Construction Cost Management: Learning from case studies*, Spon, London.

13.12 Procurement methods

Key terms: procurement; risk; change; trust; design and build; contract strategy; partnering

'Procure' is defined in the *Compact Oxford English Dictionary* as 'obtain', but originates from the Latin *procurare* which means to 'take care of, manage'. A combination of these simple definitions comes reasonably close to reflecting the essence of procurement in the construction industry.

A traditional perception of procurement used to see it just as the choice to be made between three quite different contractual approaches through which the client organisation engaged with a contracting organisation in order that the former could have something altered or built from new by the latter. While this is satisfactory as a rough working definition, it suggests a simpler world than today. Professional services have broadened out enormously over the years, such that the 'procurement' of a particular scheme may embrace everything from arrangements to purchase of the land to the ongoing management of the finished property some 30 years later.

Risk: Construction can be one of the most risky of businesses. Every new job is, in effect, a prototype, likely to throw up its own unique risks and problem issues. So, one of the key concerns of both parties will be, and should be, the identification and the effective allocation of risk. At worst, this becomes an exercise in each party trying to shift all possible risk onto the other. At best, it will call for each party to accept (and subsequently manage) those risks it is best equipped to accept, through experience, financial resources etc.

Change: One of the chief risks, it could be said, is the risk of change – changed circumstances, changed design and so on. A procurement route can be chosen specifically to accommodate change in certain ways, the consequences varying for the parties accordingly.

Trust: It is to be hoped always that parties to a contract can trust one another in their business dealings and that they will perform effectively and in a timely manner. However, only recently has a firm expectation of this been incorporated formally within contractual agreements. Mutual trust forms the 'backbone' it could be said to the partnering relationship – one of the newer facets of procurement practice.

The management of a construction project concerns, in its simplest form, managing three main factors: namely time, cost and quality. It is possible, therefore, to amplify this last statement to state that the construction industry is in the business of managing the *risks of changes* to time, cost and quality.

A client can choose from three main procurement routes – within these it can be seen how risk factors can be variously allocated.

The traditional route

The client engages their own architect, quantity surveyor and various other specialist consultants. The client's architect will be responsible for the whole design of the works on behalf of the client (with the possible exception, nowadays, of any contractor designed portions) and will manage the cost, quality and timescales involved. The contractor will generally be responsible for constructing the building according to the drawings etc. within the originally stated contract period for a cost agreed by some basis at tender stage. Usually the whole design should be finalised and measured (and set down in BoQs) before the works are sent out to obtain tenders.

Risks to the client of the traditional route

Providing that the client's consultants have produced a perfectly functional, comprehensive, top quality design before the project goes out to tender it should indeed be possible to ensure that the building is subsequently constructed to this high standard. Also, since (in the case of a contract with BoQs) every detail can be accounted for, the value of the final account should be as the tender sum.

It is thought that the factor that suffers in the case of the traditional approach is *time*. Procurement via a traditional route is not a quick process.

Management of change in the traditional route

Change normally means an improvement or addition rather than an omission – with an addition to the cost. Tender documents should be fully and finally prepared, in which case there should be no need for change – no need for the final price to increase. However, more often than not, a change *will* be required, and the client, whose design this is will have to pay accordingly.

The design and build route

The client presents the contractor with a performance specification. As the name suggests, this will detail the required scheme in broad, generic terms setting certain parameters but stipulating very little detail – (e.g. a 200 m^2 office) whereupon the contractor carries out all design work and then constructs the required building for a fixed price, all within a time limit set by the client.

This is the 'pure' form of design and build, as it was originally created – one in which all design detail and construction responsibility is left to the main contractor. There has been a tendency for clients to make their design briefs/requirements increasingly detailed and in some cases even passing employment of the architect over to the main contractor (known as 'novating'), in which case the client maintains much of the say as to the designs as they proceed.

Risks to the client of the design and build route

The contractor is responsible for the design of the building and so the client can send his project out to tender as soon as he has finalised the performance specification (usually produced with the advice and help of an architect and/or other specialists). Once the contractor has stated a time and cost for completing the works these should not change, as the responsibility for these is entirely the contractor's. It is thought that the factor that could suffer in the case of a design and build approach is that of *quality*. However, in a pure design and build scenario they have only given a performance specification – not detailed in every way as in the traditional route, and so a lot is left to the contractor, who may be tempted to 'cut corners' in order to offer a particularly keen price.

However, the opposite may be true and design could benefit. The contractor, through his own design team, may in fact come up with an innovative, exciting design.

Management of change in the design and build route

Everything about the design and about the planning and execution of the works is the contractor's responsibility. If during construction the client wants to rethink and change any aspect of the works, he could well face a severe penalty in terms of very high rates and long (costly) extensions to the contract programme. By opting for design and build the client gave the contractor total control over the quality, quantity and pace of the works.

Management based routes

The term covers a number of possible contractual/control situations, but in all cases the client appoints an individual or organisation to manage construction operations, in return for a lump sum fee or a percentage of the contract sum. In *management contracting* the lead contractor provides expertise in management and in buildability. This contractor does not take on the work or employ any of the labour or plant themselves. They will let the work as a series of packages, usually by competitive tender (though sometimes on some other basis) One advantage of this is the capacity for the client to engage the management contractor at an early point in the design stage, thus drawing on the latter's expertise in putting the whole thing together. In this case, contracts for the works packages will be between the management contractor and the works package contractors. *Construction management* operates similarly, but here, while the project manager (again, an individual, firm or contracting organisation) has the same coordinating role as that above, the contracts for the work packages are directly between the client and the works package contractors. This last may prove too onerous a responsibility for the client body unless they have prior experience.

Risks to the client under management contracts

The client, through their project manager or through the management contractor, can determine the quality of the works (through rigorous specifications, drawn details etc.) and can set an overall programme for the works, which it should be possible to enforce.

It is thought that the factor that could suffer most in the case of the management approach is *cost*. Since design may only be finalised in time for the letting of the package, there can be no actual certainty as to cost until the tender sum has been agreed for the last work package to be agreed and issued.

Management of change in the management routes

The processes and consequences for the client in this case will be much the same as for either of the above traditional or design and build routes, depending upon how a particular package has been let.

Making the decision – contract strategy

It will be seen that in following the ways of the above, there is nearly always a trade-off between time, cost and quality – one or two points on the triangle which will be stronger.at the expense of one or two others.

In practice, clients will be aware, or can be advised, of this. Generally they will be led to make a conscious choice between the three, based on the individual properties seen above. A further deciding factor may be the client's assessment of the degree of certainty about the works, and thus likelihood of change.

Newer and longer-term solutions

The procurement routes considered so far have all assumed that the relationship between the client and the contractor is, essentially, to occur on a one-off basis. So, in the above:

1 the client selects the best contractor for the job;
2 the contractor constructs the works; and
3 the contractor hands over the completed works to the client and the two parties part company.

This in turn has been made on the basis of two key presumptions:

1 that the client is financing the purchase of the building out of their own funds; and
2 there would be no cause for or benefit to be gained from a deeper or longer-term relationship between the parties.

Over recent years, two key developments in procurement processes have tried to address changes in the actual needs and circumstances of the parties, as follows:

- in terms of contractual relationships, that is *partnering;*
- in terms of project finance, that is *PFI.*

Partnering

Reference was made above to the level of trust expected/hoped for within the contractual relationship. Partly to combat the traditional adversarial nature of contractual relationships within the construction industry, a structured relationship between the key players in the construction team or 'partnering' has been developed. This relies on a willingness of those taking part to be open, sharing information as freely as they can and seeking to resolve any differences/disputes in an open constructive manner. In its earliest days, partnering was achieved through a Partnering Agreement – a legal document setting out guidelines for relationships between the parties but being unenforceable legally in its own right. In such case there would still be a standard form of contract in force. The Partnering Agreement merely suggested how parties to the contract itself should ideally behave. PPC2000 is a standard form of contract which gave legal force to the partnering relationship. In PPC2000 any number of individuals/concerns could be party to the contract (the more the better, for everybody's sake) rather than this being restricted to the usual two parties in the contract, the client and the main contractor.

A partnering relationship (whether via Agreement or via PPC2000) can be applied to a single scheme (*Project Partnering*) or to a series of schemes, stretching out over a number of years (*Serial Partnering*).

The private finance initiative

At the start of the 1990s the UK Government created a new process for financing construction works – chiefly applicable to the public sector in respect of such buildings as schools , hospitals and the like. The relationship is essentially that of mortgagor (the private sector) and mortgagee (the public sector). The private sector *provider* is responsible for raising the funding, finding the design team and the contractor (each of which may be selected by one or other of the means above) and producing the required building. The most striking feature of this approach to procurement is that the provider retains and maintains the building for a period of between 25 and 40 years, engaging in what is a long-term exercise of facilities management.

Although it seems not to be listed as a 'procurement method' in text books, it seems that it is indeed appropriate to address the issue of the private finance initiative (PFI) within this concept. Admittedly, it was primarily created by the government as a funding device, applicable whatever the immediate procurement method (in terms of traditional, design and build etc.) But, due to the 25–40 year pay-off periods involved and the continuing link between the client (in this case the government) and the provider, it presents a pattern of ownership and control that fits the broader definition of procurement.

PFI has received a significant amount of criticism, mainly in the light of the high charges made to the public sector for the long-term contracts mentioned above (see Concept 3.3). The government has introduced PF2, a revised procurement process that seeks to speed up procurement and reduce costs.

Further reading

Cartlidge, D. (2013) *Quantity Surveyor's Pocket Book*, 2nd edn, Taylor & Francis, London.
Greenhalgh, B. and Squires, G. (2011) *Introduction To Building Procurement*, Spon, London.

13.13 Contract administration

Key terms: interim valuations; valuation of variations; final account; ethics

The QS plays an important role in the management and administration of the contract signed between the contractor and the client. Many provisions of standard forms of contract relate to financial matters for which the client's QS will be responsible during the duration of the contract. In addition, the contractor's QS will have a range of contract administration duties that are somewhat different, but particularly important to the contractor in furthering their business. This is a crucial/key difference between the QSs employed by both parties – a client's QS is providing consultancy services whereas the contractor's QS is a central figure in making a business sustainable over time and profitable.

A detailed knowledge and understanding of the conditions of contract is needed by both in order to be able to apply the provisions correctly and professionally. Because of this a QS often has more knowledge about details of contract provisions than other parties and is in a position to offer advice on their use. The QS is seen as the legal expert in the construction team in administering contracts.

Most contracts place the responsibility for most of the activities on the QS acting on behalf of the employer. However, in practice a joint working arrangement is needed between the client's QS and the contractor's QS to ensure that all of the necessary work is completed to both parties' satisfaction. However, for example in the JCT suite of contracts, specific reference is made to the client's QS and not to the contractor's QS. This is because the contractor in general terms is referred to as a company and the contractor's QS acts on their employer's behalf. In fact, the main role of the contractor in terms of administration of contracts is to give notice to the client when items of contention arise and to provide information to allow the client to make judgements with regard to time and cost.

Some of the key functions of the client's QS are as follows:

- preparation of interim valuations for interim certificates;
- measurement and valuation of variations;
- preparation of final account.

The other main responsibility for the QS prescribed by most building contracts is that of ascertaining direct loss and/or expense incurred by the contractor (see Concept 13.9).

Ethics

The QS has a professional duty as a chartered surveyor to ensure that the financial administration of the contract is fair to both parties (employer and contractor). However, 'fairness' can be subjective and the QS must act within the stated conditions of the contract. The QS has no authority to vary those conditions even if the result of applying them seems to be financially penalising one or other of the parties. For example, verbal instructions must be confirmed in writing, otherwise payment should not be made. Another example, if one party is experiencing temporary cash flow problems, then payments should not be adjusted to allow for this. This ethical presumption is particularly difficult to apply for the contractor's QS who is charged with making due profit for a business – this is, in practice, a very difficult balance to achieve. For example, the contractor's QS may seek to ensure that the contractor derives any potential benefits from the provisions, for example errors/omissions in quantities. However any application for further payment can only be based on the contract provisions.

Interim valuations and certificates

In order to reduce the contractor's liabilities to finance construction work, it is normal to allow for regular periodic payments to be made to the contractor for work completed to date. Standard forms of contract set out when the payments will be made, how the amount payable will be assessed and the procedure for making payment.

Normally the contractor will be paid monthly, commencing one month after the date on which the contractor takes possession of the site, and usually continuing until at least the date of practical completion. For cash flow purposes, regular dates are agreed (and entered into the contract particulars) on which the work will be valued so that payments will follow on a regular basis.

Assessment of amounts of payment

Before each payment is made, it is usually necessary for the client's QS to prepare an interim valuation. He must visit the site and check on the progress of the work at the agreed valuation date. Although it is the client's QS's responsibility to prepare the valuation, the contractor's QS has an option to submit an application first if he so wishes.

The work is valued on a cumulative basis. At each valuation the total work carried out to date since the start of the contract and the unfixed materials and goods on site are valued as accurately as is reasonably possible. From this, retention is deducted and the total value of previous payments to give a balance payable for the last month.

The preparation of a valuation can be divided into sections as follows:

1 Preliminaries
2 Main contractors building work (as priced for the contract)
3 Variations

4 Provisional sums
   - for defined work where certain information can be provided, including the nature and scope of the work and how it is to be fixed in place
   - for undefined work where this information cannot be provided
   - contingency sums
5 Unfixed materials and goods
6 Claims for direct loss and/or expense
7 Statutory fees and charges.

Once the amount due to the contractor is calculated the client's QS must deduct retention. Items 6 and 7 above are not subject to retention and must be paid in full. The client's QS then informs the client of the amount that must be paid to the contractor through the lead consultant who is often the architect.

Financial reporting

An interim valuation is produced primarily for the purpose of calculating monthly payments to the contractor. As such it is a historical document (albeit a very up-to-date one) and does not attempt to predict what might happen in the future. A further role for the QS, beyond the contractual requirements of valuations and payments, is that of anticipating events for the remainder of the contract and maintaining a report of the likely financial outcome of the contract. Quite often a monthly financial statement is produced, alongside the interim valuation, which acts as a report to both the employer and the contractor with regard to the financial progress of the contract.

Similarly, the contractor's QS will produce a CVR report (see Concept 13.14) which makes a prudent assessment of the contractor's financial performance on the project.

Variations

Variations are defined as 'the alteration or modification of the design, quality or quantity of the works'. The definition is wide ranging, but does not include changes to the very nature of the work. Whether or not the nature of the work is being changed as the result of a 'variation' may be difficult to ascertain – it will depend on the circumstances, as shown by examples of case law.

Variations can also arise due to changes in access to the site, limitations of working space or working hours and the order in which work is to be carried out.

The variations clauses allow for changes to be made under the contract (without changing the contract itself). Variations to the contract are changes to the contract itself that can only be made by further agreement between the parties. Hence, the definition and scope of variations allowed by the contract are important to know and understand.

All variations (indeed all instructions even if they do not incorporate a variation) must be in writing. If oral instructions are given, they must be confirmed in writing by the architect. The contractor has a 'right of reasonable objection' to a variation instruction.

Valuation of variations

Variations must be valued in accordance with the provision of the particular contract. In general, a reasonable process defined in JCT contracts that serves well as good practice is given in Table 13.13.1. Dayworks, mentioned in the final row of the table, can be defined as a method of valuation based on the prime (actual) cost of all labour, materials and plant used in carrying out the work, normally with a percentage addition to the total cost of each for overheads and profit.

Table 13.13.1 Valuation of variations

Character of work	*Conditions under which work is carried out*	*Quantity of work*	*Method of valuation*
Similar	Similar	Similar	Bill rates
Similar	Different	Significantly different	Bill rates with fair allowance
Different	May also be different	May also be different	Fair rates
Work cannot be properly valued by measurement and pricing above			Daywork rates

Final accounts

Practical completion is the point in time where the contractor has substantially completed construction work and is able to hand over the building to the employer. It is a milestone that marks a number of changes to payment procedures and timing of activities, as follows:

- Half of the amount held as retention is released in the next payment to the contractor.
- Regular (monthly) interim certificates cease.
- The rectification period commences.

The final account involves the ascertainment of all changes to the contract sum that have occurred during the period of the contract. The process of receiving documents from the contractor's QS and ascertaining the final account and agreeing it with the contractor is likely to be iterative, i.e. a process of negotiation until agreement is reached.

Contractor's QS role

The above material relates to the main contract between the client and contractor. The contractor's QS additionally administers a significant number of subcontracts. As contractors subcontract 75–90 per cent of the work this is a considerable task as all of the aforementioned administrative procedures are replicated across up to 50 subcontracts which may exist on a large/complex project.

Further reading

Ashworth, A. and Hogg, K. (2007) *Willis's Practice and Procedure for the Quantity Surveyor*, 12th edn, Blackwell, Oxford.
Ramus, J. W., Birchall, S. and Griffiths, P. (2006) *Contract Practice for Surveyors*, 4th edn, Butterworth Heinemann, London.
Towey, D. (2012) *Construction Quantity Surveying: A practical guide for the contractor's QS*, Wiley-Blackwell, Chichester.

13.14 Cost value reconciliation

Key terms: management accounting; gross amount certified; payment application; internal valuation; anticipated final value

Cost value reconciliation (CVR) is the predominant method adopted by contractors to financially control individual projects. As the title suggests, the objective is to 'reconcile' the income from projects (or value) against the cost incurred in achieving that value. Whereas the CVR is usually completed by the contractor's QS – the process requires a team effort from the contractor's team. Considerable discussion is required before the CVR is issued to the rest of the management team.

The latter point is vital, as the CVR provides information for action by the company's management. CVRs from all of the contractors's projects form part of the contractor's *management accounting* process. The management accounts comprise management information that is used to review performance throughout the financial year and should not be confused with financial accounts, which are a statutory requirement. Therefore there are two basic reasons for conducting cost value comparisons in contracting organisations:

- good management practice;
- accountancy requirements.

In common with all management accounting systems, there are no legal requirements to enforce companies to monitor their performance using a set methodology. Each contractor will design their own process. However, it is beneficial to the company to follow issued guidelines which direct companies to follow certain parameters. SSAP9 (Statement of Standard Accountancy Practice) comprises a series of explanatory notes that are intended to remove inconsistencies from financial reporting procedures relating to financial accounts. It is clearly beneficial and good practice for the commercial QS to follow such guidelines in controlling their projects to increase the reliability of the company's accounts.

It is essential that management are kept informed on a regular and frequent basis of how things are going in terms of company performance – whether this is profitable or not. Within contracting organisations this amounts to comparing the actual costs to be allocated to projects against accurate assessments of the value of works carried out.

The SSAP9 guidelines advocate a conservative approach to completing annual accounts and cost–value comparisons. They divide their considerations into two sections:

- long-term projects, i.e. contracts that span more than one accounting year;
- short-term projects, i.e. those that do not.

SSAP9 therefore guides the commercial QS in how to report work in progress where a project spans an accounting period as well as emphasising the need to adopt two key accountancy principles in their analysis. First, they must be 'prudent' in their actions. This means taking a conservative view of financial transactions, for example not making an assumption that the contractor will be paid for extra/varied work completed. SSAP9 provides that profit should only be declared where there is reasonable certainty that the profit has been achieved.

Second, the principle of matching should be imposed. This means that the commercial QS must match income and expenditure together in the same period. This principle has a significant impact on the production of the CVR and the QS must adjust their figures to cater for differences in timings of interim valuations (money received) and entry into their costing systems (money paid out).

To be of most benefit to the management team it is necessary for cost–value comparisons to be produced as expeditiously as possible after interim valuations are completed and agreed on site. There is little point in producing cost information several weeks after the event, a situation that could render remedial measures ineffective.

CVRs are not just completed to monitor performance; they are an essential management tool used in many departments, in particular estimating. Feedback from project performance is required continuously to ensure that accurate and competitive estimates can be produced.

Key components of a cost value reconciliation

A CVR report contains all key management information required for the contractor's senior management to take corrective action to address problems on the project. The following figures are crucial in maintaining control over the project:

- The gross amount certified – this is the amount actually paid by the client (as ascertained by the client's QS) to the contractor and is therefore crucial in measuring performance relating to cash flow on the project.
- The payment application – this is the amount requested by the contractor. This amount should not be over stated or exaggerated (prudent) and should be fully justified by the commercial QS. This figure is the starting point for the CVR calculation.
- The internal valuation – the CVR calculation will adjust the payment application to reflect the issues addressed by SSAP9 and should implement the accounting principles of prudence and matching.
- The anticipated final value – this will require the commercial QS to anticipate future losses and unforeseen expenditures on the project to ensure that the financial status of the project is not overstated (e.g. allowing for the costs of repairing defects, making allowances for failure to complete on time).

The calculation of the internal valuation figure is where the bulk of work is required to complete the CVR appraisal. Once the internal valuation has been calculated (having made adjustments for prudence and matching) this is compared (reconciled) with the total costs for the project to give a realistic picture of performance on the project. Accordingly, the contractor will report a complete picture on project performance which will comprise an assessment of:

- cash flow
- current profit attained
- potential profit upon completion
- likely cost to complete the project.

Further reading

Potts, K. (2008) *Construction Cost Management: Learning from case studies*, Spon, London.
Towey, D. (2012) *Construction Quantity Surveying: A practical guide for the contractor's QS*, Wiley-Blackwell, Chichester.

13.15 Cash flow

Key terms: Latham Report; Egan Report; payment systems; liquidity problems

Cash flow within contracting organisations has been a significant issue for the construction industry for decades. Arguably, the most well-known judge during this time, Lord Denning, famously stated in 1971 that cash flow is the life blood of the building industry. This has been confirmed in the seminal reports by Latham (1994) and Egan (1998), which further explored the issue of cash flow in construction projects. The Latham report led directly to the Housing Grants, Construction and Regeneration Act (known as the Construction Act 1996), part of which was designed to improve the flow of cash from main contractors to subcontractors. This was achieved by implementing strict payments systems and outlawing clauses that had been written into subcontracts which imposed unreasonable and 'cash unfriendly' contractual terms such as 'paid when paid' clauses.

Nevertheless, the 1996 Act failed to make sufficient progress in solving the problems associated with cash flow and this resulted in further legislation in 2009 (The Local Democracy, Economic Development and Construction Act). The Act further tightened payment systems introduced in the 1996 Act and closed loopholes by outlawing types of contract terms that contractors had developed to avoid passing on due entitlements (e.g. 'pay when certified' clauses).

However, the term cash flow must be seen from two distinct perspectives and should not focus narrowly on the problems that exist in the contractor/subcontractor relationship. First, there must be a clear distinction made between cash flow related to projects and cash required for organisational purposes, i.e. to run a business. Second, one must not ignore the requirements for clients to efficiently fund their projects as the cash flow from client to main contractor is vital for a project. The consultant quantity surveyor will produce cash flow forecasts for the client to inform them of their monetary commitments under a contract. Cash will flow regularly from the client to the construction company over a given period (usually monthly).

Of course, a contracting organisation is faced with cash flow issues on both fronts. Whereas there is clear distinction between project and organisational finance there is a significant overlap – this is particularly true of contractors. If their business is performing well (i.e. they are securing work) there will be a requirement to finance several projects at once. This may result in pressures on the company as payment (cash) is received 'in arrears'. The most common reason for company failure in construction is insolvency – that is the company does not have enough assets to cover its debts or if it is unable to pay its debts as they fall due.

A key financial principle that companies must understand is that cash flow and profitability are two separate things. A company can be profitable, but not have positive cash flow. This principle has been reflected in the type and nature of annual financial statements provided by companies – a key document is the cash flow statement that shows the ways in which cash has been received and spent. Only in the last 30 years has the reporting of cash movements been seen as a crucial measurement of financial performance.

However, a company can have positive cash flow, but not be profitable. A business can fail if it hasn't enough cash (or working capital) in the short term to pay bills. However, a company can't stay in business indefinitely without making a profit.

It is therefore of crucial importance to manage cash flow effectively, whether it be for projects or organisational requirements. This requires effective use of cash flow forecasts that predict cash requirements over a period of time (for example, the financial accounting period of a company).

Management of cash also requires sound financial planning strategies and companies need to be aware of possible sources of finance available should the need for cash arise. There are many sources of finance available to a company and the correct source will depend on a number of factors, for example:

- the time money is needed (is it short, medium or long term);
- source of money (is it to be sourced internally or externally).

Literature varies on the definition of time – however, as a general rule short term is 1 year, medium term is up to 5 years and long term is over 5 years. In terms of source of money a key consideration is that a company retains control of internal financing, whereas a third external party becomes involved and some form of agreement is required with them to receive funding.

For example, a contractor will need to decide whether to borrow money (external) from a bank or to finance from cash earned in running the business (retained profits) – or a mixture of both in the case of multiple projects.

For contractors, the difference between cash and profit is a crucial factor. In 2012, the average profit for the largest 100 construction companies in the UK was 2 per cent . This implies that retained profit is not a regular/reliable source of internal finance for either organisational or project cash. Furthermore, financial analysis literature suggests that a well managed credit control system that includes increasing creditor payment periods and decreasing debtor collection periods, will create positive cash flows for a company. This is possibly the reason for the reluctance for main contractors to pay subcontractors (see discussion above). However, the major contractors in the UK are trying to overcome the cash flow problem by guaranteeing subcontractor payments through banks through supply chain finance initiatives. These initiatives have been supported by government and may prove to be a success and a solution to this long-running problem.

When any company decides that they need cash there are a number of considerations to take into account. They should generally match borrowing to assets held, so long-term borrowing should be used for permanent operating assets (e.g. office accommodation, plant) and short-term borrowing is appropriate for assets affected by seasons and short-term needs (e.g. liquidity problems). In addition, the company may wish to take flexibility into account. For example, short-term borrowing can postpone a commitment requiring a long-term loan. However, it is crucial to recognise that short-term borrowing requires renewing more frequently and can cause problems when the company is constantly in financial difficulty. Finally, it is commonly recognised that interest rates are usually higher on short-term debt as lenders require a higher rate of return. If frequent short-term borrowings are renewed, the financing of a company in this way can work out to be more expensive.

Further reading

Atrill, P. and McLaney, E. (2011) *Accounting and Finance for Non-specialists*, 7th edn, Prentice Hall, Harlow.

Egan, J. (1998) *Rethinking Construction*, HMSO, London.

Harris, F., McCaffer, R. and Edum-Fotwe, F. (2013) *Modern Construction Management*, 7th edn, Wiley-Blackwell, Chichester.

Latham, M. (1994) *Constructing the Team*, HMSO, London.

13.16 Benchmarking

Key terms: Egan Report; Latham Report; drivers for change; targets for improvement

Benchmarking in the construction industry came to prominence with the issue of the Egan report in 1998. The Egan report, alongside the Latham report of 1994, had a significant and far-reaching impact on the construction industry and emphasised the need for the industry to measure its performance and strive to achieve 'world class' standards.

A benchmark has two components. The main focus of the 'Egan' benchmarks relate to the first which is the establishment of a reference/measurement standard that is used for comparison with other companies. The nature of this 'comparison' will depend upon the type of benchmarking being carried out (see below). This results in a 'quantitative' measurement whereby a 'gap' between a company's performance and that of the 'best in class' can be identified. This is the essence of the technique.

However, another vital component in 'pure' benchmarking is the communication of the method of operation which achieves a particular quantitative 'score'. This allows comparison with a specific 'process' (method of working) to understand how improvements can be achieved – it is important to understand how the particular measurement or 'metric' was achieved. This allows an organisation to identify and introduce ways in which innovation can improve performance. It is clear that companies will not openly and willingly offer information about their methods of achieving a particular score – this is central to their company project strategy and part of their intellectual capital. Therefore, the construction industry relies on striving to achieve a metric that places their projects in the upper quartile (top 25 per cent) which is generally regarded as being the desired performance.

The Egan report (called 'Re-thinking Construction') established the '5-6-10' model, which encourages the industry to measure (and thereby benchmark) in key areas of performance. The '5' relates to industry-wide drivers for change which will be achieved by meeting the other key benchmarks in the model. The model further breaks down the drivers into 'project targets' which are seen to be the key areas of continuous improvement required. The 'targets for improvement' are the measurements that are regarded as a minimum requirement. Other metrics may be measured on a project-specific basis to demonstrate progress.

The achievement of the industry as a whole is published in a report entitled *UK Industry Performance Report* on an annual basis, thus providing data for comparison across the industry. Where metrics fail to reach desired level or fail to reach the previous year's results, the whole industry has the opportunity to respond. The core metrics have been extended over the years since the report was released in 2003. For example, additional metrics related to staff loss, CSCS cards (related to Health and Safety), women in construction, disabled people in construction and the age profile of the industry were introduced in 2012. It can be seen from Figure 13.16.1 that most of the metrics in the model relate to project data, but some relate to organisations (i.e. profitability and productivity).

Types of benchmarking

There are many different types of benchmarking available – the 'Egan' benchmarks could be seen as an example of external benchmarking. Other types are:

Internal

Internal benchmarking involves benchmarking operations from within the same organisation (e.g. separate strategic business units). Access to data is easier, less time and resources are needed and change may be easier to manage. However, real innovation may be limited due to the internal focus and world class performance is more likely to be found through external benchmarking – hence the use of 'Egan' targets.

Non-competitive

Competitive or performance benchmarking measures performance relating to key products and/or services of companies in the same sector. In order to protect confidentiality this type of analysis is often undertaken through trade associations or third parties.

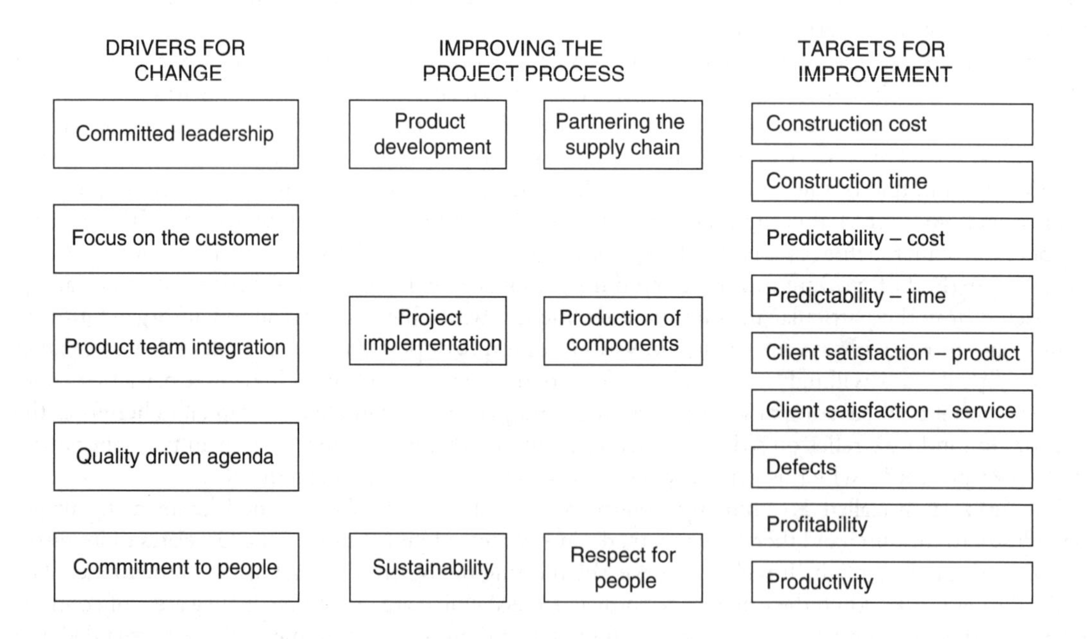

Figure 13.16.1 The benchmarking process.

Source: Centre for Construction Innovation (2004: 8).

Strategic benchmarking involves examining long-term strategies, and is used by successful high performers in order to improve a business's overall performance.

Process benchmarking attempts to improve critical processes through comparison with best performing companies performing similar work.

Functional benchmarking compares partners drawn from different sectors to find innovative ways of improving work processes. This has the potential to lead to dramatic improvements in performance in a short time.

Common pitfalls of benchmarking

It is very easy for companies to benchmarking too many things including things that are 'nice to know' rather than being fundamental to continuous improvement. Therefore, the selection of the items to benchmark is crucial – hence the focus in the construction industry on 'generic' metrics supplemented by those that are more project specific.

In general, benchmarks that are not specific and difficult to measure will not help and will fall into disuse over time. Unless initial improvements are observed companies can give up too early – the technique needs time to show signs of progression. Also, it should be noted that benchmarks should not be fixed as company priorities change over time.

Companies make common mistakes – for example, they assume that benchmarking is a 'one-off' project. To be effective, benchmarking must be seen as a continuous improvement process. However, benchmarking does *not* provide obvious/simple solutions – it provides information upon which to act. Companies should not assume it involves copying or imitating competitors (since the process is partly concerned with understanding measurement and process). Benchmarking should be seen as a process of learning about the organisation and is not a 'quick fix'. The process is often time consuming and labour intensive.

Benchmarking can convey significant advantages to a company. It helps avoid complacency and can improve efficiency and thus profits. Furthermore, the technique can improve client satisfaction levels and at the same time improve employee satisfaction by creating a more efficient, profitable and rewarding environment.

Further reading

Centre for Construction Innovation (2004) 'An Introduction to Key Performance Indicators', Constructing Excellence in the North West, Liverpool.

DETR (2000) 'KPI Report for The Minister for Construction', Published by Department of the Environment, Transport and the Regions.

Egan, J. (1998) *Rethinking Construction*, HMSO, London.

Fewings, P. (2012) *Construction Project Management: An integrated approach*, 2nd edn, Routledge, Oxford.

Latham, M. (1994) *Constructing the Team*, HMSO, London.

13.17 Value management

Key terms: value; value engineering; value planning; value review; brainstorming; functional analysis

Any phrase that includes the word 'value' needs careful interpretation as its definition is somewhat vague and depends on the reader's perception of value. Value is a very subjective judgement in one respect, but can also be seen as a complex entity made up of several different/diverse factors. For example, how scarce is the item under consideration, how useful it is, does it have value in exchange, what is its 'marginal utility'. Marginal utility is the additional satisfaction a consumer gains from consuming one more unit of a good or service. Positive marginal utility is when the consumption of an additional item increases the total utility and negative marginal utility is the opposite of this.

In construction, maximum value has been defined as:

- a level of quality from least cost;
- the highest level of quality from a given cost; or
- an optimum compromise between the two.

From the construction clients point of view, maximum value is the relationship between value and price. This is a key concept in considering value management (VM) as a technique.

There is often confusion in the definitions of related activities concerned with VM. In the UK, VM is a wider term used to describe the overall structured, team based approach to a project throughout the whole of the project lifecycle – i.e. from the concepts stage, through project definition and implementation to the operation of the building after completion on site. VM has component parts which are:

- VE: value engineering is the process of analysing the functional benefits a client required for the whole or part of the design. It is in effect a study of completed designs.
- VP: value planning is applied during the development of the brief to ensure that value is planned into the project from the beginning.
- VR: value review – monitoring the value process and feedback into subsequent areas of work.

Two important techniques employed in VM are brainstorming and functional analysis. Brainstorming is important because the ideas by which value may be increased depend on the creativity of the project participants. The technique depends upon being 'unrestrained' in approach, so there is no limitation to the strange and outlandish suggestions generated. In some ways, the more extreme the better as some creative thinking techniques can be limiting and constrained. The process depends on the ability of the facilitator of the group activity to be able to foster the correct atmosphere. It is also important to recognise that this technique intends to generate ideas and not solutions.

Functional analysis is the second important technique and intends to evaluate objects in a building from the perspective of their function in a building, i.e. what they do, rather than their form, i.e. what they are. Various alternatives that could equally provide that function are then considered. In addition, a breakdown of the functional costs can be carried out and each part individually analysed. An example of this could be a window – is the window required at all? If the window is required, the window is broken down into its constituent parts and the costs analysed – i.e. the materials, opening lights, furniture, glazing etc. are costed to determine where possible value can be adjusted. The key concept at this point is recognising that the concept of value implies that *cost* can increase as well as decrease – the technique is not about cost cutting. It is concerned with meeting the client's value requirements.

However, in the construction industry, the application of VM principles, particularly in the VE phase (evaluation of completed design) has become more of a cost cutting exercise rather than true VE. Also, in construction, due to the complex nature of buildings, the functional analysis of individual building elements would not be cost effective – hence, the reliance on VE in the post design, but pre-construction phase of a project.

VM should be a technique that is applied throughout the project life cycle. However, application of the methodology can be seen by clients as expensive as it requires the involvement of a wide range of project stakeholders over a protracted period of time – in effect the process can be limitless in terms of time, but a well used format is the 40-hour workshop. Reducing this time is often where clients can reduce costs. The success of the methodology is that a facilitator can maintain a careful structure. This is known as the 'job plan' and follows a logical sequence as follows:

1 The information phase – where project information is presented including design work completed at that point in time. This phase could include a functional analysis if deemed cost effective.
2 The creative phase – this is where the core technique of brainstorming is applied to generate ideas for the delivery of the project brief.
3 The evaluation phase – evaluating and testing the idea generated in stage 2. It is likely that most ideas will be rejected, but those that emerge from this stage will have significant support from the project team.
4 The development phase – developing ideas that are feasible and will result in enhanced value However, some ideas may well have a 'knock-on' effect and cause problems to other elements in the building resulting in the requirement for further development.
5 The ideas and presentation phase – presenting the ideas in a fully costed format with clear identification of the benefits that could be realised. The overriding objective of this phase is to gain the client's approval and support for change to the project delivery.

VM is unlikely to be cost effective for small projects, but the technique can be shortened and/or abbreviated to allow potential benefits to be realised. The technique can generate savings on project costs and the involvement of the design and construction teams at various stages can improve briefing and the design process. The key point is that unnecessary project costs can be eliminated.

Further reading

Cartlidge, D. P. (2011) *New Aspects of Quantity Surveying Practice*, 3rd edn, Spon, London.
Walker, P. and Greenwood, D. (2002) *Risk and Value Management*, RIBA Enterprises, Newcastle upon Tyne.

13.18 Risk management

Key terms: enterprise risk management; hazard risks; financial risks; operational risks; strategic risks

Risk management is more readily associated with projects – other sections in this book discuss the ways in which risk may be apportioned between the parties, for example in the choice of procurement route. The choice of traditional or design and build procurement has a key impact on the risk perceived by both parties.

However, before considering project risk, it is important to initially discuss enterprise risk – that is, the systematic approach to managing total risks a company faces in carrying out its day-to-day business. There are a growing number of companies offering associated IT solutions to assist companies in measuring and understanding the corporate risks it takes.

Enterprise risk management (ERM) is a process implemented by a company's board of directors at the very top of the management structure. Risk management must be applied in setting the strategic direction of the company across the whole enterprise in order that potential risky events are identified that might impact on the company so that the company's short- and long-term value to its stakeholders is maintained and developed.

Risks covered by enterprise risk management

The types of risks covered by ERM are as follows:

- hazard risks that have traditionally been addressed by insurers, for example fire, theft, pollution, health and pensions;
- financial risks which cover potential losses due to changes in financial markets, including interest rates, foreign exchange rates, liquidity risks and credit risk;

- operational risks which can cover a wide variety of situations, such as poor customer satisfaction, poor product development, product failure, trademark protection, corporate leadership, information technology, management fraud and information risk;
- strategic risks including factors such as changing customer preferences, technological innovation and regulatory or political change.

ERM relates to the corporate objectives of a company and emanates from the idea that 'outcomes' (i.e. actual performance) differ from the planned objectives. Failure to achieve the corporate plan can result from external factors, such as changes in the marketplace, new entrants into the market, changing client requirements, changes in the economy and money markets, and changes in the political, legal, technological, demographic and other environments that affect the company's business.

In addition, the corporate plan may not be achieved due to internal factors such as human error, fraud or systems failure.

Accordingly, it is logical that ERM is a 'top-down' process where companies will alter and adapt their strategies to lower risk. The overriding duty of the company is to its shareholders and to maintaining shareholder value. To achieve this the 'agents' (i.e. the management of the company) need to set objectives that match those of shareholders and this needs to reflect sound management of risk. In publicly quoted companies risk is a perception by the stock market and quite often high risk can lead to high rewards.

As stated above, insurance used to be the predominant method. However, companies have moved away from this due to advances in the ability to understand and prevent insurable risks (i.e. they could be managed). Sophisticated prevention and control systems have reduced the impact of corporate risks. Companies have retained risks and managed them (rather than insuring) and other mechanisms of ERM are now used. These include legal mechanisms – e.g. limited liability status, outsourcing of non-core activities such as information technology which reduces the risk of technical obsolescence. In general, more knowledge/competence of the business implies risks are lowered – of course the opposite also applies.

Link to project risks

For companies that operate in the construction industry there needs to be a clear link to projects as this is the predominant method of earning income and making profits. In ERM the 'top-down' approach required implies passing down some responsibility for management of risk to people closest to the point where risk arises, i.e. projects. This leads to the consideration of risk management at the project level which has separate methods of assessing risk, but the culmination and aggregation of total project risk needs to be managed via ERM at a corporate level as shown in Figure 13.18.1.

There are clear links between the two and yet the processes involved are somewhat different. For example, health and safety performance on site is a clear project risk, but should a serious accident

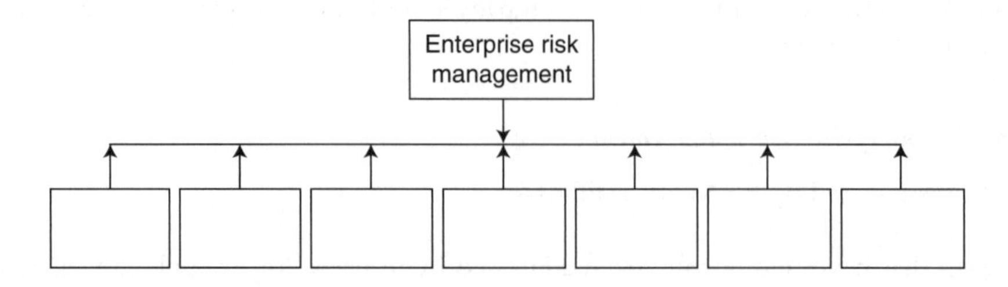

Figure 13.18.1 Enterprise risk management.

Source: Glenn Steel (2013).

occur there will a significant 'knock-on' effect on the company's ability to attract future work, particularly if the event causes a fatality.

Project risk management

There is a significant literature in this field with many separate definitions. The key issue in construction is whether or not a contractor can manage risk effectively without allowing too much within competitive bids, thus rendering the company uncompetitive. It is good practice to measure and allow for risks in a bid, but a contractor will understand that by adding money into a competitive estimate they are relying on their competitors doing the same thing, thus 'equalising' the market. However, an assessment of risk is a measure of uncertainty and is based on the probability that something will occur. However, if it *does* occur, what *impact* would it have on the project objectives and would this be a negative impact?

Certainly, the traditional view of risk has strong *negative* connotations. However, this perception is now seen as too narrow and needs to be 'balanced' with a view that with risk comes potential 'upside' opportunities. An illustration of this is in PFI procurement. The contractor is given the opportunity to reflect risk for many issues that the client is 'averse' to – i.e. the client wishes for the contractor to accept the risks on their behalf and include allowance in the costs of the project for which the public sector client pays a unitary charge or rent. It is an opportunity for the contractor to make a return by accepting and managing the client's risk effectively so risk becomes an investment opportunity.

Risk and opportunity management

The core processes involved in risk management are:

- risk identification which involves determining risks that are likely to affect the project and categorising risk characteristics;
- risk estimation including evaluation of risks and assessing possible project outcomes;
- risk response, planning and execution involving defining, developing and executing steps to enhance opportunities and responses to threats.

This leads to a risk management strategy which specifies the most important risks and how to manage them. The options for dealing with each risk are evaluated in turn against the following options:

- shrink/reduce – can be shrunk or reduced by, for example, establishing more information about a risk situation;
- accept/retain – can be accepted by a party as unavoidable and as being impossible to adopt an alternative strategy;
- distribution/transfer – may be distributed to another party, e.g. subcontractor;
- elimination/avoidance – possibly avoid the project or a particular element of the project.

The quantification of risk is the joint function of two items, likelihood (or probability) and impact. BS4778 confirms that, as it states that risk is a combination of the probability of the occurrence of a defined hazard and the magnitude of the consequences of the occurrence. This leads to the concept of a risk matrix that provides a tool for communicating the overall risk picture.

With reference to Figure 13.18.2, situation 1 is the least worrying and it may be appropriate to ignore the risk. In situation 2, risks are often allowed for by contractors by including some contingency allowance or 'risk pot' to have money available for handling the consequences of risk. Situation 3 is typically dealt with by insurances as this relates to instances that are rare but can be damaging – similar

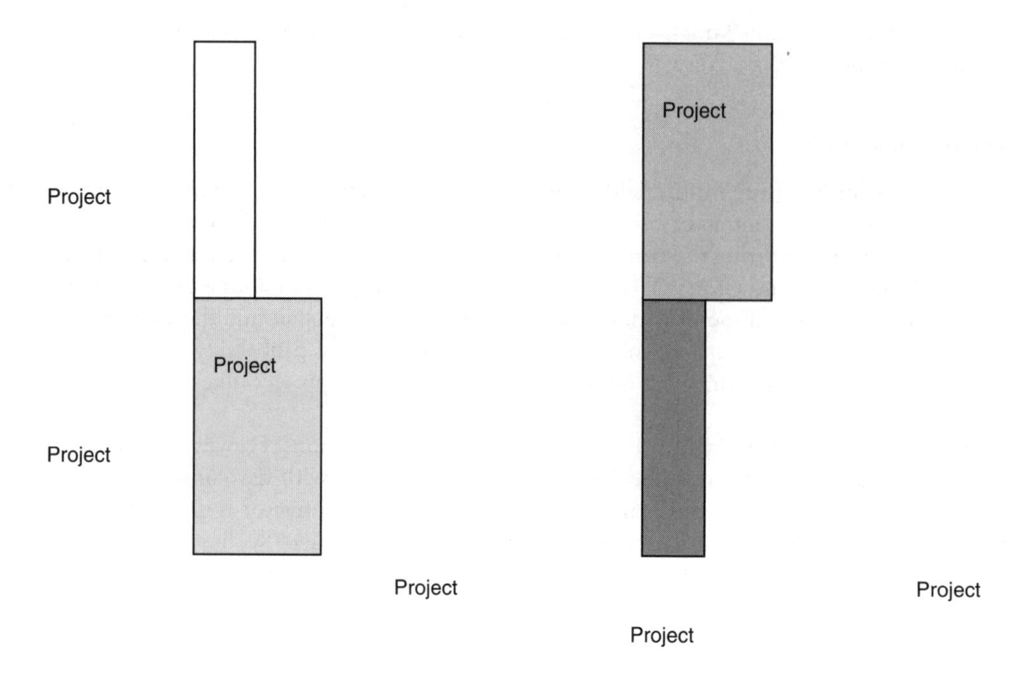

Figure 13.18.2 Quantification of risk.

Source: Adapted from Walker and Greenwood (2002: 80).

in nature to domestic house insurance. Situation 4 comprises risks of greatest concern and may be sufficient to warrant a risk avoidance strategy by refusing the project.

Further reading

Kahkonen, K. (2006) in Lowe, D. and Leiringer, R. (eds) (2006) 'Management of uncertainty', *Commercial Management of Projects: defining the Discipline*, Blackwell, Oxford, ch. 10.

Walker, P. and Greenwood, D. (2002) *Risk and Value Management*, RIBA Enterprises, Newcastle upon Tyne.

13.19 5D building information modelling

Key terms: 3-dimensional building models; BIM manager; BIM maturity; supply chain integration; time as fourth dimension; cost as fifth dimension

Building information modelling (BIM) is a fast developing technique for introducing collaborative strategies into the construction industry. BIM is central to the Government's construction strategy which includes a BIM task group. The government strategy provides for all public sector projects to be using BIM by 2016 with the objective of improved public asset performance in cost, value and carbon performance. A small sample of PFI projects has demonstrated how BIM can improve the management of projects throughout their life cycle.

At a strategic level, BIM may be the catalyst that results in the industry working collaboratively where other major initiatives have failed. The private sector is gradually realising the benefits of BIM and major engineering contractors are rapidly developing expertise in the field.

What is BIM?

BIM uses software to create a 3-dimensional model of a building that can be changed to test a proposed building and the professional design team can visualise the way the real building would look should certain designs and changes be implemented. At the design stage of the project, it is necessary for the whole design team to 'pool' their designs into a single 3D model so that all relevant design information is shared, thereby allowing an integrated and collaborative approach to be adopted. The traditional mode of operation in the industry makes the sharing of a consultant's 'intellectual property' a difficult hurdle to overcome, but compulsory for the technique to be successful. One of the main questions posed as the methodology develops is who takes control of the model once complete? Some organisations are publishing BIM 'protocols' to assist in such decisions (e.g. RICS). The role of a (so called) BIM manager will be a crucial function in the development of BIM.

Designers can use BIM to explore alternative concepts, conduct value engineering and optimise their designs. Contractors can use a BIM model to 'rehearse' construction, coordinate drawings and prepare shop and fabrication drawings. Owners can use the model to optimise building maintenance, renovations and energy efficiency, as well as to monitor life cycle costs. BIM enables collaboration among designers, constructors and owners in ways the construction industry has never known before.

Levels of BIM maturity

BIM can be used at different levels of integration and sophistication. In the UK construction industry. Level 2 is the current standard that major contracting organisations have reached. However, the UK is behind the progress made in other countries. For example, Scandinavia and North America are ahead of the UK. In 2007, Finland and Denmark mandated BIM use on all public sector projects.

The recognised levels of BIM maturity are as follows:

- Level 0 – unmanaged CAD, in 2D, with paper (or electronic paper) data exchange.
- Level 1 – managed CAD in 2D or 3D format with a collaborative software used to provide a common data environment.
- Level 2 – a managed 3D environment held in separate discipline 'BIM' tools with data attached. This level of BIM can utilise 4D construction and/or 5D cost information. This level is recognised as providing a significant challenge to industry incumbents. The Government Construction Strategy report provides for industry to achieve Level 2 BIM by 2016.
- Level 3 – a fully integrated and collaborative process enabled by 'web services' and compliant with emerging Industry Foundation Class (IFC) standards. This level of BIM will be able to utilise 4D construction sequencing, 5D cost information and 6D project life cycle management information.

Benefits of BIM

There are many possible uses of BIM on construction projects. However, to exploit the full advantage of BIM's capabilities, a collaborative approach needs to be adopted with the full integration of the supply chain and the information that they provide.

The advantages of the 3D model can be seen as the following:

- Model walkthroughs allow problems to be resolved.
- Traditionally designed drawings must be coordinated to assure that different building systems do not clash and can actually be constructed in the allowed space. BIM allows clash detection to be achieved before design completion.
- Project visualisation allows the client and planning authorities to observe the appearance of the final building.

- The level of construction information in a BIM model means that prefabrication can be used with greater assurance.

The fourth dimension of BIM relates to time and applying 4D would allow improvements in construction planning and management e.g. the contractor can visualise crane locations and traffic access routes and visualisation of design changes.

The fifth dimension of BIM relates to cost that allows automated quantity take-offs to generate essential financial management information. In addition, cost data can be added to each BIM object to enable the model to automatically calculate a rough estimate of material costs.

The sixth dimension features facilities management which allows life cycle cost management by turning the model into an 'as built' model, which is provided for the owner.

Barriers to implementation

BIM has the potential to improve the process of designing, constructing and operating buildings. However, despite the clear vast potential there are barriers to its implementation as follows:

- Different BIM models in the initial stages means there will not be a common technology in use.
- Current technologies and levels of BIM adoption do not yet allow seamless coordination between different BIM models.
- Some firms may not be able to afford BIM and therefore traditional drawings will still have to be produced to communicate down the supply chain. It is also likely that smaller subcontractors will not have access to the digital BIM model on site.
- The ownership of the BIM model data is not clear at the moment and different legal problems may be faced in relation to each type of information, e.g. exposure to design liability.
- Provisions on the input of data, limitation of liability and defining the role of any BIM 'model manager' will be crucial in the adoption of BIM.
- The risk profile for construction projects and project participants will change with use of a BIM model and a collaborative, integrated approach. This implies that contract conditions will need to be changed to fall in line.
- BIM methodology will completely change the way that designers approach the design process. It is likely that more hours will be spent on the design and less in production.
- BIM places increased reliance upon information technology. With this reliance comes the need for specific support and security precautions.
- BIM models will require significant investment in hardware, software and staff training across the industry.

Further reading

Isurv (2013) 'What is BIM?' available at: www.isurv.com/site/scripts/documents_info.aspx?documentID=7216&categoryID=1246 (accessed 13/12/2013).

13.20 Expert witnesses

Key terms: tribunal arbitration; official enquiry; Civil Procedure Rules; expert immunity

There are two kinds of witness. They are witnesses of fact and expert witnesses (EWs). An EW produces expert evidence. Once material facts in issue have been identified the parties aim to prove (or disprove) those facts to the satisfaction of the court.

An expert can be anyone with knowledge of, or experience in, a particular field or discipline beyond that to be expected of a layman. An EW is an expert who makes this knowledge and experience available to a court (or other judicial bodies, e.g. tribunals, arbitrations, official enquiries, etc.) to help it understand the issues in a case and thereby reach a sound and just decision

Moreover, an EW is paid for the time it takes to:

- form an opinion; and, where necessary,
- support that opinion during the course of litigation.

It's very important to keep these definitions clear and in focus. If an individual strays from acting as an EW into advising the client – and thus becomes an expert adviser – this significantly changes their legal position.

English law has developed detailed rules to govern the use of expert evidence in civil proceedings. The basic requirements that affect the use of expert evidence are set out in the Civil Procedure Rules 1998 Rule 35 – Experts and Assessors, implemented in April 1998.

The Royal Institution of Chartered Surveyors (RICS) has published a practice statement (*Surveyors Acting as Expert Witnesses*, third edition, 2011) on surveyors acting as EWs.

The English legal system

Surveyors will often find that they are the first point of contact for a client with a problem and need to understand the framework in which they are operating in order to give clients the best advice about the options that are available to them for dealing with a potential dispute. Surveyors should never forget that the best advice they can give a client contemplating resolution of a dispute is to take early legal advice.

There are two types of cases within the English legal system – criminal cases and civil cases, which involve an individual, business or organisation taking action against another, usually to obtain money or other legal redress.

The system remains adversarial, although the Civil Procedure Rules have introduced reforms emphasising that legal action is to be regarded as a last resort and introducing requirements to undergo mediation processes before, if such processes should fail, full court appearances.

What is expert evidence?

The fundamental characteristic of expert evidence is that it is opinion evidence. Generally speaking, lay witnesses may give only one form of evidence, namely evidence of fact. To be practically of assistance to a court, however, expert evidence must also provide as much detail as is necessary to allow the court to determine whether the expert's opinions are well founded. It follows, then, that it will often include:

- factual evidence supplied in the expert's instructions that requires expertise in its interpretation and presentation;
- other factual evidence that, while it may not require expertise for its comprehension, is linked inextricably to evidence that does;
- explanations of technical terms or topics;
- hearsay evidence of a specialist nature, e.g. as to the consensus of medical opinion on the causation of particular symptoms or conditions; as well as
- opinions based on facts presented in the case.

Expert evidence is needed when the evaluation of the issues requires technical or scientific knowledge only an expert in the field is likely to possess.

Pre-1999 position

Prior to the publication of the Civil Procedure Rules 1998, there were considerable misgivings in the courts as to the function of EWs in the courts. This came about because each party to a dispute appointed its own expert and a tendency developed for EWs to present their client's case in the best possible and uncritical light. Hence, lawyers would refer to 'hired guns' and attack experts for their poorly concealed bias and lack of impartiality. Mr Justice Laddie stated: 'An expert should not consider that it is his job to stand shoulder to shoulder through thick and thin with the side that is paying his bill.'

Immunity of the expert (negligence) – prior to 2011

The protection of witnesses from civil claims arising out of their evidence has (subject to certain exceptions) been embedded in the English legal system since at least 1873. The immunity of advocates (an advocate can be defined as 'a person who puts a case on someone else's behalf') at trial was considered and removed by the House of Lords in *Hall v Simons* (2002) but this case reaffirmed the existence of the immunity afforded to EWs. So prior to 2011 an expert who failed to make sure that their oral evidence in court was properly prepared could not be sued for negligence.

Such immunity was founded on the principle that evidence that is put before a court, whether by way of report or oral testimony, should be given freely and without fear of reprisal. The effect of this is that, in contrast to other work they might carry out, EWs were not exposed to claims for breach of contract and/or negligence in respect of the evidence that they give to the court. Immunity from being sued was argued to be one of the reasons why experts were perhaps willing to give evidence in difficult or controversial cases.

If such immunity was removed some experts may no longer choose to give evidence for fear of later being sued because the result is not to their client's satisfaction. However, in relation to civil proceedings, indications started to come from the judiciary that, where professional EWs offer their services in return for payment, it would be increasingly difficult to justify the continued retention of immunity from claims for breach of contract and/or negligence.

Expert immunity – after 2011

The court in the case of *Jones v Kaney* reviewed the immunity of the EW. As mentioned above, there had been much disquiet in courts about the role of the EW and the perceived support by the expert of the 'bill payer' rather than providing expert opinion on complex issues in dispute.

In the case of *Jones v Kaney*, the Supreme Court resolved the matter and EWs can now be sued for providing negligent expert evidence just as they can be sued for negligently providing any other service. It is, therefore, vital that anyone acting as an EW obtains personal indemnity insurance to protect themselves against the risk of claims from their clients.

An expert who is not seen to be independent and even-handed in his evidence will be of little value to his client and may injure his client's case. The real cost of a biased expert may be many times the cost of their fee.

Further reading

Horne, R. and Mullen, J. (2013) *The Expert Witness in Construction*, John Wiley & Sons, Chichester.
RICS (2011) *Surveyors Acting as Expert Witnesses*, 3rd edn, RICS, Coventry.

14 Regeneration

Julie Clarke, Hannah Furness, Paul Greenhalgh, Rachel Kirk and David McGuinness

14.1 Defining urban regeneration

Key terms: urban regeneration; neighbourhood renewal; area-based initiatives; multiple deprivation

Regeneration is a term that is widely misused and misunderstood. It means different things to different people, and can range from large-scale activities that promote economic growth, to neighbourhood interventions that improve people's quality of life. Fundamentally, it is a holistic process of reversing economic, social and physical decline in areas where market forces alone will not suffice. Holistic regeneration may be conceptualised as a stool, the three legs of which represent physical, economic and social aspects, without any one of which the stool will fall over.

Urban regeneration as an idea encapsulates both the perception of a town or city in decline and the hope of renewal, reversing trends in order to find a new basis for economic growth and social well-being. It may be defined as a comprehensive and integrated vision and action that leads to the resolution of urban problems and that seeks to bring about a lasting improvement in the economic, physical, social and environmental condition of an area that has been subject to change (Roberts and Sykes, 2000).

Physical regeneration is work on the physical fabric of an area where such work forms part of a strategy to promote social, physical and economic improvements in a given locality, rather than just redevelopment driven solely by market forces (Commission for Racial Equality, 2007). It is necessary to clearly distinguish physical regeneration activity from redevelopment; the latter has arguably been occurring ever since urban settlements existed. It is the process of recycling or reusing land that has already been developed, and may involve demolition, land remediation and reclamation, rebuilding, rehabilitation, conversion and change of use etc. To be considered to be physical regeneration, a project must have some public sector input or intervention that results in it making a contribution beyond the delivery of profitable or worthwhile property development.

Urban policy is the collection of policies and programmes introduced by, sometimes, successive governments, to tackle social, economic, physical and environmental problems in urban areas. Urban policies are often targeted at areas of greatest need, as identified by a weighted combination of socio-economic indicators of poverty and deprivation. Area-based initiatives (ABIs) are spatially targeted programmes of intervention intended to reverse the process of decline, alleviate poverty and promote a process of transformation in the fortunes of an area. Such initiatives are long term, taking anywhere from 10 to 20 years to deliver measurable improvements.

Further reading

Commission for Racial Equality (2007) *Regeneration and the Race Equality Duty*, CRE, London.

Communities and Local Government (2008) *Transforming Places, Changing Lives: A framework for regeneration*, DCLG, London.

Roberts, P. and Sykes, H. (2000) *Urban Regeneration: A handbook*, Sage, London, ch. 2.

14.2 Development corporations and regeneration agencies

Key terms: Urban Development Corporations; Urban Regeneration Companies; development agencies

Development corporations have been used in a variety of countries to pursue urban renewal and redevelopment. Examples include in the US where most major cities have a development corporation to administer funding programmes (see separate concept); in Germany there are Stadtebauliche Redevelopment Agencies; in the Netherlands the Rotterdam Development Agency was responsible for facilitating the Kop van Zuid project; in France Agences d'Urbanisme et Aménagement du Térritoire have been responsible for large infrastructure projects, for example in Lille (see Couch *et al.*, 2003). While there is considerable variation between the institutional arrangements and performance of development corporations around the world, they do have some common characteristics:

- remit to pursue redevelopment of land and property;
- assembling of development sites (often using their own compulsory purchase powers);
- use of capital funding to 'pump prime' development sites;
- reclamation and remediation of derelict and contaminated 'brownfield' land;
- masterplanning of strategic developments;
- investment in infrastructure, services and utilities;
- de-risking of development to lever in private sector investment and development;
- operating at arm's length from central or local government.

The UK has a long and rich history of the use of development corporations to pursue the above activities. The Scottish and Welsh Development Agencies were created in the mid 1970s with remits to pursue economic development with a particular focus on attracting inward investment. In England the New Town Development Corporations had long been established to build out the new towns' programme of the 1960s and 70s. The step change occurred with the 1980 Local Government Planning and Land Act by the Conservative Government, which gave the Secretary of State the power to create Urban Development Corporations (UDCs) to secure the regeneration of their (urban development) areas by:

- bringing land and buildings back into effective use;
- encouraging the development of existing and new industry and commerce;
- creating an attractive environment;
- ensuring that housing and social facilities are available to encourage people to live and work in the area.

(Section 136 Local Government Planning and Land Act 1980)

Fourteen UDCs were created in four phases between 1980 and 1990. Phase 1 created the pioneering London Dockland and Merseyside Development Corporations, partly in response to the urban riots that had swept like wildfire across the country the previous year. The second and largest phase, in 1987, created Teesside, Tyne and Wear, Trafford Park, Black Country and Sheffield Corporations; phase 3, a year later, comprised smaller corporations in Central Manchester, Leeds and Bristol; the final phase in 1990 created corporations in Birmingham and Plymouth. UDCs were also designated in Laganside (Belfast) and Cardiff Bay; there were none in Scotland. It may be observed that the pattern of designation is dominated by the large metropolitan cities that were almost exclusively controlled by Labour authorities.

UDCs were given wide ranging powers to acquire, hold, manage, reclaim and dispose of land and property, carry out building, run a business and do anything necessary or expedient for the purposes incidental to those purposes. They could take over land held by public bodies (vesting) or use their compulsory purchase powers to acquire land from private individuals against their will. They not only had statutory planning powers, but took responsibility for building control functions, provided mortgages, loans, grants and could even take over housing authority and public health functions. UDCs were controversial for a variety of reasons, not least because they were generously funded at a time when local authorities were being rate capped by the Conservative administration. They took over the local planning authorities' development control powers for their urban development area; they were not accountable to the general public, only to the Secretary of State for the Environment, who appointed the Chief Executive, Chair and board members; they pursued their remit in such a narrow and single-minded way that the local population were often the last to benefit from their activities; and finally, they gave lie to the notion that the basis for urban revitalisation lies in the market and entrepreneurship.

UDCs were successful at delivering end results in terms of the remediation of brownfield sites, investment in highways and other infrastructure, construction of commercial and industrial floorspace and residential units, often in mixed use projects. This was only achieved at considerable expense (London Docklands Corporation alone was estimated to have cost the taxpayer in excess of £3 billion at 1996 prices) and was sometimes at the expense of the local residents who were often the last to benefit from the 'trickle-down' effect.

Most UDCs were wound up in 1998, by which time they were regarded by many as rather anachronistic. The UK has entered a new era of consensual and holistic regeneration pursued through a partnership between public and private sectors and local communities. The replacement for the UDC was the Urban Regeneration Company (URC), the model for which is compared with the UDC model in Table 14.2.1.

Table 14.2.1 Urban Development Corporations and Urban Regeneration Companies

Origin	1980 Local Government Planning and Land Act, Part XVI	Urban Task Force 1999 Urban White Paper 2000 Guidance and Principles 2001
Statutory	Yes	No
Number	14 (none in Scotland)	24 (across the UK)
Life-span	5–17 years	In theory 10–15 years
Aims	1 To bring land and buildings back into effective use 2 Create environment conducive for private sector investment 3 Encourage development of new commerce and industry 4 Ensure housing and social facilities are available	1 Coordinate and deliver holistic and sustainable regeneration 2 Engage the private sector in an agreed physical and economic regeneration strategy 3 Work with key partners to deliver employment opportunities 4 Work with key partners to link activity to local communities
Staffing and use of consultants	Medium sized in-house establishments with moderate use of consultants	Small in-house teams with great reliance on use of external consultants
Politics and governance	Usurped local authorities	Local authorities a key partner
Appointment of board/exec	By Secretary of State	By local authority and Regional Development Agency; approved by SoS
Relationship with LA	Completely autonomous	A key partner but operating at arms length
Funding	Generously funded	No dedicated funding; rely on partners for funding/resources
Planning powers	Development and building control powers in Urban Development Area	Rely on Local Planning Authority to use powers to assist their work
CPO powers	Full powers used extensively	Key partners exercise on their behalf (RDA, English Partnerships and local authority)
Vesting of publicly owned land	Common	Control of land often retained by local authority or other public body

Perhaps because of their consensual modus operandi and lack of funding and powers, the URCs were not as dynamic and successful as their forbears. Many have subsequently been wound up due to lack of funding or have morphed into economic development companies.

Further reading

Couch, C., Fraser, C. and Percy, S. (2003) *Urban Regeneration in Europe*, Blackwell, Oxford.

Deas, I., Robson, B. and Braidford, M. (2000) 'Re-thinking the Urban Development Corporation experiment: the case of Central Manchester, Leeds and Bristol', *Progress in Planning* 54(1): 1–72.

Foster, J. (1999) *Docklands: Cultures in conflict, worlds in collision*, Routledge, London.

Imrie, R. and Thomas, H. (2000) *British Urban Policy: An evaluation of Urban Development Corporations*, Sage, London.

National Audit Office (1993) *The Achievements of the Second and Third generation Urban Development Corporations*, HMSO, London.

Parkinson, M. (2008) *Make No Little Plans: The regeneration of Liverpool City Centre 1999–2008*, Liverpool Vision, Liverpool.

14.3 Neoliberal urban policy

Key terms: laissez-faire; Urban Development Corporations; Enterprise Zones; Michael Heseltine

The election of Margaret Thatcher as Prime Minister in 1979 ushered in a new era of urban policy, apparently based on neoliberal ideologies of deregulation and competition that manifested themselves in relaxation of planning controls and a more laissez-faire regulatory approach to encourage private sector enterprise and entrepreneurialism. Somewhat ironically, the first two significant tools of regeneration introduced by the Thatcher Government were on the one hand positively Stalinist (Urban Development Corporations), while also causing significant distortion in the operation of the free market (Enterprise Zones – EZs).

The prevailing situation for the introduction of UDCs and EZs was a malaise in Britain's inner cities, culminating in 1981 in the worst rioting the UK had seen for decades, most notoriously in Toxteth in Liverpool, Handsworth in Birmingham, St James' in Bristol and Brixton in London. The Government had to be seen to act decisively and Michael Heseltine, the Secretary of State for the Environment, grasped the nettle, symbolically landing in a 'war-torn' Toxteth by helicopter and promising to tackle Britain's failing inner-cities.

The main pieces of legislation used to introduce these new 'property-led' tools of regeneration were the Local Government Planning and Land Act and Finance Acts of 1980, Part XVIII, Schedule 32 and Part XVI, Schedules 26–9 of which contained provisions for EZs and UDC respectively. Fourteen Development Corporations were created across the UK with the exception of Scotland, in four phases between 1980 and 1990. These were complemented by the designation of 32 EZs across the whole of the UK between 1980 and 2006. The Conservative Government originally regarded EZs as the remedy to Britain's urban problems but by the mid-1980s it became apparent that this fiscal tool lacked the political traction to really make much difference at the local level, provoking the Government to change tack and embark on creating a second round of UDCs in most major cities in England. It was no coincidence that all the cities in which these QUANGOs (quasi-autonomous non-governmental organisations) were created were controlled by Labour local authorities.

The 1980s is now regarded as one of the most dynamic and controversial periods in British urban policy, in part due to the plethora of policy initiatives that were introduced during a short space of time (see Table 14.3.1).

Table 14.3.1 1980s policy initiatives

Year	Initiative
1980	Local Government Planning and Land Act and Finance Act gives powers to the Secretary of State for the Environment to create UDCs and designate EZs
1981	Inner city riots 13 EZs designated London Docklands and Merseyside Development Corporations established Urban Regeneration Grant introduced Simplified Planning Zones established
1982	Derelict Land Act and Grant introduced 14 more EZs designated Urban Development Grant introduced
1983	Conservative Government re-elected
1984	First Garden Festival in Liverpool New Assisted Areas
1985	5 City Action Teams created Inner City Enterprises set up Estate Action established
1986	8 Inner City Task Forces established Second Garden Festival in Stoke on Trent
1987	5 more UDCs created 8 more Task Forces established
1988	Action for Cities Launched City Grant introduced to replace Urban Development and Regeneration Grants 4 more 'mini' UDCs created Third Garden Festival in Glasgow 2 more City Action Teams
1989	Local Government Housing and Finance Act which legislated for the creation of City Challenge Partnerships 3 more Task Forces established

In practice this experimental policy cocktail brought about greater centralisation of political power and funding, encouraged the usurping of local authorities through the introduction of QUANGOs, and manifested itself in limited-life, property-led, area-based initiatives that favoured capital-led projects and encouraged greater private sector involvement. The consequences of this short-term approach was that the economic base of inner city areas continued to erode as there was actually *less* money going into them due to rate capping. What financial incentives that did exist tended to benefit investors and developers, increasing private sector investment through large-scale 'flagship' developments that caused significant levels of displacement and relocation. The lack of an overarching strategy meant that policies were not 'joined up' and were thus failing to relieve deep-seated poverty, unemployment and deprivation (see Atkinson and Moon, 1994).

Even before the decade was out, serious concerns were being expressed about the direction and impact of urban policy in the UK, most notably by the Archbishop of Canterbury's Commission on Urban Priority Areas, which published *Faith in the City: A call for action by church and nation* in 1985, which was critical of the way that the urban poor had been marginalised in society and questioned whether the political will existed to set in motion a process to tackle this social exclusion. By the end of the 1980s it was apparent to most commentators that Government urban policy was not working. The Government's own Audit Commission produced a report in 1989 famously describing urban policy as a 'patchwork quilt of complexity and idiosyncrasy', its performance being undermined by a fragmented and incoherent array of policies. The notion of trickle down, on which many of the policies were predicated, was increasingly regarded with scepticism and doubt. Most damningly, it appeared that the people who were in most need of help were least likely to benefit from the policies.

There was an urgent need for change and it came in the shape of Michael Heseltine, who returned to the DoE after some years in the political wilderness and immediately performed a complete '*volte-face*' by re-orientating urban policy to concentrate again on local solutions for local people in partnership with locally accountable and democratically elected bodies – namely local authorities.

Further reading

Archbishop of Canterbury's Commission on Urban Priority Areas (1985) *Faith in the City: A call for action by church and nation*, Church House, London.

Atkinson, R. and Moon, G. (1994) *Urban Policy in Britain: The city, the state and the market*, Macmillan, London.

Brownhill, S. (1990) *Another Great Planning Disaster*, Paul Chapman, London.

Healey, P., Davoudi, S., O'Toole, M., Tavsanoglu, S. and Usher, D. (eds) (1993) *Rebuilding the City: Property-led regeneration*, Spon, London.

Imrie, R. and Thomas, H. (1993) 'The limits of property-led regeneration', *Environment and Planning C* 2(1): 87–107.

Lawless, P. (1989) *Britain's Inner Cities*, Paul Chapman, London.

MacGreggor, S. and Pimlott, B. (1991) *Tackling the Inner Cities*, Clarendon, Oxford.

Robson, B. (1988) *Those Inner Cities: Reconciling the social and economic aims of urban policy*, Clarendon, Oxford.

Smyth, H. (1994). *Marketing the City: The role of flagship developments in urban regeneration*, Spon, London.

Turok, I. (1992) 'Property-led urban regeneration: Panacea or placebo', *Environment and Planning A* 24(3): 361–379.

14.4 Compact cities and urban sprawl

Key terms: compact city; urban sprawl; densification; urban development; suburbanism

The compact city movement originated in northern Europe during the inter-war period as a response to urban sprawl and suburban development. Although the term 'compact city' creates the impression of a single, coherent concept, this is somewhat misleading as there are a variety of notions of a compact city and many of different models of varying significance and influence that seek to promote more compact and sustainable urban development. However, there is one core principle that is common to all models: increased intensification of urban land use predominantly by way of residential densification. In order to successfully counter urban sprawl, densification strategies need to go hand in hand with a policy of urban containment.

Urban sprawl is the excessive spatial growth of cities, characterised by a widely dispersed population in low-density residential development, rigid separation of homes, shops, and workplaces, a lack of distinct and thriving city/town/suburban centres and poor interconnectivity between localities. According to Downs (1999) sprawl has ten traits:

1. Unlimited outward extension of new development.
2. Low-density residential and commercial settlements.
3. Leapfrog development beyond established settlements.
4. Fragmentation of powers over land use among many small localities.
5. Dominance of transportation by private automotive vehicles.
6. No centralised planning or control of land uses.
7. Widespread strip commercial development.
8. Great fiscal disparities among localities.
9. Segregation of specialised types of land uses in different zones.
10. Reliance mainly on trickle-down to provide housing to low-income households.

A significant characteristic of urban sprawl is the inefficient use of land and resources, the consequences of which are environmental degradation and pollution, particularly from vehicle emissions due to increased dependence on automobiles, the loss of rural land and the impairing of urban vitality and diversity. Thus, compact urban form should reduce congestion, pollution, carbon emissions and development pressure on greenfield land.

Compact urban form highlights the value placed upon proximity and ease of contact between people. It gives priority to the provision of public areas for people to meet and interact, to learn from one another and to join in the diversity of urban life (DETR/UTF, 1999). Compact city models typically involve planning for new housing at higher densities than may have been previously considered appropriate. It also involves raising the densities of existing areas of housing through infill development or retrofitting of current structures and plots. Table 14.4.1 presents residential densities measured by dwellings per hectare (ha) and persons per ha for a range of different UK housing types and locations.

Proponents of compact city models believe that by developing at higher densities, urban areas may avoid the problems associated with low density housing and urban sprawl. A compact urban form typically reduces the distance that infrastructure and utilities need to cover, which in turn reduces the cost of providing such infrastructure (this may be compared with 'ribbon development' that is symptomatic of suburban development). Other benefits of urban densification are improved security, a reduction in derelict land. Thus, a compact city should encourage and promote urban regeneration, the revitalisation of town centres, construction at higher densities, mixed use development, enhanced public transport provision and the concentration of urban development at public transport nodes, consequently reducing urban sprawl.

Table 14.4.1 UK residential density gradient

UK residential density gradient	*Units/ha*	*Persons/ha*
Low density detached – Hertfordshire	5	20
Milton Keynes average 1990	17	68
Average density of new development in UK 1981–91	22	88
Minimum density for a bus service	25	100
Private sector 1960s/70s – Hertfordshire	25	100
Inter-war estate – Hertfordshire	30	120
Private sector 1980s/90s – Hertfordshire	30	120
Hulme – Manchester 1970s	37	149
Average net density London	42	168
Ebenezer Howard – Garden city 1898	60	240
Minimum density for a tram service	60	240
New town higher density low rise – Hertfordshire	64	256
Sustainable urban	69	275
Victorian/Edwardian terraces – Hertfordshire	80	320
Central accessible urban	93	370
Holly Street – London 1990s	94	376
Holly Street – London 1970s	104	416
Hulme – Manchester 1930s	150	600
Islington – 1965	185	740

Source: Commission for Architecture and the Built Environment (2005).

NB: an average dwelling size of 4 bed-spaces has been assumed throughout this table although it should be noted that this is higher than the average household size in the UK.

According to Gordon and Richardson (1997) there are three generic approaches to delivering more compact urban forms:

1 the 'macro' approach whereby average densities are high across the extent of an urban area;
2 the 'micro' approach which focuses on achieving higher densities at the neighbourhood level;
3 the 'spatial structure' approach, which recognises the role of the city centre as the core around which all other elements of the city revolve.

Compact cities are therefore places that reduce fuel consumption by reducing distances travelled by residents, particularly by private motor vehicles, through the provision of good public transport and encouraging walking and cycling. Such cities also provide residents with improved accessibility to a wide range of employment and services through mixed use development. With an increased number of residents per hectare, investment in frequent and reliable public transport becomes more financially viable, and the excessive journey times associated with sprawling low density cities that people endure commuting between their homes and places of work may be reduced or avoided.

Further reading

Bruegmann, R. (2006) *Sprawl: A compact history*, University of Chicago Press, Chicago, IL.
CABE (2005) *Better Neighbourhoods: Making densities work*, CABE, London.
DETR/UTF (1999) *Towards an Urban Renaissance*, Spon, London.
Downs, A. (1999) *Some Realities about Sprawl and Urban Decline*, Brookings Institution, Washington, DC, p.1.
Ewing, R., Schmid, T., Killingworth, R., Slot, A. and Raudenbush, S. (2008) 'Relationship between urban sprawl and physical activity, obesity and morbidity', *Urban Ecology*, 567–582.
Gordon, P. and Richardson, H. W. (1997) 'Are compact cities a desirable planning goal?' *Journal of the American Planning Association*, 63: 95–106.
Layard, A., Dvoudi, S. and Batty, S. (eds) (2001) *Planning for a Sustainable Future*, Spon, London.
Jenks, M. and Burgess, R. (eds) (2000) *Compact Cities: Sustainable urban forms for developing countries*, Spon, London.

14.5 Shrinking cities

Key terms: industrial decline; depopulation; managed decline; rustbelt cities

From the 1950s to the end of the twentieth century many of the UK's industrial towns and cities outside London suffered significant depopulation often due to deindustrialisation and the demise of traditional heavy industries such as coal mining, steel production and shipbuilding. More recently, a sharp decline in the population of some British towns and cities has been linked to loss of a core employer due in part to increased global competition, the mobility of manufacturing firms, technological advances, cuts in defence spending or social changes such as where people choose spend their holidays. Towns and cities that have lost population have not done so in a uniform way; inner city areas have usually been most acutely affected, often as a consequence of suburbanisation and competitions from new towns, market towns and commuter settlements. Such large-scale migration is often along ethnic lines and has been termed 'white flight' by some commentators.

Shrinking cities may be defined as significant urban conurbations (min. population of 10,000 people) that have suffered continual population decline over a sustained period of time (more than two years) and exhibit signs of structural economic crisis. Shrinking cities are a global phenomenon and much academic research has been focused on the significant European and North America industrial cities of the nineteenth and twentieth centuries that have suffered loss of their key industrial employers and severe economic restructuring resulting in depopulation as mobile unemployed labour migrates to alternative towns and cities in search of new employment opportunities. Such population loss can lead to serious economic, social and environmental consequences, including failing housing markets, land abandonment, the failure of retail centres, vandalism and crime. For local authorities/municipalities, population loss is a serious threat to their economic stability as depopulation results in a decline in tax revenue (from both domestic and commercial rates), which has a corrosive impact on urban infrastructure, housing, schools, social services, the environment, amenities, etc. Towns and cities, once they are in a downward spiral of decline, find it very difficult to reverse their fortunes.

Notable examples of shrinking cities are Detroit and Cleveland in the United States, both casualties of the increased global competitiveness and success of developing countries (e.g. southeast Asian car industry), and the former East German cities of Dessau and Liepzig, which have struggled in terms of economic competitiveness since German reunification. The most notorious US example of the decline of a 'single sector' city is Detroit, previously known at 'Motor City' and home to Motown Records, which has been decimated by the decline in motor vehicle manufacturing resulting in the city recently being pronounced bankrupt and placed into special measures by the US administration. Prominent UK examples of shrinking towns/cities include Liverpool, Blackpool and Middlesbrough.

It should be noted that depopulation of Western conurbations, towns and cities in the mid to late twentieth and early twenty-first centuries contrasts graphically with wider global trends of urbanisation in developing countries, during an era in which urbanisation has continued inexorably such that now well over half the world's population live in cities.

Further reading

Bernt, M. (2009) 'Partnership for demolition: The governance of urban renewal in East Germany's shrinking cities', *International Journal of Urban and Regional Research*, 33(3): 754–769.

Couch, C. and Cocks, M. (2011) 'Underrated localism in urban regeneration: The case of Liverpool, a shrinking city', *Journal of Urban Regeneration and Renewal*, 4(3): 279–292.

Hollander, J., Pallagst, K., Schwartz, T. and Popper, F. (2009) 'Planning shrinking cities', *Progress in Planning*, 72(4): 223–232.

14.6 The urban renaissance

Key terms: Urban Task Force; urban renaissance; gentrification; compact city

The Urban Task Force (UTF) was established in 1998 by the Deputy Prime Minister of the newly elected Labour Government, John Prescott. The UTF, under the chairmanship of the architect Richard Rogers, comprised a group of experts from the public and private sectors, complemented by a number of working groups. Their mission was to:

> identify causes of urban decline in England and recommend practical solutions to bring people back into our cities, towns and urban neighbourhoods. It will establish a new vision for urban regeneration founded on the principles of design excellence, social wellbeing and environmental responsibility within a viable economic and legislative framework.
>
> (DETR, 1999)

The Government's motivation for pursuing the urban renaissance agenda was driven by increasing concern about depopulation of major conurbations in the UK outside London (see Table 14.6.1) and represented a concerted attempt by the British urban movement to pursue a set of long-term and holistic principles that sought to place disadvantaged communities and neglected neighbourhoods back at the centre of the policy arena.

Table 14.6.1 Population change in UK conurbations 1991–2001

	Population 1991 (000s)	*Population 2001 (000s)*	*Change 1991–2001 (%)*
London	6829	7308	7.0
West Midlands	2619	2570	–1.9
Greater Manchester	2553	2512	–1.6
West Yorkshire	2062	2084	1.1
Clydeside	1728	1666	–3.6
Merseyside	1438	1366	–5.0
South Yorkshire	1288	1266	–1.7
Tyne and Wear	1124	1078	–4.1

Source: ONS (1991).

To pursue their mission, the UTF visited not only English cities but also cities in Europe, such as Barcelona and Amsterdam, and in North America. They published their landmark report *Towards an Urban Renaissance*, also referred to as the 'Rogers Report', in 1999, that set out their urban vision for England in five sections (containing 105 separate recommendations):

1 The sustainable city.
2 Making towns and cities work.
3 Making the most of our urban assets.
4 Making the investment.
5 Sustaining the renaissance.

A central premise of the report is that increasing the intensity of activities and people within an area is crucial to creating sustainable neighbourhoods. The report recommended revising planning guidance to promote greater residential densification (see Concept 14.4) by discouraging local authorities from using arguments of excessive density and over-development as reasons for refusing planning permission, creating a planning presumption against excessively low density urban development and providing advice on use of density standards linked to design quality. Such recommendations sought to create a policy, legislative and financial regime that would encourage both developers and local planning authorities to pursue well-designed higher density brownfield development.

One of the more contentious components of the UTF report was its analysis and recommendations around delivery of new housing on brownfield land, the Government target for which at the time was 60 per cent of all new housing to be built on previously developed land. The UTF, in hindsight, wrongly concluded that the target would be difficult to achieve and sustain when in fact it was exceeded within a year and would ultimately peak at 80 per cent. However, the debate around the potential of brownfield land to accommodate new housing (the Government at the time anticipated that around 4 million units would be needed by 2016), continued beyond the publication of the UTF report, becoming the focus for Kate Barker's reports (Barker Review of Housing Supply and Barker Review of Land Use Planning) and provoking Peter Hall's striking statement/caveat in Richard Rogers's independent report 'Towards a strong Renaissance in 2005'.

The UTF report was unashamedly pro-urban, with a bias towards the built environment and physical regeneration, but was of undoubted influence on the Government's Urban White Paper (UWP) published the following year. 'Our towns and cities – the future: delivering an urban renaissance' (DETR, 2000), was the first urban white paper for nearly a quarter of a century. Its scope was broader and more comprehensive than the UTF report, drawing on the work of the Social Exclusion Unit and the State of English Cities reports. Fundamentally though, the UWP echoed the UTF's call for a move back to the city and presented a new vision of urban living in England. While the UWP adopts most of the policy reforms recommended by the UTF, it does reject calls to significantly increase public expenditure for urban regeneration or to interfere with the workings of the free market, instead making some marginal adjustment to fiscal arrangements that ultimately failed to deliver any demonstrable improvements to delivery. Ultimately, the Task Force Report is a more considered and detailed document but the UWP managed to mediate some of the worst of its middle class excesses.

Further reading

DETR (2000) 'Our Towns and Cities – the future: Delivering an urban renaissance', DETR, London.

DETR/UTF (1999) *Towards an Urban Renaissance*. Spon, London.

Imrie, R. and Raco, M. (2003) *Urban Renaissance? New Labour, community and urban policy*, Policy Press, Bristol.

Rogers, R. (2005) 'Towards a Strong Renaissance: An independent report of the Urban Task Force', available at: www.urbantaskforce.org/UTF_final_report.pdf (accessed 20/11/2013).

14.7 Enterprise Zones

Key terms: displacement; capital allowances; simplified planning; rates holiday

Enterprise Zone (EZ) regimes vary between countries, but essentially they are designated areas that benefit from a combination of government subsidies, tax breaks, simplified planning and reduced bureaucracy. When EZs were first introduced in the UK in 1980, they were seen as a property driven initiative designed to encourage the physical and economic regeneration of relatively small areas (Department of Environment, 1995); similar models having operated in France and the US for many decades. The policy evolved from the concept of Freeports espoused by Peter Hall in the late 1970s. Geoffrey Howe, Margaret Thatcher's first Chancellor of the Exchequer, seized on the idea and adapted the model to deliver a package of fiscal and other benefits targeted at developers, landlords, investors and occupiers. The power for the Secretary of State for the Environment to create EZs was contained in the Local Government, Planning and Land Act 1980 and provision for the fiscal measures was contained in the Finance Act (Section 74). In all, 32 EZs were designated across the UK, 22 in England, five in Scotland, three in Wales and two in Northern Ireland. They were designated in four tranches and each had a 10-year life span.

EZs in the UK comprised a package of eight measures, the three most significant of which were 100 per cent capital allowances in respect of investment in buildings and fixed plant and machinery, a rate-free period for the life of the zone, and a simplified planning regime. The rate-free period was intended to reduce operating costs for occupiers although half of this benefit was clawed back by landlords by way of increased rents. Capital allowances allow investors to deduct the price of purchasing an EZ investment from their taxable income, effectively subsidising the purchase price of EZ property by the rate at which investors would usually pay tax. This incentive, in combination with the higher rents that landlords were able to charge as a consequence of the tenants paying no rates during the life of the zone, resulted in higher investment yields. At their peak, EZs encouraged the creation of unit trusts to allow wealthy individual investors to reduce their tax bills by investing in the EZ trusts. In some locations the capital allowances were extended beyond the life of the zones using golden contracts that allowed the build out of undeveloped sites long after the zones had officially expired.

The simplified planning regimes established in the zones, which provided fast track planning permission for any development that conformed with the regimes established by the zone authority, encouraged the development or regional shopping centres, such as Merry Hill in the West Midlands and the Metrocentre in Gateshead. Such large-scale developments would normally have had to go through a protracted planning process that usually resulted in an expensive and time-consuming public planning inquiry. The necessity for this was negated by simplified planning. Ultimately, EZs were a powerful supply-side stimulant that encouraged the construction of millions of square metres of commercial and industrial floorspace in the zones in a relatively short space of time. Whether this was the right space in the right place at the right time to encourage sustainable economic growth is debateable.

EZs were subject to more monitoring and evaluation than any other property-led regeneration initiative, probably because they proved to be a very expensive policy tool that was paid for by the Treasury. The early EZs covered large swathes of inner urban land (e.g. Tyneside and Isle of Dogs) and were found not to be as effective as they might have been for four reasons: first, by the time derelict land had been reclaimed and infrastructure put in, the zone could be halfway through its life; second, some of the inner urban areas were not where modern business wanted to be located; third, private landowners whose land was designated as EZ received windfall profits for doing nothing; and fourth, the early zones included within their boundaries many firms that pre-dated the designation of the zones; these firms received 10-year rates holidays just by being there. Later zones were often designated in more accessible locations, on vacant greenfield sites in public ownership.

As a result of the powerful cocktail of financial incentives and simplified planning, EZs encouraged a concentration of property development, investment and occupation activity in particular locations, to the detriment of other areas. They created a two-tier property market, with an on-zone market characterised by high rents, yields, supply and demand, and an off-zone market of stagnating or reducing values due to a lack of investment and diminished investor and occupier demand. EZs also caused displacement by attracting local occupiers to relocate to the zones from nearby.

Problems also arose when zone incentives expired. Occupiers found themselves not only paying inflated 'on-zone' rents on leases with upward-only rent reviews, but simultaneously having to cope with the re-imposition of business rates calculated at inflated rateable values. Vacant premises, that had not been occupied during the life of the zone, continued to overhang the market. Landlords that had benefitted from the generous capital allowances were able to offer heavily subsidised rents and other occupier incentives to attract tenants, undermining the viability of competing developments.

In November 2010, George Osborne announced the designation of a 'new breed' of Enterprise Zones in England. The new EZs were to benefit from business rate discount worth up to £275,000 per business over a 5-year period, simplified planning by way of Local Development Order powers or Simplified Planning Zone, retention of new and additional business rates within the zone by the local authority and introduction of superfast broadband (DCLG, 2010). The announcement provoked a flurry of publications comparing the proposed new EZs with the old style ones, with the new zones being branded EZ 'Lite', compared to their EZ 'Max' predecessors. In total, 24 areas in England were given EZ designation, with Wales announcing five EZs of its own. A notable difference between old and new zones is the sectoral focus of many of new zones, for example renewable energy, automotive manufacture, aerospace and advanced manufacturing.

Because of these side effects, EZs were generally regarded as an outmoded tool of regeneration. It was therefore something of a surprise when in November 2010, five months after the forming of a Coalition Government, the chancellor of the exchequer, George Osborne, announced the designation of a 'new breed' of Enterprise Zones. The new EZs were to benefit from business rate discount worth up to £275,000 per business over a 5-year period; simplified planning by way of Local Development Order powers or Simplified Planning Zone; retention of new and additional business rates within the zone by the local area to be invested locally; and introduction of superfast broadband (DCLG, 2010).

The announcement provoked a flurry of publications comparing the proposed new EZs with the old style ones (see Sissons and Brown, 2011; Larkin and Wilcox, 2011, Higton, 2011). Eleven areas were identified as recipients of the first wave of EZ designations with four named locations in the vanguard. Notionally, all areas were to be covered by a local enterprise partnership (LEP). LEPs missing out on the first wave of designations were invited to bid in a second round for a further ten; ultimately another 13 locations were selected, resulting in a total 24 new EZs. Two further

EZs announced for Lancashire and Hull and Humber LEPS in response to proposed closure of BAE plants and the Scottish Government and Welsh Assembly have made similar announcements of the intention to create EZs. Time will tell whether the new 'Lite' EZs will perform with anything like the vigour of the old EZ 'Max'.

Further reading

Department of the Environment (1995) *Final Evaluation of Enterprise Zones*, London, HMSO.

Higton, A. (2011) *Enterprise Zones in the UK since 1980*, South West Observatory, Bristol.

Jones, C., Dunse, N. and Martin, D. (2003) 'The property market impact of British Enterprise Zones', *Journal of Property Research*, 20(4): 343–369.

Larkin, K. and Wilcox, Z. (2011) *What Would Maggie Do? Why the Government's policy on Enterprise Zones needs to be radically different to the failed policy of the 1980s*, Centre for Cities, London.

Office of the Deputy Prime Minister (2003) *Transferable Lessons from Enterprise Zones*, HMSO, London.

Segade, O. and Barrett, S. (2012) *Enterprise Zones: Only one piece of the economic regeneration puzzle*, London Chamber of Commerce, London.

Sissons, A. and Brown, C. (2011) *Do Enterprise Zones Work? An Ideopolis policy paper*, The Work Foundation, London.

14.8 Partnership working

Key terms: networked governance; collaborative working; effective partnership working; barriers to partnership working

As a concept, partnership involves 'an agreement between two or more independent bodies to work collectively to achieve an objective' (Audit Commission, 2005: 7). Partnerships and partnership working are key to delivering regeneration. Such partnerships involve diverse agencies in diverse arrangements, incorporating different extents of collaboration; they can range from a contractual obligation, to an arrangement for funding, to adopting a whole new way of working that has implications for all the partners involved. While the theory of effective partnership working is clearly established in the literature, extensive evidence exists to demonstrate the challenges of effective partnership working in practice.

The wide range of stakeholders engaged in regeneration activity – from central and local government to private, statutory and third sector organisations, to local residents and communities – reflects the context of complex networked governance (see for example Stoker, 1998), which requires collaboration and partnership working. A growing recognition of the complexity of the problem to be addressed in declining areas, and an understanding of the failures of regeneration in the past, as well as the need to consider alternative sources of funding for regeneration, have all supported the rise to prominence of partnership working (see for example Shaw and Robinson, 2010). A coordinated, joined up, multi-agency approach, working with communities as partners is seen as key to delivering sustainable outcomes.

Effective partnership working can bring many advantages for the stakeholders, reflecting the notion of collaborative gain, defined by the Improvement Service (2008, p. 3) as:

> a situation where partnership working brings about 'added value' benefits, which would not have been achieved by the individual partner organisations operating on their own ('More than the sum of the parts'). These benefits might include, for example, enhanced outcomes for service users/ communities or efficiency gains for the partners themselves.

Ensuring a 'fruitful partnership' is achieved (Audit Commission, 1998) requires various criteria to be met. Hudson and Hardy (2002) outline the following six principles for successful partnership working:

- Acknowledgement of the need for partnership
- Clarity and realism of purpose
- Commitment and ownership
- Development and maintenance of trust
- Establishment of clear and robust partnership arrangements
- Monitoring, review and organisational learning.

Such principles highlight the challenges facing partnership working in practice. As well as potential benefits there are multiple potential barriers facing partnerships operating in the context of complex networked governance, reflecting the diverse vested interests, different organisational cultures, and uneven power relations among the various stakeholders (see for example the evaluation of New Deal for Communities (DCLG, 2010) or Coulson (2005). In practice, partnership working can be characterised by conflict and tensions; trust, open and effective communication, and a willingness to compromise, for example, can be difficult to achieve and can take considerable time. Partnership working can be expensive, unrewarding and time consuming if not done effectively (Audit Commission, 1998), however where successful partnership working is evident the benefits can be considerable for all involved. Recognising the importance of partnerships (and also the challenges of the process of partnership working in practice), a number of toolkits/frameworks to evaluate and improve partnership working are available for partner agencies to use (see for example Audit Commission, 2009).

Further reading

Audit Commission (various) e.g. 'A fruitful partnership' (1998); 'Governing partnerships' (2005); 'Building better lives' (2009); all available at: http://archive.audit-commission.gov.uk/auditcommission/nationalstudies/pages/default.aspx.html (accessed 20/11/2013).

Coulson, A. (2005) 'A plague on all your partnerships: Theory and practice in regeneration', *International Journal of Public Sector Management* 18(2): 151–163.

DCLG (2010) *What Works in Neighbourhood-level Regeneration? The views of key stakeholders in the New Deal for Communities Programme*, HMSO, London.

Hudson, B. and Hardy, B. (2002) 'What is a successful partnership and how can it be measured', in: C. Glendinning *et al.* (eds) *Partnerships, New Labour and the Governance of Welfare*, Policy Press, Bristol, ch. 4.

Improvement Service (2008) available at: www.improvementservice.org.uk/collaborative-gain/ (accessed 20/11/2013).

Shaw, K. and Robinson, F. (2010) 'UK urban regeneration policies in the early 21st century: Continuity or change?', *Town Planning Review*, 81(2): 123–150.

Stoker, G. (1998) 'Governance as theory: Five propositions', *International Social Science Journal*, 50(155): 17–28.

14.9 Funding and finance for regeneration

Key terms: regional development agencies; public–private partnerships; local enterprise partnerships; Regional Growth Fund

The majority of funding for regeneration comes from public sector bodies such as central, regional and local government, other regional, local and national bodies such as regional development agencies, and via the European Commission. In addition, funding is also available from other bodies such as the Heritage Lottery Fund. In 2007, approximately 80 per cent of grant aid available for regeneration projects came from the public sector.

Sources of private sector finance for regeneration and economic development include banks, insurance companies, and other lending institutions such as finance houses.

A number of different 'pots' of public sector funding are available, depending on the nature and scope of the project. Funding can be the full funding of the project or funding of a large proportion of the costs, e.g. site remediation works, which would otherwise make the project unviable for the developer.

Single pot funding was distributed and administered by the nine regional development agencies (RDAs) (now defunct) across England, whose remit was to provide funding to encourage economic development and regeneration in the areas of business efficiency, investment, competitiveness, employment, skills and sustainable development. The funding came from contributing central government departments, which at the time were Department of Trade and Industry (DTI), Department for Communities and Local Government (DCLG), Department for Education and Skills (DES), Department for Environment, Food and Rural Affairs (Defra), and Department for Culture, Media and Sport (DCMS). Each RDA was allocated a set annual budget, determined by economic need and performance, which varied region by region. These budgets were significant: in 2007/8 the RDAs' total budget under the Single Programme was £2.2 billion which was available to the RDAs to spend to meet regional priorities identified in their regional economic strategies and corporate plans, such as promoting business growth, development and innovation, brownfield land reclamation, learning and skills, infrastructure, communities, sustainability, culture, and regional promotion and tourism.

Public sector grants are usually awarded as one-off single payments towards a project that has a robust business case that meets defined criteria and provides evidence that the project will deliver required outputs, outcomes and impacts, such as creating jobs and businesses. In addition, a condition of many public sector grants is that the project must secure 'match' funding from another funding body, or from the private sector. This 'leverage' of additional finance demonstrates 'added value'. Another typical condition of a public sector grant is that the award must be spent within a specified timescale, usually within a financial year, so that the proposed outcomes and impacts are delivered and realised. Depending on the nature of the proposed project, other conditions might include a demonstration of 'deliverability'; strategic fit with existing policy and priorities; and quality and sustainability considerations such as environmental sustainability, quality of place and social impacts. Overall, proposed projects must demonstrate that they are fulfilling a specific economic need and delivering economic impact.

Common public sector funding mechanisms are private finance initiative (PFI) and public–private partnerships (PPP). Both are used by the public sector to develop buildings that will deliver services. PFI was introduced in the UK in 1992 as a mechanism for procuring public buildings without the need for public borrowing to support the capital cost. The private sector therefore develops the building on behalf of the public sector, which then pays back the costs as a revenue stream over the term of the agreement. Many services such as hospitals, schools and prisons have been delivered in this way, with the public and private sectors entering into a PPP, whereby the private sector contributes its expertise and funding in return for guaranteed repayments from the procuring public body. The public sector is therefore effectively paying a rate of interest relative to the capital cost of the building.

Since the onset of the credit crunch in 2007 and subsequent recession, public sector funding for regeneration has been significantly reduced following central government efficiency savings and austerity measures that were put in place to stabilise the economy. The removal of the regional tier of government and the demise of the regional development agencies and other public sector funding bodies and the move towards 'local' policy has meant that traditional public sector funding for regeneration projects has all but disappeared. This, in addition to the lack of finance available from banks and other lending institutions for development and regeneration has meant that new mechanisms of funding and finance have to be considered. The government has responded to this by giving local authorities more power to support growth through development by shifting the focus of regeneration away from a subsidised regional strategy to a more local and independently financed model. Public sector funding is now being channelled through a series of programmes that are designed to lever in private sector investment, which is seeing a move to more innovative and collaborative methods of funding and finance. New mechanisms include Local Enterprise Partnerships (LEPs), Growing Places Fund, Regional Growth Fund, European funding, tax incremental financing (TIF) and the Homes and Communities Agency (HCA). From the private sector, finance is becoming available through, for example, traditional equity funds setting up debt funds or moving into bridging finance or senior debt; however, it is still extremely restricted due to the perceived high risk of property-led regeneration by private sector lending institutions.

Some of the current available funding mechanisms for regeneration are summarised below.

LEPs and the Regional Growth Fund

LEPs were created under the Localism Act 2011 to encourage the public and private sectors to come together to influence and shape local policies for economic development based on local need. Established LEPs are able to bid for funding from the Regional Growth Fund allocated by central government to stimulate the economy within LEP areas. £1.4 billion was allocated in the first round with a further £1 billion allocated in the second. In addition, the Growing Places Fund (GPF) has allocated £730 million to be used for transport and infrastructure improvements.

European funding

Funding for projects has been allocated by the Joint European Support for Sustainable Investment in City Areas (JESSICA) and the European Regional Growth Fund. In addition, the European Investment Bank offers investment via JESSICA Urban Development Funds that have been established in the UK.

Tax incremental financing

TIF enables local authorities to keep up to 50 per cent of increases in business rates that result from a development to invest in the local area, or to borrow against, rather than those rates being transferred back to central government to be redistributed. TIF 'zones' therefore act as a mechanism for local authorities to lever in additional funding.

Homes and Communities Agency

The HCA has now taken control of the land and property portfolios of the disbanded RDAs, amounting to approximately 10,000 hectares of brownfield land. It has a £550 million fund under the Get Britain Building programme to kick-start stalled residential development sites across the UK. In addition, there is a £225 million 'large sites' fund to stimulate development of unviable sites, and a £190 million public land infrastructure fund.

Private sector investment

Under the RGF programme, the private sector is expected to 'match' public sector funding and provide up to 70 per cent of the total investment required to deliver the economic, social and structural improvements identified. Pension funds are now starting to invest in lower risk infrastructure and regeneration projects, mainly via debt, mezzanine and bridging finance, and by working collaboratively with local authorities and LEPs; the private sector is assisting in refinancing regeneration and development projects that were previously seen as unviable and undeliverable.

Further reading

BPF (2007) 'Funding regeneration – a guide to public sector grants', available at: www.bpf.org.uk/en/files/bpf_documents/regeneration/Funding_regeneration_final_printed_version.pdf (accessed 02/12/2013).

GVA (2013) Alternative Funding Development Spring 2013, available at: www.gva.co.uk/WorkArea/DownloadAsset.aspx?id=*15032394537* (accessed 02/12/2013).

Isaac, D. O'Leary, J. and Daley, M. (2010) *Property Development Appraisal and Finance*, 2nd edn, Palgrave Macmillan, Basingstoke, ch. 6.

14.10 Brownfield land

Key terms: brownfield land; previously developed land; National Land Use Database

Transforming the vast areas of brownfield land that lie across the country is one of the biggest challenges collectively facing government, communities and businesses. Failure to recycle brownfield land may potentially lead to further erosion of the countryside and natural resources; perhaps more importantly it will represent a missed opportunity to revitalise urban and rural areas blighted by centuries of industrial activity (LGA, 2002).

Brownfield land comprises land or premises that have been previously used or developed and are not currently fully in use, although they may be partially occupied or utilised. It may be vacant, derelict or contaminated. Fundamentally, therefore, a site is not available for immediate and full use without some intervention (Alker *et al.*, 2000). For some, the term brownfield land is viewed as a synonym for contaminated land, which ignores the potential for land to be derelict without being contaminated; whereas other people use the term in a more generic way to mean non-greenfield.

More recently, the term 'previously developed land' (PDL) has gained some traction, defined as land that was occupied by a permanent structure including the curtilage of the developed land and any associated fixed surface infrastructure. The specific focus of this concept is vacant, derelict and underused PDL, which is regarded as land so damaged by previous industrial or other development that it is incapable of beneficial use without treatment (ODPM, 2004).

There have been many initiatives aimed at reclaiming brownfield land, one of the earliest of which was garden festivals, introduced in West Germany to help with post-war reconstruction. The first Garden Festival was held in Hanover in 1951, and the concept has since been adopted in the UK, Japan, Singapore and Canada. Another common approach is a grant regime that subsidises the cost of reclaiming derelict and contaminated land, for example the Derelict Land Grant, which operated in England in the late twentieth century. A third approach is to create quasi-public agencies that pursue brownfield reclamation and development, for example development corporations and regeneration agencies that were tasked with bringing brownfield land back into productive use.

Historically, derelict land surveys were conducted to record the quantity, location and condition of brownfield land in England and Wales by the Department of the Environment. In 1998, English Partnerships established a National Land Use Database (NLUD) to provide data on vacant and derelict sites and other previously developed land and buildings that may be available for development in England and provide statistics on the number, type and planning status of previously developed land.

NLUD recorded that, in 2009, there were 61,920 ha of PDL in England, down 3 per cent from 2008, an estimated 33,390 ha (54 per cent) of which were vacant or derelict; the remaining 28,530 ha were in use but with potential for redevelopment (HCA, 2011). Disappointingly, NLUD surveys ceased after 2009, when English Partnerships merged with the Housing Corporation to form the Homes and Communities Agency; the Scottish Derelict and Vacant Land Survey continues.

Further reading

Alker, S., Joy, V., Roberts, P. and Smith, N. (2000) 'The definition of brownfield', *Journal of Environmental and Planning and Management*, 43(1): 46–49.

HCA (2011) *Previously Developed Land that may be Available for Development*, HCA, Warrington.

LGA (2002) *Something Old, Something New: A report of the LGA Inquiry into the Development of Brownfield Land*, LGA, London.

ODPM (2004) *Previously Developed Land that may be Available for Development*, ODPM, London.

14.11 Contaminated land

Key terms: Environmental Protection Act Part IIA; polluter pays; suitable for use; pollutant linkage; remediation

All industrial societies have, in the past, allowed land to become derelict and contaminated. Various industrial practices, such as industry, mining and waste disposal, have led to toxic or noxious substances being in, on or under land. Land in this condition can pose a serious threat to health and the environment, including pollution of water (DETR, 2000).

Land is considered to be contaminated if it contains substances in sufficient quantities or concentrations that are likely to cause harm directly or indirectly to humans and the environment. Contaminated land may give rise to statutory nuisance, contain unlawfully deposited regulated waste, cause pollution of controlled waters, contravene environmental regulations or give rise to private nuisance in common law. It may typically require measures to be undertaken in order to make it suitable for reuse.

The UK Government introduced a legislative framework to address the country's legacy of derelict and contaminated land. The primary piece of legislation is the Environmental Protection Act 1990 which established the principle of 'polluter pays' and the 'pollutant linkage' test. In addition, a 'suitable for use' approach has also been introduced that seeks to balance the policy goal of remediating contaminated land with concerns about the prohibitive cost of doing so.

The planning regime in the UK establishes procedures for dealing with contaminated land in relation to the future use of land but is a voluntary process with a vague definition of unacceptable harm. In contrast, the Part IIA regime deals with contamination in relation to current uses of land, adopting a clear statutory definition of contamination and significant harm and imposing a compulsory regulatory process. Part IIA of the EPA 1990 follows the 'polluter pays principle' such that the cost of remediation will normally lie with the person(s) who caused or knowingly permitted the contamination. Therefore, any person, organisation or business might be liable to remediate contaminated land if they caused or knowingly permitted the contamination. However, if this person(s) cannot be identified then the owner or occupiers of the land will be responsible. Lenders, investors, insurers and professional advisers are normally excluded from liability for remediation (DETR, 2000).

Part IIA of the Environmental Protection Act 1990 charges local authorities with identifying land in their area that is contaminated, determining what needs to be done by way of remediation and serving remediation notices on individuals or companies that the law deems to be responsible for that contamination. It recognises that harm to health and the environment arises not from the mere presence or contaminating substances in land, but from their movement along a 'pathway' to where they can cause damage to a 'receptor'. S78A defines harm as being to the health of living organisms or other interference with ecological systems of which they form part and, in the case of humans, includes harm to their property.

The principle underlying Part IIA is that land should be decontaminated to a level that is suitable for the use for which the land is intended. Contaminating substances should therefore not be ignored when there is a change in the use of land because the 'acceptable' level of contamination is different depending on the end use. This common sense approach was adopted following experience in the US and the Netherlands, where the 'clean as green' approach of restoring land to its pre-development condition was found to be prohibitively expensive in the long term.

Remediation is a process that should address the issues identified by desk top study and site investigation. It addresses risks arising from a pollutant linkage, makes the site suitable for its actual or intended use and assists with the preparation of the site for general development. It is the process by which the risk of harm or pollution arising from substances in land is eliminated or reduced to an acceptable level. This may take place by reference to a particular use of land, or alternatively, to all possible uses of that land (RICS, 2010).

Remediation may comprise:

- removing the substance or substances from the source, thus rendering it harmless;
- severing the 'pathway'; or occasionally by
- removing the receptor or reducing its vulnerability.

The choice of remediation technique will depend on the following factors: time, cost, ground conditions, contamination risk, the nature of the development and actual or intended use, the need for long-term environmental monitoring and residual risks.

Further reading

DETR (2000) 'Contaminated Land'. Circular 02/2000, DETR, London.

HM Government (2012) 'Environmental Protection Act 1990: Part 2 A – Contaminated Land Statutory Guidance', Defra, London.

RICS (2010) *Contamination, the Environment and Sustainability: Implications for chartered surveyors and their clients*, 3rd edn. GN13/2010. RICS, London.

14.12 Gap funding

Key terms: grant funding; negative residual; state aid

Gap funding is a government or public sector grant to fund what is called the negative residual, which is the difference between the cost and value of a property development project that is non-viable, e.g. the costs outweigh the value, thus it would not go ahead without gap funding. For example, if the total development costs (including land) of a development is £800,000 and its gross development value is £1 million, the negative residual is £200,000 and the scheme will not progress as the developer is making a loss. If a grant of £200,000 is made available to bridge the negative residual, together with a reasonable allowance for the developer to make a profit, then the project may go ahead. Grants were targeted at Assisted Areas where development viability was a problem, for example office and industrial development in northern towns and cities. In locations with higher end values, where residual values were positive then no grant would be needed even if it was available. In some instances a grant was made available based on pessimistic end values but with a clawback arrangement so that if higher end values were achieved on completion then some or all of the grant may be repaid. It is possible to calculate a gearing or leverage ratio of public to private investment, which for the example above would be 1:4 (£200,000:£800,000).

Gap funding operated across the UK for over two decades, the first example of which was the Urban Development and Urban Renewal Grants introduced in 1982, which were merged in 1988 to create City Grant, later transferred to English Partnerships in 1993 and renamed the Partnership Investment Programme (PIP) which operated between 1994 and 1999.

The continued use of gap funding in the UK was thrown into turmoil in 1999 when the European Commission classified PIP as state aid under Article 92 (1) of the Treaty of Rome. Their decision was provoked by concerns that the grant regime was anti-competitive and had been used to bail out the ailing Rover car plant at Longbridge. Despite the EC's decision being regarded by many as perverse, the Government decided to pull the PIP programme with immediate effect.

As a consequence of the decision there was a hiatus in physical urban regeneration activity in England, with the loss of a significant level of public sector investment in property-led regeneration with which to lever in greater levels of private investment, hampering the operation of the newly established regional development agencies that had just inherited the grant programme from English Partnerships. In response to the adverse decision, a number of alternatives to gap funding were considered, which were to allow a modest level of funding in assisted areas up to the European state-aid limits, pursue direct development funded by the public sector, revolving investment fund or competitive procurement. Eight new EC approved schemes were launched in England, none of which adequately replaced PIP, supporting as they did only 10 per cent to 20 per cent of schemes previously qualifying for gap funding.

In 2007 a new regime was introduced to support investment to bring back into productive use land or buildings that are derelict, contaminated, under-used or vacant and are suitable for use or conversion for business purposes. In order to be eligible, the investment had to be on land situated within an assisted area and owned by the prospective developer or by a relevant public body. The beneficiary of the aid was the developer who owned the land on which the investment was to be made and the grant was calculated with reference to the gap between the projected eligible costs of the initial investment and the projected market value of the material assets.

Categories of eligible costs included:

- market value of the land;
- cost of preparing the land for development;
- cost of providing services and infrastructure;
- costs of constructing and refurbishing buildings;
- a development fee (developer's profit);
- professional fees;
- reasonable costs of marketing and letting and disposing of units.

While the new scheme was similar in many ways to the original gap fund, the momentum of the programme had stalled during the intervening seven years and the private sector had little appetite to get involved in a time-consuming bureacratic process that offered relatively modest levels of funding (due to State Aid limits). As a consequence, grant funding in the UK has all but disappeared. The concept of gap funding is now regarded by many as an outmoded method of regeneration funding, an addictive 'fix' that was administered to regions with weak and failing economies. The new methods of regeneration funding, such as TIF business rates retention and revolving infrastructure funds are now predicated on borrowing against the future uplift in values facilitated by investment in infrastructure rather than the a public sector grant used to lever in private sector investment.

Further reading

Department for Communities and Local Government (2007) 'Regional Investment Aid Scheme for Speculative and Bespoke Development', Commission Regulation (EC) No. 1628/2006 on the application of Articles 87 and 88 of the Treaty to national regional investment aid, DCLG, London.

European Commission (2006) *State Aid Control and Regeneration of Deprived Urban Areas*, DG Competition, Brussels.

House of Commons (2000), Environment Transport and Regional Affairs Committee – 'The Implications of the European Commission Ruling on Gap Funding Schemes for Urban Regeneration in England', House of Commons, London.

Needham, B., Adair, A., Van Geffen, P. and Sotthewes, M. (2003) 'Measuring the effects of public policy on the finances of commercial development in redevelopment areas: Gap funding, extra costs and hidden subsidies', *Journal of Property Research*, 20(4): 319–342.

Royal Institution of Chartered Surveyors (2000) *Alternatives to the Partnership Investment Programme*, RICS, London.

14.13 Community engagement

Key terms: community; engagement terminology; the participation continuum; barriers to engagement

It is generally held within the context of public policy, irrespective of ideological belief, that engagement with communities is a 'good thing'. These communities can be ones of faith, interest, employment, education, health etc. However within the policy of regeneration these communities are communities of place.

Engagement with communities in regeneration developed as a reaction to the top-down regeneration initiatives of the 1970s and 1980s, typified by the UDCs. Centralised state-determined policy decided how, where and in what form regeneration was to be delivered. These projects, certainly in their early stages, offered very little to disadvantaged communities. Economically driven with an agenda to transform areas physically, the needs of poorer communities were peripheral at best (Tallon, 2010). Post-1990, approaches to the role of community in regeneration began to undergo a change. Policies such as City Challenge and the Single Regeneration Budget saw a shift away from a professionally dominated approach to delivering regeneration to one where the views of the community were actively sought (Robinson *et al.*, 2005).

The concepts involved in community engagement are, however, difficult to reach a consensus on and a failure to do so can lead to policy failure.

Community as a concept is nebulous. Within the context of regeneration it is spatial, expressed at the neighbourhood or estate level. However, policy makers would be wrong to assume that a community is homogenous, expressing one single viewpoint. Communities are made up of many differing groups differentiated by age, household type and stage in the life cycle, tenure, economic status, faith, ethnicity etc. This in itself can make reaching a consensus about issues and responses problematic.

The same can be said of the notion of engagement. There are a multitude of phrases that are used by policymakers to describe interaction with communities. These phrases are often used in such a manner that they often substitute for each other without recourse to their full meaning. There are numerous issues that need to be considered to ensure the most appropriate form of engagement is used. Not least of which is the question – what is the purpose of that engagement? Traditionally, consideration of participation processes was around how to do it well. However, as community engagement began to form a much more important element of regeneration policies, the debate was opened up to determine what was the motivation behind the decision to engage? Summarised briefly, the three rationales are:

- maintaining a consensus approach to ensure decisions are within the status quo (organisation led);
- bargaining to ensure decisions are agreed collectively but are contained within what is considered to be acceptable change (balance in decision making); and
- increasing the consciousness of different groups to agree collective priorities and achieve radical change (community led) (Hague, 1990).

Determining what the rationale is behind engagement with the community will determine where on the participation continuum any engagement will sit (Figure 14.13.1).

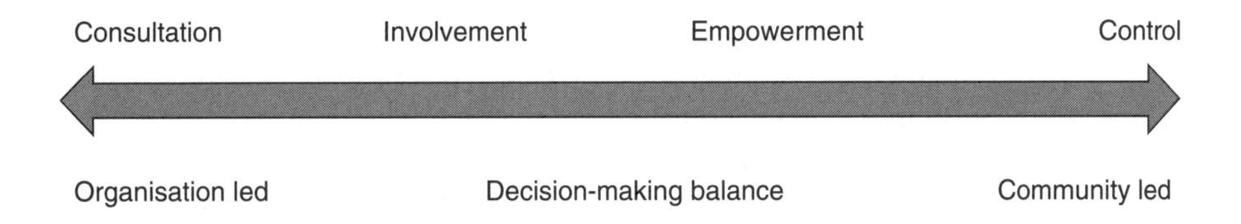

Figure 14.13.1 The participation continuum.

Emphasis post-1997 in terms of community engagement and regeneration saw a much greater emphasis on community governance which saw a move towards the participation of residents in decision making about their community. Communities that could articulate their concerns, determine the most appropriate outcomes, ensure effective implementation and see successful change were seen as being highly empowered (Robinson *et al.*, 2005). This was demonstrated in the New Deal for Communities programme and more recently in current ideas around community self-determination, i.e. the localism agenda. However, community engagement at this level of sophistication, in the form of formal partnerships and active citizenship, has many difficulties to overcome.

Individuals from communities who find themselves in formal partnerships are expected by the other partners to represent the community. However, as the very nature of community is varied they find themselves with an impossible task that is limited by their own experiences and aspirations. The community themselves expect their varying views to be articulated but time and support to allow for meaningful interaction between representatives and the community is often lacking (JRF, 2000). The scale of interaction in formal partnerships can be daunting with time-consuming demands being placed on a few individuals, leading to the breaking of 'community stars' or their absorption into the partnership's agenda (JRF, 2000). There is a steep learning curve for what amount to unpaid volunteers and the pressure for results, particularly among local politicians, can negate the process of meaningful engagement. Lack of representativeness is often cited as a failing in regeneration schemes that causes delays but adds little value to the process or its outcomes (Ball, 2004).

Support can be put in place for those actively involved, such as training, skills development, mentoring etc. Engagement with the community over regeneration initiatives needs to be multifaceted to ensure all views are heard. Expectations of and on community representatives on partnerships needs to be determined, collectively at the outset of any regeneration project.

Further reading

Ball, M. (2004) 'Cooperation with the community in property led urban regeneration', *Journal of Property Research*, 21(2): 119–142.

Hague, C. (1990) 'The development and politics of Tenant Participation in British Council Housing', *Housing Studies*, 5: 4.

JRF (2000) *Community Participant's Perspectives on Involvement in Area Regeneration Programmes*, Joseph Rowntree Foundation, York.

Robinson, F., Shaw, K. and Davidson, G. (2005) '"On the side of Angels": Community involvement in the governance of neighbourhood renewal', *Local Economy*, 20(1): 13–26.

Tallon, A. (2010) *Regeneration in the UK*, Routledge, Abingdon.

14.14 Gentrification and abandonment

Key terms: gentrification; abandonment; displacement; polarisation; urban renaissance

> Gentrification is the transforming of a working class or vacant area of the central city into a middle class residential or commercial use.
>
> (Lees *et al.*, 2010)

It occurs when low income residents of inner-city areas are replaced by new higher income residents, from other areas of the city, in a spatially concentrated manner. The definition hinges on economic, social and population changes that cause physical changes to neighbourhoods. It should be noted that while physical change may be the most apparent manifestation of gentrification, it is not essentially part of the process, but more of an end result. Gentrification involves the exploitation of the economic value of real estate and the treatment of local residents as objects rather than subjects of upgrading.

Even though population movement is a common feature of cities, gentrification is specifically the replacement of a less affluent group by a wealthier social group (Berg *et al.*, 2009).

Established models of urban land use, such as Burgess's Concentric Zone and Hoyt's Sector Theory, recognise that residential land use is segregated in terms of income and social class that change over time through a process of invasion and succession due to economic and population growth. Both theories describe how inner city areas are abandoned by high income households and infilled, usually at high density, by lower income housing. Conversely, gentrification may be regarded as the reverse of this process, whereby high income residents re-invade previously abandoned inner city areas.

Abandonment occurs when owners and occupiers of property have no incentive for continued ownership or occupation and are willing to surrender their title, without compensation, due to the absence of any effective demand for (re)use (paraphrasing Marcuse, 1985). As a result, property rental and capital values collapse due to there being little or no demand for property. This is often accompanied by stigmatisation of an area resulting in blight (link to economic theory?). Gentrification, conversely, results in rapid increase in land and property values which is to the detriment of residents on low incomes who are unable to afford inflated rents and values and become displaced.

The benefits of gentrification are that it usually improves the quality of housing in an area, revitalising areas of the city and increasing the tax base though domestic rates and local taxation on economic activity. It is regarded by some as a means of solving social malaise, not by providing solutions to unemployment and poverty (see Concept 14.1) but by transferring or displacing problems elsewhere in a city and is sometimes mistakenly regarded by policymakers as a solution to abandonment.

The reciprocal relationship between gentrification and abandonment may be explained by adopting a dual market theory of housing, in which gentrification happens in one market and abandonment in another. Essentially they are two sides of the same coin, sometimes existing side by side, and are mutually reinforcing, each contributing to and accentuating the other, as populations move in opposite directions. Thus, gentrification results in displacement and social polarisation on both sides, and in a vicious circle in which low income groups are continuously under pressure of displacement and high income groups seek to wall themselves within gentrified neighbourhoods. Far from being a cure for abandonment, gentrification worsens the process (Marcuse, 1985).

The role of real estate speculation in the gentrification is significant and needs to be acknowledged. Property investment speculation is a strong accompaniment and necessary ingredient to the process of gentrification, and the behaviour of speculators and investors in the property market generally is the single most sensitive indicator of the type of change that is likely to occur in a neighbourhood.

Notably, despite reading like a gentrifiers' charter, the term 'gentrification' is never used in either the 1999 Urban Task Force Report or the 2000 Urban White Paper (see Concept 14.6), which is remarkable, given the fact that the many of the principles and characteristics of urban renaissance are consistent with those typically associated with that of gentrification. The UTF report not only heralded the urban renaissance in England but, arguably, launched a new phase of state-led gentrification in English towns and cities, resulting in a doubling of apartment building in England between 2001 and 2005. According to the DCLG, the number of apartments constructed in England rose from a quarter of all new dwellings in 2001–2, to half of all new dwellings in 2005–6, since when apartment construction has returned to around a quarter of new dwellings by 2012–13, mainly as a consequence of the credit crunch and collapse in housing development in the UK during the subsequent recession. Despite increased supply of residential units in British towns and city centres, problems of housing affordability and access, which continue to characterise housing markets in many parts of the UK, have been exacerbated by the capping in housing benefit payments, effectively pricing out low income groups from more expensive residential areas in towns and cities, resulting in increased social polarisation.

Further reading

Berg, J., Smith, N., Breznik, M. and Uitermark, J. (2009) *Houses of Transformation: Intervening in European gentrification*, NAi Publishers, Rotterdam.

Communities and Local Government (2013) 'House Building: March Quarter 2013, England', Housing Statistics Release, DCLG, London.

DETR (2000) *Our Towns and Cities – the Future: Delivering an urban renaissance*, DETR, London.

Lees, L., Slater, T. and Wyly, E. (2010) *The Gentrification Reader*, Routledge, Abingdon.

Marcuse, P. (1985) 'Gentrification, abandonment and displacement: Connections, causes and policy response in New York City', Urban Law Annual. *Journal of Urban and Contemporary Law*, 28: 195–240.

Urban Task Force (1999) *Towards an Urban Renaissance: Final report of the Urban Task Force*, Spon, London.

14.15 Social enterprise

Key terms: social mission; corporate social responsibility; business model; regeneration

Social enterprises are businesses. They trade, are revenue generating, make a profit and pay reasonable salaries to staff. They do not function purely through volunteering, the receipt of grants or donations. However, social enterprises are businesses with a particular social mission. That mission is to ensure, as a consequence of their activities and income generation, that their clearly stated social and environmental objectives are met. Social enterprises are not to be confused with the corporate social responsibility strategies adopted by many private sector businesses. In this case businesses engage with social and environmental actions to minimise any negative impacts and to generate positive images of themselves as a company. Their responsibilities are primarily to their investors.

Social enterprise businesses are set up to make a difference and they do this through reinvestment of surpluses in accordance with their social mission. They achieve this in lots of different ways: creating jobs for those who cannot get into mainstream employment; reinvesting in community projects; protecting the environment; providing vital services.

Social enterprise businesses are varied in terms of both their activities and the organisational model they adopt. Social enterprises can be:

- community interest companies (CICs) – a legal form especially created for the use of social enterprises to ensure that social enterprise businesses cannot deviate from their social mission or be sold for profit;
- registered with the Industrial and Provident Society (IPS) – this is the model usually adopted by cooperatives and mutual societies. This business model ensures that the social enterprise is democratically controlled and all members are actively involved in decisions;
- a company limited by guarantee (CLG) – this is the most common legal model used by businesses. Social enterprise businesses often choose to use this because it allows for flexibility in terms of how the organisation is governed and also it can make accessing finance simpler. However, a social enterprise business would have to include details of its social mission and policy regarding reinvestment of profits in its memorandum and articles of association;
- charities – some social enterprise businesses will, as part of their organisation's structure, have a charitable arm. Registration with the Charities Commission places certain restraints on an organisation's activities; however, it does provide an opportunity for social enterprise and other businesses to take advantage of different forms of tax relief. Where retention of surpluses is essential to the success of a social enterprise, a charitable arm could assist.

There are many different types of businesses operating as social enterprise companies. They also operate at local, regional, national and international levels. They can produce goods or deliver services. As a consequence there are many different types, from those that are instantly recognisable such as:

- Divine Chocolate – 45 per cent of the business is owned by cocoa growers.
- The Big Issue – works to support vendors who are homeless, vulnerable because of their housing or in danger of becoming homeless.
- The Eden Project – works on projects worldwide that bring about social, economic and environmental regeneration.
- Jamie Oliver's Fifteen – offering training and employment to unemployed young people.

To those that are less well known outside their local communities:

- The Salmon Youth Centre, Bermondsey – delivering an extensive range of activities aimed at challenging young people to widen their aspirations.
- Growing Well, Kendal – provides local people with mental health problems support in a farming environment.
- Greenworks, London – recycles surplus office furniture and sells it at reduced costs to local small businesses.

Social enterprise businesses have contributed to regeneration projects in a number of ways; 39 per cent of social enterprises are concentrated in the most deprived communities (13 per cent standard small and medium enterprises – SMEs) (Social Enterprise UK, 2011). Social enterprise businesses can support regeneration objectives through the creation of employment, the generation of local economic activity and the provision of services and facilities. Home Baked in Anfield, Liverpool is both a community land trust (CLT) that aims to provide affordable homes for local people through its CLT, and a cooperatively run bakery offering training and job opportunities, again for local people. This contribution to the local economy of disadvantaged communities is significant. Social enterprise businesses when asked in 2011 about their economic development in a time of poor or non-existent growth in the general economy reported the following:

- 58 per cent of social enterprises reported growth last year (28 per cent SMEs);
- 57 per cent of social enterprises predicted growth next year (41 per cent SMEs) (Social Enterprise UK, 2011).

In addition, social enterprise businesses can also be a conduit through which regeneration occurs. An example is SKINN based in what was the industrial heartland of Sheffield. SKINN is a not for profit organisation that is working with local residents and businesses to improve the area of Shalesmoor, Kelham Island and Neepsend. Faced with multiple issues around the environment, public space and negative developments in the area, SKINN came together to provide a coordinated approach to the strategic planning of regeneration in the area.

Further reading

Social Enterprise UK (2011), available at: www.socialenterprise.org.uk (accessed 20/11/2013).

14.16 Area-based initiatives

Key terms: regeneration; trickle down; neighbourhood renewal; Urban Development Corporations; New Deal for Communities; multiple deprivation

Area-based initiatives (ABIs) are an approach where regeneration policies are brought to bear on a defined and specific area of a conurbation, city or town. The area in question will typically exhibit high levels of multiple deprivation, usually the result of underlying adverse economic, social and environmental conditions. ABIs first came to the fore through the Urban Programme that was launched in 1968 but only became mainstream government policy in the 1980s under Margaret Thatcher's Conservative administration, driven by a dominant neoliberal ideology. They believed that finite resources would be better spent on spatially, financially and time constrained initiatives, concentrated on specific areas of need rather than being thinly spread across a region or urban area. Such an approach was also politically motivated as it allowed the usurping of usually Labour controlled (inner) city authorities by QUANGOs that were imposed by and accountable to Central Government rather than the local electorate. The approach was closely linked to the 'trickle-down' paradigm which suggested that the positive benefits generated by ABIs would not only 'trickle down' to the most needy and deprived members of society but would also spill over into neighbouring areas, creating a virtuous trickle down of wealth and opportunities.

Area-based initiatives have been developed under a series of guises over the past four decades; examples include the Urban Programme, Urban Development Corporations, City Challenge, Single Regeneration Budget Challenge Fund and New Deal for Communities. Often the financial approach to regeneration pursed by ABIs is that of 'pump priming', using public sector resources up front, to lever in private sector investment often through infrastructure and building projects. Most ABIs are time limited but the scale of projects involved in ABIs can range from the refurbishment of a single building to the redevelopment a whole neighbourhood. ABIs have also been employed to deliver flagship regeneration projects such as iconic cultural attractions, business and industrial parks, retail and leisure centres, and heritage and conservation areas, all of which seek to enhance the economic competitiveness and attractiveness of an urban area.

Tony Blair's New Labour government, elected in 1997, launched a plethora of ABIs in the form of 'action zones' and neighbourhood initiatives, across a range of sectors including health, education, crime, employment, in an attempt to eradicate social exclusion, the most prominent of which was New Deal for Communities. NDC was a well-resourced ABI that aimed to tackle inequalities faced by 39 of the poorest urban areas in England over a 10-year time frame. It should be noted that previous ABIs were often awarded through a competitive bidding process, e.g. City Challenge and Single Regeneration Budget (SRB); however, New Labour chose to use 'need', as represented by the Index of Multiple Deprivation, as the key determinant of which areas would receive resources.

One criticism of ABIs is that they often simplify the complex process of regeneration by focusing on hard (physical) outputs and ignoring the softer, more complex aspects of holistic regeneration; for example, many of the property-led regeneration initiatives of the 1980s which resulted in iconic flagship projects that did little to relieve deep seated deprivation and poverty – trickle down does not work! Some early ABIs were criticised for lack of engagement with the wider community and, under New Labour, the complexity and lack of coordination between individual ABIs was highlighted as a failing. By 2010, when the Coalition Government came to power, most ABIs were reaching expiry, but the Coalition chose to speed up the process under the auspices of debt reduction and austerity. Subsequently, a report by the Work Foundation (2012) reported that for the first time in 40 years there would be no ABIs in England tasked with reversing decline in some of England's most deprived neighbourhoods.

Further reading

CLES (2010) 'Area based initiatives – do they deliver?', available at: www.cles.org.uk/wp-content/uploads/2011/01/Area-Based-Initiatives-do-they-deliver.pdf (accessed 20/11/2013).

Lawless, P. (2004) 'Locating and explaining area-based urban initiatives: New deal for communities in England', *Environment and Planning C: Government and Policy*, 22: 383–399.

Matthews, P. (2012) 'From area-based initiatives to strategic partnerships: Have we lost the meaning of regeneration?', *Environment and Planning C: Government and Policy*, 30(1): 147–161.

Shaw, K. and Robinson, F. (1998) 'Learning from experience? Reflections on two decades of British urban policy', *Town Planning Review*, 69: 49–63.

Work Foundation (2012) *People or Place? Urban policy in the age of austerity*, Work Foundation, Lancaster.

14.17 Tax incremental financing

Key terms: finance; investment; regeneration; property development; infrastructure; business rates

Tax incremental financing (TIF) is a financing tool that allows local authorities to borrow against the predicted future growth in local business rates arising from a new development. This money is used to fund the infrastructure improvements and enable the development scheme to proceed. TIF is therefore a way for the public sector to fund infrastructure investment, to drive regeneration and unlock economic growth, by borrowing against the future additional tax revenues that the infrastructure investment unlocks. TIF is thus a form of PPP, based on the premise that in areas affected by blight, property taxes tend to be relatively low. When an area is redeveloped, property values rise and commercial activity increases, which causes an incremental increase in the tax revenues generated. In particular, TIF focuses on improving infrastructure in blighted areas to facilitate new development and increase property values. In essence, it is a mechanism of paying for growth with growth.

TIF originated in the USA in the 1950s and its use grew rapidly in the 1980s at a time when federal government funding for redevelopment was being cut and responsibility for urban policy was being transferred to lower-level government. In the UK, enthusiasm for TIF has grown in the prevailing post-credit crunch economic conditions.

Two preconditions are typically required to ration its use:

1 *Blight test*: demonstration that urban regeneration is required in order to prevent or remove 'blight'. US cities have their own statutory criteria of what constitutes a 'blighted' area, such as the presence of unsafe buildings, prevalence of depreciated property values, environmental contamination and inadequate infrastructure.
2 *'But for' test*: demonstration that private funds are insufficient and the development would not reasonably be anticipated or realised without the adoption of the TIF.

TIF schemes should therefore only be approved in situations where private sector developers have deemed a project viable were it not for the cost of infrastructure provisions. By implication, proposals deemed unviable for a range of issues as well as infrastructure problems, would not acquire TIF status. Other preconditions could include an assessment of value for money, that the scale and quantum of investment should be material to the local authority and that private sector investment would be leveraged through the deployment of TIF.

There are three financing models employed by TIF schemes:

1 Local authority prudential borrowing powers – in the UK this would be from the Public Works Loan Board.
2 Municipal bonds – this involves local authorities issuing bonds that are sold to the capital markets to raise money to pay for infrastructure.

3 'Pay as you go' developer financing – this involves no upfront borrowing by the local authority; instead, the private development company pays for all the infrastructure costs up front and then is repaid by the local authority using the uplift in taxation.

With core funding for regeneration in England only a third of what it was a few years ago, and the Coalition Government pursuing a localism agenda that sees responsibility for urban regeneration residing with local authorities and LEPs, TIF may be the most effective way of using public finances to encourage urban regeneration schemes at a time when developers and developments are suffering from a chronic lack of finance.

Further reading

Core Cities Group and British Property Federation (2010) *A Rough Guide to Tax Increment Financing*, CCG and BPF, London.

Greenhalgh, P., Furness, H. and Hall, A. (2012) 'Time for TIF? The prospects for the introduction of tax increment financing in the UK from a local authority perspective', *Journal of Urban Regeneration and Renewal*, 5(4): 367–380.

Hutchinson, N., Liu, N., Adair, A., Berry, J., Harran, M. and McGreal, S. (2012) *Tax Increment Financing – An opportunity for the UK?*, RICS Research, London.

Squires, G. (2012) 'Dear Prudence: An overview of tax increment financing', *Journal of Urban Regeneration and Renewal*, 5(4): 356–379.

15 Residential property

Julie Clarke, Rachel Kirk and Cara Hatcher

15.1 The private rented sector

Key terms: tenancies; rents; Local Housing Allowance; condition in the private rented sector; trends in the private rented sector; yields

The private rented sector is defined as consisting of properties, owned by individuals or companies, that are rented out for the purpose of making a profit from a wide range of different households from across the socio-economic spectrum. The proportion of households whose head is classified as in a professional or managerial role renting in the private sector has increased from 11 per cent in 1980 to 33 per cent in 2009. Conversely the proportion of skilled manual households fell from 19 per cent to 11 per cent over the same time period (Pawson and Wilcox, 2012).

There are different types of tenancies that exist in the private rented sector. The main tenancy types are as follows:

- *Regulated* – a regulated tenancy gives a tenant greater protection from eviction than other private sector tenancies. It applies to tenancies prior to January 1989.
- *Assured* – this is a legal tenancy that gives you the right to live in your accommodation for a defined period of time. This may be a fixed term such as 6 months or a periodic tenancy that rolls on and can be weekly or monthly. It came into effect in January 1989 and was the default tenancy until February 1997.
- *Assured shorthold* – the majority of tenancies in the private sector are assured shorthold with a fixed term of 6 to 12 months. This tenancy came into effect in February 1997.

Rents charged within the private sector are determined by the market – what a landlord believes a tenant will pay. For those needing assistance with their rent payment a Local Housing Allowance is in place to determine the amount of assistance available. Local Housing Allowance is set at the level of the cheapest 30 per cent of private rents where the dwelling is located.

The condition of stock in the private rented sector is very variable. In 2010 over a third of private rented dwellings, under the Housing and Health Safety Rating System were not deemed decent (Pawson and Wilcox, 2012). This percentage had almost halved in the period from 1996 when it stood at 62.4 per cent. However this lack of fitness is concentrated among long-term and more vulnerable tenants; 54 per cent of long-term tenants live in dwellings with substantial long-term repairs needed (CLG, 2011).

The number of private rented sector dwellings within the UK housing stock has undergone considerable change over time. At the end of the First World War the private rented sector made up over 90 per cent of the housing stock of the United Kingdom. From that point on there had been, until the turn of the century, a steady decline in the proportion of private rented dwellings (see Figure 15.1.1).

Since the start of the 2000s this downward trend has reversed so that by 2009 the percentage of dwellings in England rented from the private sector reached 17.4 per cent (Pawson and Wilcox, 2012). The more than doubling of the private rented sector between 1991 and 2011 can be explained by a number of factors:

1 Significant investment in the 1990s and early 2000s by institutional investors in the buy-to-let market for rent to professional households. Willingness to invest is dependent on yields. The scale of the yield available determines the willingness of institutional investors to invest in the sector. Historically, yields have been seen as high because of the sustained rise in house prices. However, the housing market downturn from 2007 saw the rates of yield diminish and, as a consequence, the number of buy-to-let investors fell. Recent years however have seen, for the reasons outlined in 3 and 4 below, increased demand for private rented properties and as a consequence average gross income yields were averaging 5.8 per cent in 2012 (Daly, 2012).
2 The buy-to-let trend had stabilised by 2012 but there was still continued growth in the size of the private rented sector. This can be attributed to the depressed housing market. Many home owners are reluctant or unable to sell and so they have turned to renting their property as an alternative.
3 A further contributory factor is the reduced availability of mortgage products and a loan-to-value ratio that has resulted in the need for larger deposits. This has, until April 2013, seen the number of mortgage approvals and the number of first time buyers decline. A consequence of this has been the rise of what the media has termed 'generation rent'.
4 A final compounding factor in the growth of the private rented sector has been the stimulus for demand created by a reduction in the number of new social housing homes for rent under the Government's National Affordable Homes Programme.

Further reading

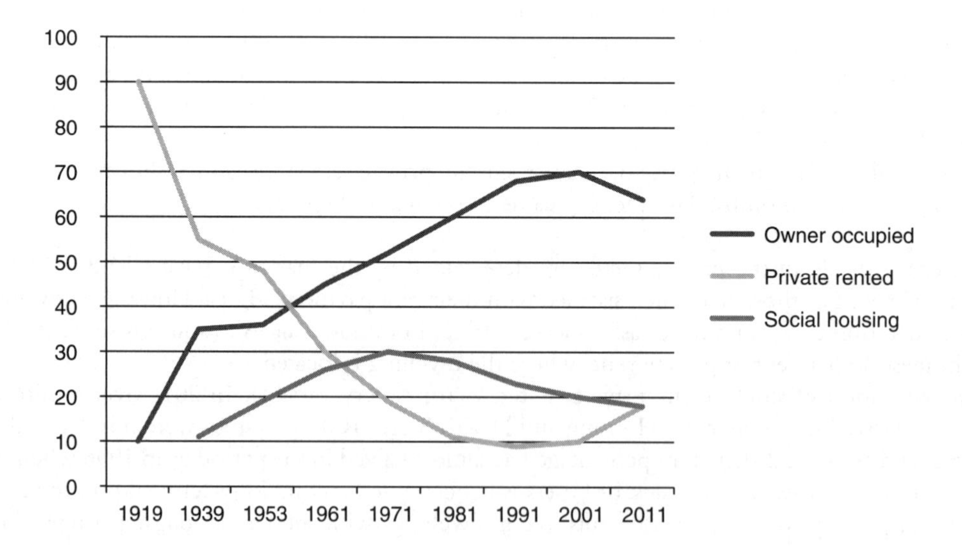

Figure 15.1.1 Changes in the tenure of dwellings 1919–2009.

Source: Adapted from Holmans (2005: 143); ONS (2009).

CLG (2011) *English Housing Survey 2009/10*, Bulletin 4.
Daly, J. (2012) 'Higher Yields Attract Investors', Residential Property Focus Q2, available at www.savills.co.uk/research_articles/141564/141008-0 (accessed 20/11/2013).
Holmans, A. (2005) *Historical Statistics of Housing in Britain*, Cambridge Centre for Housing and Planning Research, Cambridge.
ONS (2009) *Social Trends: Housing*, ONS, London.
Pawson, H. and Wilcox, S. (eds) (2012) *UK Housing Review 2011/12*, Chartered Institute of Housing, Coventry.
Rugg, J. and Rhodes, D. (2008) *The Private Rented Sector: Its contribution and potential*, University of York, York.

15.2 The social housing sector

Key terms: trends in the social housing sector; stock condition; Decent Homes Programme; household diversity

The social housing sector is defined as housing provided at below market rent on a secure basis to households unable to meet their housing needs in the private sector. The ability to provide housing at sub-market rents is due to two factors: the provision of social housing is subsidised; and the organisations providing and managing social housing are non-profit making. Landlords providing and managing social housing include housing associations (also called Registered Providers), local authorities and, on behalf of local authorities, arm's-length management organisations (ALMOs).

Over the course of the last century the proportion of social housing stock has fluctuated greatly from 11 per cent in 1939, peaking at 30 per cent in 1971 before declining to current levels of 18 per cent (see Figure 15.2.1). The role of social housing has also changed. The reasons for this are varied.

The advent of the Right to Buy policy in 1980 and a move away from funding the development of new council housing saw the beginning of a rapid decline in the number of council properties. By 2012, 1.77 million council homes had been sold under the Right to Buy (Pawson and Wilcox, 2013). By 1991 the number of council dwellings had reduced by a third to 19.8 per cent of total housing. This reduction was further compounded when an increasing number of local authorities opted to transfer their housing stock to become either new housing associations or to join an existing housing association. This was known as large scale stock transfer and was motivated primarily by a desire to invest in existing stock and develop new homes.

By 2011 local authorities owned only 7.6 per cent of the properties in the United Kingdom. This contrasts with the growth of housing associations from a starting point of owning 1 per cent of properties in 1939 to owning 10 per cent of the housing stock in 2011. This increase, gradual at first, gained in momentum from 1974 when housing associations begin to receive substantial government funding in the form of Housing Association Grants to build new homes. Despite reduction in grant levels over subsequent years, the vast majority of new social housing starts are housing association. In 2004/5 housing associations accounted for all but one hundred of the 19,070 social housing new build starts. By 2011/12 although the number of local authority new starts had reached 2,470, housing association new build was 33,410. The formation of a coalition government in 2010 saw the capital programme for the National Affordable Homes Programme reduced by 63 per cent. Government funded contributions to the provision of social housing reduced to approximately 20 per cent of build costs with the remainder being funded through reserves or private sector borrowing (Pawson and Wilcox, 2013).

The condition of social housing stock has improved significantly since the introduction of the Decent Homes Standard in 2000. In 1996 non-decent homes in the social sector accounted for 52.6 per cent of the stock. This figure had fallen to 20 per cent by 2010 (Pawson and Wilcox, 2012). The programme of Decent Homes investment aimed to ensure that by 2010 all properties in the social housing sector met the following:

- the statutory minimum standard for housing;
- a reasonable state of repair;
- reasonably modern facilities and services; and
- a reasonable degree of thermal comfort.

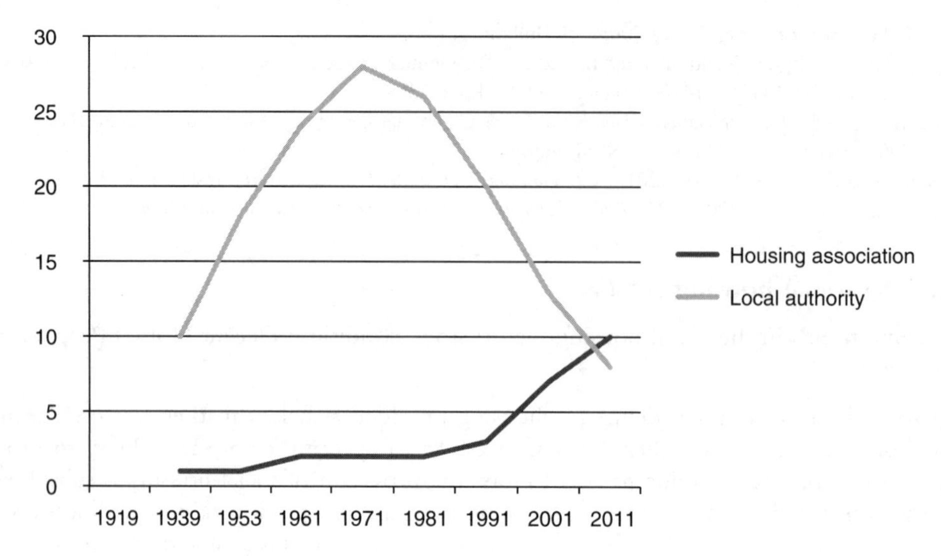

Figure 15.2.1 Changes in the tenure of dwellings 1919–2011.

Source: Adapted from Holmans (2005: 143); Pawson and Wilcox (2013: Table 17b).

The last three decades have seen a significant change in the characteristics of tenants of social housing. The diversity of households within the social housing sector has decreased since the 1970s. A long-term characteristic of those households leaving the sector has been for them to be in employment and in the 25–45 age group (Perrie and Capie, 2008). The proportion of households who are classed as economically inactive renting from the social housing sector has increased, from 47 per cent in 1980 to 62.5 per cent in 2009. Conversely the number of households whose head is classified as in a professional or managerial role has fallen from 5 per cent in 1980 to 1 per cent in 2009 and for those classified as skilled manual from 23 per cent in 1980 to 5 per cent in 2009 (Pawson and Wilcox, 2013). Social housing has become a tenure characterised by a concentration of younger and older households, economic inactivity and low incomes. This residualisation of the tenure has, as a consequence, concentrated households with multiple issues into concentrated geographic locations.

Further reading

Holmans, A. (2005) *Historical Statistics of Housing in Britain*, Cambridge Centre for Housing and Planning Research, Cambridge.

ONS (2009) *Social Trends: Housing*, ONS, London.

Pawson, H. and Wilcox, S. (eds) (2012) *UK Housing Review 2011/12*, Chartered Institute of Housing, Coventry.

Pawson, H. and Wilcox, S. (eds) (2013) *UK Housing Review 2013*, Chartered Institute of Housing, Coventry.

Perrie, J. and Capie, R. (2008) *Who Lives in Social Housing?*, CIH/Housing Corporation, Coventry.

15.3 Owner occupation

Key terms: freehold; leasehold; flying freehold; owner occupation; first time buyers

Owner occupation can be simply defined as a property in which the owners reside. The average number of current owner occupiers in the UK is 64 per cent (ONS, 2013). London, however, has a current owner occupier rate of only 49.6 per cent (ONS, 2013).

There are a variety of properties available for owner occupation within the residential sector including houses, flats and bungalows. These can be owned freehold or leasehold. A fee simple absolute in possession, commonly known as a freehold, refers generally to a house or bungalow that can be owned for an indefinite period of time and can grant the ownership to their next of kin into perpetuity. The owner occupier also has the definitive right to alter their property without having to negotiate with a landlord. A flying freehold is a less common type of ownership but readily occurs in Right to Buy properties whereby an owner occupier may share the common rights to a shared access, for example. This will be explored later in other forms of occupation.

A leasehold property refers to a property that is bound or held by the terms of a legal document, referred to as a lease. This document is drafted to include the relevant restrictions and covenants on a property. Generally, flats and apartments are held under such leases comprising of 99 years or 125 years. In terms of owner occupation of a leasehold property there is less freedom to alter and reside in the property without due consideration of others and importantly the terms of the lease. In some cases, a house can be held under a leasehold restriction. In many cases, an owner occupier will attempt to purchase the freehold, at a premium, from the landlord. If the landlord wishes to dispose of the freehold they must provide the leaseholder(s) with the first right of refusal whereby they have the right to purchase the freehold.

In terms of owner occupation rates in England and Wales, there has been a significant decline in owner occupation via a mortgage since 2001. There are a variety of reasons why this may have occurred including inflated house prices, restrictions on mortgage/bank lending and a stark reduction in wages. This has subsequently led to a surge in demand for rental properties rather than owner occupation. The changing situation is illustrated in Figure 15.3.1.

As illustrated in Figure 15.3.2, there is a striking reduction in first time buyers in occupation in the UK. This will ultimately have an effect on the residential property market as a whole and this is driven by the current economic climate whereby restrictions on bank lending, reduced consumer confidence and lower wages has seen a steady decline in owner occupation.

At present, owner occupation has suffered from bank lending restrictions making affordability and availability of mortgages difficult, resulting in a spike of demand in the rented sector.

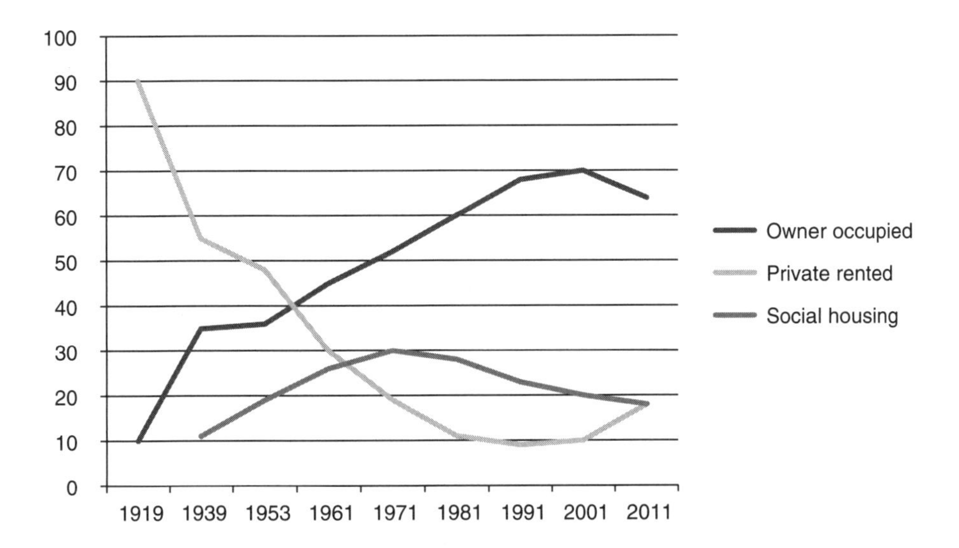

Figure 15.3.1 The rise and fall of owner occupation.

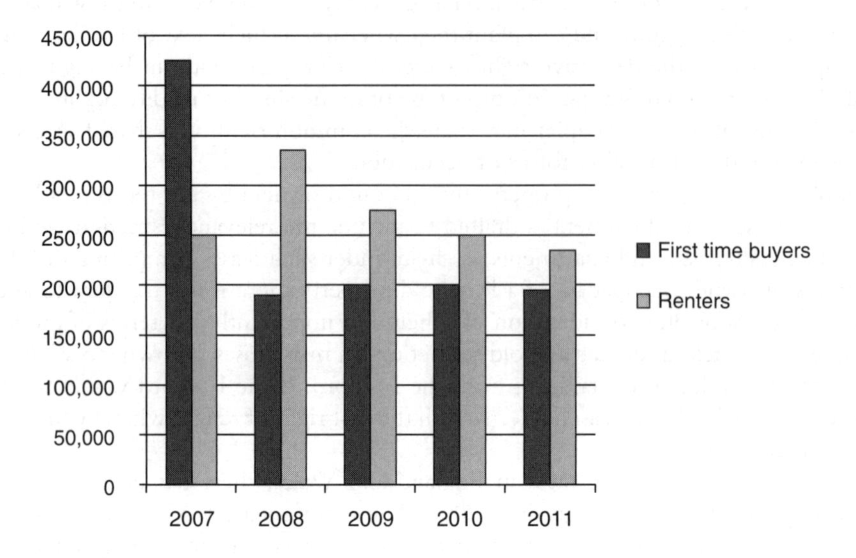

Figure 15.3.2 First time buyer activity versus growth in private renting.

Source: Adapted from Savills (2013).

Further reading

Mackmin, D. (2008) *Valuation and Sale of Residential Property*, 3rd edn, Estates Gazette Books, London.

ONS (2013) 'A Century of Home Ownership and Renting in England and Wales'. Available at: www.ons.gov.uk/ons/rel/census/2011-census-analysis/a-century-of-home-ownership-and-renting-in-england-and-wales/short-story-on-housing.html (accessed 20/11/2013).

Savills (2013) 'Residential Property Focus Q2 2013'. Available at: http://pdf.euro.savills.co.uk/residential---other/rpf-q213.pdf (accessed 20/11/2013).

15.4 Housing tenure – other forms of ownership

Key terms: shared ownership; shared equity; rent to buy

There are various other forms of owner occupation whereby the resident does not have sole ownership of the property. There are various private companies that act as part purchasers for clients wishing to own their own property.

Due to the recession, the first time buyer particularly has found difficulty in obtaining finance in order to purchase their own property so many companies and house builders have started to offer shared equity or help to purchase property.

Shared ownership

Many shared ownership schemes are governed by the government and under social housing. However, there are a variety of private shared ownership schemes in the market place.

Gov.uk (2013) states 'Shared ownership schemes are provided through housing associations. You buy a share of your home (between 25% and 75% of the home's value) and pay rent on the remaining share.' The benefit of this is that a person can own a percentage and pay a small rent until they are able to own the whole amount. This is popular in new build properties at present and the government has launched a help to buy scheme. The help to buy scheme is available where:

- the household earns £60,000 a year or less;
- you're a first time buyer;
- you rent a council or housing association property.

(Gov.uk, 2013)

The FirstBuy scheme works in conjunction with the Homebuy Agency and house builders 'will enable an eligible buyer to purchase a brand new property, funded by an affordable mortgage (Firstbuy, 2013).

Shared equity

The purpose of shared equity schemes works like a mortgage but instead of obtaining a loan from a bank or building society, a company will keep a share of the equity in the property, usually 30 per cent. This 30 per cent is on the value only and legally, the property is 100 per cent owned by the person living in the property. The issue with shared ownership is that a rent needs to be paid on the 30 per cent which can be more expensive in the long term. The situation is illustrated in Figure 15.4.1.

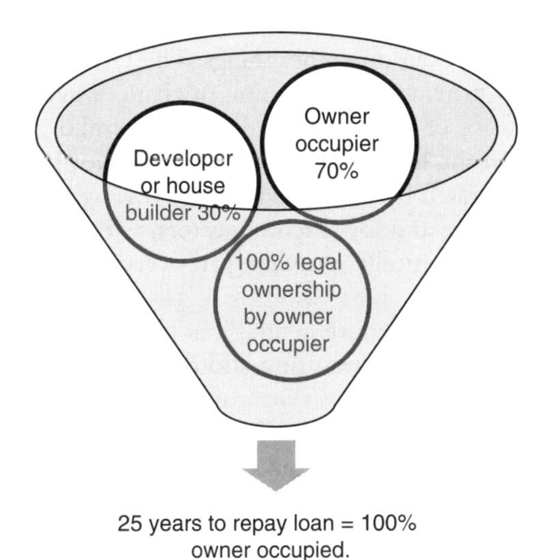

Figure 15.4.1 Shared equity scheme.

Rent to buy

Rent to buy offers first time buyers and tenant buyers an opportunity to buy their own property without the need for a deposit. This type of purchasing was introduced following the Government Homebuy Scheme. The purchaser acts as a tenant in the first instance but can own up to 100 per cent of the property. This is still in its infancy and is not a traditional form of property ownership; however, it may become popular if the market remains difficult for mortgage lending.

Further reading

Gov.uk (2013) 'Affordable home ownership schemes', available at: www.gov.uk/affordable-home-ownership-schemes/shared-ownership-schemes (accessed 20/11/2013).

Firstbuy (2013) 'Firstbuy scheme', available at: www.firstbuyscheme.org.uk/firstbuy.php (accessed 20/11/2013).

Mackmin, D. (2008) *Valuation and Sale of Residential Property*, 3rd edn, London, Estates Gazette Books.

15.5 Affordability in housing

Key terms: affordability; affordable housing; house prices to income ratio; mortgage costs to income ratio; rent to income ratio; residual income approach

Essentially, housing affordability is about the relationship between a household's income and the cost of housing – whether it be house prices, mortgage costs or rents. The related notion of affordable (or unaffordable) housing is a relative, shifting concept and is therefore difficult to define; it is open to varying interpretation and can be political in its application.

As housing has become increasingly expensive (reflecting rising demand from households outstripping the supply of housing), and earnings have increased significantly more slowly, issues of affordability in housing markets have become more prevalent –whether in terms of access to home ownership for first time buyers or addressing the housing needs of lower income households. Such debates about affordability require measures of the extent and nature of the problem, something which is not straightforward.

The most common measures of affordability consider the ratio of house prices to income (see for example DCLG, 2013). The focus on house prices can be seen as quite a crude measure of affordability as, for many people, housing costs are determined by mortgage payments which are influenced by availability of credit and interest rates among other things. The UK Housing Review Affordability Index (Pawson and Wilcox, 2013) provides an illustration of changes over time using mortgage cost to income ratios. See also the Council of Mortgage Lenders (www.cml.org.uk) for wide-ranging data and debate on affordability issues in the home ownership market. Affordability issues are also significant in the rental sector, as demonstrated by the UK Housing Review's (Pawson and Wilcox, 2013) comparison of rents (in both the private and social rented sectors) with earnings. Comparisons of such data can be made over time and geographically, providing an overview of the extent and distribution of the problem.

Other analyses of affordability are more tenure neutral, but potentially more complex and difficult to apply. The residual income approach to measuring affordability – i.e. considering the income a household has remaining once their housing costs are paid –provides an alternative to the various housing costs to income ratio measures. This approach recognises the particular issues facing lower income households and the greater proportionate impact of their housing costs given the smaller overall income in the first place. However, this approach also raises the difficult question of defining what is the minimum level of resources that a household requires to meet their non-shelter needs (see for example, Whitehead *et al.*, 2009).

Affordability, in its many facets, is a central aspect of the housing policy agenda. The Barker Review of Housing Supply (2004) in the UK argued that the development of additional housing stock was key to improving affordability. A major policy implication of the affordability debate in the UK is the need to provide more affordable housing, i.e. below market price housing. Policy interventions intended to boost affordable provision have included the use of the planning system as a mechanism to secure additional affordable housing, support for low cost and shared home ownership initiatives and government subsidised development of affordable housing. The latest Affordable Homes Programme 2011–2015 (Homes and Communities Agency, 2011) provides some grant funding for a range of housing associations, ALMOs, councils and private landlords to develop affordable housing, but at a much reduced rate which requires the developers to raise a significant proportion of the cost themselves through rental income. To facilitate this, the Government in 2011 introduced 'a new, more flexible form of social housing, Affordable Rent' – a key aspect of this model is that the affordable rent homes are available to tenants at 'up to a maximum of 80 per cent of market rent' (www.homesandcommunities.co.uk/ourwork/affordable-rent). The implications of this policy shift for increasing rents raises a question that underpins this discussion of affordability – how affordable is affordable?

Further reading

Barker Review (2004) 'Review of housing supply. Delivering stability: securing our future housing needs'. Final Report – Recommendations, HM Treasury, London.

DCLG (2013) Department of Communities and Local Government Statistical data set – Housing Market Series, DCLG, London, available at: www.cml.org.uk (accessed 20/11/2013).

Homes and Communities Agency (2011) *Affordable Homes Programme 2011–2015*, available at: http://www.homesandcommunities.co.uk/affordable-homes

Pawson, H. and Wilcox, S. (eds) (2013) *UK Housing Review 2013*, CIH, Coventry.

Whitehead, C., Monk, S., Clarke, A., Holmans, A. and Markkanen, S. (2009) *Measuring Housing Affordability: A review of data sources*, Cambridge Centre for Housing and Planning Research, University of Cambridge.

15.6 Homelessness

Key terms: statutory homelessness; main homelessness duty; priority need criteria; temporary accommodation; rough sleeping; hidden homelessness

The meaning of 'home' with all its psychological as well as physical connotations highlights the multifaceted and complex nature of what being without a home – homelessness – can mean (see for example, Somerville, 1992). Homelessness in its broadest sense is a wide-ranging and potentially contested concept and can include circumstances such as people forced to leave home due to domestic abuse, households living in overcrowded conditions, as well as those sleeping rough. Homelessness is intrinsically difficult to define and to measure.

Official homelessness statistics provide a measure of statutory homelessness. In England statutory homelessness is defined as 'those households which meet specific criteria on priority need set out in legislation, and to whom a homelessness duty has been accepted by a local authority' (DCLG, 2013a). Such legal definitions can vary – for example in 2003 Scotland moved away from priority need categories and towards a national strategy for preventing homelessness. In England and Wales the Homelessness Act 2002 required local authorities to develop a Homelessness Strategy, incorporating the provision of advice and assistance to help households meet their housing needs and avoid homelessness (i.e. not just those identified as priority homeless). This preventative approach has become an increasingly significant aspect of homelessness policy (Pawson and Wilcox, 2013).

In England statutory homeless legislation was first introduced in the 1977 Housing (Homeless Persons) Act. This has since been updated in the Housing Act 1996 and Homelessness Act 2002. Such legislation placed a legal obligation on local authorities in England to ensure suitable accommodation is available for the household where a main homelessness duty is owed; this requires several criteria to have been met.

Applicants will be accepted where they are eligible for assistance (households recently arriving in, or returning to, the country may not qualify for statutory assistance), are not intentionally homeless and whose circumstances mean that they meet priority need criteria. Priority groups as outlined in the legislation include pregnant women, households with dependent children, those who have lost their home through natural disaster such as flood, or are vulnerable because of old age, mental illness or physical disability. The Homelessness Act 2002 extended the priority need criteria to include 16–17-year-olds and those aged 18–20 who had previously been in care, as well as introducing additional 'vulnerable' considerations such as people fleeing their home due to violence or the threat of violence, and people leaving care, custody or the armed forces. Local authorities can also consider whether the applicant has a local connection to their area and may refer to an alternative local authority where this is not the case.

Where a main housing duty is accepted the local authority has an obligation to provide suitable and settled accommodation – this may be in their stock or through referral to a social housing organisation or in the private rented sector. Where appropriate accommodation is not available local authorities may provide temporary accommodation – for example in bed and breakfast hotels – to meet their duty to the homeless household in the meantime.

Since 2009 there has been an increase in the number of homeless households accepted by local authorities in England (i.e. households owed the main homelessness duty); Government statistics show that in 2012/13 there were 53,540 acceptances, a 6 per cent increase compared to the previous year (DCLG, 2013b). However, pressure groups, such as Shelter (www.shelter.org.uk), argue that statutory measures provide an underestimation of the scale and nature of homelessness. While attempts, however methodologically difficult, are made to assess the number of people who are sleeping rough (see DCLG, 2013a, rough sleeper counts), the notion of hidden homeless – 'non-statutory homeless people living outside mainstream housing provision' (Reeves and Batty, 2011) – highlights the invisible nature of aspects of homelessness. The hidden homeless can include, for example, single people who are sleeping on friends' sofas ('sofa surfing') or people living in illegal accommodation such as 'beds in sheds' (Pawson and Wilcox, 2013).

Further reading

DCLG (2013a) 'Homelessness data: notes and definitions', available at: https://www.gov.uk/homelessness-data-notes-and-definitions (accessed 27/03/2014).

DCLG (2013b) 'Series Homelessness Statistics', available at: https://www.gov.uk/government/collections/homelessness-statistics (accessed 27/03/2014).

Pawson, H. and Wilcox, S. (eds) (2013) 'Commentary: Housing needs, homelessness, lettings and housing management', in *UK Housing Review 2013*, CIH, Coventry.

Reeves, K. and Batty, E. (2011) *The Hidden Truth about Homelessness: Experiences of single homelessness in England*, Crisis, London.

Somerville, P. (1992) 'Homelessness and the meaning of home: Rooflessness or rootlessness?', *International Journal of Urban and Regional Research*, 16(14): 529–539.

15.7 Housing management – allocating property (social housing)

Key terms: waiting lists; allocations policy; choice-based lettings

Allocation is the task of matching a household with a property. The allocation of rented housing to prospective tenants is a fundamental role of the landlord or their agent irrespective of tenure. Allocations need to be made quickly and effectively to minimise rent loss, maximise rental income and prevent properties falling into disrepair or becoming vandalised.

The allocation of social housing is determined by need and all organisations allocating social housing must have regard to the law which states those households deemed as having a 'reasonable preference' must be given priority. Examples include those at risk of becoming homeless or are homeless, those facing violence or those with a medical condition that makes their existing housing unsuitable.

Local authorities are obliged to have in place a register of housing applicants called a housing waiting list. In some areas this is a combined list from which Housing Associations will also select applicants. This approach has been followed for a number of reasons:

- It avoids the need for applicants to complete multiple forms to join the waiting list of different landlords with housing in the same area.
- The avoidance of duplication ensures that estimates of housing need in an area are more accurate.
- Common waiting lists and common application forms are usually the result of cooperation between landlords in an area where landlords have agreed a shared set of priorities for allocating housing. By agreeing a shared set of priorities it ensures that whenever a vacancy arises, irrespective of landlord, the household in greatest need is allocated the property.

However some housing associations continue to have their own application form, waiting list and allocations policy. This is common for the large national associations and smaller specialist providers. It is for this reason that it is difficult to estimate with accuracy the number of people who are seeking accommodation. Statistics from government identify that waiting lists held by local authorities have increased by over a third (Table 15.7.1) in the ten years since 2002.

Social landlords must have a procedural system to determine who is allocated property and this procedure must be published. This is called an allocations policy.

Historically allocations policies have included the following:

- Date order – housing occurs in strict date order but as landlords have to have regard to the law date order tends to be used within other systems.
- Points schemes – an applicant is considered against a scheme where differential points are awarded for different factors such as overcrowding, households living separately, medical conditions.
- Group schemes – depending upon their primary housing need an applicant is allocated to a group, examples of which are medical cases, management moves, homeless households. The landlord will usually then apply a quota of their allocations to each group.
- Combined schemes such as points and date order were also commonly used.

Recently concerns were raised by applicants about the complexity and lack of transparency of many allocations policies. While landlords themselves found the traditional approaches time consuming, resource intensive and inefficient. Information collected about the applicant, often many years before an allocation is made, can quickly become out of date leading to wasted time verifying data in order to allocate a property. Refusal rates under these systems were high leading to properties standing empty for longer.

The government in 2000 introduced its Housing Green Paper, 'Quality and Choice – a Decent House for All'. This paper challenged organisations to rethink how they allocated housing to address these concerns. The result was the introduction of choice-based lettings (CBL) for the majority of social landlords.

CBL schemes are different from traditional needs-based waiting lists. They allow applicants who are registered to express an interest in any vacant property for which they are matched in terms of type and size. Properties are advertised widely via shops, newspapers, websites etc. Need still plays an important role in the allocation of vacant social properties and those registered are allocated into a priority band. Evaluation of CBL found that applicants welcomed the choice, control and transparency offered by the new system. For social housing landlords they found that properties were refused less often and tenants stayed longer in the property than under the previous systems.

In England in the year 2011–12 the social housing sector made almost 300,000 allocations of property (Table 15.7.2). The management of this process forms a fundamental part of a landlord's housing management function.

Table 15.7.1 Local authority waiting lists in England

Waiting lists	*2002*	*2012*
Unitary Authorities	165,975	420,406
London Boroughs	266,789	380,301
Metropolitan Districts	291,74	483,499
Shire Districts	408,836	564,330
Total	1,133,342	1,848,536

Source: Department for Communities and Local Government.

Table 15.7.2 Allocations of social housing 2011–12

	New tenants	*Existing tenants*	*Total allocations*
Local authority (including ALMOs)	89,100	51,800	140,900
Housing associations	23,000	135,000	158,000

Source: Department for Communities and Local Government.

Further reading

Thornhill, J. (2010) *Allocating Social Housing: Opportunities and challenges*, Chartered Institute of Housing, Coventry.

15.8 Housing management – rent collection and recovery (social housing)

Key terms: rent setting; housing benefit; rent collection; recovering rent arrears

Rent is the payment made to a landlord for the occupation of a property. Rent setting, collection and the recovery of unpaid rent are important elements of the management of any tenancy irrespective of tenure.

Rents within the social housing sector as well as paying for the management and maintenance of a tenants home also covers the provision of a range of services such as welfare benefit advice, into work advice, community involvement and tackling anti-social behaviour. In addition rents contribute towards the payment of capital borrowing costs allowing new homes to be built.

Rents within the social housing sector are not set by the market but do have a relationship to it. Since 2002 rent setting has been the subject of a rent formula aimed at establishing a convergence of rents between all social housing providers within a local area by 2012. The formula was based on the Retail Price Index (RPI) figure from the September preceding the increase and a maximum increase of +0.5 per cent and £2.

$$\text{RPI} + 0.5\% + £2$$

The 'formula rent' at the point of convergence is determined by three factors: value, size and local average earnings. An issue for social housing landlords with the rent convergence formula has been the wide variation in RPI over the last decade with the highest figure being 5.6 per cent in September 2011 which, including the additional £2, resulted in an average rent increase for tenants of 8 per cent in April 2012. Concerns have also been expressed when RPI falls to very low levels which results in small rent increases impacting upon organisations' abilities to service debt and raise finance. This formula has been continued following the change of government in 2010 and the deadline for convergence has been extended to 2016.

Households on a low income or in receipt of welfare benefits may be entitled to claim housing benefit to cover some or all of the rent costs and some service charges such as the upkeep of communal areas and the services of a caretaker. It does not cover energy, food or care costs. Approximately two-thirds of tenants in social housing claim housing benefit.

2013 saw significant changes to housing benefit policy for social housing tenants. Tenants of working age in receipt of housing benefit but who are deemed to be under occupying their property have had their housing benefit reduced. Under occupation by one bedroom results in a 14 per cent reduction in benefit while under occupation by two rooms will see housing benefit reduce by 25 per cent. Universal Credit is another significant change coming into effect during 2013–14. Under this move working age recipients of welfare benefits will receive one single payment that brings together all of the benefits they are in receipt of. As part of this measure tenants can no longer opt to have their housing benefit paid direct to their landlord but must actively make that payment once their Universal Credit is paid once every four weeks. Social housing landlords have raised concerns about the consequent increase in rent arrears as the direct result of these two changes to housing benefit payments. In addition rent payment transactions and recovery action will increase in numbers with the consequent increase in associated costs.

Table 15.8.1 gives details of the number of housing benefit payments and average payment by tenure in England.

A variety of methods are used by social landlords to collect rent:

- Door to door collection, where a rent collector called on a regular basis, was the traditional method of collecting rent. This has declined due to the cost of collection and the vulnerability of rent collectors to robbery.
- Most landlords have some form of office-based collection where tenants can come to the office to pay their rent. This has the benefits of being less costly, safer and rent accounts are updated immediately a payment is made.
- Tenants of some landlords can pay their rent via post offices, Allpay and other similar services. This method tends to be associated with higher arrears and delayed updating of rent accounts but is useful for more rural communities where access to a rent collection point is difficult.
- Landlords prefer either bank payments such as direct debits or online payment facilities whereby tenants can pay direct from their bank accounts via the landlord's website. This is less costly and easier to administer. However a major issue is the number of tenants without bank accounts.

Social housing rent arrears are difficult to accurately state due to the demise of the Tenant Services Authority and the Audit Commission. However, estimates are that 5 per cent of all rent due is owed to social landlords. CIPFA reported in 2012 that local authority rent arrears in England totalled £202,731,000.

Rent arrears recovery is an important function of housing management. Tenants get into arrears for a range of reasons and the reason for non-payment can influence how the debt is recovered. Reasons given for non-payment include problems with housing benefit, other debts, illness, low income, domestic problems, etc.

Prevention of rent arrears – organisations ensure tenants at tenancy sign up are aware of the consequences on non-payment. It is also important to ensure tenants income is maximised through the completion of a welfare benefits check and that appropriate housing benefit forms are completed.

Early action – organisations have in place policies that ensure when a payment is missed tenants are contacted immediately. The format of the contact can vary but is usually one of the following; letter, visit, text message or email. However, the intent is to establish the reason for the missed payment to minimise it happening again.

Table 15.8.1 Housing benefit recipients and average weekly payment in England

Local authority tenants		*Housing association tenants*		*Private tenants*	
Number of recipients (000s)	Average weekly housing benefit paid (£)	Number of recipients (000s)	Average weekly housing benefit paid (£)	Number of recipients (000s)	Average weekly housing benefit paid (£)
1,196	74.17	1,626	83.69	1,455	109.90

Source: Pawson and Wilcox (2013).

Legal action – as arrears increase organisations will then choose to follow a legal process that could ultimately lead to recovery of the property. However, before this, social landlords will have to serve a Notice of Intention to Seek Possession. At this stage further discussions can take place to put in place a payment plan acceptable to the organisation and within the means of the tenant. Failure to keep to an agreement can result in the landlord seeking a possession order from the court. This is costly and time consuming. In almost all cases a suspended possession order will be granted that formalises the payment plan.

All rent arrears recovery action is dependent upon the quality, accuracy and timeliness of the landlord's management information systems and the use of new and developing technologies to advise tenants that payments are due or have been missed.

Further reading

DCLG (2006) *Guide for Effective Rent Arrears Management*, DCLG, London.
Pawson, H. (2010) *Rent Arrears Management Practices in the Housing Association Sector*, Tenant Services Authority, London.
Pawson, H. and Wilcox, S. (eds) (2013) *UK Housing Review 2013*, Chartered Institute of Housing, Coventry.

15.9 Housing management – repairing property (social housing)

Key terms: repair categories; responsive repairs; cyclical repairs; planned repairs; stock condition survey; major works; Decent Homes Standard

Repairs to a landlord's stock are a fundamental aspect of housing management. It is essential that stock remains in good condition not only to protect its value as an asset but also because house conditions can impact upon the quality of life of tenants. To this end landlords will have in place a process to ensure that repairs are carried out effectively (in a timely manner) and efficiently, providing good value for money.

Within the context of social housing, repairs to housing stock represents a significant proportion of the revenue expenditure of landlords. In addition for most tenants this is their primary form of contact with their landlord once their tenancy has commenced and one upon which they will make a judgement on their landlord's ability.

Repairs to the stock of social landlords can be categorised in the following way:

- responsive repairs;
- cyclical works;
- planned works;
- major repairs.

These categories can be further subdivided.

Responsive or day-to-day repairs are unplanned and reported by tenants. Most landlords will subdivide that into emergency repairs that require an immediate response such as a broken window that requires boarding up or an older person's faulty heating system in winter. Other repairs are classed as routine in that they are not urgent but require action. These can include leaking radiators, heating faults or a fence blown down in the wind. Within this routine repair category landlords may determine a further set of response times depending on urgency. So in the case of the examples above, a heating fault would be deemed to be more urgent, generally, than a fallen down fence. Most landlords have moved towards appointment systems for responsive repairs in response to tenant feedback.

Cyclical works are carried out in a repeating, predetermined cycle. The duration may differ depending on organisational priorities and legal requirements. Gas servicing needs to be carried out annually. However, painting repairs and painting of woodwork may be carried out every five years. This ability to plan should increase efficiencies and reduce costs.

Planned maintenance work usually arises out of a comprehensive assessment of the stock condition. This will determine at which point components need to be replaced such as windows, roofs, heating systems, kitchens etc. Each of these planned works will have a different cycle so for example a whole heating system might be replaced every 15 years and a roof every 50 years. However depending on the design, original methods of construction, building materials used, and maintenance over time these timescales will vary enormously. This is where a stock condition survey can help a landlord understand the peculiarities of their stock especially where there are non-traditional design and build schemes. This type of work is usually organised in large contracts covering a specific geographical area. All components in the designated contract area will be replaced irrespective of condition. While this may appear self-defeating in terms of value for money it is more expensive to replace individual components on a responsive basis once the contract has ended.

Major works occur when a landlord undertakes extensive improvements to a property in one go. Properties can have many if not all of the following components replaced in one contract over a few weeks – electrics, heating systems, windows, kitchens, bathrooms. As you would expect, there is a great deal of upheaval and ideally the work should be carried out without the tenant in the property. However, moving a tenant out (decanting) for the duration of the work into another property is costly. Costs incurred are loss of rent from the decant property, moving costs, delays to the contract if decants are delayed. Most organisations will attempt to carry major works around the tenant with support in place through contract liaison officers. Decanting will occur where tenants are vulnerable.

The introduction of the Decent Homes Standard introduced in 2000 set a minimum standard for the condition of social housing stock. The standard stated that all housing must:

- be above the statutory minimum standard;
- be in a reasonable state of repair;
- provide reasonably modern facilities and services;
- provide a degree of thermal comfort.

This resulted in a significant amount of investment in social housing stock, dramatically improving its condition.

Irrespective of the method of repair, social landlords regularly collect performance data around customer satisfaction in order to improve services.

Further reading

Joseph Rowntree Foundation (2002) 'Britain's housing 2022', available at www.jrf.org.uk/publications/britains-housing-2022-more-shortages-and-homelessness (accessed 20/11/2013).

15.10 Housing management – managing tenancies (social housing)

Key terms: anti-social behaviour; prevention; tenancy agreement; diversionary activities; support; ASBOs; family intervention projects; enforcement

Tenancy management has, in recent years, become an increasingly important aspect of housing management in social housing. Tenant expectations about the quality of where they live coupled with increased emphasis from the sector regulator, the Homes and Community Agency, on tenancy standards has resulted in a greater emphasis on tenancy management. In parallel with this has been a growth in incidents of anti-social behaviour (ASB). ASB can range from occasional but annoying neighbour nuisance such as noise late at night to, sustained harassment and criminal activity. Most minor cases of ASB can be resolved through dialogue and mediation. Criminal activity is referred to the police.

For the more problematic cases of ASB housing managers have had to develop a range of responses and work with a diverse range of partners in order to achieve successful resolution. The Joseph Rowntree Foundation (JRF) in 2005 used the British Crime Survey as a source of data to identify some of the characteristics associated with ASB. They found:

- ASB was an acute problem but for a minority of people;
- acts of ASB were concentrated in deprived urban areas (34 per cent in inner city areas saw ASB as a problem);
- 27 per cent saw rowdy teenagers as the worst form of ASB they experienced;
- empirical research in the case study areas saw drug and alcohol misuse, neighbour disputes and 'problem families' as contributing to ASB.

Responses by social landlords can be categorised as:

- preventive;
- supportive;
- enforcement.

Preventive measures adopted by social landlords begin with the tenancy agreement. This contractual agreement between tenant and landlord varies depending upon the type of tenancy on offer. However, most tenancy agreements will place a responsibility on the tenant, their household and visitors not to cause a nuisance or annoyance either within the property or in the broader estate and neighbourhood. This broadening out of the responsibilities beyond the tenant and their property was in response to much of the ASB being caused by household members and their friends.

Many landlords offer new tenants introductory or probationary tenancies. These tenancies do not offer security of tenure for the first 12 months and enable landlords to recover possession of the property more easily should breaches of tenancy occur. As part of receiving a new tenancy most social landlords spend time with tenants at sign-up explaining their responsibilities and the consequences of breaching their tenancy, including what constitutes ASB.

Other preventive actions landlords can take is rapid responses to minor issues, such as graffiti, petty vandalism and neighbour disputes to prevent escalation. In areas where ASB is a noticeable problem many social landlords have introduced neighbourhood wardens whose role is to work with the community to assist in the prevention and reporting of ASB. Neighbourhood wardens along with housing officers and others such as youth workers also deliver diversionary activities, especially during the school holidays and summer nights, that offer young people on an estate positive activities. This helps avoid unintended youth ASB.

A second strand of the responses to ASB is supporting households to change their behaviour. Depending upon the severity of the ASB, different approaches to support will be adopted. Support measures often involve housing officers working with other agencies to deliver a more holistic solution. Where young people are involved in ASB the landlord along with the police and school may ask the family to sign up to an Acceptable Behaviour Contract (ABC). This sets out the acceptable and unacceptable behaviours and the responsibilities of the young person and their parents/carers and is signed by all parties. A breach of an ABC can be used as evidence when going to court for possession of the property.

Landlords can also seek injunctions or specifically Anti-Social Behaviour Orders (ASBOs). ABSOs seek to ensure individuals modify their behaviour or keep away from areas where they have caused problems in the past. ASBOs can be sought for both tenants and individuals who are not tenants but whose behaviour is deemed as being unacceptable. Breaches of an ASBO can lead to the individual being jailed. ASBOs have been criticised for being over-used and used in inappropriate situations, such as expecting someone with severe mental health problems to change their behaviour, and so their value has diminished over time.

Where problems are more deeply entrenched and have occurred over a long period family intervention projects can provide support to households either in their home or removed from the area in a residential placement with family support officers. This is usually only used where the behaviour has deteriorated so much normal relations between the family and their neighbours has irretrievably broken down and there is no possibility of them returning. Families are appointed an individual support worker who coordinates the involvement of other agencies such as educational welfare, health services, the police and probation services, youth offending teams and schools.

The final set of actions available to social landlords involves enforcing the tenancy agreement. This requires the landlord to provide evidence of serious and persistent breaches of the tenancy. Often those affected feel very vulnerable and are unwilling to give evidence against their neighbours. When this is the case landlords have a number of options available to them such as professional witnesses and the use of mobile closed circuit surveillance. The landlord can then either seek a possession order suspended on the guarantee of adherence to the tenancy agreement or absolute possession. The number of possession cases granted has increased in recent years as the result of more robust evidence collection.

Further reading

Shelter (2007) 'Tackling ASB', available at: http://england.shelter.org.uk/professional_resources/policy_and_research/policy_library/policy_library_folder/back_on_track_a_good_practice_guide_to_addressing_anti-social_behaviour (accessed 20/11/2013).

15.11 Housing management – allocating property (private rented sector)

Key terms: availability; affordability; housing stock

The private rented sector consists of a wide variety of housing stock, ranging from studio apartments to detached country estates. The allocation of this housing stock is very different to social housing, with less of an onus on fundamental need and more on demand for a particular type of property.

The private rented sector is currently a growing market place and there is a demand for rental property in this sector from students, young professionals, families and individuals and also the Department of Social Security (DSS).

Some of the housing stock in the private rented sector is allocated to social housing needs and the DSS will provide rental payments on behalf of these tenants.

Allocation of property

The allocation of property in the private rented sector is fundamentally through the market system and based on affordability. If a tenant can prove they can afford the rent through their salary and references, they will be provided an assured shorthold tenancy on the property for usually 6 months.

The rent is payable every month and includes a tenant's deposit at the start of the tenancy and any associated management costs.

In terms of the DSS, they will pay the rent and associated costs on behalf of the tenants and therefore will have their own allocation policy to adhere to regarding number of beds, location and affordability.

Schemes to assist private rented tenants

There are a number of schemes to assist people in establishing funds for a property in the private rented sector. Many local councils offer assistance to tenants by providing a paper guarantee, which can be used to guarantee a month's rent if they have problems with payments.

This type of scheme is beneficial to a variety of tenants including:

- homeless priority groups, whereby offering a property in the private rented sector would prevent homelessness;
- homeless households with priority needs;
- single people or couples who are homeless.

The Localism Act 2011 has allowed 'local authorities to discharge their homelessness duty with an offer of accommodation in the private rented sector without the applicant's consent' (Gov.uk, 2013). However, there are a number of issues with this use of legislation, including a lack of suitable property and the possible increase in homelessness that this could lead to.

There are two distinct users of the private rented sector – the privately guaranteed and referenced individual users and those that have the support of the local councils and DSS who assist or make rental payments.

Further reading

Gov.uk (2013) 'The Private Rented Sector'. Available at: www.publications.parliament.uk/pa/cm201314/cmselect/cmcomloc/50/50.pdf (accessed 14/11/2013).

Mackmin, D. (2008) *Valuation and Sale of Residential Property*, 3rd edn, Estates Gazette Books, London.

15.12 Housing management – rent collection and recovery (private rented sector)

Key terms: rent collection; non-payment; recovery

Many landlords in the private rented sector collect rent themselves, or through a managing agent. Rent is generally paid on a monthly basis either directly to a landlord or if through a managing agent into a client account for that property.

The RICS *UK Residential Property Standards*, fifth edition (October 2011), known as the *Blue Book*, reiterates the importance of the lease in providing information on the service charge. It does, however, emphasise the importance of the management surveyor especially when dealing with recovery costs.

The *Blue Book* states 'There are no statutory rights for landlords to recover any costs or to collect service charges in advance; the rights are purely contractual, thus the lease is paramount.' (RICS, 2011).

The Tenancy Deposit Scheme

At the start of a new assured shorthold tenancy, a deposit is paid which usually amounts to one calendar month rental payments. The Tenancy Deposit Scheme (TDS) is run by the Dispute Service Ltd. It is insured by the government. It has two main roles, which are protecting deposits and resolving deposit disputes. Tenancy deposit protection applies to all deposits for assured shorthold tenancies that started in England or Wales on or after 6 April 2007.

The deposit for the tenancy is placed in a fund that is protected and is independent from the landlord and therefore as long as the tenant does not damage the property, they will receive their whole deposit back. The legal requirements are contained in sections 212–215 of, and Schedule 10 to, the Housing Act 2004.

In terms of protection for the tenant, the Housing (Tenancy Deposits (Prescribed Information) Order 2007 is a protection for the tenant and includes information on the landlords contact details, the deposit amount and how the deposit can be used. A certificate is provided to the tenant so they are aware of their deposit contribution.

Recovery of rent

The recovery of the non-payment of rent is a complex issue and can be a lengthy process in terms of recovery, not only of rent but possession of the property.

There are two main routes accessible to private landlords if they wish to regain possession of their property under the Housing Act 1988, allowing for 2 months written notice, which are:

- *Section 21* – gives a landlord an automatic right of possession without having to give any grounds (reasons) once the fixed term has expired;
- *Section 8* – allows a landlord to seek possession under certain grounds, one of which includes rent arrears.

These involve attending court proceedings which, depending on the specific case can take from 14 days to 42 days, if there are exceptional hardship circumstances. However, the landlord must have already given 2 months written notice before either notice will be accepted.

Further reading

Gov.uk (2013) 'Improving the rented housing sector', available at: www.gov.uk/government/policies/improving-the-rented-housing-sector--2/supporting-pages/private-rented-sector (accessed 14/11/2013).

Gov.uk (2013) 'Gaining possession of a privately rented property let on an assured shorthold tenancy', available at: www.gov.uk/gaining-possession-of-a-privately-rented-property-let-on-an-assured-shorthold-tenancy (accessed 14/11/2013).

Parliament.uk (2013) 'Private Rented Sector', available at: www.parliament.uk/business/committees/committees-a-z/commons-select/communities-and-local-government-committee/inquiries/parliament-2010/private-rented-sector/ (accessed 14/11/2013).

RICS (2011) *UK Residential Property Standards*, 5th edn, RICS, London.

15.13 Housing management – repairing property (private rented sector)

Keys terms: repair; obligation; responsibility; landlord and tenant

The repairs to a residential property are linked to the health and safety issues that surround the management of residential property. The role of the landlord or the managing agent is to protect not only the tenant but also the property.

The tenant's responsibility

The tenant has the responsibility to maintain the property to a fair standard and in the condition at the start of their tenancy. In order to ensure that the tenant maintains the property in a condition that is acceptable to the landlord, the landlord or landlord representative is usually allowed to enter the property, with prior written permission, in order to inspect the condition of the property periodically. At the start of the tenancy, the landlord will undertake a condition report on the property which details any issues with the property, any furnishings or fixtures that are part of the property and therefore should remain at the end of the tenancy, and the extent of the property including any grounds and outbuildings.

The tenant's obligations can be found in the tenancy agreement they have signed and the tenant must ensure that they adhere to these. The managing agent should always refer to the tenancy agreement if a dispute about repairs arises.

The tenant in many cases must ensure they maintain the garden area, any furniture (which must be fire resistant) and any equipment supplied by the landlord such as white goods. The tenant must ensure the property is kept clean, free from damage, generally maintained; for example with the changing on light bulbs, fuses, unblocking of sinks and protection of the heating system through sensible and proper use.

If an instance where the tenant has left the property in a condition below that required by the tenancy agreement, the tenant's deposit money can be used to rectify the damage or maintenance issue caused.

The landlords obligations

As with the tenant, the landlord is required to keep the property in a well maintained order which must be equal to the condition at the start of the tenancy. However, if the property was not in good order at the beginning of the tenancy there may be an obligation for the landlord to upgrade the property to a habitable standard.

The tenancy agreement, however, will not determine that a landlord must significantly improve the property simply to suit the requirements of the property. The landlord is not required to carry out repairs to the property until they are made aware of the defect by the tenant (Landlord and Tenant Act 1985). This notice can be verbally or in writing; however, in writing is preferable as proof of the defect may be necessary if a dispute arises. The landlord is usually not responsible for carrying out repairs to the plasterwork or skirting boards unless the damage has occurred directly due to an exterior defect that the landlord was responsible for, such as a dampness problem.

The Landlord and Tenant Act 1985, section 11 – repairing obligations in short leases for the landlord

Section 11 states that in a short lease such as an assured shorthold tenancy, the landlord has the responsibility:

> '(a) to keep in repair the structure and exterior of the dwelling-house
> (b) to keep in repair and proper working order the installations in the dwelling-house for the supply of water, gas and electricity and for sanitation
> (c) to keep in repair and proper working order the installations in the dwelling-house for space heating and heating water' (Landlord and Tenant Act 1985).

It is important that both the tenant and the landlord adhere to their repairing obligations which change from tenancy to tenancy. Therefore, it is important that the managing agent or surveyor is aware of the responsibilities of each party in the agreement to ensure that the obligations are undertaken when required.

Further reading

Gov.uk (2013) 'Private renting', available at: www.gov.uk/private-renting/repairs (accessed 20/11/2013).
Landlord and Tenant Act 1985.

15.14 Housing management – managing tenancies (private rented sector)

Key terms: management; private rented sector; tenancies

The management of residential property can be diverse and complex and there is a fundamental need to understand the legal obligations of tenancies that a property management might encounter in the profession. These tenancies all have unique requirements as a managing agent; you will need to abide by these rules and regulations as a managing agent and ensure your landlord does also. There are three types of tenancy that a residential property management surveyor may come across in the management of tenancies and residential properties. These are:

- the regulated tenancy;
- the assured tenancy;
- the assured shorthold tenancy.

The regulated tenancy

A regulated tenant or tenancy consists of:

- a tenant of residential accommodation only;
- the tenant has exclusive use of living accommodation;
- the tenancy is not the market rent each month;
- the tenancy began prior to 15 January 1989.

The regulated tenant has the right to reside in the property until they choose to leave or they die. The tenant pays well below market rent for the property and this has a fundamental negative impact on the value of the property because the tenancy runs with the sale of the property.

The assured tenancy

For an assured tenant or tenancy:

- the rent is payable to a private landlord or their representative such as a managing agent;
- they have the right to reside in the property free from disturbance from the landlord or any other person associated with the landlord;
- the tenancy began between 15 January 1989 and 27 February 1997 and was not defined as an assured shorthold tenancy when signed.

The assured tenant has a right to automatically renew their tenancy. The can pay market rent unlike the regulated tenancy.

The assured shorthold tenancy

This is the most common form of residential tenancy and is utilised in most residential tenancies. They usually consist of 6 months tenancy agreements and include:

- a tenancy that commenced either on or after 28 February 1997;
- the rent is market rent and is directly payable to the landlord or their representative (the property manager);
- the landlord has to provide 24 hours' notice before they or their representative can access the property.

The assured shorthold tenancy was introduced to provide a fairer tenancy agreement for the landlord as well as the tenant. The landlord can regain possession of the property after the 6 month agreement but must give the tenant 2 months' written notice to allow the tenant the opportunity to find and move into another property; however the tenant only has to give one months' notice.

When valuing or managing residential properties, there is a need to understand the type of tenancy you are dealing with. The tenancy in place can determine the impact on value whether that be a negative impact or little impact at all; in managing a property, you need to be aware of your ability to enter a property or your legal obligations to the tenant to maintain their quiet enjoyment of the property they rent.

Further reading

Gov.uk (2009) 'Regulated tenancies', available at: www.gov.uk/government/publications/regulated-tenancies (accessed 20/11/2013).

Gov.uk (2009) 'Assured tenancies and assured shorthold tenancies', available at: www.gov.uk/government/publications/assured-and-assured-shorthold-tenancies-a-guide-for-landlords (accessed 20/11/2013).

15.15 Housing support – independent living

Key terms: self-determination; deinstitutionalisation; aids and adaptations, lifetime homes

Conceptually, independent living is a set of beliefs or philosophy: 'it is about disabled people having the same level of choice, control and freedom in their lives as any other person' (Office for Disability Issues, 2008). Independent living challenges what is perceived as a socially constructed view that has traditionally medicalised disability, portraying it as a medical problem that requires specialist help and intervention, i.e. encouraging dependency rather than self-determination and independence. Empowerment of the individual underpins the concept of independent living. Whereas historically people with mental and physical disabilities may have been placed in institutions or some form of residential care, there has been a long-standing policy support for deinstitutionalisation and the provision of care within communities and independent living.

Access to housing, the design of housing, and the types of support available to people in their own home, are key factors in facilitating independent living, enabling people to choose to live in communities rather than in residential care if they wish to. Practical support for people to live in their own homes can incorporate many different aspects. It can include, for example, the provision of floating support through support workers going into people's homes to enable people with disabilities to live independently. Physically, the current housing stock is poorly equipped to deal with the needs of disabled people and the increasing number of people in later life. Support for independent living can include the provision of aids and adaptations within houses. These range from grab rails being installed, to specialist equipment such as bath seats being provided, to the replacement of a bath with a level access shower, for example.

A more radical way of sustaining independent living, and achieving a housing stock that is more flexible and adaptable to diverse needs, is to consider the design of new housing development – something that is encapsulated by the notion of Lifetime Homes (see for example www.jrf.org.uk). Lifetime Homes are designed so that they can respond to the changing needs of individuals and families as their circumstances change over their lifetime and remain accessible. The Lifetime Homes standard is made up of 16 criteria – including for example width of available parking and doorways being sufficient to facilitate use of buggies and wheelchairs; height of window sills and window handles meaning they are accessible; a wheelchair accessible WC on the entrance level with potential for accessible shower (see www.lifetimehomes.org.uk. for more details). This standard has been incorporated into the Code for Sustainable Homes (see Concept 16.5). Meeting all the criteria is mandatory for Code level 6.

In spite of a clear policy commitment to the principle of independent living, its implementation in practice presents considerable challenges as it requires significant investment as well as cultural change.

> We know that people's housing plays a critical role in helping them to live as independently as possible, and in helping carers to support others more effectively. However, people told us during the *Caring for our future* engagement that there were not enough specialised housing options for older and disabled people.
>
> (DoH, 2012)

Further reading

DoH (2012) White Paper, 'Caring for our future: reforming care and support', HMSO, London.

Homes and Communities Agency at www.homesandcommunities.co.uk/ourwork/care-support-specialised-housing-fund (accessed 20/11/2013).

Office for Disability Issues (2008) available at: http://odi.dwp.gov.uk/odi-projects/independent-living-strategy.php (accessed 05/06/2014).

15.16 Housing support – specialist supported housing

Key terms: vulnerable adults; supported housing schemes; foyers

For some people, their particular needs and circumstances may mean that they require a form of supported housing that specialises in providing integrated support services and accommodation. Such circumstances might include health-related needs such as people with physical or learning disabilities or mental health problems or drug/alcohol dependency issues. Specialist supported housing can also be provided for people whose experiences or changing circumstances make them vulnerable or create additional needs, for example women and children fleeing domestic violence, refugees and asylum seekers, people leaving prison and teenage parents. The Homes and Communities Agency identifies 15 categories of vulnerable groups in relation to supported housing (see www.homesandcommunities.org.uk for further information).

For a variety of reasons, such individuals and households may face a number of challenges and difficulties in obtaining and sustaining their own tenancy or home. Moreover, the often multifaceted and complex nature of the types of problems such people are facing means they require support as well as accommodation. Supported housing schemes are necessarily diverse in nature, offering a variety of accommodation and types of support. For example, they can be purpose built residential schemes incorporating accommodation units and communal facilities, or they may be clusters of housing where additional support is provided for the residents, or they can include hostel settings. Support services can include practical assistance and social support to enable vulnerable people to live as independently as possible; they can include personal development or training in terms of life skills (for example budgeting or cooking) or provision of education, training and employment opportunities; or the support might include counselling and advice, as well as many other types of support.

An example of a specialist supported housing initiative is the foyer movement. In the 1990s foyers were introduced in the UK as a response to the combination of youth unemployment and youth homelessness, drawing on the foyer approach developed in France. The intention was to address the 'no home–no job–no home' cycle by providing young homeless people aged 16–25 with accommodation in a secure environment and also support in terms of learning and skills, personal development, and training and employment opportunities. Now there are over 120 foyers across the country each providing both housing and direct services for young people (see www.foyer.net).

Supported housing schemes are provided by a range of organisations, including local authorities, social landlords, charity and voluntary organisations or social enterprises. Supported housing may be an organisation's specialist function or it may be an aspect of an organisation's wider remit. Organisations may also focus on a particular issue, e.g. mental health problems, or they may provide housing support for a range of client need groups. The combination of housing and specialist support for vulnerable adults provides a vital service in communities. It has been argued that the preventative approach of such schemes is beneficial and cost effective. However, while there are significant needs, there are significant challenges in terms of funding and developing such services. Financially, there is a complex economy for housing, support and care. As public funding becomes more limited, charitable, voluntary and private resources and alternative models of development and provision are increasingly being explored.

Additional reading

Homes and Communities Agency at www.homesandcommunities.org.uk (accessed 20/11/2013).
National Housing Federation at www.housing.org.uk (accessed 20/11/2013).

15.17 Housing an older population

Key terms: demographic changes; three phases of old age; specialist housing; staying independent

The population of the United Kingdom is ageing. This raises many issues, one of which is how are we going to meet the housing needs of an older population. Old age is complex and the aspirations of older people, in terms of their housing, differ enormously.

An ageing population is seen as a national challenge for housing, health and social care. Statistics from government indicate the following trends:

- 60 per cent of projected household growth up to 2033 will consist of households where the main householder is over 65.
- By 2016 this will mean an extra 2.4 million older households.
- Older people occupy a third of all homes.
- 3.7 million older people live alone.
- The 75+ age group is growing faster than any other.
- In 2000 90 per cent of older people lived in mainstream not specialist housing.

These statistics, however, mask a range of other trends that need to be addressed in terms of investment in provision and support services:

- It is predicted that over one million people will suffer from dementia by 2025 rising to 1.7 million by 2051.
- The number of older people with a learning disability is not accurately known but is expected to increase between 37 per cent and 59 per cent.
- The number of disabled older people will double to 4.6 million by 2041.
- Over the next 30 years the population is predicted to increase in the following age bands:
  - 65+ by 76 per cent an increase of 7.3 million
  - 75+ by 95 per cent an increase of 4.4 million
  - 85+ by 184 per cent an increase of 2.3 million.
- These figures also hide variations geographically as the population in rural areas is growing and ageing faster than that in urban areas.

(DCLG/DoH/DWP, 2007)

It is also essential that we do not see old age as homogenous but recognise that there are significant variations between what is commonly defined as the three stages of old age. Many of the most important distinctions between groups of older people are about phases of life rather than chronological age groups and importantly each stage has differing needs.

The three phases of older life are identified as:

- older workers – people who either still have jobs or are still actively seeking work;
- third agers – people who have retired from work and can reorganise their lives around leisure, family responsibilities, non-vocational education or voluntary work;
- older people in need of care – people whose lives are substantially affected by long-term illness or disability.

These three groups are not wholly distinct, and people can move between them in all directions, and the types of housing and levels of services that these groups will require will be different, making decisions about investment in provision and support more complex.

Priorities for housing older people are seen as supporting independence for as long as possible and the offering of high quality housing choices.

In terms of specialist housing for older people it has traditionally been a combination of older persons' bungalows usually in the social or owner occupier sector; sheltered accommodation developed post war; or residential care. Bungalows are in short supply, sheltered accommodation is dated in terms of size, style and communal facilities and residential care is perceived very negatively as an option of last resort.

Recently there has been a growth in the development of retirement communities for sale by leasehold. Developers in this niche market include Retirement Villages and McCarthy & Stone. This is an expensive option for a significant number of older people particularly in relation to on-going costs such as service and management charges.

In 2012 the Joseph Rowntree Foundation (JRF) published a report that stated that the specialist housing on offer did not reflect the choices most older people make. They also stated that while three quarters of older people are owner occupiers only a quarter of specialist housing is for sale. Those properties traditionally built by the social housing sector because of pressures on funding and scheme viability tend to have only one bedroom. Older people aspire to a minimum of two bedrooms for many reasons. They concluded that there has been slow progress in developing different housing options for older people. Specialist retirement developers are seen as having a limited model that for many is not affordable. General house builders, who have the opportunity to develop mixed age housing, were seen as not designing with older people in mind as a target market. In 2013 JRF concluded that there was insufficient investment in specialised housing options and that this lack of investment was compounded by the economic downturn in the housing market.

The government has in the light of this committed, via the Department of Health, £160 million of capital funding for specialist housing providers to bring forward proposals for development of specialist housing to meet the needs of older people (and adults with disabilities) outside of London in the 5 years from 2013–14. This fund is to be managed by the Homes and Community Agency.

Other trends within housing and older people are focused around keeping people independent in their own home through direct support such as care workers; assisted technologies such as community alarm systems and remote health monitoring; handy person services carrying out small-scale repairs and other tasks such as hanging curtains; advice services such as First Stop providing free independent advice on housing, care and finance; and adaptations. The latter process for those in the private and owner occupied sectors is managed by a network of Home Improvement Agencies. The agencies advise on the type of improvements and adaptations needed, assist with applications for grants and loans; help to find reputable tradespeople; and oversee the work. Home Improvement Agencies are not-for-profit organisations supported by the government and local authorities.

Further reading

DCLG (2013) 'Providing Housing and Support for Older and Vulnerable People', available at: www.gov.uk/government/policies/providing-housing-support-for-older-and-vulnerable-people (accessed 20/11/2013).

DCLG/DoH/DWP (2007) *Lifetime Homes, Lifetime Neighbourhoods: A national strategy for housing in an ageing society*, DCLG, London.

Joseph Rowntree Foundation, available at: www.jrf.org.uk/work/workarea/housing-with-care-older-people (accessed 20/11/2013).

Joseph Rowntree Foundation (2012) *Older People's Housing: Choice, quality of life and under occupation*, JRF, York.

16 Sustainability

Graham Capper, John Holmes, Ernie Jowsey, Sara Lilley, David McGuinness and Simon Robson

16.1 Sustainable development

Key terms: inter-generational equity; intra-generational equity; environmental limits; mega-cities; ecological footprint

A definition of sustainable development that is often used is: 'development that meets the needs of the present without compromising the ability of future generations to meet their own needs'. This comes from the United Nations World Commission on Environment and Development report of 1987 (often known as the Brundtland Report after its chairperson). It incorporates the concept of inter-generational equity or fairness to people in the future.

The concept of sustainable development also requires intra-generational equity or fairness to the current generation in the form of relief of poverty and decent living standards in both the developed and the developing world. The built environment and especially world cities play a key role in achieving sustainable development and inter- and intra-generational equity.

Sustainable development has a number of practical features that can be implemented directly in policies for the built environment. They include:

- *Environmental limits* should not be exceeded e.g. absorption of waste, protection against radiation, overuse of resources, air quality etc. The precautionary principle should be adopted where there is any doubt.
- *Demand management* should follow an acceptance of environmental capacity limits, e.g. instead of building roads or airports in anticipation of more demand (which is then self-fulfilling) the increase in demand should be managed. This is done for energy through conservation and efficiency measures rather than building new power stations.
- *Environmental efficiency* – this can be increased by:
  – increasing durability
  – increasing technical efficiency
  – avoiding overuse of renewable natural resources (faster than replenishment)
  – closing resource loops by reuse, recycling and salvage
  – reducing primary non-renewable resource use.
- *Welfare efficiency* means gaining the greatest human benefit from each unit of economic activity. Environmental issues cannot be separated from social issues and the built environment should protect and enhance health.
- *Equity* for people in the current generation (intra-generational equity) because poverty leads to environmental damage.

According to the United Nations, 70 per cent of the world population will be urban dwellers by 2050. Mega-regions have developed where urbanisation has spread to join cities together to form continuous urban sprawl, often with unbalanced development, income inequalities and slum shanty towns. The largest of the mega-regions is Hong Kong–Shenhzen–Guangzhou in China, which is home to about 120 million people. Other mega-regions have developed in Japan and Brazil, and are developing in India, west Africa and elsewhere in the developing world.

The UN-Habitat report in March 2010 described the process of urbanisation and 'endless cities' as unstoppable but generally positive because these regions are driving wealth. The top 25 cities in the world account for more than half the world's wealth and the five largest cities in India and China account for 50 per cent of the wealth in those countries. Migration to cities for economic reasons continues at a fast pace and much of the wealth in rural areas comes from people in urban areas sending money back. The process of urban sprawl, however, is wasteful, adds to transport costs, increases energy consumption, requires more resources and causes the loss of prime farmland.

The biggest threat to sustainable cities comes from inequalities of income. The more unequal cities become, the higher the risk that economic disparities will result in social and political tension. Cities that are prospering the most are generally those that are reducing inequalities (UN Habitat 2012/13).

The ecological footprint of cities is the land required to feed them, to supply their water, to supply their timber and other products and to reabsorb their carbon dioxide emissions by areas covered with growing vegetation. For London this is at least 50 million acres (which equals the Great Britain land area) or 125 times its surface area of 400,000 acres. If the population of the world lived at the standards of USA citizens, then three planets would be needed. A typical North American city with a population of 650,000 would require 30,000 square kilometres of land – in comparison, a similar size city in India would require 2,800 square kilometres (Wackernagel and Rees, 1996). The high levels of consumption are associated with large amounts of waste and pollution which are causing health problems and climate change. Changing this is a central objective of sustainable development policies. The built environment can be made more sustainable by achieving targets to reduce its ecological footprint.

Further reading

Jowsey, E. (2011) *Real Estate Economics*, Palgrave Macmillan, Basingstoke, ch. 14.

UN (2009) 'World Economic Situation and Prospects 2009', Doc. No. E.09.II.C.2, available at: www.un.org/esa/policy/wess/wesp.html (accessed 03/12/2013).

UN-Habitat (2010) *State of the World's Cities 2010/2011, Bridging the Urban Divide*, available at: http://sustainabledevelopment.un.org/content/documents/11143016_alt.pdf (accessed 05/06/2014).

UN-Habitat (2012/13) *State of the World's Cities 2012/2013*, available at: http://www.unhabitat.org/pmss/listItemDetails.aspx?publicationID=3387 (accessed 24/03/14).

Wackernagel, M. and Rees, B. (1996) *Our Ecological Footprint*, New Society Publications, Canada.

World Commission on Environment and Development (1987) *Our Common Future*, 'The Brundtland Report', Oxford University Press, Oxford.

16.2 Biomass

Key terms: renewable energy; energy policy; renewable heat incentive

Biomass is the catch-all term used to describe energy primarily generated from plant based sources rather than fossil fuels and can include:

- wood and wood residues (chunks, sawdust, pellets, chips)
- agricultural residues (straw, chaff, husks, animal litter and manure)
- energy crops (hybrid poplars, switchgrass, willows)
- municipal solid waste (MSW).

Biomass is considered to be a renewable energy source, as opposed to the burning of fossil fuels. If the biomass material comes from sustainable sources, such as miscanthus grass or short coppice willow (which has a three-year rotation), it may be deemed to be carbon neutral because the carbon released from the plant material has already been fixed from the atmosphere through the process of photosynthesis. However, environmentalists have raised concerns that as demand increases for biomass it is likely that sufficient supply can only be achieved through harvesting of mature trees, which will cause a considerable carbon debt.

Biomass can be converted into direct heat energy through burning or this heat energy can be used for electricity generation through a steam turbine. Biomass has a significant role in achieving the UK's target of 15 per cent renewable energy by 2020. There are plans to convert coal-powered power stations to biomass but this is a finely judged investment decision primarily based in the UK on carbon taxes and government renewable energy subsidies. For example, in 2011 work began on the conversion of the 900 MW Tilbury power station to biomass but in 2013 the plan was shelved due to uncertainty about the renewables subsidies. Meanwhile, Drax, the UK's largest power station, which supplies 7 per cent of the country's electricity and burned 25,000 tons of coal per day in 2012 has committed to being predominantly biomass by 2017. This is likely to involve importing 7.5 million tons of biomass from the USA and Canada which will need 4,600 sq miles of forest to maintain continuous supply. Despite the carbon emissions involved in harvesting, processing and transporting the biomass material Drax claim that their biomass generation has half the CO_2 emissions of a modern gas-fired power station.

Prior to the discovery of North Sea gas in the 1970s the default heat source was coal; practically this meant the delivery and storage of bulky, dirty fuel, manual or automatic feeding of boilers and the disposal of ash. The discovery of natural gas and the installation of gas boilers and fires made the whole process of keeping warm cheaper and cleaner. There has been significant research and development of gas-fired technology to produce highly efficient (95 per cent) gas boilers which are readily understood by the design and facilities management professions. The advent of biomass effectively returns heating to a Victorian (but low carbon) age with the consequent need to relearn storage and delivery strategies. Biomass heating installations in residential or commercial applications will have the same basic requirements: a boiler, an automatic feed system for the biomass fuel, biomass storage area and easy access for bulk delivery and disposal of ash. Compared to gas boiler installations, there are considerably more moving parts and maintenance implications.

In urban residential applications biomass will not be competitive with gas-fired heating and is only likely to be installed in small-scale district heating schemes in social housing – for example the replacement of a boiler installation in a block of flats. In rural areas, where occupiers are dependent on expensive oil or electricity as the primary heating source, biomass may be an attractive alternative provided the residential property has sufficient space for the relatively large boiler, biomass feed hopper and space to store the timber pellets which will normally be delivered in 15–20 kg bags. It should be noted that timber pellets require considerable processing of the source timber which has to be finely chopped and re-formed into pellets to a specific moisture content to allow for automatic boiler feeding and minimising of ash production.

In commercial applications there has been a significant growth in installations over the past decade, particularly to satisfy town and country planning conditions which have required 10 per cent of energy to be supplied from renewable sources. In these instances compliance has been readily achieved by installing a small biomass boiler as an adjunct to the main gas boiler installation. Anecdotal evidence suggests that in many instances these biomass boilers are not normally switched on. However, as the economics of gas/biomass fuels changes in future, biomass may become the primary fuel source.

Apart from these token installations there is a substantial body of experience in the UK on the successful commercial application of biomass technology. One of the better examples is Queen Margaret University Edinburgh. The university was built in 2008 as a low-carbon campus for 5,000 students. It illustrates the best aspects of the biomass technology. The boiler installation and long-term fuel supply is a package deal between the university and the Baccleuch Estate (largest landowner in the UK). The main heat source is an

Austrian Kohlack boiler of 1,500 kW with two small gas boilers for summer use. The biomass installation provides 90 per cent of the heat and hot water demand for the university and produces 75 per cent less carbon than the equivalent gas installation. The fuel is chipped timber from the Baccleuch Estates in Scotland and can be supplied at up to 60 per cent moisture content as the boiler preheats the fuel. The short supply line and minimal processing needed reduces the embedded carbon in the fuel supply.

A driver of demand for biomass heating has been the UK Government's Renewable Heat Incentive (RHI). This will provide a subsidy to a variety of renewable heat technologies, biomass, heat pumps and solar collectors. The payment applies to installations completed after 2009 and will continue to be paid for 20 years. It is likely that this subsidy will encourage the installation of more biomass heating schemes. However, in 2013 the Government announced that it was planning to restrict RHI to power stations, hence the uncertainty described earlier.

The wholesale adoption of biomass could have unintended consequences, for example, delivery of biomass by truck (rather than gas by underground pipe) will have an impact on traffic density. Security of supply and cost must also be considered; the cost of biomass has been predicted by the American Energy Information Administration to be double the cost of natural gas by 2017 because of the reduction of gas cost due to fracking in the USA.

In the UK, biomass has been adopted as a renewable heat source; however, the technology is more risky in terms of design, security of supply and ongoing maintenance compared to the default gas installation. Future energy costs are difficult to predict – carbon taxes and subsidy regimes make accurate calculation impossible. Moreover, the conventional wisdom that biomass is carbon neutral is clearly flawed when supply chains are taken into consideration; however, coal is imported from Austria and LPG from Qatar, so the question is: which is the least worst choice.

Further reading

Committee on Climate Change (2011) 'Bioenergy Review', available at: www.theccc.org.uk/publication/bioenergy-review/ (accessed 24/10/2013).

MacKay, D. J. C. (2009) *Sustainable Energy without the Hot Air*, UIT Cambridge, Cambridge.

16.3 Building Research Establishment Environmental Assessment Method

Key terms: sustainable buildings; environmental assessment; environmental impact

The Building Research Establishment Environmental Assessment Method (BREEAM) is, as the name suggests, an environmental assessment method devised and administered by BRE Global, a brand owned by the BRE Trust, a charitable not for profit organisation. The BRE grew out of the Building Research Station which was established by the government in the 1920s to carry out research into building materials and set standards.

BREEAM is one of a number of environmental standards in existence around the world, in the US LEED (Leadership in Energy and Environmental Design), in Australia, Greenstar and CASBEE (Comprehensive Assessment System for Building Environmental Efficiency) in Japan. Almost every developed country has devised an environmental assessment method, often with a methodology linked to the specific climate conditions or environmental issues in the region. The International Institute for Sustainable Development listed just under 900 methodologies in 2012.

The aims of BREEAM are listed by BRE as:

- to provide market recognition to low environmental impact buildings;
- to ensure best environmental practice is incorporated in buildings;
- to set criteria and standards surpassing those required by regulations and challenge the market to provide innovative solutions that minimise the environmental impact of buildings;

- to raise the awareness of owners, occupants, designers and operators of the benefits of buildings with a reduced impact on the environment;
- to allow organisations to demonstrate progress towards corporate environmental objectives.

When it was devised in the 1990s there was a need for an independent environmental assessment system as property developers could claim to be producing 'green' buildings on the most spurious of grounds. BREEAM takes a comprehensive approach to environmental sustainability allowing a project to gain 'credits' under a wide range of criteria. The credit criteria and the weighting of the credits are established by a panel of stakeholders, the criteria are modified regularly (every two or three years) to keep them ahead of legislation and current practice. The main headings considered are:

- management (of the construction process);
- health and well-being (of the building occupants);
- energy (relating to energy consumption as measured by the energy performance certification (EPC) metering and low- or zero-carbon technologies);
- transport (relating to the location of the building and carbon generated by commuting or visitors);
- water (consumption and metering);
- materials (to encourage the selection of low environmental impact construction materials linked to the Green Guide to Specification);
- waste (to discourage wastage of materials in the construction process);
- land use and ecology (to encourage the use of brownfield sites, the preservation of ecological features and enhancement of biodiversity);
- pollution (to reduce pollution effects across a wide range of issues including light, noise and watercourses).

As demand for environmental accreditation has increased so BRE Global have devised BREEAM methodologies for a broad range of building types, beginning with offices, then going on to cover industrial buildings, education, prisons, courts and retail. Nonstandard buildings can be covered by an 'other buildings' BREEAM which BRE will devise on a bespoke basis. Buildings can be designated as Pass, Good, Very Good, Excellent or Outstanding; the Outstanding designation has been added since 2008 to recognise the most sustainable buildings which will comprise only the top 3 per cent of developments.

BRE Global train and licence BREEAM assessors through what is effectively a franchise system. BRE Global are ISO 9001 accredited and have rigorous QA procedures to monitor the quality of assessor reports and adherence to technical standards.

Despite being devised in 1990 BREEAM accreditation was relatively rare until 2000 when the Government made a policy decision that they would not occupy offices unless they had a Very Good BREEAM rating. This policy was generally ignored until about 2005 when the requirement was adopted by regional development agencies when approving grants, and planning authorities began to impose BREEAM ratings as a condition of granting planning permission. These developments forced design teams to engage with the BREEAM process and as a result there has been a considerable improvement in certain aspects of the construction process – for example, the Considerate Constructors scheme and the availability of timber from sustainable sources. Initially BREEAM accreditation was based on the building design at the procurement phase but now accreditation can be achieved at the design stage and the post construction stage to ensure that the aspirations of the design team are achieved in practice.

As BREEAM has matured so the detail of the BREEAM manuals have grown and the time needed to collect the evidence needed. Design teams and clients can be very selective as to which credits they will attempt to gain – some credits are relatively easy to gain while others, such as the responsible sourcing of materials, require extensive time-consuming evidence collection. In some instances the provision of sub meters to measure substantial energy uses and sub tenancies are installed to gain the credit, but not used by office managers when the building is occupied.

BREEAM has become institutionalised in the UK but as recently as 2011 research by Fuerst and McAllister found no statistical evidence of enhanced appraised property value. BREEAM's effect is to provide the developer and occupier the assurance that buildings are built to an accepted environmental standard; however, research shows that this is no guarantee that the building will be economical to run or pleasant to work in.

Further reading

Fuerst, F. and McAllister, P. (2011) 'The impact of Energy Performance Certificates on the rental and capital values of commercial property assets', *Energy Policy*, 39: 6608–6614.

www.bsria.co.uk/news/breeam-or-leed/ (accessed 20/11/2013).

www.bre.co.uk/greenguide/podpage.jsp?id=2126 (accessed 20/11/2013).

16.4 Code for Sustainable Homes

Key terms: environmental benchmarking; market differentiation; building regulations; zero carbon homes

The Code for Sustainable Homes has its origin in Ecohomes, the residential environmental benchmarking scheme devised by the Building Research Establishment (BRE) in 2000. Ecohomes was effectively 'nationalised' when Defra commissioned the BRE to revise the scheme to become a compulsory rating and certification system rather than a voluntary label. The Code is now the national standard for the sustainable design and construction of new homes. The Code aims to reduce carbon emissions and create homes that are more sustainable. It set design principles for energy, materials and water usage and to improve health, reduce pollution and minimise waste. The Code measures the sustainability of a new home against nine categories of sustainable design, rating the 'whole home' as a complete package.

The Code's nine elements of design and construction are:

- Energy and CO_2 emissions
- Water use
- Materials
- Surface water runoff
- Waste
- Pollution
- Health and well-being
- Management
- Ecology.

Similar to BREEAM, homes can be classified in two stages: at the design stage and after a post construction review. The buildings can be classified from level 1 to 6 stars where level 6 represents a zero carbon home. Code level 3 has become the minimum standard for social housing funded by the Homes and Communities Agency (HCA) and their counterparts in Wales and Northern Ireland.

The code provisions deal with fundamental design and specification issues – for example, fittings to reduce energy and water consumption and the selection of materials of low environmental impact. In addition, credits are awarded for features that help the occupant to lead a more sustainable lifestyle – for example, an outdoor clothes drying line, cycle storage shed and rainwater storage butts. The Code sets minimum standards for energy and water use at each level. It is intended to provide valuable information to home buyers, and offers builders a tool with which to differentiate themselves in sustainability terms.

The Code may be described as a 'top-down' scheme, enforced as a condition of funding by HCA and occasionally by local authorities as a condition when selling land to house builders. It is noticeable, however, that the Code does not appear to be affecting the marketability of properties. A review of new homes for sale by house builders and existing homes being sold by estate agents shows that the Code is not being used as a sales feature. Legislation insists that the energy performance certificate (EPC) should be featured in the sales particulars in an attempt to introduce energy performance as an issue in the sales negotiation. In fact, homes are bought with reference to location, neighbourhood, size, age, garden, public transport and school facilities. For the vast majority of home buyers the energy performance of their home or indeed the provision of a rainwater butt is not a deal breaker.

In July 2007 the Government's *Building a Greener Future* policy statement announced that all new homes would be zero carbon from 2016. There were to be clear carbon reduction milestones along the way, specifically in 2010 and 2013. The 2010 Building Regulations required a 25 per cent reduction in emissions; however, the 2013 revision has watered down the proposed reduction from the intended 25 per cent to just 6 per cent. This has been in response to an election commitment not to put additional burdens on house builders during the current parliament. In addition, the requirement for zero carbon energy for the home has been amended to exclude electrical appliances. Consultation on 'allowable solutions', i.e. what technology is deemed to be zero carbon and to what extent carbon emissions may be mitigated off site has yet to be agreed. This reduction in ambition and lack of clarity has left the house building industry in some disarray; some developers are working on designs to achieve the zero carbon target but the lack of certainty has undermined the drive to innovate for many. It may be that the 2016 target is postponed to try to ensure that the additional cost of zero carbon is only imposed when the market has fully recovered from the 2008 recession.

Further reading

DCLG *Code for Sustainable Homes*, available at: www.gov.uk/government/policies/improving-the-energy-efficiency-of-buildings-and-using-planning-to-protect-the-environment/supporting-pages/code-for-sustainable-homes (accessed 20/11/2013).

DCLG *Code for Sustainable Homes Case Studies*, Vols 1, 2 and 3, DCLG, London.

16.5 Combined heat and power

Key terms: cogeneration; sustainable buildings; local energy generation

The conventional means of generating electricity in most developed countries is to have a centralised power station using gas, coal or oil to produce steam that passes through as turbine to produce electricity that is then distributed around the region using a high voltage network. There are losses at each stage, in converting the fuel into steam, in the generation system and in the distribution network. The Department for Energy and Climate Change quote end user efficiency figures of 35 per cent for coal and 48 per cent for gas generation. How much better would it be if the electricity could be generated locally so that the waste heat that would otherwise disappear up a cooling tower could be used to heat a building, and the electricity used on site to avoid losses in the distribution system?

The solution is called combined heat and power (CHP), sometimes referred to as cogeneration. It has been used extensively in continental Europe, particularly when hooked up to district housing schemes. Although the theory looks indisputable, the practicalities at the local and individual level can be problematic and need careful scrutiny and design.

CHP systems can be classified as large-scale custom built CHP for industrial applications, often using steam or gas turbines for outputs of 1–100 MWe. Small-scale packaged CHP systems of 60 kW to 1.5 MWe will use a reciprocating engine while a micro-CHP, designed for individual homes, uses a sterling engine as its primary source.

In more detail, a modern packaged CHP system will comprise a reciprocating engine, similar in scale to a truck or a ship (depending on the scale of the building), fed by gas or oil, this works at its optimum power to produce electricity which is fed directly into the host building. The waste heat, which in a truck or ship would be removed to a radiator and out through the exhaust system, is carefully harvested through heat exchangers to heat water for building heating and domestic hot water. To work effectively, the host building needs a fairly constant demand for heat and electricity; for this reason they are most effective in hospitals and swimming pools, both of which have a constant and substantial demand for hot water.

In substantial applications such as hospitals CHP has provided a dual benefit of reducing energy costs and CO_2 emissions. Some hospital trusts have also entered into private finance initiative (PFI) schemes with energy supply companies (ESCOs); the ESCO supplies and maintains the CHP system in return for an inflation-linked energy cost over 25 years. The risk of installing and maintaining the capital equipment is transferred to the ESCO but the risk to the Hospital Trust is to ensure that they have judged their demand for heat and energy correctly and they are not left with an energy contract for a hospital they may wish to run down or close during the period for the PFI contract.

At the domestic scale, micro CHP, which comprises a wall-mounted, gas-fired Sterling engine, will generate electricity when heating. It is similar in size to a large wall-mounted gas boiler and thus is an easy replacement for an existing gas boiler. It would be best installed in a utility room or garage as the engine gives off a mechanical hum. Typically, micro CHP systems will generate 1Kw of electricity, the homeowner will gain the main benefit if they can use the electricity when the heating is being used, and this may mean doing the ironing or laundry in the evening rather than during the day. Currently in the UK micro CHP is eligible for the Feed-in Tariff of 10p for each kWh of energy generated and a further 3p for each kWh exported to the grid, provided the CHP system and installer are certified under the Micro Generation Certification Scheme. While the sterling engine technology is proven over decades, micro CHP has not been installed widely. The public may be wary of an unfamiliar technology and the difficulty of securing maintenance services from local plumbers.

CHP has been successfully embraced by large-scale industrial, leisure and health undertakings. Domestic installations have had limited application – perhaps because of the high efficiency of conventional gas boilers and competition and promotion of PV panels for those who want to make an investment in renewable technology.

Further reading

Department of Energy and Climate Change 'CHP Focus', available at: http://chp.decc.gov.uk/cms/ (accessed 20/11/2013).

Energy Saving Trust, 'Choosing a renewable technology' available at: www.energysavingtrust.org.uk/Generating-energy/Choosing-a-renewable-technology (accessed 20/11/2013).

16.6 Electric vehicles and electric vehicle infrastructure

Key terms: battery electric vehicles; hybrid electric vehicles; electric vehicle charging infrastructure; electric vehicle charge points

The Climate Change Act (2008) established a long-term framework and targets to reduce the UK's greenhouse gas emissions by at least 80 per cent, compared to 1990 levels, by 2050. Transport accounts for approximately a quarter of domestic carbon dioxide (CO_2) and other greenhouse gas emissions in the UK and is contributing to declining air quality. Reducing greenhouse gases from transport will help the long-term goal of reducing the UK's greenhouse gas emissions by at least 80 per cent compared to 1990 levels by 2050. Ultra-low-emission vehicles – any vehicle that emits extremely low levels of carbon emissions compared with current conventional vehicles, such as electric, plug-in hybrid and hydrogen powered cars and vans – help cut down greenhouse gas emissions and air pollution on our roads.

A battery electric vehicle is a vehicle that is run on an electric motor, powered by a rechargeable battery, charged via the mains electricity. They are entirely run on electricity, and with no petrol or diesel used, the vehicles do not produce any tailpipe emissions. Most EVs (electric vehicles) offer a full battery range of about 100 miles, for example the Nissan Leaf travels 124 miles per full charge. The distance/range of the EV when fully charged largely depends on how the car is driven, dependent on driving style, road condition, weather, heating/air conditioning used on the journey etc. Driving the car in the most efficient way maximises the car's range and ensures driver satisfaction.

A hybrid electric vehicle is a vehicle powered by a combustion engine with varying levels of electrical energy storage captured when braking and stored in a battery or supercapacitor.

A hydrogen fuel cell electric vehicle is a vehicle driven by an electric motor powered by a hydrogen fuel cell that creates electricity on board.

Plug-in hybrid electric vehicles (PHEVs) combine petrol or diesel engines with a battery and electric motor; they can be plugged into the mains electricity to charge the battery, and provide a much longer driving range on electric-only power. The batteries have a greater storage capacity than an existing hybrid, making it possible to drive considerably further using the electric motor only. A PHEV is a plug-in version of a full hybrid, usually with a larger battery and a greater electric driving range. In addition to capturing energy when braking, the on-board battery can be charged from an external source when the vehicle is not in use.

Hybrid cars use both electric motors and a conventional petrol or diesel engine to drive the vehicle. A series hybrid is a plug-in hybrid where the wheels are driven exclusively by an electric motor with an additional internal combustion engine connected in series. The engine runs at optimum efficiency to power an on-board generator to charge the battery. 'Range extenders', which use a small combustion engine to charge the battery to enable longer-distance journeys, are a type of series hybrid.

In the UK, the Office of Low Emission Vehicles (OLEV) is the unit that brings together the work of three government departments; the Department for Transport (DfT), the Department for Business, Innovation and Skills (BIS), and the Department of Energy and Climate Change (DECC). OLEV supports the early market for ultra-low emission vehicles (ULEV) and is responsible for:

- grants to reduce the upfront cost of new ULEVs;
- a programme of research through the Technology Strategy Board and supporting wider green growth opportunities;
- encouraging UK businesses to seize commercial opportunities in the ULEV sector;
- raising awareness, developing and strengthening the capability of ULEV manufacturing, demonstration and the associated UK supply chain;
- a nationwide recharging infrastructure strategy, including funding to eight pilot areas under the Plugged-in Places programme;
- contributing to the development of new carbon dioxide emissions standards (OLEV, 2013).

Government support in the form of the Plug-in Vehicle Grant is available to reduce the higher initial cost of purchasing an electric vehicle. This provides a subsidy of:

- 25 per cent, up to £5,000, towards the cost of an electric car;
- 20 per cent, up to £8,000, towards the cost of an electric van.

At the end of September 2013, 5,702 claims had been made through the Plug-in Car Grant scheme (OLEV, no date).

Cenex – the UK's first Centre of Excellence for low carbon and fuel cell technologies – is a delivery agency, established with support from the Governments Department for Business, Innovation and

Skills, to promote UK market development in low carbon and fuel cell technologies for transport applications. Cenex is known for its research into the market dynamics for low carbon vehicles of all types as well as its analysis work on real-world operation and user behaviour for electric vehicles and vehicles running on natural gas, bio-methane and hydrogen (CENEX, no date).

EV charging infrastructure

There are electric vehicle charge points available to use at accessible public locations. Examples of the charge point locations include on-street parking, businesses and organisation's car parks, hotel car parks, airports, shopping centres, supermarkets, restaurants, cafés, as well as at existing petrol stations. There are three main types of EV charging infrastructure in the UK:

1 Standard (3 kW) points can be used to top up an EV battery in a couple of hours and charge a battery from flat to full in 6 to 8 hours.
2 Fast (7–46 kW) points are able to top up batteries in 30 minutes, and charge from flat to full in less than 4 hours.
3 Rapid/Quick (50–250 kW) chargers will act as an emergency to be utilised when drivers have a near-empty battery. This is due to the high level of infrastructure required for the technology to operate safely, as result these will be located in strategic off-road locations.

OLEV provides funding to the Plugged-in Places programme (2011–13) which offers match-funding to consortia of businesses and public sector partners to install electric vehicle charging points. There are eight Plugged-in Places: East of England; Greater Manchester; London; Midlands; Milton Keynes; north-east England; Northern Ireland and Scotland (OLEV, no date). By the end of March 2013, over 4,000 charge points had been provided through the eight Plugged-in Places projects. About 65 per cent of these Plugged-in Places charge points are publicly accessible. Using data provided by charge point manufacturers, it is estimated that non Plugged-in Places organisations may have also installed about 5,000 charge points nationwide (OLEV, no date).

Buyers of electric vehicles need to know that that there is sufficient electric vehicle charge point infrastructure available to make them feel confident that they are able to easily charge the electric vehicle. This will help to alleviate range anxiety, one of the main barriers to uptake of EVs. Range anxiety is the driver's fear that the electric vehicle battery will run out of power before they reach their destination or a charge point. There are a number of different resources for EV drivers to locate EV charge points, including websites, mobile phone applications and in-car technology such as satellite navigation systems.

Further reading

CENEX, available at: www.cenex.co.uk/ (accessed 21/11/2013).

Department of Energy and Climate Change (2008) 'The Climate Change Act', available at: www.decc.gov.uk/en/content/cms/legislation/cc_act_08/cc_act_08.aspx (accessed 21/11/2013).

Department of Energy and Climate Change (2011) 'The Low Carbon Plan: Delivering our Low Carbon Future', Crown Copyright.

Department of Energy and Climate Change, available at: www.gov.uk/government/organisations/department-of-energy-climate-change (accessed 21/11/2013).

Department for Transport, available at: www.gov.uk/government/organisations/department-for-transport (accessed 21/11/2013).

OLEV, available at: www.gov.uk/government/organisations/office-for-low-emission-vehicles (accessed 21/11/2013).

Plugged in Places programme, available at: www.gov.uk/government/publications/plugged-in-places (accessed 21/11/2013).

Reducing greenhouse gases and other emissions from transport, available at: www.gov.uk/government/policies/reducing-greenhouse-gases-and-other-emissions-from-transport#actions (accessed 21/11/2013).

16.7 Energy policy and the built environment

Key terms: energy efficiency; energy policy; housing; built environment; renewable; energy use

The UK Government Energy White Paper (DTI, 2007) introduced a commitment to reduce carbon dioxide emissions by 60 per cent by 2050 in the UK. The UK's legally binding target under the Kyoto protocol is to cut greenhouse gas emissions by 12.5 per cent below 1990 levels by 2008–12. The 2008 Climate Change Act established a target of a 34 per cent reduction in carbon emissions by 2020 and at least 80 per cent by 2050. Five-year carbon budgets set the trajectory to 2050 and in the first carbon budget period (2008–12) the limit of greenhouse gas emissions is 3,018 $MtCO_2e$. The provisional figures for 2011 suggest that both the international (Kyoto) and domestic targets for reducing greenhouse gases are being met. The 2011 UK emissions are provisionally estimated to be approximately 570 $MtCO_2e$ (26 per cent below the baseline).

In 2009 buildings accounted for about 45 per cent of the UK's total carbon emissions: 27 per cent from domestic buildings and 18 per cent from non-domestic buildings. Much of these emissions come from space heating and hot water provision as a result of burning fossil fuels to heat buildings, and generating the electricity that powers lighting and appliances. In order to achieve such ambitious targets to reduce carbon dioxide emissions, the UK government has recognised the need to target the domestic housing sector. The energy consumed in homes accounts for more than a quarter of energy use and carbon dioxide emissions in Great Britain. More energy is used in housing than either road transport or industry and housing represents a major opportunity to cut energy use and CO_2 emissions (DECC, 2011) and therefore mitigate the impacts of climate change. The energy we use for heating and powering our non-domestic buildings is responsible for around 12 per cent of the UK's emissions, three-quarters of which comes from private businesses, with the remainder from public buildings. In addition, energy use for cooling is more significant in the commercial sector than for residential buildings.

Since 1990, government policies, including Warm Front, the Energy Efficiency Commitment and the Carbon Emissions Reduction Target have dramatically accelerated the deployment of cavity wall and loft insulation: a simple method for improving the energy efficiency of buildings and contributing to decreased carbon dioxide emissions and fuel bills for householders. In 2010, over 400,000 existing homes received cavity wall insulation (DECC, 2011), and this increased by 31 per cent between April 2008 and April 2012, meaning 11.4 million of 19 million homes with cavities were insulated (DECC, 2012). Over a million lofts were insulated in 2010 (DECC, 2011) and the number of homes with loft insulation (at least 125 mm) increased by 47 per cent between April 2008 and April 2012, meaning 14.5 million of the 23.4 million homes with lofts are insulated to this level (DECC, 2012a). New buildings standards mean that a house built today demands less energy for space heating required by a house built before 1990, as well as new condensing boiler standards (DECC, 2011). Since legislation was introduced in 2005 mandating the installation of condensing boilers in all but special applications, installation rates have increased to over 1.5 million a year which in turn has saved 4.1 $MtCO_2e$ alone. This has led to savings for many householders (approximately £95 off their energy bills this year) and at least £800 million for the UK as a whole.

Further UK government energy policies relevant to the built environment include:

Green Deal

The Green Deal, launched early 2013 and first outlined in the Energy Act 2011, is the Government's energy efficiency policy to overcome barriers to improving the UK's building stock and provide a means to aid households and businesses to improve energy efficiency with no upfront additional cost. Individuals are able to pay back the costs of installing the measures through the energy savings that they have made on their energy bills (DECC, 2011). This will promote a 'whole house' approach – offering a package of measures and ensuring the needs of the property are assessed as a whole. Micro generation technologies may be eligible for the Green Deal.

Private rented buildings are one of the most difficult sectors to improve. While tenants benefit from more energy efficient buildings, the landlords decide whether to pay to make the changes. The Green Deal will help tackle this split incentive. From 2016, domestic private landlords will not be able unreasonably to refuse their tenants' requests for consent to energy efficiency improvements. In addition, the Energy Act 2011 contains provisions for a minimum standard for private rented housing and commercial rented property from 2018, and the Government intends for this to be set at EPC band E. Use of these regulation-making powers is conditional on there being no net or upfront costs to landlords, and the regulations themselves would be subject to caveats setting out exemptions. If these powers are used, the Government envisages that landlords would be required to reach the minimum standard or carry out the maximum package of measures fundable under the Green Deal and Energy Company Obligation (even if this does not take them to band E).

Energy Company Obligation

The Energy Company Obligation (ECO) was introduced in January 2013 to reduce the UK's energy consumption and support people living in fuel poverty, for those most in need and for properties that are harder to treat. The ECO funds energy efficiency improvements worth around £1.3 billion every year. This will play an important role in supporting the installation of solid wall insulation, and also in providing upfront support for basic heating and insulation measures for low-income and vulnerable households. The costs of ECO are assumed to be spread across all household energy bills in Britain. It will place a duty on energy companies both to reduce emissions through undertaking solid wall insulation and to tackle fuel poverty by installing central heating systems, replacing boilers, and subsidising cavity wall and loft insulation.

Feed-in Tariff

The Feed-in Tariff (FiTs) scheme was introduced April 2010, under the Energy Act 2008. People are able to invest in small-scale low-carbon electricity, in return for a guaranteed payment from an electricity supplier of their choice, for the electricity they generate and use, as well as a guaranteed payment for unused surplus electricity they export back to the grid (DECC, 2012b). At the end of quarter 1, 2012, 247,953 FiTs installations were confirmed, 69 per cent of the total installed capacity were in the domestic sector. The majority of installations were solar PV, with other technologies installed being micro-CHP, anaerobic digestion, hydro and wind (DECC, 2012b).

Renewables Obligation and Renewable Heat Incentive

The Renewables Obligation (RO) aims to support the use of large-scale renewable electricity generation and the Renewable Heat Incentive (RHI), introduced in March 2011, supports generation of heat from renewable sources at all scales and provides long-term financial support to renewable heat installations, to encourage the uptake of renewable heat (DECC, 2012b). The RHI Premium Payment scheme managed by the Energy Saving Trust, provided £15 million funding to households to help them to buy renewable heating technologies (DECC, 2013).

Smart meters

Government aims for all homes and small businesses to have smart meters by 2020 and energy suppliers will be required to install smart meters. The Government is mandating the provision of in-home displays for domestic customers ensuring that consumers have the information and advice to make changes that will cut carbon and energy bills (through its consumer engagement strategy) The in-home display (IHD) shows how much energy consumers are using and what it will cost, encouraging controlled energy use and energy savings and money savings. More specifically, smart meters will give consumers: near real-time information on energy use, expressed in pounds and pence; the ability to manage their energy use, save money and reduce emissions; an end to estimated billing – people will only be billed for the energy they actually use, helping them to budget better; and easier switching – it will be smoother and faster to switch suppliers to get the best deals.

CRC Energy Efficiency Scheme

The CRC Energy Efficiency Scheme (or CRC Scheme) is designed to improve energy efficiency and cut emissions in large public and private sector organisations. It requires participants to buy allowances for every tonne of carbon they emit. It is expected to reduce non-traded carbon emissions by 16 million tonnes by 2027, supporting our objective to achieve an 80 per cent reduction in UK carbon emissions by 2050. The CRC scheme applies to emissions not already covered by Climate Change Agreements (CCAs) and the EU Emissions Trading System (EUETS). Organisations that use more than 6,000 megawatt-hours (MWh) of electricity settled on the half-hourly market in a qualification year are required to participate, and must buy allowances for every tonne of carbon they emit (DECC, 2013).

National Planning Policy Framework

The National Planning Policy Framework outlines how developments should be planned to reduce carbon emissions from buildings. Local planning authorities are required to make sure that new developments are energy efficient, while new homes are required to be zero carbon from 2016, which may be extended to other buildings from 2019. EPCs have been improved so that they are more informative and user friendly. Green Deal is mentioned, to enable people to pay for home improvements over time using savings on their regular energy bills. The Code for Sustainable Homes is introduced which provides a single national standard for the design and construction of sustainable new homes (DECC, 2011).

Building Regulations

The Government is committed to improvements in new-build standards through changes to Part L of the Building Regulations in England and their equivalents within the devolved administrations. Increased standards of insulation were introduced by the changes to the Building Regulations in 2006 and 2010. In October 2010, the new regulations in England and Wales introduced a 25 per cent improvement on 2006 carbon emissions standards for new buildings, while regulation in Scotland delivered a 30 per cent reduction on their 2007 standards. It is intended that changes in 2013 will be an ‘interim step on the trajectory towards achieving zero carbon standards’ for new homes by 2016. The changes will come into effect on 6 April 2014. Domestic new buildings will have to be 6 per cent more efficient than current requirements and for non-domestic new buildings the requirement will be 9 per cent. Despite these requirements and the perception that they have and do demand onerous standards of both designers and constructors (the 2013 standards require a 44 per cent efficiency improvement over the 2006 baseline), the low number of new house completions means that the impact of such changes is likely to be limited (DECC, 2011).

Further reading

DECC (2011) 'The Carbon Plan: Delivering Our Low Carbon Future'. Crown Copyright.

DECC (2012a) Statistical Release: Experimental Statistics. Revised estimates of Home Insulation Levels in Great Britain: April 2012. Available at: https://www.gov.uk/government/uploads/system/uploads/attachment data/file/49407/5457-stats-release-estimates-home-ins-apr2012.pdf

DECC (2012b) 'UK Energy in Brief', available at: https://www.gov.uk/government/uploads/system/uploads/attachment_data/file/65898/942-uk-energy-in-brief-2012.pdf (accessed 19/11/2013).

DECC (2013) 'Domestic Renewable Heat Incentive: The first step to transforming the way we heat our homes', available at: https://www.gov.uk/government/uploads/system/uploads/attachment_data/file/212089/Domestic_RHI_policy_statment.pdf (accessed 20/11/2014).

Department of Energy and Climate Change, Committee on Climate Change, Department for Environment, Food and Rural Affairs and Department for Transport (2013) 'Reducing the UK's greenhouse gas emissions by 80% by 2050', available at: www.gov.uk/government/policies/reducing-the-uk-s-greenhouse-gas-emissions-by-80-by-2050 (accessed 21/11/2013).

Department for Communities and Local Government (2013) 'Improving the energy efficiency of buildings and using planning to protect the environment', available at: www.gov.uk/government/policies/improving-the-energy-efficiency-of-buildings-and-using-planning-to-protect-the-environment (accessed 21/11/2013).

DTI (2007) 'Meeting the Energy Challenge: A White Paper on energy', available at: http://www.berr.gov.uk/files/file39387.pdf (accessed 19/11/2013).

16.8 Environmental impact assessment

Key terms screening; scoping; baseline; impact prediction and assessment; mitigation; alternatives; decision making

> A process for identifying the likely consequences for the biophysical environment and for man's health and welfare of implementing particular activities and for conveying this information, at a stage when it can materially affect their decision to those responsible for sanctioning the proposals.
>
> (Wathern, 1990)

This definition of environmental impact assessment (EIA), although dated, encapsulates the essence of the process. The key issue is that decision-makers, often planning officers, committees and inspectors are fully apprised of the likely environmental impacts of a development proposal prior to permission being granted or refused.

A key milestone in the evolution of EIA was the National Environmental Policy Act (NEPA) which was enacted in the US in 1969. NEPA established the purpose of EIA and established the basic procedures. Since 1969 the majority of developed countries, and many developing countries, have introduced some form of EIA for major projects.

The main stages in the EIA process are as follows. Although illustrated as a linear process in Figure 16.8.1, iteration of some of the stages may be required.

The first stage of the process is to determine whether an EIA is required, this is known as *screening*. The decision will be made following a review of the type and scale of the development with reference to various criteria.

Scoping is the process of deciding what should be included in the EIA. This is an important part of the process to ensure that resources are focused on the most significant issues and that the report produced will be useful and relevant to the ultimate decision-makers.

A *baseline study* is a description of conditions existing at a point in time against which subsequent changes can be detected through monitoring.

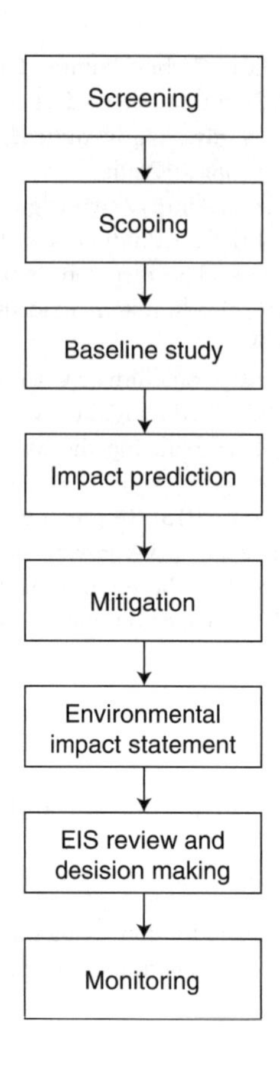

Figure 16.8.1 The environmental impact assessment process.

An attempt must be made to predict the *impact* the proposed development will have on all the scoped issues. This must be done in a quantifiable way and should also include a prediction of the time frame in which the impacts are likely to occur.

Following the prediction of which impacts will occur it is necessary to *assess* their *importance* and *significance*. It is critical that this assessment is carried out in an overtly unbiased way as the assessment will be scrutinised by experts, the public, pressure groups and ultimately the decision-makers.

Mitigation measures are taken to either remove or reduce impacts. Mitigation may also include enhancing the development in some way to offset rather than reduce the impact.

An *Environmental Statement* must then be written including a description of the development, a section on mitigation measures explaining how any significant adverse effects will be avoided, reduced or remedied, an assessment of the main effects the development is likely to have on the environment, a summary of the main alternatives considered (designs and sites) and a justification as to why the selected scheme was chosen, and a non-technical summary.

In the UK the EIA should be submitted with the planning application to the competent local planning authority. They will then consult on the application and make a decision based on all the material considerations. Material considerations include a wide variety of matters that relate to the use and development of land. Material considerations will include all relevant planning policies and the environmental statement if an environmental impact assessment has been required.

If planning permission is granted, it will generally be subject to conditions or a planning agreement. The conditions or agreement may contain measures to mitigate impacts identified in the EIA.

Further reading

'Town and Country Planning (Environmental Impact Assessment) Regulations 2011', available at: www.legislation.gov.uk/uksi/2011/1824/introduction/made (accessed 20/11/2013).

Wathern, P. (1990) *Environmental Impact Assessment: Theory and practice*, Routledge, Abingdon.

16.9 Ground/air source heat pumps

Key terms: renewable energy; sustainable buildings; local energy generation; ground source heat pump (GSHP); air source heat pump (ASHP)

Heat pumps can best be described as a refrigerator mechanism applied to a building. A refrigerator takes heat out of the ice box or cold space and transfers the heat to the (warm) coil on the back of the fridge. It achieves this by using a refrigerant which typically (R134A) boils at –24 degrees centigrade. The refrigerant liquid is allowed to evaporate in the ice box, this evaporative process draws heat out of the local space (you can test this by wetting your finger and blowing on it, the evaporation of the moisture on your finger will make it feel cold – this is latent heat of evaporation, which you may recall from school physics lessons). The refrigerant gas is then taken through a compressor that increases its pressure and temperature, the gas is allowed to condense which liberates the heat and warms up the refrigerator coil, the liquid then goes through an expansion valve which lowers its pressure and allows it to evaporate, thus starting the cycle again.

The theoretical attraction of a heat pump is that it transfers free heat from one place to another using only electricity to drive the compressor (and fans or pumps). The process has a multiplier effect whereby the process of the refrigeration cycle may produce three or four times the heat energy which would be produced by running the electricity though a radiant electric fire. The heat source may be the outside air in which a fan passes volumes of air over the evaporator. Alternatively if there is enough space, heat may be drawn from the ground by means of bore holes or pipes (termed a slinky) laid in trenches in the ground. Ground temperature remains relatively stable more than 1m below ground level (10–15 degrees in the UK) so it provides a stable source of heat to be exchanged. The slinkies draw in heat from the ground which has been warmed by the sun; deep bore holes draw heat from the ground by migration of heat from the surrounding geology. It should be noted that neither system is truly 'geothermal', which refers to heat sources deep underground, up to a mile in the UK (or at surface level in Iceland). In the UK a slinky will produce 1kW for a 10m length trench, thus a medium-size detached house will need enough garden or local land to accommodate 80 metres of trench which if doubled back must be at least 5 metres apart, an area of land not normally available in a UK domestic garden.

GSHPs are more are more efficient than ASHPs because they draw on a more stable and reliable heat source, but they are much more expensive to install due to the drilling or trench digging. Air source heat pumps are drawing heat from the air which may range from –5 to + 30 in the UK. Paradoxically, they are much less efficient at low temperatures so they are best in summer when there isn't any need for space heating, and least effective in winter when there is most demand for heating.

Heat pumps produce low grade heat, up to 55 degrees centigrade. This is relatively cool for space heating; in homes, radiators will need to be larger than conventional gas-fired heating, taking up more wall space. The ideal heat transfer system for heat pumps is an under-floor heating system but this is tends to restrict installations to new build properties. The low temperature also means that the water heating system must use an immersion heater to periodically heat the water up to 70 degrees as a precaution against legionella.

A field trial by the Energy Savings Trust produced some very interesting results, the efficiency or multiplier effect of the heat pump installations is measured as the coefficient of performance (CoP). Installations surveyed produced CoPs of 1.2 to 2.2 for ASHP (mode 1.6) and 1.6 to 3.4 (mode 2.2) for GSHP; this is much lower than the nominal CoP advertised by manufacturers. The overall conclusions were that heat pumps may be an effective alternative to heating by oil, coal or LPG but could not compete with a modern gas boiler fuelled by mains gas. In this study of early installations there were many instances of poor design and installation, the report also reported problems of occupiers failing to understand the operation of the system.

Further reading

Energy Saving Trust (2010) 'Getting warmer, a field trial of heat pumps', available at: www.energysavingtrust.org.uk/Organisations/Technology/Field-trials-and-monitoring/Field-trial-reports/Heat-pump-field-trials (accessed 20/11/2013).

Energy Saving Trust (2012) 'Detailed analysis from the first phase of the Energy Saving Trust's heat pump field trial', available at: www.gov.uk/government/publications/analysis-from-the-first-phase-of-the-energy-saving-trust-s-heat-pump-field-trial (accessed 20/11/2013).

16.10 Life cycle assessment of buildings

Key terms: environmental impact; scoping; inventory; impact assessment; interpretation

Life cycle assessment (LCA) is a tool that can be used to assess the environmental impacts of a product, process or service from design to disposal, that is, across its entire life cycle, from raw material extraction through materials processing, manufacture, distribution, use, repair and maintenance, and disposal or recycling. The procedure involves:

- compiling an inventory of relevant energy and material inputs and environmental releases;
- evaluating the potential impacts associated with identified inputs and releases;
- interpreting the results to help make a more informed decision in order to reduce environmental impact.

The term life cycle is used because measuring environmental impact requires the assessment of raw-material production, manufacture, distribution, use and disposal, including all intervening transportation steps necessary or caused by the building's existence. The procedures of LCA are part of ISO 14000 environmental management standards. There are four phases of LCA: a scoping exercise; then a life cycle inventory; then impact analysis and finally interpretation of results.

Scoping starts with an explicit statement of the goal and scope of the study that sets out the context of the study, and explains how and to whom the results are to be communicated. This is a key step and the ISO standards require that the goal and scope of an LCA be clearly defined and consistent with the intended application.

Life cycle inventory (LCI) analysis involves creating an inventory of flows from and to the environment for a product system. Inventory flows include inputs of water, energy and raw materials, and releases to air, land and water. To develop the inventory, a flow model of the technical system is constructed using data on inputs and outputs. The input and output data needed for the construction of the model are collected for all activities within the system boundary, including from the supply chain.

Once the inventory is complete the next phase is impact assessment. This evaluates the significance of potential environmental impacts based on the LCI results. This involves selecting and classifying impact categories, measuring impacts in common units that can then be aggregated to give an overall environmental impact.

Life cycle interpretation is a systematic technique that identifies, quantifies, checks and evaluates information from the results of the LCI and the LCA. This provides a set of conclusions and recommendations for the study, including identification of significant issues; evaluation of the study considering completeness, sensitivity and consistency; conclusions, limitations and recommendations. A key purpose of performing LCI is to determine the level of confidence in the final results and communicate them in a fair, complete, and accurate manner.

For a building LCA, the process would typically involve assessing:

- production of building materials;
- transport of materials to construction site;
- actual construction of building including additional materials, building processes and transport of workers and equipment to site;
- operation of building over its lifetime;
- disposal of building materials at end of building life.

Research by Bribián *et al.* in 2009 in a study based on Spanish property revealed that embodied energy can represent more than 30 per cent of the primary energy requirement during the life span of a single house of 222 m^2 with a garage for one car. The contribution of the building materials decreases if the house does not include a parking area, since this increases the heated surface percentage. Usually the top cause of energy consumption in a residential building is heating, but the second is the building materials, which can represent more than 60 per cent of the heating consumption.

Further reading

Bribián I. Z., Scarpellini S. and Usón A. A. (2009) 'Life cycle assessment in buildings: State-of-the-art and simplified LCA methodology as a complement for building certification', *Building and Environment*, 44(12), December: 2510–2520, available at: www.bre.co.uk (accessed 20/11/2013).

16.11 Retrofit

Key terms: sustainable buildings; low carbon technology; carbon reduction commitment; Green Deal

Retrofit has become a catch-all term to describe the process of upgrading of homes and commercial buildings to reduce their environmental impact.

The UK government has ambitious plans to reduce carbon emission by 80 per cent by 2050. As buildings account for approximately 50 per cent of carbon emissions, Building Regulations have been successively amended to lower energy demand in new buildings with the target of achieving zero carbon homes by 2016 and zero carbon commercial buildings by 2019. However, in England, for example, 52 per cent of homes were built before 1960, Victorian, Edwardian and interwar homes had no provision for thermal insulation and thus without extensive modifications will continue to consume energy for space and water heating, jeopardising carbon reduction targets.

The consideration of home retrofit may be considered under two headings: the technical evaluation of which interventions are effective; and the policy options available to ensure that energy efficient retrofit takes place. As regards physical improvements, the payback period, how long it takes for the energy savings to repay the capital cost, may be estimated reasonably accurately. However, this is not the only consideration, occupiers will also consider whether the improvement will improve their comfort balanced against the disruption and inconvenience caused by having the work done. For example, under-floor insulation is cost effective but the disruption caused by retrofitting in an existing home makes it an unlikely first choice for homeowners.

When triangulating capital cost, energy savings and ease of installation, most homeowners will choose the following:

- upgrading lights to compact fluorescent or LED lighting – DIY job, cost effective;
- loft insulation – most existing homes have 100 mm but 300 mm will be cost effective;
- draught proofing – cheap, easy to install and improves comfort;
- heating controls – hot water cylinder thermostats, room thermostats, thermostatic radiator valves – half a day for a competent plumber.

After these simple steps things begin to get more complicated: for the 21 million homes heated by gas in the UK (83 per cent) the next step will be to install a modern gas boiler. Gas boiler technology has improved considerably in the past two decades; old conventional gas boilers will typically be 80 per cent efficient, modern condensing boilers can achieve 95 per cent efficiency. Replacing a boiler and hot water storage cylinder with a combi-boiler which only heats water as it is needed will also improve carbon outputs and energy costs.

For those off the mains gas network, most will use electricity for space heating, often 'night storage' radiators using off-peak electricity. It is tempting to replace these with air source heat pumps but although this offers a notional multiplier effect the space heating will be working during the day when electricity is at its most expensive. If space is available in the home, biomass may be a better solution and has the advantage of being notionally carbon neutral.

After these measures, retrofit becomes more intrusive and expensive; if the home has cavity walls then insulating them with blown fibre is cost effective, especially at the subsidised rate available due to the energy companies' carbon reduction commitment. In a solid wall home upgrading the insulation value is comparatively expensive; if applied to the outside of the home it will change the appearance, usually not for the better, and if applied inside it will reduce the size of rooms and be very disruptive during the retrofit process.

Although wall insulation is more cost effective many homeowners will choose to replace single glazed windows with double glazed; this creates a more comfortable environment and reduces the incidence of condensation. Replacing timber frames with uPVC will avoid the need for regular repainting although they are often less aesthetically appealing. As regards policy drivers to facilitate retrofit, the Government has imposed energy saving commitments on the energy supply industries which has translated into subsidised loft and cavity wall insulation; this has been very effective to the extent that in many cities all lofts and cavities have been filled in eligible households.

The main driver towards retrofitting is the Government's Green Deal, launched in January 2013 – it is intended that it will facilitate retrofitting for energy efficiency at no upfront cost to the homeowner. The scheme requires homeowners to establish which energy efficiency measures are cost effective through a paid-for survey by an approved surveyor. The work is then carried out by approved contractors and paid for by a loan, the repayments of which are financed by the energy cost savings. This appears, on the face of it, to be a fool proof means of retrofitting the homes of householders who don't have much capital or easy access to finance. The scheme got off to a shaky start; nine months after the launch, although there have been over 7,000 surveys carried out, fewer than 400 households had signed agreements under the scheme measured against a DECC target of 10,000. The scheme administration appears complex, has a relatively high interest rate and, because the loan and repayment obligation is applied to the property rather than the individual, could be a disincentive when it comes to selling the property.

The main success in retrofit has come from the social housing sector, taking advantage of a succession of Homes and Community Agency grants for registered social landlords (RSLs), which have had the technical resources and expertise to bid for funding and execute retrofit on a large scale to such an extent that now virtually all social housing achieves the Government's 'decent homes' standards and many homes have been installed with renewable technologies.

Although RSLs have been very proactive in retrofitting their properties, it has been difficult for the Government to incentivise homeowners to make radical interventions to effectively reduce domestic energy consumption. Despite significant increases in energy cost and little prospects of reduction in the future, the inertia of homeowners has been difficult to overcome when it comes to retrofitting energy efficiency measures unless the grant regime is excessively generous, as found in the first iteration of the FiT for PV installations.

Further reading

Low Energy Building database, available at: www.retrofitforthefuture.org (accessed 22/10/2013).

National Audit Office (2010) 'The Decent Homes Programme', available at: www.nao.org.uk/report/the-decent-homes-programme/ (accessed 22/10/2013).

16.12 Sustainability appraisal

Key terms: strategic environmental assessment; cost–benefit analysis; environmental audit.

Sustainability appraisal (SA) is an umbrella term that covers a range of assessment methods. It involves being clear about objectives and the alternative methods of achieving them by estimating the costs and benefits of each option. A SA is mandatory in the UK under Section 39(2) of the Planning and Compulsory Purchase Act 2004. The purpose of a SA is to promote sustainable development through the integration of social, environmental and economic considerations in project planning. A key element of this is the testing of projects or developments against the SA objectives in order to identify likely positive impacts and also determine whether any negative impacts could arise. A range of assessment techniques for policies and plans are available but the most useful are shown in Table 16.12.1.

Local authorities are also required to conduct an environmental assessment in accordance with the requirements of European Directive 2001/42/EC (the Strategic Environmental Assessment Directive). Although the requirements for carrying out SA and strategic environmental assessment (SEA) differ, both are satisfied through the SA process.

SEA is a process of environmental assessment that leads to an environmental report that should help inform policy choices. SEA enables city level policies to be evaluated in terms of their environmental implications. Relationships between the city and surrounding areas can also be considered. Individual housing, industry or transport projects are often considered in terms of their environmental impact, but their effects on other sectors may be ignored if some form of assessment is not undertaken at a strategic level (Jowsey and Kellett, 1996). SEA is a form of assessment of the environmental effects that arise from policies, plans and programmes before individual projects are authorised. An SEA identifies the significant environmental effects that are likely to result from the implementation of a project or plan. These effects can then be considered together with financial, technical, political and other relevant factors.

Table 16.12.1 Environmental assessment techniques

	Strategic environmental assessment	
	Environmental impacts of policy/plan City–region relationship Integration of policy	
Environmental impact assessment	*Cost–benefit analysis*	*Environmental audit*
Potential impacts of new developments Utilises environmental standards and thresholds	Monetary appraisal of private and social costs of development proposals and of alternative proposals	Existing land uses and processes identified and environmental targets set

Source: Adapted from Jowsey and Kellett (1996).

The process of producing a report has a number of stages (Therivel, 2004):

Stage 1 Identify the SEA objectives, indicators and targets that the plan or proposal can be tested against.
Stage 2 Describe the environmental baseline (using data on the current environment) – and what environmental and sustainability issues to consider.
Stage 3 Identify links to other relevant strategic actions.
Stage 4 Identify sustainable alternatives to the plan proposed (if there are any).
Stage 5 Prepare a 'scoping report' based on stages 1–4 and showing how to proceed.
Stage 6 Predict and evaluate the impact of alternative proposals.
Stage 7 Write the SEA report, establishing guidelines for implementation.
Stage 8 Consult with all those affected by the proposals and address any concerns.
Stage 9 Monitor the environmental/sustainability impacts of the strategic action.

Following this procedure should facilitate public participation and make the strategic action more transparent. The SEA should improve the project or plan and promote urban sustainability. The downside is, of course, its cost in terms of time and resources, and this can be considerable.

Further reading

Jowsey, E. and Kellett, J. (1996) 'Sustainability and methodologies of environmental assessment for cities', in C. Pugh (ed.) *Sustainability, the Environment and Urbanisation*, Earthscan, London, pp. 197–227.
Therivel, R. (2004) *Strategic Environmental Assessment in Action*, Earthscan, London.

16.13 Sustainable urban drainage systems

Key terms: urban drainage; flood alleviation; natural drainage

The UK has recently experienced a number of extreme weather events; Meteorological Office statistics indicate that heavy rainfall has become more common since 1960. Extreme rainfall is defined as a the volume of rain that may be expected every 100 days – in recent years these volumes of rain have been occurring every 70 days, leading to a greater instance of flooding. Climate change scientists note that as the atmosphere warms it can contain more moisture, a 0.7 degree Celsius increase in temperature means that the atmosphere can contain 4 per cent more moisture; this can precipitate out as rain – often in more intense bursts than has been experienced in the past. To check the risk of flooding in any locality in the UK the Environment Agency has a flood map available at www.environment-agency.gov.uk/homeandleisure/floods/31650.aspx.

Intense downpours can overwhelm old drainage systems and flood through sewerage treatment plans causing pollution of rivers and coastline. More visible is the localised fluvial flooding when the drainage capacity is overwhelmed and rainwater accumulates in watercourses, roads, parks and fields, spilling over to flood homes and businesses. In some instances local flood effects can be traced back to poor maintenance of rainwater gullies in roads or poorly maintained rivers, which allows debris, such as fallen trees disturbing the flow of the river when in spate.

The effect of extreme rain events can be exaggerated by the intensity of development in cities, whereby hard surfaces, roofs, roads and paths rapidly discharge rainwater into the surface water drainage system or local watercourse. Some local authorities, especially on London, are discouraging the paving over of front gardens for parking as it reduces the area of permeable land available to absorb rainfall.

To mitigate the effects of these extreme rainfall events the UK has been in the forefront of developing sustainable urban drainage systems (SUDS) to reduce the incidence of flooding and also help the rainwater replenish the aquifer rather than running into rivers and the sea.

The first priority is to try to retain rainwater on the site onto which it falls. This can be achieved by a variety of design features. One highly visible and effective means of attenuating rainwater is to cover buildings with a sedum or 'green' roof rather than a hard surface. Sedum are low maintenance flowering plants, commonly known as 'stonecrops'. In a green roof application they can be planted on a thin layer of soil over a root barrier and waterproofing membrane. As well as improving the biodiversity of the roof as a host to insects, green roofs will increase the durability of the roof by increasing insulation and reducing thermal movement of the waterproof membrane. The major advantage is the ability of the green roof to retain rainwater; it has been found that a well-designed green roof can retain 75 per cent of rainwater, subsequently releasing it back into the atmosphere by evaporation or transpiration. The extra weight of the soil and plant covering has to be accommodated in structural design of the building. The capital cost of a sedum roof is likely to be double the cost of a conventional roof; it is also likely that an irrigation system and access for maintenance will be needed. Figure 16.13.1 shows a sedum roof covering a nursery building. The shallow pitch will help slow rainwater discharge. The photograph was taken after the summer flush of colourful flowers had died back.

Another simple means of retaining rainwater on the site is the provision of permeable hard surfaces; open jointed concrete block will allow rainwater to percolate through the paving rather than rapid run-off to surface water drainage or adjacent stream. Figure 16.13.2 shows open jointed concrete paving in a car park that will allow rainwater to percolate through to the aquifer.

Some commercial and residential developments can retain rainwater on site by constructing a buffer pond; although this will inevitably take up land, it can be positioned in such a way that it occupies space that otherwise is undevelopable. A well-designed water feature can considerably enhance the value of a development as well as improve biodiversity on site.

Rainwater harvesting has some applications; in the residential setting rainwater butts can retain water for subsequent garden watering. In commercial applications, rainwater harvesting has limited application and value – in most office development the demand for water is limited and the capital investment and subsequent maintenance obligation rarely make such an investment worthwhile. There are also real risks involved; in one unfortunate instance the rainwater harvesting tank shared a basement location with the IT server. When a pump failed the flooding of the basement caused tens of thousands of pounds worth of damage to the IT installation.

Figure 16.13.1 A sedum roof.

Figure 16.13.2 Open jointed concrete paving.

Having reduced the volume of water leaving the site to a minimum, the next stage is to convey the water off site. The traditional means of water conveyance has been sealed pipes – fired clay in the past, UPVC currently. The aim in a SUDS is to disperse the rainwater into the soil. A novel innovation has been a 'swale' – a landscaped trench that will contain the water, in large volumes, and allow it to disperse into the ground as it travels (assuming a permeable soil structure). Water can also be conveyed in an infiltration trench (or French drain) – a trench filled with rubble that provides a 'route of least resistance' that will convey the rainwater without containing it and again allowing infiltration into the subsoil.

SUDS will also include detention basins and flood plains as a final line of defence; these are areas of land that can be flooded without causing damage to infrastructure and buildings. Detention basins and flood plains will allow excess rainwater to disperse over the landscape to avoid it flooding rivers which will be walled in between quaysides and buildings as they run through towns and cities.

The Environment Agency is spending in the region of £600 million per year on flood defences. They have to make difficult decisions on competing flood defence schemes, assessing costs and benefits to achieve the optimum value for money, while negotiating with the insurance industry. The more attenuation work that can be carried out through site-based SUDs, the less major investment will be needed downstream where the risks and costs are infinitely greater.

Further reading

Defra (2011) 'National Standards for Sustainable Urban Drainage', available at: www.gov.uk/government/consultations/implementation-of-the-sustainable-drainage-provisions-in-schedule-3-to-the-flood-and-water-management-act-2010 (accessed 20/11/2013).

16.14 Solar power photovoltaics

Key terms: low carbon technology; renewable energy; Feed-in Tariff; retrofit.

Photovoltaic (PV) technology has a long history; the photovoltaic effect was first observed by Edmond Becquerel in 1839 but it was not until the need for electrical power in spacecraft that the technology was refined. The first solar powered satellite was launched in 1958 and the first earth-based commercial application was a solar powered lighthouse in Japan in 1966.

It isn't essential to know how a PV panel works (who knows how a TV works?) but the technology is interesting and gives an insight into the operational parameters. Very simply, PV cells comprise a wafer thin silicon crystal of which half has been treated to make it electron deficient and the other half treated to make it electron rich. When photons of light strike the cell they liberate electrons from the electron rich side of the crystal which flow to the electron deficient side of the crystal; they leave holes which, when the cell is connected in a circuit, allows the freed electrons to fill the holes on the enriched half of the crystal, thus providing a current and usable electricity. In addition to these 'conventional' PV panels, thin film solar cells have been developed that can be applied to glass or cladding panels, thus creating an integrated design for commercial applications.

To gain the optimum power from PV panels they must be fixed to gain the maximum light from the sun; this means predominantly south facing and fixed at an angle of 30–35° from the horizontal. This can be achieved when fixed on frames on a flat roof such as an office block, but when retrofitted to pitched roof they will be fitted at the same pitch as the existing roof which may be other than the optimum. Slate and tiled roofs are normally designed at a minimum of 35°, while in the 1970s and 1980s speculative builders would construct roofs at 22.5° with interlocking concrete roof tiles because they were relatively cheap.

Care must be taken to avoid shading of the panels as it has a disproportionate effect on power output. Trees, chimneys and dormer windows can all shade a PV array; even if a small section of the panel is in shade, the performance of the whole panel will be significantly reduced. This is because PV panels consist of a number of cells wired together into a series circuit. When the power output of a single cell is reduced by shading, the power output for the whole series is reduced to the level of the current passing through the shaded cell. Therefore, a small amount of shading can significantly reduce the performance of the entire system.

PV panels generate direct current; electrical appliances use alternating current, so to make the electrical energy useable on site or to export it to the grid it must be put through an inverter to convert from DC to AC. PV panels are normally guaranteed for 20 to 25 years with an assumption that the power output will decline to about 80 per cent of new condition over that time. The inverter will last for 10 to 15 years before it must be replaced.

At UK latitudes PV generated electricity is not viable without a substantial subsidy. In early commercial applications the Department for Trade and Industry provided capital grants to encourage PV installations which would develop the design and installation expertise. A prominent early example was the recladding of the Cooperative Insurance Services (CIS) offices in Manchester. The cost of the PV cladding was £5.5 million, which was subsidised by a grant of approximately £1 million by the North West Development Agency and the UK Major Photovoltaic Demonstration Programme. The project was not financially viable in terms of the electricity generated but the project went ahead to gain experience in installing the system and measuring the actual output against design outputs. The CIS have a long tradition of high quality architecture and environmental awareness, and the PV installation was a demonstration of their commitment to a low carbon future.

The current means of subsidising PV installations on both residential and commercial property is by means of a Feed-in Tariff (FiT). The Government imposes a levy on all electricity bills to provide a subsidy to PV installations (and other low carbon technologies). A current estimate for a residential installation on the Energy Saving Trust calculator, indicates that a 3.4Kw PV installation on a south-facing roof in the north of England would cost £6,700 to install and receive a total income of £11,000 over 25 years. To be eligible for the FiT, installation must be carried out by a contractor certified under the Micro Generation Certification Scheme and the home must have an EPC rating of band D or better. If the property has an EPC rating of E, F or G the homeowner has the option of improving its energy performance or accepting a lower rate of FIT over the lifetime of the contract.

The cost of PV panels and installations has reduced considerably over the past 5 years as Chinese manufacturers have moved into the market. The Chinese success is such that German manufacturers are approaching the European Union to invoke anti-dumping legislation to protect their PV manufacturing industry. As costs diminish, PV electricity is achieving parity with fossil fuel-generated electricity in locations with ample sunlight. In addition to the cost advantage, PV provides almost zero carbon electricity with a low maintenance package. There are, however, aesthetic issues to be considered as residential retrofit installations on historic buildings can look inappropriate and may be detrimental to subsequent saleability.

Figure 16.14.1 shows retrofit PV panels and solar water panels on a refurbishment of a Victorian terraced house. Figure 16.14.2 shows building integrated thin film PV allied to the south-facing external wall of an office building. Figure 16.14.3 shows the interior view of the same building.

Figure 16.14.1 Retrofit photovoltaic panels and solar water panels.

Figure 16.14.2 Thin film PV.

Figure 16.14.3 Interior view thin film PV.

Further reading

Carbon Trust (2011) 'A place in the sun, lessons learned from low carbon buildings with photovoltaic electricity generation', available at: www.carbontrust.com/media/81357/ctg038-a-place-in-the-sun-photovoltaic-electricity-generation.pdf (accessed 21/10/2013).

16.15 Solar water heating

Key terms: renewable energy; solar collector; heat exchanger; solar water heating

Solar water heating is a relatively simple renewable technology that has been exploited for decades, particularly in southern Europe where solar water heating is routine – for example, in Greek holiday villas. Solar water heating can be used in residential buildings and non-domestic buildings where there is a high demand for hot water – for example, leisure centres.

The components of solar water heating comprise a solar collector, preferably on a south-facing pitched roof, a hot water storage cylinder, a pump, connecting piping and control gear to manage the system.

The solar collector absorbs radiation from the sun and converts it into heat energy to be transferred to a heat exchanger in the hot water cylinder by a fluid, usually water or anti-freeze. There are two types of solar collector, flat plate collectors and evacuated tubes. Flat plate collectors are slightly cheaper and simpler to build; they comprise a dark coloured panel, not unlike a radiator, mounted in a structural box and covered by a glass sheet to provide a measure of thermal insulation. Evacuated tube collectors are much more sophisticated devices: glass tubes are mounted in parallel rows with dark coloured absorber tubes in the centre; these tubes can collect solar radiation more effectively, because, as the name suggests, the glass tube is evacuated to form a vacuum that reduces the conduction of heat away from the tubes, thus achieving higher temperatures.

Flat plates are more readily incorporated into a pitched roof; in new build they can be installed in the plane of the roof pitch, in a retro fit they will normally be installed on pegs, above the pitch of the roof. Evacuated tube collectors have a more 'industrial' or 'hi-tech' appearance that some homeowners may find alien. In summer conditions both systems produce a similar output but in the winter when the air temperature is colder and sunlight is weaker evacuated tubes should in theory be more efficient.

Solar radiation heats the transfer fluid in the collectors and this is pumped to a heat exchanger in a hot water cylinder. To be most effective the hot water storage cylinder should be larger than normal; a typical domestic cylinder will be 450 mm diameter × 1000 mm high with a 100 litre capacity; for a solar water system the cylinder will preferably be 300 litres capacity with twin coils, one for the solar system and another for the backup system, be it a water boiler or an electric immersion heater. It is important to have a control system that gives precedence to the 'free' solar water rather than the back-up system, otherwise the occupier will be paying to heat water and there will not be any space in the cylinder for the 'free' solar water when the sun is shining.

The Energy Saving Trust (EST) conducted an evaluation of residential solar water installations and published a report: 'Here Comes the Sun: a field trial of solar water heating systems' in 2011. The survey of 88 installations between April 2010 and April 2011 produced some very interesting results. It found that the theoretical advantage of evacuated tubes was not observed in practice, possibly because the flat plate collectors have a larger working area than evacuated tubes. They found that a well-installed system could provide 60 per cent of a household's hot water demand, but the average was 40 per cent. The worst installation only produced 9 per cent and the best 98 per cent. An important issue is the demand for hot water in the home – if there are appliances that heat cold water as part of their function, e.g. electric showers and cold feed clothes and dish washers, these will comprise a significant element of the demand for hot water but will not be displaced by a solar water system.

The most disappointing element of the EST report was the economic considerations. It found that solar systems typically produced a saving of £55 per year when replacing gas and £80 per year when replacing electric immersion heating (2011 prices). With installations currently costing from £3,000 to £6,000 this gives an optimistic payback period of over 35 years! Clearly, energy costs are increasing much faster than inflation and payback will improve as energy costs increase. The economics will also improve if the Government brings in the much delayed Renewable Heat Incentive payments (which are currently only paid to commercial installations). As regards carbon saving, the EST report noted that the carbon saving was similar to that achieved by installing draft proofing around all the doors, windows and skirting boards.

Solar water heating is a reliable technology that is difficult at present to justify on economic or carbon saving terms; however, it is highly visible and can be a demonstration of green credentials by social housing providers and commercial developers. Figure 16.15.1 shows a large-scale evacuated tube installation on the roof of a city centre library. Most libraries would not have enough demand for hot water to justify such an installation but in this instance the library public toilets are heavily used and hot water usage is significant.

Figure 16.15.1 Large-scale evacuated tube installation.

Further reading

The Energy Saving Trust (2011) 'Here Comes the Sun: A field trial of solar water heating systems', available at: www.energysavingtrust.org.uk/Publications2/Generatingenergy/Field-trialreports/Here-comes-the-sun-a-field-trial-of-solar-water-heating-systems (accessed 15/10/2013).

16.16 Wind turbines

Key terms: load factor; planning issues; Renewable Obligation Certificates

The UK has probably the best wind resources in Europe and wind turbines do now contribute a significant proportion of the UK's commitment to the production of renewable energy.

Most wind turbines seen in the UK have a horizontal axis and are technically known as 'HAWTs' (horizontal-axis wind turbines). They have the main rotor shaft and an electrical generator at the top of a tower and are pointed into the prevailing wind to generate power. Less common and typically a lot smaller are 'VAWTs', where the axis is vertical.

Modern commercial turbines can range in potential peak output from 2 MW to 7 MW with larger turbines suited to offshore applications. The power generated by a turbine is determined by a combination of the area that an individual turbine blade sweeps and the prevailing wind speed. All wind turbines work best with a high, constant wind speed and a wind that flows smoothly, i.e. one that is not interrupted by hills, trees or buildings. A power law demonstrates that maximum wind speeds occur at higher turbine heights. These factors have resulted in larger wind turbine blades on increasingly high towers.

As a resource, wind is intermittent and therefore turbines do not run at maximum (or peak) capacity at all times. The load factor (the ratio of average power to peak power) is typically 30 per cent for modern turbines on a suitably windy onshore site.

In many rural areas of the UK, applications for planning permission for wind turbines are a major issue for local authority planning officers. For a developer, gaining permission for wind farms can be a long process, likely to go to an appeal and guaranteed to raise considerable local issues. The success rate of wind farm applications compares unfavourably with most other major applications for development.

There are permitted development rights for some small domestic-scale wind turbines. For large-scale wind 'farms' the planning process will depend on the exact size of the proposed development. In England, planning applications for renewable energy projects, such as onshore wind turbines, above 50 megawatts (MW) are treated as nationally significant infrastructure projects (NSIPs) and under the rules provided in the Planning Act 2008 require 'development consent'. With modern individual wind turbines having an output of between 2 and 3 MW this equates to wind farms of 15–20 turbines. Projects below 50 MW are dealt with at local authority level in accordance with policies set out in the National Planning Policy Framework (NPPF).

Irrespective of size, most development will normally start with a desktop scoping study to establish the suitability of the terrain and the likely wind speed across a potential site. This is often followed by an application for planning permission for an anemometer mast – a lightweight steel pylon-like structure with anemometers for measuring wind speeds at differing heights. Remote monitoring of the data from the mast enables a more accurate assessment of likely wind speeds and therefore potential return to developers.

A developer will have to decide how intensively a potential wind farm site could be developed, but for technical and practical purposes individual wind turbines are typically spaced 400–600 m apart. Other considerations include proximity to residential dwellings and potential noise issues, access, ecological issues, historic environment impacts, landscape and visual amenity, traffic and transport, and the not inconsiderable cost of the connection to the electricity grid. Given these considerations, larger wind farms can deliver financial economies of scale but they don't significantly increase the total power output because larger turbines have to be spaced further apart.

The return on investment to a developer comes from a combination of the electricity sold to the national grid along with various subsidy arrangements for renewable energy provided by the government. In 2002 the government introduced the Renewable Obligation (RO), which requires electricity suppliers to source a proportion of the electricity that they provide to customers from renewable sources. Renewable generators, such as wind farm developers, receive Renewable Obligation Certificates (ROCs) for each MWh of electricity generated and these certificates are traded on the open market.

Further reading

MacKay, David J. C. (2009) *Sustainable Energy – without the hot air*, UIT Cambridge, Cambridge, chs. 4 and 10.

17 Taxation

Ernie Jowsey and Rachel Williams

17.1 Direct taxes

Key terms: income tax; corporation tax; capital gains tax; inheritance tax; progressive taxation; tax relief

Direct taxes are so called because the taxpayer makes payment direct to the revenue authorities – Her Majesty's Revenue & Customs (HMRC) or the local authority. Usually each individual's tax liability is assessed separately. A direct tax, such as income tax, is assessed on and collected from the individual intended to bear it. An indirect tax is not levied directly upon the person on whom it ultimately falls e.g. VAT.

The most important direct taxes are:

- income tax – on incomes with higher percentage rates on higher incomes above a threshold (which means that income tax is a progressive tax);
- corporation tax – on company profits;
- capital gains tax – levied at income tax rates on any capital gain above a threshold amount;
- inheritance tax – on legacies (and lifetime gifts) above a threshold amount;
- other taxes including national insurance contributions, stamp duties, motor vehicle duties, petroleum revenue tax, council tax and uniform business rate.

Direct taxes have several advantages including: low costs of collection and convenient methods of payment such as PAYE; certainty of liability to the tax; they can be levied at progressive rates; and they can be used to redistribute income. Their disadvantages include: a disincentive effect on effort and enterprise (because the rewards are reduced); and if rates are high, a tendency for people to do all they can to avoid the tax.

While direct taxes may affect incentives to effort and risk-bearing investment, they are basically neutral in their effects on items of expenditure. There are some effects, however, on real estate, as follows:

- Interest relief on income tax for home-owners is a subsidy to home ownership.
- UK real estate investment trust dividend income is received net of basic rate income tax (20 per cent in 2013).
- There is 'rent a room' relief on the income of an individual from letting furnished accommodation which is part of his/her only or main residence. The exemption from income tax applies up to £4,250 (in 2013).
- In the computation of corporation tax (income tax in the case of an unincorporated business), agricultural and industrial buildings enjoy capital depreciation allowances. Commercial and residential buildings, however, can only offset actual repair and maintenance costs against tax. As a result, agricultural and industrial buildings tend to have a shorter life irrespective of technical

considerations, while landlords of residential properties are deterred from making improvements to buildings with a limited life.

- To avoid the break-up of agricultural estates that occurred in the past in order to pay inheritance tax upon the death of a major owner, there is now no tax payable on agricultural land. The tax advantage of leaving farm land rather than other assets influences demand, making farms more attractive to the wealthy.
- Since owner-occupied houses are exempt from capital gains tax (CGT), the owner receives a further 'subsidy' compared with other asset holders, thereby increasing demand for houses as compared with other assets such as equities, unit trusts and expensive antiques, which are subject to CGT.

Further reading

Harvey, J. and Jowsey, E. (2007) *Modern Economics*, Palgrave Macmillan, Basingstoke, ch. 36.
Jowsey, E. (2011) *Real Estate Economics*, Palgrave Macmillan, Basingstoke, ch. 22.

17.2 Income tax

Key terms: progressive tax; tax relief; MIRAS; home ownership subsidy

Income tax in the UK is a progressive tax because it has higher rates above certain thresholds. Taxes can be classified according to the proportion of a person's income that is deducted. A regressive tax takes a higher proportion of a poorer person's income than of a richer person's. Indirect taxes, which are a fixed sum irrespective of income (e.g. television licences), are regressive. A proportional tax takes a given proportion or percentage of one's income. A progressive tax takes a higher proportion of income as income increases (see Figure 17.2.1).

Justification for taxing the rich at a higher rate than the poor rests on the assumption that the law of diminishing utility applies to additional income, so that an extra £50 affords less pleasure to the rich person than to the poor person. So, taking from the rich involves less hardship than taking from the poor. Generally, this can be accepted as true, but we can never be sure, simply because there is no absolute measure of personal satisfaction.

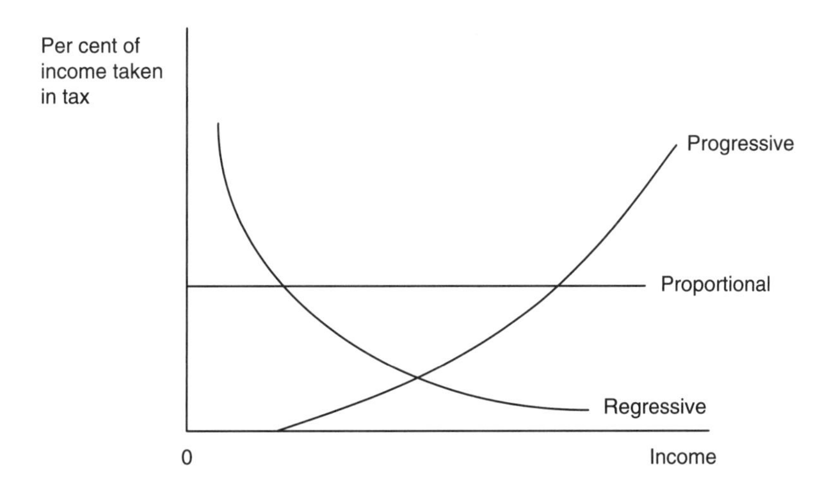

Figure 17.2.1 Regressive, proportional and progressive taxes.

Income tax interacts with real estate in some areas:

- Interest relief on income tax for homeowners is a subsidy to house buyers that started in 1803. Until 1950, when only 10 per cent of households were homeowners, the cost to the Exchequer was small. But the subsequent increase in ownership led to the qualifying loan being limited to £25,000 in 1974. This was raised to £30,000 in 1983 with the introduction of 'mortgage interest relief at source' (MIRAS), under which the borrower paid the lender interest less the tax relief. This homeownership subsidy peaked at £8 billion in 1991, but concern over government borrowing meant that the rate was subsequently reduced in steps to 10 per cent. On 6 April 2000, when MIRAS was abolished, the Exchequer saved only £1.4 billion a year. Over the previous years, however, this subsidy, by favouring owner occupiers, effectively moved housing tenure from the rented to the owner-occupied sector.
- UK real estate investment trust dividend income is received net of basic rate income tax (20 per cent in 2010).
- There is 'rent a room' relief on the income of an individual from letting furnished accommodation which is part of his/her only or main residence. The exemption from income tax applies up to £4,250.

Further reading

Harvey, J. and Jowsey, E. (2007) *Modern Economics*, Palgrave Macmillan, Basingstoke, ch. 36.
Jowsey, E. (2011) *Real Estate Economics*, Palgrave Macmillan, Basingstoke, ch. 22.

17.3 Corporation tax

Key terms: company taxation; profits; depreciation allowances; dividends

Corporation tax is a tax on the taxable profits of limited companies and other organisations including clubs, societies, associations and other unincorporated bodies. There are currently two rates of UK corporation tax, depending on the company or organisation's taxable profits:

- the lower rate – known as the 'small profits' rate;
- the upper rate – known as the 'full' rate or 'main' rate.

In the computation of corporation tax agricultural and industrial buildings enjoy capital depreciation allowances. Commercial and residential buildings, however, can only offset actual repair and maintenance costs against tax. As a result, agricultural and industrial buildings tend to have a shorter life irrespective of technical considerations, while landlords of residential properties are deterred from making improvements to buildings with a limited life.

Private landlords can make savings on tax by forming a company and paying corporation tax rather than income tax. Corporate tax is currently charged at 23 per cent, falling to 20 per cent in 2015. The individual rate of income tax can be up to 45 per cent and the capital gains tax rate is 28 per cent. There are some drawbacks with this, however, for example to take money out of a company structure might require a dividend payment, and that is liable to income tax; and annual accounts must be completed for companies. If profits were extracted, the additional income tax payable on receipt would eliminate much of the benefit, but if the profits were retained in the company there would be no additional taxes. Forming a company to hold properties has the added advantage of conferring limited liability should any claim be made against a property.

Further reading

www.gov.uk/renting-out-a-property/paying-tax (accessed 25/11/2013).

17.4 Inheritance tax

Key terms: death duty; tax threshold; agricultural relief; US estate tax

The origin of inheritance tax was Death Duty (or Estate Duty) which was introduced in 1894. This tax was payable when an estate was transferred on the death of the owner. Inheritance tax was introduced in 1986 and comes out of the deceased's estate before the inheritance is passed on. It applies to transfers on death, gifts made within 7 years of death and lifetime gifts. Some gifts are exempt, such as small gifts of less than £250 per year and marriage gifts.

Inheritance tax is due when a person's estate (their property and possessions) is worth more than £325,000 when they die. This amount is known as the inheritance tax threshold. The rate of inheritance tax is 40 per cent on anything above the threshold. The rate may be reduced to 36 per cent if more than 10 per cent of the estate is left to charity. Inheriting a property from a spouse or civil partner means the receiver is an 'exempt beneficiary' and normally won't have to pay inheritance tax as long as they are domiciled in the UK.

Inheritance tax could result in the break-up of agricultural estates in order to pay inheritance tax upon the death of a major owner. In order to avoid this, there is now no tax payable on agricultural land. This relief applies to the agricultural value of the asset only and for the purposes of agricultural relief, a farmhouse, cottage or building must be proportionate in size and nature to the requirements of the farming activities conducted on the agricultural land or pasture in question. For a woodland estate, the value of the timber (but not the land) is excluded from the estate, but if the timber is sold, inheritance tax is then due. The woodland might also qualify for agricultural relief or business relief if it's part of a working farm or business.

Of course the tax advantage of leaving farm land rather than other assets influences demand. A person who wishes to pass on his wealth to an heir may purchase and enjoy the amenity of owning a rural estate for at least two years and can, if he so wishes, carry on the actual farming through a contractor or similar device. Moreover, since exemption from inheritance tax is given for gifts and bequests to certain national institutions such as the National Gallery and the National Trust, it tends to decrease the land available for holding in private hands.

In the United States an inheritance tax differs from an estate tax, which is a tax levied on an entire estate before it is distributed to individuals. The Federal government currently has an estate tax for estates in excess of $2 million. Estate taxes are imposed on the total value of the property at death. Some states have an estate tax, some have an inheritance tax, and some have both. Inheritance tax is imposed on the transfer of assets, including real estate, at death and the rate is dependent on the relationship between the deceased and the inheritor.

Further reading

Jowsey, E. (2011) *Real Estate Economics*, Palgrave Macmillan, Basingstoke, ch. 22.

17.5 Indirect taxes

Key terms: specific tax; ad valorem; VAT; inelastic demand; hidden tax; tax evasion; tax avoidance; inflation

Indirect taxes are levied on spending on goods and services. They are 'indirect' because the revenue authority (HMRC in the UK) collects them from the seller, who tries to pass the burden on to the consumer by including the tax in the final selling price of the good. Indirect taxes can be *specific*, meaning they are a fixed sum regardless of the price of the good, or *ad valorem*, where they are a given percentage of the price of the good or service.

There are three categories of indirect tax:

1 Customs duties levied at EU rates on goods imported from non-EU countries.
2 Excise duties on certain domestic goods and services, e.g. beer, whisky, petrol, cigarettes and gambling.
3 Value Added Tax (VAT, see Concept 17.6).

Because governments tend to have very great revenue needs they have to use indirect taxes so that direct taxes are not extremely high. And where goods have inelastic demand (meaning they continue to be bought even after a rise in price) the revenue yield is fairly certain and easy to calculate. Indirect taxes are known as 'hidden' taxes because they are part of the purchase price that is not usually identified. They are also cheap to collect and difficult to evade. Of course they can be avoided by not buying the good or service in question.

The disadvantages of indirect taxes include:

- They are regressive – the poor will pay a larger proportion of their income in taxes on spending than the rich (who spend a smaller part of their total income).
- They can add to inflationary pressure if rates of duty are increased – this is particularly true of VAT, which is applied to most goods and services in the UK except food, housing and children's clothing.
- They result in greater loss of satisfaction to consumers than direct taxes because their expenditure is rearranged to reflect altered prices. Then of course resources are not allocated as efficiently as possible.

Further reading

Harvey, J. and Jowsey, E. (2007) *Modern Economics*, Palgrave Macmillan, Basingstoke, ch. 36.
Jowsey, E. (2011) *Real Estate Economics*, Palgrave Macmillan, Basingstoke, ch. 22.

17.6 Value Added Tax

Key terms: VAT; option to tax; waive VAT exemption; transfer as a going concern (TAGC).

VAT is a form of indirect taxation which applies to the supply of goods and services within the UK. The rates of VAT are; zero rate, reduced rate 5 per cent and standard rate 20 per cent. In a real estate context it is of particular importance to the supply of land (sales and lettings) and to construction work. The VAT treatment of some real estate related transactions and activities are considered in Table 17.6.1.

As noted in Table 17.6.1, except for new commercial buildings the supply of land does not give rise to liability to pay VAT. Commercial property owners can however, elect to pay VAT which is sometimes referred to as the option to tax or waiver of the exemption. When one first encounters the notion of someone choosing to pay tax it can be quite difficult to comprehend. The way a VAT registered business operates is that they must account to HMRC for the difference between the amount of

VAT they charge on their sales (output tax) and the amount they are charged on their purchases (input tax). If you sell goods or services that are exempt from VAT then you cannot reclaim the VAT on your purchases as illustrated below in a case where a property owner has refurbished an office building.

If property owner does not elect
They pay building contractors £100,000 + VAT
Cost of works to owner = £120,000

If property owner elects
They pay building contractors £100,000 + VAT
They sell the property for £600,000 charging VAT on the purchase price.
Input tax = £20,000; Output tax = £120,000.
Property owner accounts to Customs & Excise for £100,000
Cost of works to owner = £100,000

Another way of thinking about it is that the tax is being passed along the supply chain (see Table 17.6.2). This could be an issue for the tenant if they are making exempt or zero rated supplies such as insurance companies and banks.

Even if the election to pay VAT has been exercised the tax will not be charged if the property is being disposed of as part of a sale of a business (a transfer as a going concern – TAGC).

Table 17.6.1 Value Added Tax rates

Activity	*VAT status*	*Rate of VAT (if applicable) (%)*
Construction of a new house	Taxable supply	0
Sale of existing commercial building or lease of commercial building	Exempt – but can opt to waive the exemption	If exemption waived 20
Sale of new commercial buildings	Taxable supply	20
Construction of new commercial building or alterations to existing building	Taxable supply	20
Sale of land of buildings/assignment of lease as part of the sale of a business as a going concern	Not a taxable supply	

Source of rates: HM Revenue & Customs, 2012.

Table 17.6.2 Input and output tax

Input tax	*Output tax*
Office block owner refurbishes the property and is charged VAT on building materials and services by building contractor	If the owner disposes of property without opting to tax then there would be no output tax to offset the input tax If the owner elects to pay VAT then when the property is sold it will be subject to VAT
Property investors buy the refurbished property paying VAT on the purchase price	Property investors let the property and receive VAT on the rent from their tenant (a firm of surveyors)
The tenant pays VAT on its rent	The tenant charges VAT on its services
The tenant's clients pay VAT on services	If the client is a registered VAT business they may be able to offset but if not then it will be a cost to the business or consumer

Further reading

HMRC (2012) 'VAT', available at: www.hmrc.gov.uk/vat/ (accessed 25/11/2013).

17.7 Stamp Duty

Key terms: Stamp Duty Land Tax; stamp duty; Stamp Duty Reserve Tax

Stamp Duty is a tax paid by a buyer on the purchase of land and shares. There are three types of stamp duty:

- Stamp Duty on transfers of shares for more than £1,000;
- Stamp Duty Reserve Tax (SDRT) on electronic share transactions; and
- Stamp Duty Land Tax (SDLT) for transfers of real estate in the UK.

The amount of duty payable on share transactions is 0.5 per cent of the purchase price.

In order to identify the amount of SDLT payable it must first be identified whether the property is freehold or leasehold and whether it is residential or not. The current rates of SDLT for the sales of freehold property are set out in Table 17.7.1.

It can be argued that SDLT:

- discriminates against investment in property as opposed to shares which pay only 0.5 per cent;
- depresses overall property values;
- by increasing purchase costs, reduces market liquidity;
- where more accommodation is required, makes it cheaper to build on to one's current house rather than move to a higher-priced house;
- has a disproportionate effect on south of England properties, where property prices are highest.

NB: Lease duty is also payable at 1 per cent on the capital value of leases (found by capitalising the rent payable over the term of the lease at the discount rate of 3.5 per cent) less the threshold of £150,000.

Table 17.7.1 Stamp Duty Land Tax rates

Rate (%)	*Residential*	*Non-residential*
0	£0–125,000	£0–150,000
1	£125,000–£250,000	£150,000–£250,000
3	£250,000–£500,000	£250,000–£500,000
4	£500,000–£1,000,000	£500,000
5	£1,000,000–£2,000,000	
7	Over £2,000,000	
15	Over £2,000,000 for acquisitions by certain persons including corporate bodies	

Source: Rates sourced from HMRC (no date).

If the property purchase price is above a payment threshold then SDLT is charged at that rate on the whole amount. So, for example, if a house was sold for £250,500 then SDLT at a rate of 3 per cent would be charged on the entire purchase price. It is therefore not very common to find properties being sold for just over a SDLT threshold as, in the example above, the difference between buying the property for £250,000 (SDLT = £2,500) and £250,500 (SDLT = £7,515) would be £5,015 in additional tax. This is also an important factor when determining what price to market a property at.

SDLT is only payable on land transactions, so any element of the purchase price that is attributable to chattels, e.g. carpets and furniture (see Concept 10.3 regarding fixtures and chattels) will not be chargeable to tax. There are sometimes attempts to reduce the tax payable by artificially apportioning the amount of the overall purchase price that is payable for the chattels but this is tax evasion and therefore not legal.

There are calculators available on HM Revenue & Customs' website to determine the amount payable on other more complicated transactions including the grant of a new lease. A SDLT return will have to be made within 30 days of 'the effective date of the transaction' (normally the date of completion of the lease or the purchase). In some instances businesses opt to acquire real estate through a property owning company rather than selling the land itself and there is a special type of relief for transfers of land between group companies (stamp duty group relief). As has been noted earlier, the rate of stamp duty on shares is half a per cent and therefore less than the rate of SDLT on properties over £125,000/£150,000. There are, however, a number of anti-avoidance measures that are designed to prevent this relief being used as part of arrangements whose main purpose is to avoid tax.

In the past there have been special provisions put in place to try and promote market activity and regeneration. For example, until April 2013 Disadvantaged Area Relief applied to certain parts of the country in which a higher threshold was applied for the payment of SDLT (HMRC, no date) and from March 2010 to March 2012 first time buyers were exempt from paying SDLT on homes costing less than £250,000.

Further reading

HMRC (no date) 'Stamp Duty Land Tax (SDLT): The Basics', available at: www.hmrc.gov.uk/sdlt/index.htm (accessed 25/11/2013).

17.8 Mansion tax and annual tax on enveloped dwellings

Key terms: wealth tax; double taxation; administration problems

A mansion tax does not exist in the United Kingdom although versions of it are in place in other parts of the world, for example, in Ireland. As the name implies it is a tax that applies only to the most valuable homes.

It is the policy of the Liberal Democrat party to introduce a mansion tax (Liberal Democrats, 2013). Their proposal is that the tax would be charged at 1 per cent of the amount by which the value of a residential property exceeds £2 million. For example the owner of a property worth four million pounds would pay £20,000 (1 per cent of £2 million) per annum. They are also debating whether or not an individual taxpayer's property holdings should be added together so that mansion tax is payable on the cumulative value of the portfolio.

Proponents of mansion tax argue that assets perpetuate privilege – most professionals inherit money – taxing wealth is better than taxing incomes as it doesn't distort incentives for employment or profit-making – while under-taxed property encourages speculation and under-occupation. It is true that the richest 10 per cent of population in the UK own 44 per cent of all wealth and the current tax system does not seem to be redressing that. Inheritance tax is frequently avoided (see Concept 17.4).

The proposal has provoked criticism in some quarters and has been described as stemming from the 'politics of envy'. In particular, concerns have been raised about the difficulties of administering such a tax and double taxation (as the householder will also be paying council tax). David Cameron has ruled out introducing a mansion tax if re-elected on the grounds that it is not sensible for a country which wants to attract wealth creation. The Labour Party made mansion tax the subject of a vote in March 2013, which they lost, but they have not to date committed to adopting mansion tax as one of their manifesto policies.

Annual tax on enveloped dwellings/annual residential property tax

The new annual residential property tax (ARPT) came into effect on 1 April 2013. This is an annual charge that will be levied upon high-value UK residential properties owned by companies (enveloped by companies). The charge will be calculated by reference to the value of the property, as follows:

Property value	*Annual charge 2012/2013*
£2m – £5m	£15,000
£5m – £10m	£35,000
£10m – £20m	£70,000
Over £20m	£140,000

Further reading

Knight, Frank (2013) 'Taxing High Value Homes – Mansion Tax', available at: http://my.knightfrank.co.uk/research-reports/taxation-of-prime-property.aspx (accessed 25/11/2013).

Liberal Democrats (2013) 'Liberal Democrats Policy Consultation: Taxation', available at: www.libdems.org.uk/siteFiles/resources/docs/conference/2013-Spring/114%20-%20Taxation.pdf (accessed 25/11/2013).

www.hmrc.gov.uk/ated/basics.htm#1 (accessed 25/11/2013).

17.9 Council tax

Key terms: local government; rates; council tax bands; regressive taxes

Local authorities need to raise finance in order to provide local services. It is thought to be better if some of their revenue is raised through local taxation, rather than being financed by central government – this is so that voters can see the results of local democracy via their influence on spending. Sources of revenue for local authorities are: uniform business rates (see Concept 17.10), central government grants, user charges and fees and the council tax. The amount by which the area authorities' spending exceeds revenue from the other sources has to be covered by levying a council tax on households.

Council tax is very similar to the old rating system, but instead of a notional annual letting value, for which market evidence was deficient, it is based on the 1 April 1991 capital value of the property – what it would have sold for on that date. The council tax band of a property is not related to its current market value. This is because, by law, council tax valuations are based on the price a property would have fetched if it had been sold on 1 April 1991. For Wales the Valuation date is 1 April 2003. The Valuation Office Agency (VOA) determines the council tax bands of the 23.4 million domestic properties in England and Wales. To allow for the difficulties involved in making a precise valuation of each individual dwelling, values are divided into eight bands, A to H, with all households in the same band paying the same amount of tax, but increasing upwards to H, whose payment band is treble that of band A and double that of D. There is a 50 per cent discount for unoccupied dwellings and second homes. There is no provision for a general revaluation, but a future sale or a change in the locality, such as a new nearby motorway, may afford grounds for a revaluation (Jowsey, 2011).

Council tax is regressive because it is not proportional to either property values or incomes and the rate charged is higher on lower value properties than it is on higher value properties (note that this does not mean that the actual amount payable is higher on lower value properties).

For single residents there is a 25 per cent discount, but people exempt from the council tax are ignored for the purpose of determining the single-person discount. Exempt people include: students, student nurses, apprentices, youth trainees, those on income support, the severely mentally handicapped and elderly dependent relatives.

In Scotland a similar system operates to that in England and Wales, but the Scottish Government believes that the council tax is fundamentally unfair and should be abolished and replaced with a fairer local income tax based on ability to pay (Scotland.gov.uk).

Further reading

Jowsey, E. (2011) *Real Estate Economics*, Palgrave Macmillan, Basingstoke, ch. 22.

17.10 Rating and uniform business rates

Key terms: rate poundage; multiplier; rateable value; effect on rents; effect on investment

Uniform business rates (UBR) are the taxes paid by those who occupy, or have the right to occupy, commercial property in the UK. HM Revenue & Customs sets these rates through the VOA, and they are determined by rental levels – what the premises could be let for less rates and taxes, insurance and maintenance expenses.

Surveyors working in rating help clients to minimise their business rates while complying with legislation, or work for the VOA to set the rates businesses must pay. Rating involves property inspections, client meetings, database management, report writing and negotiating skills.

Business rates are easy and cheap to collect and they ensure that firms contribute to local expenditure on collective services, such as police, roads, street lighting and open spaces. They are tax deductible from profits. Taxes on commercial property, however, such as the UBR can also reduce investment in property relative to plant and machinery which is not subject to such a tax.

In 1990 the response of the government to the weaknesses of the business rate was to take away the power of the local authority to levy it. A new rating list was compiled by the valuation officer appointed for each charging authority (the district borough or city council), using an unchanged formula of a notional rental value – what the premises could be let for less rates and taxes, insurance and maintenance expenses.

A rate (now known as 'the multiplier') is set for England and Wales by the Department of Communities and Local Government. If the rateable value of a business property, in England, was £20,000, and the local authority was calculating the 2013/14 business rates bill, it would multiply it by 47.1p to get a total for the year of £9,420. If the business is entitled to any form of rate relief, this sum is then adjusted to reflect that, making a final total for the rates bill. The multiplier usually changes each year in line with inflation.

In England, since 1 April 2005, a small business rate relief scheme has been in operation. Eligible businesses who apply to their local authority to be part of the scheme can have their liability calculated using the small business multiplier (46.2p in 2013/14). Dependent on their rateable value, they may also be eligible for a further discount on their bill. The City of London is able to set a different multiplier from the rest of England (slightly higher in 2013/14).

Revaluations of the rating list take place every five years, the next being in 2015. At each revaluation the UBR is reset to ensure that in real terms the revenue from business rates is unchanged following the revaluation. The revenue from the new rate is paid into a national pool and then distributed to local authorities in proportion to their adult populations. There are separate pools for England and Wales.

It is doubtful whether businesses, while still relying on local authorities for services, will find their needs treated with more respect now that they are paying a national rather than a local tax. Indeed, when businesses paid the local tax, their Chambers of Commerce were listened to and could exert influence. Thus accountability could be less under the UBR.

Increased UBR can affect rents. Since the firm has to sell in competition with other firms, it is likely that it will have to bear most of the increase. In the short period, only if the demand for a firm's product is completely inelastic can the whole of a rate increase be passed on to the consumer through higher prices. In the long term when a rent review is due, a firm may be able to pass some of the higher cost on to the landlord by a rent reduction, depending on the relative elasticities of demand and supply.

In his Autumn Statement of 2013, the Chancellor of the Exchequer cut the scheduled increase in business rates in April 2014 from 3.2 per cent to 2 per cent and also said pubs, restaurants and small shops would receive a £1,000 discount on their rates for the next 2 years. Businesses that move into vacant shops on Britain's struggling high streets will, in addition, enjoy a 50 per cent discount on their rates for 18 months, while rates relief for small businesses will be extended for another year.

Further reading

Jowsey, E. (2011) *Real Estate Economics*, Palgrave Macmillan, Basingstoke, ch. 22.

17.11 Land value tax

Key terms: site value; potential value; betterment; external costs and benefits

Land value taxation (LVT – sometimes called site value rating, or site value taxation when used in the context of local taxes) is a method of raising public revenue by means of an annual *ad valorem* tax on the rental value of land. The site value is the value of the unimproved land used in its most productive use, regardless of its current use. It equates to the sale value of the vacant freehold plot. Advocates say LVT would replace, not add to, existing taxes and, if properly applied, could support a range of social and economic initiatives, including housing, transport and other infrastructural investments. There is currently no LVT in the UK.

The value of every parcel of land in a country would be assessed regularly and the land value tax levied as a percentage of those assessed values. 'Land' means the site alone, not counting any improvements. The value of buildings, crops, drainage or any other works that people have erected or carried out on each plot of land would be ignored, but it would be assumed that all neighbouring properties were developed as at the time of the valuation; other things being equal, a vacant site in a row of houses would be assessed at the same value as the adjacent sites occupied by houses. The valuation would be based on market evidence, in accordance with the optimum use of the land within the planning regulations. If the current planning restrictions on the use were altered, the site would be reassessed. Information on recent transactions makes valuations easier, but if such information is scarce, valuers' estimates can be subjective and open to challenge.

LVT can be justified for economic reasons because if it is implemented properly, it will not deter production, distort market mechanisms or otherwise create welfare losses the way other taxes do. Income taxation discourages work and sales, and value-added taxes discourage consumption, capital gains taxes discourage investment, and real property taxes discourage building and improving property.

LVT does not have distortionary effects on output, because it is payable regardless of whether or how well the land is actually used. Land is effectively in fixed supply, so an increase in the tax rate on land value will raise revenue without damaging incentives for owners to invest in and use their land. Because the supply of land is inelastic, market land rents depend on what tenants are prepared to pay, rather than on the expenses of landlords, and so LVT cannot be passed on to tenants. The only direct effect of LVT on prices is to lower the market price of land.

The situation is shown in Figure 17.11.1, where the tax comes from the producer surplus or economic rent, with no effect on the amount of land supplied.

The idea behind LVT as the basis of assessment is that the tax falls only on the land element of real property, that is, the open market value of the site, on the assumption that it is currently available for its most profitable use. Thus, compared with a net annual value (NAV) or capital value base, the buildings are not taxed, but *potential value* is taxed (Figure 17.11.2).

As the basis of a property tax, LVT has many advantages. First, it has strong moral backing in that it is closely associated with taxing 'betterment'. The argument is founded on Ricardo's theory of economic rent: since land is fixed in supply, its value is determined solely by demand; LVT is a tax on this demand-determined value and is thus a means by which 'betterment' can be returned to the community. Moreover, because sites are, for spatial reasons, fixed in supply, the tax falls entirely on economic rent with no effect on supply.

Second, LVT should improve the efficiency of land use. Because site-value is the sole tax base, it is in effect a lump-sum tax which is levied irrespective of how the site is used, the value of the building on it, or whether the buildings are improved. That is, LVT is neutral as regards the type of use, intensity of use and improvements. There is thus an incentive to develop sites to their most profitable use, since the burden of the given tax would then be spread over higher gross receipts. Even if speculators continued to hoard land – vacant central sites and agricultural fringe land – LVT would ensure that they had to pay towards the cost of public services provided. Furthermore, the improvement of existing buildings would be encouraged. Thus LVT should speed up the renewal of inner-city areas. Council tax, and the national non-domestic rates levied on businesses penalise landowners who contribute to their communities by developing their land, while the owners of derelict, community-damaging sites pay little, if any, tax. With LVT landowners have to pay whether or not they are using the land productively. This provides a powerful incentive for bringing the land into productive use.

Third, LVT could have benefits for housing. In the short period, the rate burden would tend to shift to central sites and away from suburban houses. Figure 17.11.3 shows that under LVT (a) rates are payable on one-half of the original base, whereas (b) rates are payable on only one-quarter. In the long period, the more intensive use of central sites should, given no change in the demand for land resources, reduce the demand for and thus the price of peripheral land, the main source for new housing (Figure 17.11.3).

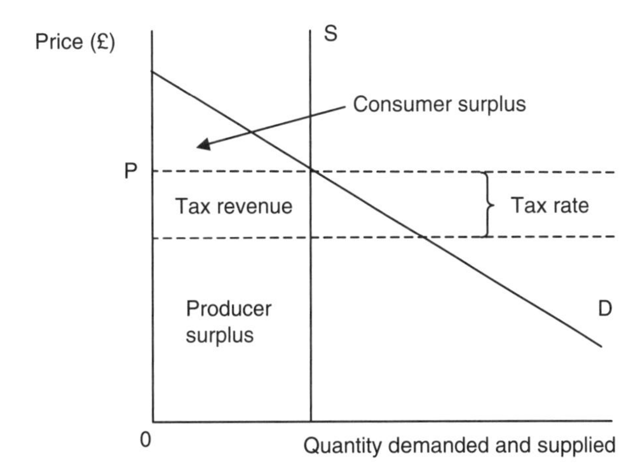

Figure 17.11.1 The effect of a land value tax.

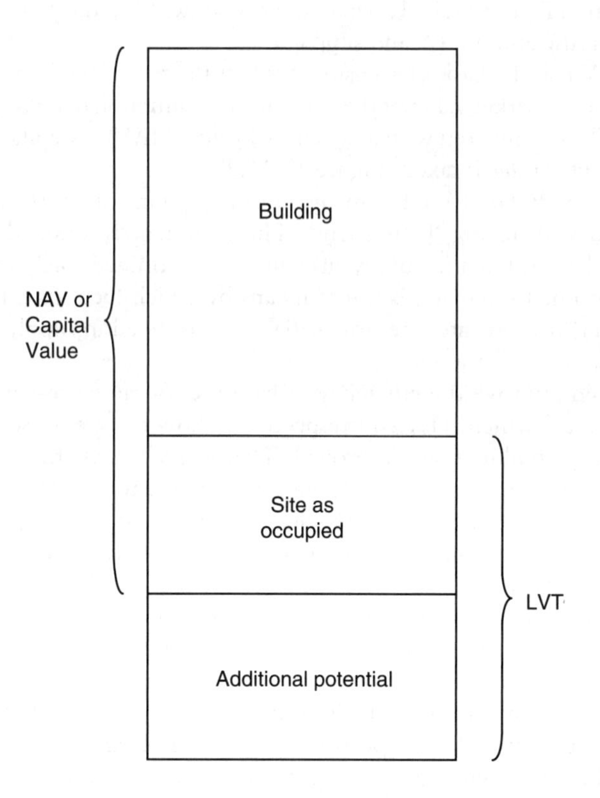

Figure 17.11.2 Comparison of the incidence of NAV, capital value and LVT on buildings and land.

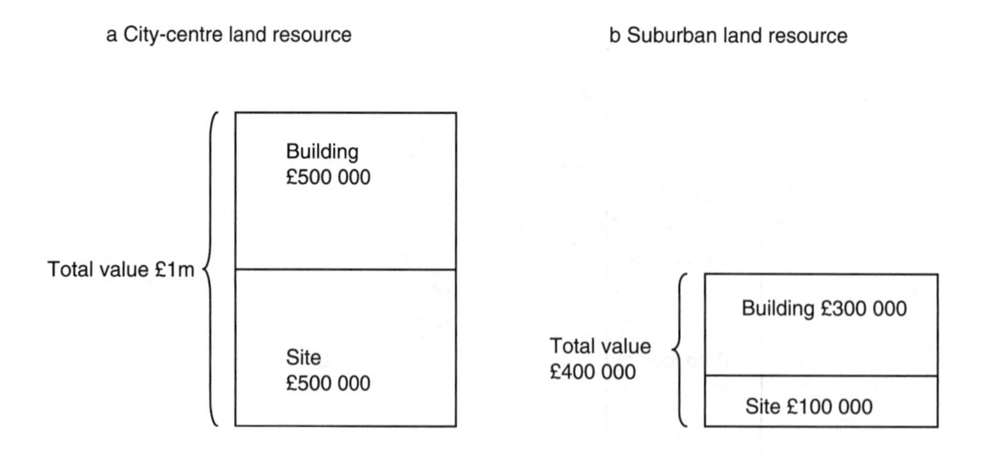

Figure 17.11.3 The incidence of land value tax or site-value rating.

Fourth, LVT should produce external benefits. Given no change in the demand for land resources, its impetus towards the redevelopment of central sites would reduce city sprawl. On the other hand, this could be accompanied by external costs – increased city-centre traffic congestion and the loss of open space, such as large private gardens and centrally situated recreational facilities, which could be taxed out of existence. Planning regulation, however, could partly deal with such costs.

Fifth, LVT would reduce 'fiscal zoning' since, in the long run, the movement of business and people to 'low-rate' areas would be self-defeating in that it would simply raise site values there.

Sixth, LVT should promote objectivity in making planning decisions. Because the NAV system rates both buildings and land, the local authority has a built-in reason for approving a proposed development through the higher rateable value that would result. With the building element removed by LVT, planning decisions can follow a consistent policy based solely on environmental considerations.

Supply-side policy attempts to increase production and the supply of goods by decreasing costs, such as by lowering taxes and eliminating excessive regulations and barriers to trade. A tax shift, away from taxing production to taxing land values, is supply-side policy, since it removes the excess economic burden of taxation.

Nevertheless, though LVT has advantages, it faces objections on principle and difficulties in implementation. The uncertainties in assessing site-values would give rise to considerable challenge by owners and a transition period of several years might have to elapse before it could become fully operative. In the meantime it might not provide a predictable basis for financing local services.

LVT has been introduced in a variety of forms in several countries around the world. Denmark, Hong Kong and Taiwan have used land values to raise revenues. On a more local scale, cities in parts of Australia, South Africa, New Zealand and North America have adopted forms of LVT.

In the United States many local authorities operate a so-called split-tax system, in which buildings and land are taxed separately. Some weight the tax towards buildings and others towards land. It seems that the more it is biased towards land, the more this benefits the local economy.

Further reading

Jowsey, E. (2011) *Real Estate Economics*, Palgrave Macmillan, Basingstoke, ch. 22.

18 Valuation

Lynn Johnson and Becky Thomson

18.1 Income cash flows

Key terms: implicit; rack rented; under-rented; over-rented

Traditional investment valuation centres on the *implicit* reflection of income growth and risk through the use of an all-risks yield (ARY). The alternative approach is the use of a discounted cash flow. With regard to traditional investment valuation, the method applied depends on the type of cash flow.

Rack rented property

When an investment property is *rack rented*, the current rental income equates to market rent and the valuation technique employed is relatively straightforward. The rental income is simply capitalised using years purchase (YP) in perpetuity at an appropriate ARY.

Case study 1

The self-contained office building in Figure 18.1.1 extends to approximately 185.80 sq m (2,000 sq ft) and was let in August 2013 on a new 10-year FRI lease at a market rent of £29,000 per annum (£156.08 per sq m; £14.50 per sq ft).

Similar investments on Metro Riverside with similar unexpired lease terms are achieving ARYs of 10 per cent. The lease has been agreed at market rent and therefore the cash flow over the life of the lease will be as shown in Figure 18.1.2.

Figure 18.1.1 Unit 3, Metro Riverside, Gateshead.

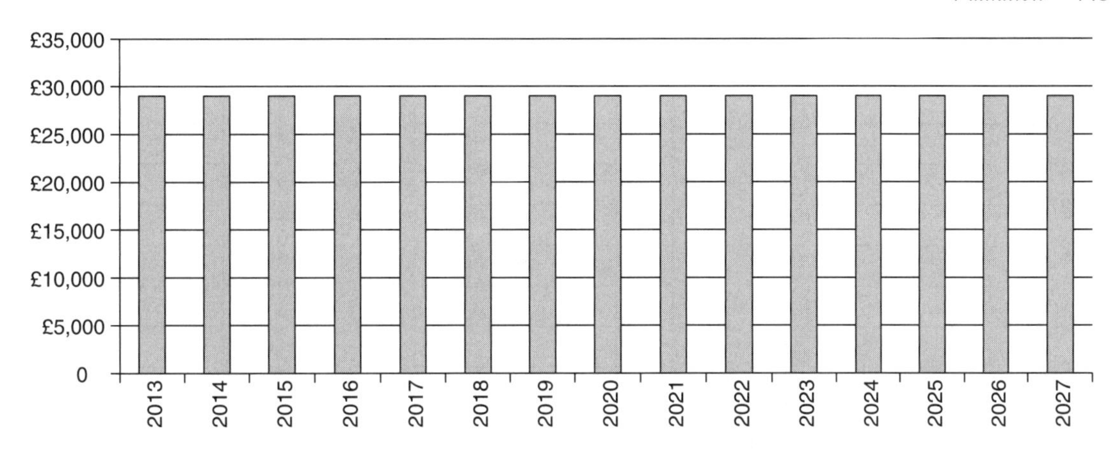

Figure 18.1.2 Lease cash flow.

We can therefore assume that the market rent is received into perpetuity:

Market rent	£29,000 per annum
YP in perp @ 10%	10.0000
Market value	*£290,000*

The rack rented approach only applies when properties are let at full market rent and this occurs when:

a) the property has recently been let;
b) there has recently been a rent review to market rent;
c) market rents have not changed since the rent was last agreed.

However, some rents are not equal to market rent and are likely to either increase or decrease at some point in the future, i.e. at a rent review or lease expiry.

Under-rented property

Under-renting occurs when the current passing rent received is *below* market rent. This is usually due to the fact that the passing rent has been agreed some time ago (either at the start of the lease or at a previous rent review) and over that time there has been rental growth.

Case study 2: Unit 1 Metro Riverside, Gateshead

Using the same property from Case Study 1, we can study a straightforward under-rented cash flow. In this scenario, a 10-year FRI lease was agreed in August 2011 at a rent of £26,000 per annum (£139.94 per sq m; £13.00 per sq ft) subject to 5-yearly rent reviews.

Unit 3 is in the same development and is the same size and specification as the subject property. Market rent was recently agreed at £29,000 per annum (£156.08 per sq m; £14.50 per sq ft) so an increase is justified for the subject property at the next rent review in 2016, and therefore the investment is currently under-rented (see Figure 18.1.3).

Note that we are only interested in valuing the cash flows from the present into the future. We are not interested in the cash flows that have gone before, as those sums have been paid and are not due to be received. An investor wants to know the value of the income that is still to be received.

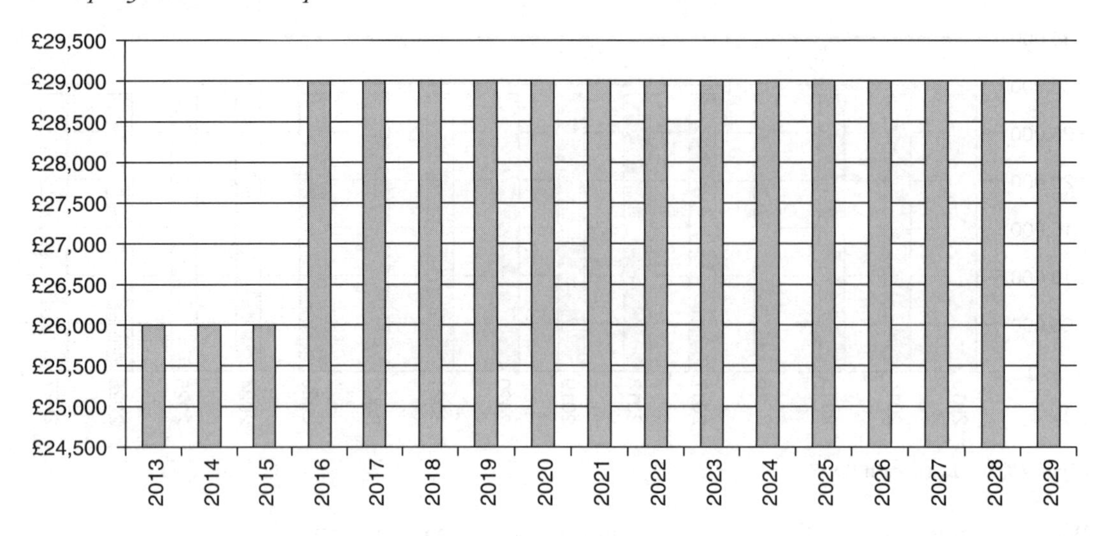

Figure 18.1.3 Under-rented property.

Over-rented property

Over-renting occurs when the passing rent is *above* market rent. This is commonplace in the current market, where rents which were agreed five or more years ago are now inflated over market rent due to a decrease in rental values.

Case study 3: Unit 1, Metro Riverside, Gateshead

Using the same property again, we can assume a new 10-year FRI lease was agreed in August 2010 at a rent of £33,000 per annum (£177.61 per sq m; £16.50 per sq ft) subject to 5-yearly rent reviews.

The market rent for Unit 3, Keel Row House is £29,000 per annum (£156.08 per sq m; £14.50 per sq ft) and therefore no increase is justified at the rent review in 2015 as the investment is over-rented (see Figure 18.1.4).

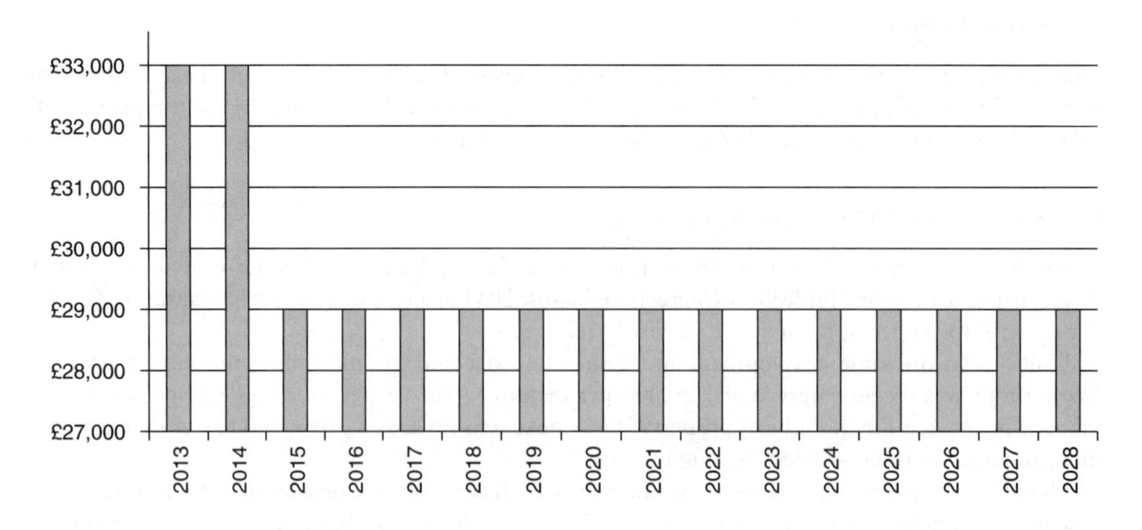

Figure 18.1.4 Over-rented property.

Further reading

Sayce, S. *et al.* (2006) *Real Estate Appraisal*, Wiley-Blackwell, Oxford.

18.2 Term and reversion

Key terms: term and reversion; passing rent; market rent; ARY; perpetuity; market value

The term and reversion method was derived in the 1950s when inflation was negligible and fixed incomes below market rent were more secure. The method considers two blocks of income received over time – the *term* when the passing rent is received until the first reversion (i.e. a rent review or lease expiry) and the *reversion* which is a future block of income received after the term at a higher *market rent*.

Case study 1: Bowburn Industrial Estate, Durham

A 1,579 sq m (17,000 sq ft) detached industrial unit on a popular estate in Durham is owned freehold and let on a 10-year FRI lease from 2010. The current passing rent is £71,000 per annum and there is a rent review in Year 5 (2015). After careful analysis of the market and comparables we know that the market rent is £76,000 per annum.

ARY for similar types of property (i.e. good location with good unexpired term and potential for rental growth) are in the order of 9 per cent.

Cash flow graph

The term comprises two years of income at £71,000 per annum. Due to the rent review mechanism in the lease, the freeholder has the opportunity to increase the income to market rent, so the reversion starts from the rent review in 2014 when the income increases to £76,000 per annum. We then assume that the £76,000 per annum will be received in perpetuity (see Figure 18.2.1).

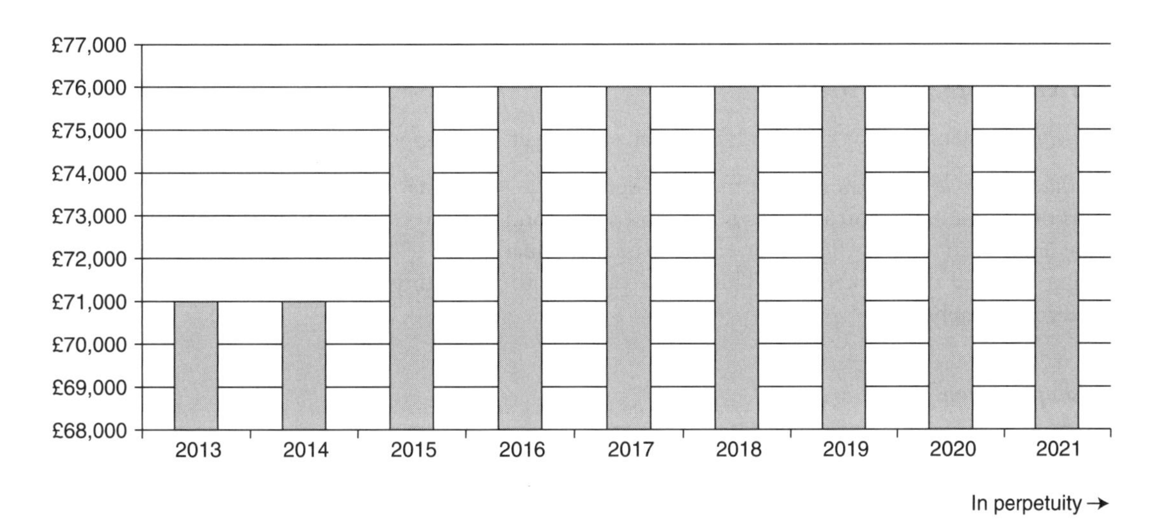

Figure 18.2.1 Term and reversion.

Valuation

Term	
Passing rent	£71,000 per annum
YP @ 9% for 2 years	1.7591
Market value of the term	*£124,897*
Reversion	
Market rent	£76,000 per annum
YP in perp @ 9%	11.1111
PV of £1 @ 9% for 2 years*	0.8417
Market value of the reversion	*£710,752*
Market value	*£835,649*

*The term is 2 years, so we have to defer the reversionary income for 2 years. It is also possible to do this using YP in perp deferred.

Reversion	
Market rent	£76,000 per annum
YP in perp @ 9% deferred 2 years	9.3520
Market value of the reversion	*£710,752*

Further reading

Isaac, D. and O'Leary, J. (2013) *Property Valuation Techniques*, 3rd edn, Palgrave Macmillan, Basingstoke.

18.3 Hardcore/layer method

Key terms: hardcore or layer method; core; top slice/layer; all-risks yield; passing rent; market rent

The *hardcore* or *layer* method is the more modern of the traditional basis of valuation. There are two layers of income – the bottom layer is the most secure and this is received into perpetuity. The *top slice* or *top layer* of income is the riskier element of the valuation, as it comprises the difference between passing rent and the market rent. This is also valued into perpetuity, but at a higher yield due to the greater risk attached.

Yield application

In practice it is perceived that the core income is the more secure, and a lower yield is applied. The top slice is the future income over and above the rent currently received, so the reversionary yield is therefore slightly higher. The level of risk applied to the top layer depends on the uncertainties and risks associated with the investment.

Case study 1: Unit 9a, Chollerton Drive, North Tyneside

The subject property comprises a 1970s semi-detached industrial unit with a steel portal frame, insulated metal cladding and a pitched metal clad roof incorporating translucent sections. The gross internal area is 893.0 sq m (9,620 sq ft).

The property was let to a local textiles company in 2010 on a new 15-year FRI lease at a rent of £24,000 per annum. The rent is subject to 5-yearly rent reviews. Comparable evidence suggests that market rent is £27,500 per annum and an appropriate ARY is 10 per cent.

Valuation

The ARY has been applied to the core income while 0.5 per cent has been added to the yield for the layer/top slice, due to the shorter unexpired term.

Core	
Passing rent	£24,000 per annum
YP in perp @ 10%	10.0000
Market value of the core	*£240,000*
Layer	
Market rent	£27,500 per annum
Less passing rent	£24,000 per annum
Incremental rent	*£3,500 per annum*
YP in perp @ 10.5%	9.5238
PV of £1 @ 10.5% for 2 years*	0.8190
Market value of the layer	*£27,299*
Market value	*£267,299*

*There are two years remaining until the next rent review.

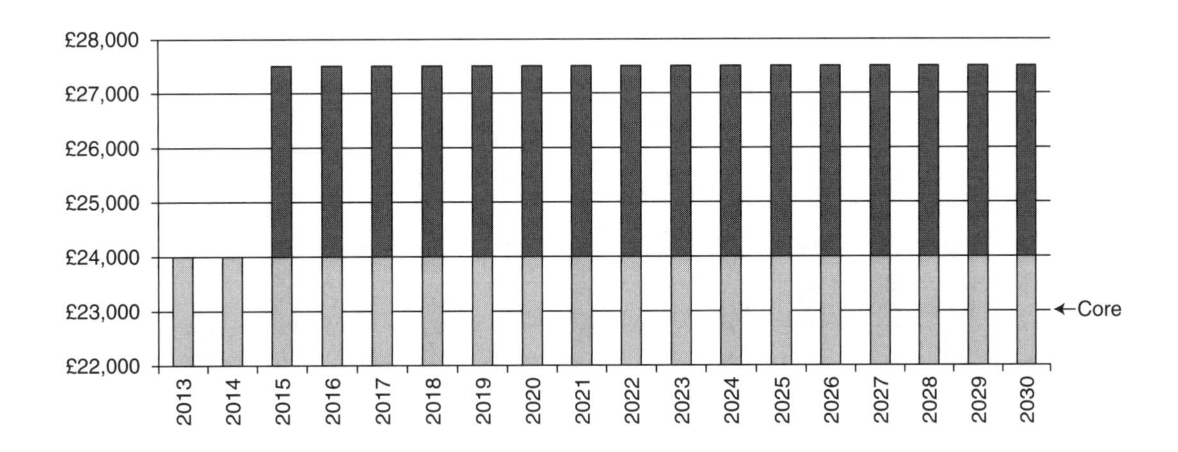

Figure 18.3.1 Cash flow 1.

Valuation with more than two layers

When considering properties with stepped rental increases, we have to consider each tranche of income similar to the term and reversion method.

Case study 2: Dean Street, Newcastle upon Tyne

The subject property comprises a mid-terraced Grade II listed property (Figure 18.3.2). It is a three-storey brick building with a slate-covered mansard roof. The ground floor retains the original ornate sandstone façade, while the upper floors have been rendered and painted.

In 2013, a new 10-year FRI lease was agreed with a local clothes store on a stepped rental basis. The initial rent was agreed at £22,000 per annum, rising to £23,500 in years 2 and 3, and then to the market rent of £25,000 per annum in year 4. An appropriate ARY of 8 per cent has been determined by comparable evidence. The valuation will include three tranches of income, and we need to think about *when* the landlord will receive the increases.

Figure 18.3.2 Dean Street, Newcastle upon Tyne.

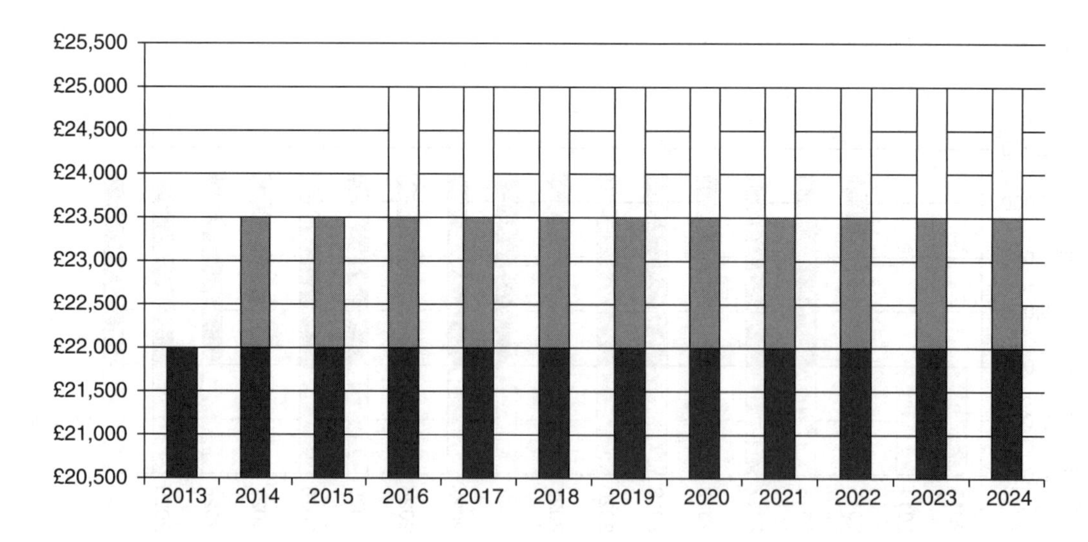

Figure 18.3.3 Cash flow 2.

Valuation

The ARY has been applied to the core income which is below market rent and therefore considered secure. The yield has been increased by 0.25 per cent as the stepped rent is introduced, to account for the associated uncertainties and risks.

Core	
Passing rent	£22,000 per annum
YP in perp @ 8%	12.5000
Market value of the core	*£275,000*
Middle layer	
£23,500 per annum – £22,000 per annum	£1,500 per annum
YP in perp @ 8.25%	12.1212
PV of £1 @ 8.25% for 1 year*	0.9238
Market value of the middle layer	*£16,796*
Top layer	
£25,000 per annum – £23,500 per annum	£1,500 per annum
YP in perp @ 8.5%	11.7647
PV of £1 @ 8.5% for 3 years**	0.7829
Market value of the layer	*£13,816*
Market value	*£305,612*

*The landlord has to wait one year before receiving the middle layer
**The landlord has three years before receiving market rent

The hardcore method is favoured in practice, and computer software programs such as Argus Investor tend to default to the hardcore method.

Further reading

Isaac, D. and O'Leary, J. (2013) *Property Valuation Techniques*, 3rd edn, Palgrave Macmillan, Basingstoke.

18.4 The all-risks yield

Key terms: yield; investment valuation; rent; comparable evidence; tenant covenant

A yield is the rate of return that a purchaser is seeking from a potential property investment (Shapiro *et al.*, 2012). The ARY is the yield most commonly used when valuing a property for investment purposes, and it implicitly reflects all of the risks and prospects attached to a particular property investment. This means that any anticipated future movement in value is reflected in the proposed purchase price by the adjustment of the yield by the valuer.

In order to establish the ARY, we need to use transactional evidence from the relevant property sector. The ARY is expressed as a percentage and is calculated as follows:

$$ARY = \frac{Market\ Rent}{Market\ Value} \times 100$$

Example 1

A fully let office block in central Newcastle with a recently reviewed rent of £160,000 per annum has just been sold to a pension fund for £2m. Calculate the ARY:

$$ARY = \frac{£160,000}{£2,000,000} \times 100$$

Therefore, ARY = 8%

Example 2

It is important to note that the better the investment, the more investors will be prepared to pay for the same rental income. If an investor were to offer £2.5m rather than £2m for the same investment as that in Example 1, the yield changes:

$$ARY = \frac{£160,000}{£2,500,000} \times 100 = 6.4\%$$

Therefore, a better property investment commands a lower ARY.

The quality of a property investment depends on the security of both the initial capital invested and the rental income received; the liquidity of the property as an asset (i.e. buoyancy of the market) and the costs of transfer (Shapiro *et al.*, 2012). Therefore, external factors play a key part in assessing the security of a property investment when analysing comparable evidence for an ARY:

- *The economy*
  The economic climate is vital to the property investment market and how it performs. For example, if an economy is emerging from a recession then there is uncertainty in the market as a whole. Fewer investments come onto the market, as investors are waiting for the return of better market conditions and a chance to secure a better return on their investments.

- *Growth*
  An investor wants to make sure that the investment will grow in terms of market rent and market value. An investment with little prospect of rental growth (i.e. a property that is currently over-rented) is not attractive in the market and hence will affect the ARY achieved.

- *Legislation*
  Changes in taxation, planning policy and law can all have an effect upon a property investment. Does the property comply with current legislation? How much will an investor have to spend to achieve compliance and let the property?

- *The tenant*
  The quality of property investments can be affected significantly by tenant covenant, i.e. the financial worth of a tenant. If you are valuing a property with a strong tenant covenant, you need to be looking for similar tenant covenant strength in your comparables. Additionally, the status of occupation (i.e. single occupier or multi-tenanted) must also be taken into consideration as these factors impact upon ARY.
- *The lease*
  As mentioned earlier, lease terms can have a great impact on the viability of a property investment and the yield it will achieve. If a landlord has just agreed a new 10 or 15 year lease with a tenant, then this will appeal to a potential purchaser as the rental income is guaranteed for the next 10 or 15 years. Therefore, he will pay more for the investment and the ARY will drop. Conversely, an investor would be very wary if a lease is coming to an end and this would be reflected in what he would be prepared to pay, so the ARY would be higher.
- *Rent review*
  Rent reviews are common in commercial leases and they are normally 'upwards only' to encourage rental growth. If market rent has dropped below passing rent at the rent review, then the rent will simply remain the same rather than go down. Upwards/downwards reviews do exist but are less common. When searching for comparables, the rent review pattern should be the same.
- *Breaks*
  Breaks can have an impact on the yield as they affect the security of income received by the investor. A tenant may not even exercise their break, but its presence is unnerving to investors as this could potentially convert into a vacant period or void. An ARY would ordinarily be increased to account for this.

When analysing comparable evidence for a yield, the valuer needs to take all of the above factors into consideration and make the necessary adjustments to the ARY in order to reflect any differences between the comparable and the subject property.

Further reading

Blackledge, M. (2009) *Introducing Property Valuation*, Routledge, London, pp. 34–53.
Shapiro, E., Mackmin, D. and Sams, G. (2012) *Modern Methods of Valuation*, 11th edn, Estates Gazette Books, London, pp. 67–78.

18.5 Over-rented property

Key terms: rent reviews; over-supply; tenant strength; unexpired term; degree of over-renting; hard-core/layer method; short cut discounted cash flow

Over-renting occurs when a property is let at a rent higher than its market rent. This happens when either market rent has fallen since the property was let and upwards-only rent reviews have maintained the high historic figure; or the property was let at a headline rent with incentives and upwards-only rent reviews:

Example 1

An office building in Edinburgh's New Town was let on a 20-year lease at a rent of £50,000 per annum. By the first rent review, the market rent of the building had fallen to £40,000 per annum and the building had become over-rented.

- The £40,000 per annum market rent is sometimes referred to as the *core income*.
- The £10,000 per annum difference between the core income and the rent payable is called the *overage*.

Any valuer acting on behalf of a property investor has to deal with the risk of both the core income and the overage for an over-rented property investment. There are three areas that must be examined to assess the risks:

1. Tenant strength:
   - receipt of the overage depends on the lessee continuing to fulfil the lease covenants;
   - if the lessee leaves the property and the freeholder is forced to try and re-let, he will only be able to achieve market rent (or core income);
   - a proven multinational tenant will mean very little risk of default and security of income.
2. The unexpired term of the lease:
   - most investors want at least an average of 8–10 years unexpired on the leases of the occupying tenants within their property investment.
   - if only a short average unexpired term remains, the investor may suffer a reduction in income at expiry as he will only be able to re-let the property at the market rent;
   - if the market rent has not grown sufficiently to overtake the passing rent by the lease end, the tenant is likely to leave for cheaper alternative premises.
3. The degree of over-renting:
   - both of the above issues become less of a problem if the over-renting is slight;
   - if there is a large degree of over-renting, then default creates a large reduction in income.

With regard to the methods used for valuing over-rented property, the hardcore (or layer) method was the first method adopted for valuing over-rented freehold interests, when the problem first arose in the early 1990s.

When applied to over-rented property, the hardcore approach values market rent into perpetuity as the core income (at an appropriate ARY), and the overage as the top slice (at a higher yield to reflect additional risk) (see Figure 18.5.1). Note that the overage yield is based on an *equated yield approach*, i.e. gilts yield, plus a property risk premium, plus in excess of 2–3 per cent, to reflect overage risk.

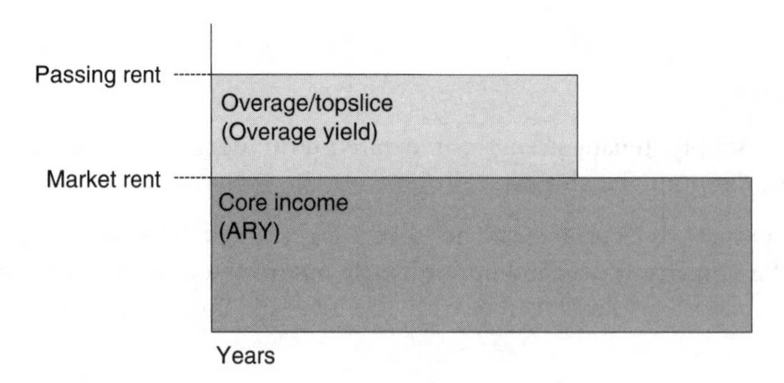

Figure 18.5.1 Overage/topslice.

Example 2

You have been asked to value an office building that was let 6 years ago on a 20-year FRI lease with 5-yearly upward-only rent reviews. The rent passing is £575,000 per annum and market rent is £500,000 per annum.

The ARY for similar property is around 6.5 per cent and the appropriate overage yield is 9.75 per cent (the gilt yield (4.75%) + property risk premium (3%) + over-renting risk (2%)).

Core income

Market rent	£500,000 per annum
YP into perpetuity @ 6.5%	15.3846
	7,692,308

Topslice

Overage	£75,000 per annum
YP single rate @ 9.75% for 14 years	7.4682
	560,112

Market value	*£8,252,420*
	Say £8.253m

The hardcore method assumes the overage will last until the lease end, but in reality the market rent is likely to increase during the remainder of the lease term (Figure 18.5.2).

The point at which market rent rises above passing rent is known as the *breakthrough* (Figure 18.5.3).

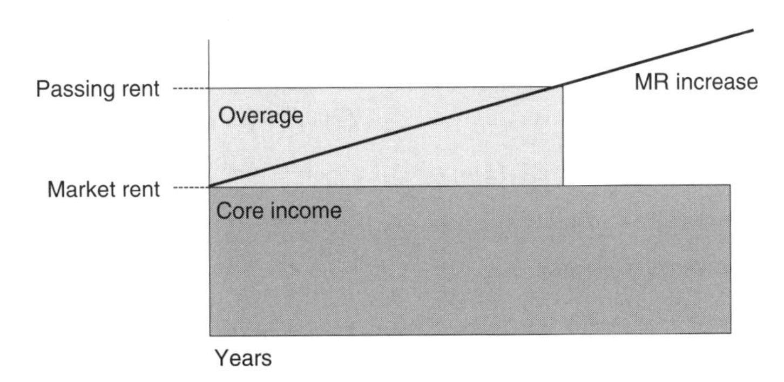

Figure 18.5.2 Market rent increases.

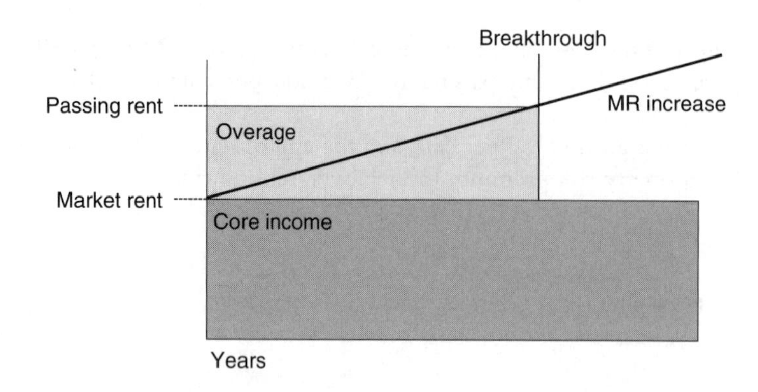

Figure 18.5.3 The breakthrough.

An alternative approach is to use a short-cut discounted cash flow which provides us with an opportunity to estimate when 'breakthrough' occurs. A discounted cash flow is different from other investment valuation methods as it *explicitly* adds rental growth to annual rent (using Amount of £1). In order to discover the point at which market rent *breaks through*, we add rental growth at each reversion opportunity.

Example 3

You have been asked to value an office building which was let 6 years ago on a 20-year FRI lease with 5-yearly upward only rent reviews. The passing rent is £575,000 per annum and market rent is £500,000 per annum.

The ARY for similar property is around 6.5 per cent, an appropriate equated yield is 7.75 per cent and rental growth is predicted at 1.75 per cent per annum.

Estimate market rent at the second rent review (4 years' time)

Market rent	£500,000 per annum
Amount of £1 @ 1.75% for 4 years $(1.0175)^4$	1.0719
	£535,930

Estimate market rent at the third rent review (9 years' time)

Market rent	£500,000
Amount of £1 @ 1.75% for 9 years $(1.0175)^9$	1.1690
	£584,500

At the third rent review breakthrough occurs – the market rent is expected to exceed the passing rent of £575,000 per annum.

Term	
Passing rent	£575,000 per annum
YP single rate @ 7.75%* for 9 years	6.3124
	£3,629,630
Reversion	
Breakthrough market rent	£584,500 per annum
YP in perp @ 6.5%	15.3846
PV of £1 @ 7.75% for 9 years	0.5108
	£4,593,266
Market value	*£8,222,896* *Say* *£8.223m*

* As we are explicitly adding rental growth, we cannot then use an ARY (which implicitly reflects growth) to value income. Therefore, in a discounted cash flow we use an equated yield.

Further reading

Baum, A. E. and Crosby, N. (2007) *Property Investment Appraisal*, 3rd edn, Wiley Blackwell, London; the hardcore/layer approach (pp.103–107); the discounted cash flow approach (pp. 145–149); a case study (pp. 283–291).

Blackledge, M. (2009) *Introducing Property Valuation*, Routledge, London, pp. 362–369.

18.6 Analysing tenant incentives

Key terms: net effective rent; straight line; discounted; discounted cash flow; landlords perspective; tenants perspective

Lease incentives can take many forms, but among the most common are rent free periods in excess of 3 months, break clauses, staggered rents, caps on future rent reviews or cash payments reverse premiums. Landlords prefer to grant concessions (such as prolonged rent free periods) rather than simply accepting a reduced rent.

In a market such as that of today, most new lettings are at headline rents subject to some sort of incentive. Valuers involved in rent reviews or valuations in these markets need to analyse headline rents and incentive packages in order to find comparable evidence for market rents.

The question is, what rent would have been agreed by the parties to the lease if the letting had not been subjected to incentives? This rent is known as the *net effective rent*.

Net effective rent

There are several different methods of analysing transactions to ascertain a net effective rent, and four approaches are identified by the RICS under UK Guidance Note 6 (2014) Analysis of Commercial Lease Transactions.

1 rent and value apportionment not assuming the time value of cash flows (*straight line* or *simple* method of analysis);
2 rent and value apportionment assuming the time value of cash flows (*discounted* analysis);
3 comparison by reference to effect on investment value (the assumption that market evidence for similar properties will also be over-rented in a weak market, and therefore evidence will already reflect the market's perception of risk);
4 a DCF approach (assumptions are explicitly stated and different prospective cash flows mapped out).

Some of these methods can be approached from two sides, particularly when the objective is to analyse a comparable letting to inform a rent review negotiation:

- *landlord's perspective* – where the valuer will seek a method of analysis that gives a high net effective rent, near to the headline rent;
- tenant's *perspective* – where the valuer will seek a method of analysis that gives the lowest net effective rent possible.

On the other hand, when the objective of rental analysis is to estimate market rent for the purposes of valuing a similar property, the valuer needs an unbiased method.

Simple analysis of net effective rent (straight line approach)

Example 1

The third floor of a prime office building in Newcastle achieved a peak rent of £125,000 per annum in 2008.

By 2012, after a period of weakened economic activity, there were several vacancies in the office building and a letting was agreed for the identically sized fourth floor at £130,000 per annum on a 15-year lease with 5-yearly upwards-only rent reviews. The lease is subject to 2 years rent free.

What is the net effective rent (i.e. market rent) of the fourth floor? Start by analysing the comparable fourth floor letting from the landlord's perspective. The landlord wants to produce a high net effective rent to support his stance in negotiating a rent review on another office building in this location. He can argue that the inducement is to persuade the tenant to pay the headline rent for the whole of the 15 year lease, through the mechanism of the upwards only rent review. Therefore the effect of the inducement should be spread over the entire 15 year lease:

Landlord's perspective
Total rent (allowing for inducements) payable over the lease term:

13 years @ £130,000 per annum	£1,690,000
Spread over 15 years	£1,690,000 ÷ 15
Net effective rent	*£112,667 per annum*

The tenant, however, will suggest the value of the rent free should be analysed only over the period to the first rent review, i.e. 5 years:

Tenant's perspective

Total rent (allowing for inducements) payable over the first 5 years:	
3 years @ £130,000 per annum	£390,000
Spread over the period to the first rent review	£390,000 ÷ 5
Net effective rent	£78,000 per annum

These two opposed ways of analysing the net effective rent give two very different outcomes. They serve the needs of the landlord's and the tenant's surveyors in arguing a rent review on a similar building, but are not helpful to a valuer wishing to find evidence of market rent. In this situation, the valuer will have to arrive at a view of the appropriate figure somewhere within this range, based on evaluation of the circumstances.

Discounted analysis of net effective rent

Lettings with inducements can also be analysed using the same principle as the simple approach, but this time including discounting to allow for the timing of rents:

Example 2

The third floor of a prime office building in Newcastle achieved a peak rent of £125,000 per annum in 2008.

By 2012 after a period of weakened economic activity, there were several vacancies in the building and a letting was agreed for the identically sized fourth floor at £130,000 per annum on a 15-year lease with 5-yearly upwards-only rent reviews. The lease is subject to 2 years rent free and the current ARY for similar property is around 8%.

What is the net effective rent (i.e. market rent) of the fourth floor?

Instead of simply multiplying the rent by the number of years it is paid, we can replace this with YP single rate to account for the timing of the rents:

Landlord's perspective

Total rent (allowing for inducements) payable over the lease term:

Passing rent	£130,000 per annum
YP single rate @ 8% for 13 years	7.9038
PV of £1 @ 8% for 2 years	0.8573
	£880,870
Spread over 15 years:	
YP single rate @ 8% for 15 years	8.5595
Net effective rent	*£102,911 per annum (£880,870 ÷ 8.5595)*

As before, the tenant will suggest the value of the rent free should be analysed only over the period to the first rent review, i.e. 5 years:

Tenant's perspective

Total rent (allowing for inducements) payable over the lease term:

Passing rent	£130,000 per annum
YP single rate @ 8% for 3 years	2.5771
PV of £1 @ 8% for 2 years	0.8573
	£287,215
Spread over the period to the first rent review:	
YP single rate @ 8% for 5 years	3.9927
Net effective rent	*£71,935 per annum (£287,215 ÷ 3.9927)*

Using DCF to analyse net effective rent

In some cases, a valuer will analyse a headline rent to find net effective rent so that it can be used to estimate market rent for another property being valued to market value. In such cases an unbiased estimate of market rent is required, and the valuer may resort to using the extreme landlord and tenant perspectives to arrive at a compromise figure.

This is effectively recognition that both extreme views (i.e. writing off the inducement over the whole lease or writing it off over the period to the first review) are incorrect. It recognises that the headline rent should be written off for some period in between the two, where the headline rent is likely to be overtaken by the growing market rent. In that case, the compromise is justified in theory, but is inexactly calculated.

Advocates of the DCF growth explicit approach to analysing net effective rent seek to investigate exactly when market rent will *break through* headline rent, and the effect of choosing it as the basis for writing off the inducements.

We will apply the DCF approach to the Newcastle prime office letting:

Example 3

The third floor of a prime office building in Newcastle achieved a peak rent of £125,000 per annum in 2008.

By 2012, after a period of weakened economic activity, there were several vacancies in the building and a letting was agreed for the identically sized fourth floor at £130,000 per annum on a 15-year lease with 5-yearly upwards-only rent reviews. The lease is subject to 2 years rent free. The current ARY for similar property is around 8 per cent, a suitable equated yield is 10 per cent and a reasonable rate of rental growth is 2.5 per cent per annum.

What is the net effective rent (i.e. market rent) of the fourth floor?

Initially, we need to estimate a point at which the market rent will pass through the headline rent. In this example, we could trial the breakthrough point at the second rent review.

We assume that the landlord is indifferent whether he receives:

Scenario A: market rent on standard lease terms from Day 1

or

Scenario B: headline rent of £130,000 per annum with 2 years rent free.

Therefore, the value of the two scenarios should be the same and we can calculate the value of each scenario and put the two values together in order to find the market rent:
Scenario A

Market rent	£X per annum
YP in perp @ ARY of 8%	12.5
Market value	*12.5X*

Scenario B

Table 18.6.1 Scenario B summary – Assuming breakthrough at year 10

Years	*Income (per annum)*	*Amount of £1 @ 2.5%*	*YP single rate @ 10%*	*PV of £1 @ 10%*	*Net present value*
1–2	–	–	–	–	–
3–10	£130,000	–	5.3349 (8 years)	0.8264 (2 years)	£573,139
Exit value *(Breakthrough)*	Re-let @ MR	1.2801 (10 years)	12.5 (YP in perp)	0.3855 (10 years)	6.1685MR
Market value					£573,139 + 6.1685 MR

On the basis that both scenarios are worth the same to the landlord:

£573,139 + 6.1685MR (Scenario B) is equal to £12.5MR (Scenario A)		
£573,139	=	12.5MR – 6.1685MR
£573,139	=	6.3315MR
MR	=	£573,139 ÷ 6.3315MR
Market rent		£90,522 per annum

However, the assumption that the market rent, growing at 2.5 per cent per annum, breaks through the headline rent by the second rent review must be double checked. Does it cross, if it starts at £90,522 and grows at 2.5 per cent per annum?

$$(1.025)^{10} \times £90{,}522 \text{ per annum} = £115{,}876 \text{ per annum}$$

The market rent would only grow to £115,876 per annum by the second rent review so it would not exceed the headline rent of £130,000 per annum, and it's clear the trial assumption regarding breakthrough was wrong.

We must try again with the crossover rather later, say, when the lease ends in Year 15. This involves modifying the valuation of Scenario B (Scenario A's value remains unchanged):

Scenario B

Table 18.6.2 Scenario B modified summary – Assuming breakthrough at year 15

Years	*Income (per annum)*	*Amount of £1 @ 2.5%*	*YP single rate @ 10%*	*PV of £1 @ 10%*	*Net present value*
1–2	–	–	–	–	–
3–15	£130,000	–	7.1034 (13 years)	0.8264 (2 years)	£763,132
Exit value *(Breakthrough)*	Re-let @ MR	1.4483 (15 years)	12.5 (YP in Perp)	0.2394 (15 years)	4.3340MR
Market value					£763,132 + 4.3340MR

On the basis that both scenarios are worth the same to the landlord:

£763,132 + 4.3340MR (Scenario B) is equal to £12.5MR (Scenario A)		
£763,132	=	12.5MR – 4.3340MR
£763,132	=	8.1660MR
MR	=	£763,132 ÷ 8.1660
Market rent		£93,452 per annum

Checking that the new breakthrough assumption is correct:

$(1.025)^{15} \times$ £93,452 per annum = £135,346

This accords with the assumption that breakthrough occurs at the end of the lease. The result does, as expected, fall between the extreme landlord and tenant views.

DCF methods can be used in this way to arrive at an **unbiased** analysis of net effective rent, theoretically suitable as the basis for deriving Market Rent.

Further reading

RICS Valuation – Professional Standards (2014) 'UKGN 6 analysis of Commercial Lease Transactions', RICS, London.

18.7 The discounted cash flow approach to valuing property investments

Key terms: rent reviews; equated yield; explicit rental growth; all-risks yield; net present value

When valuing property for investment purposes, a valuer can either use a traditional approach or a discounted cash flow. In traditional investment valuation, the rent reverts to market rent at the first opportunity and is then capitalised into perpetuity, despite the fact that the rent will probably increase at a later point. The reversion is also at today's market rent – we make no explicit assumptions regarding rental growth as this is reflected in the ARY.

The alternative to traditional investment valuation is to use a discounted cash flow (DCF) that recognises the importance of *all* income streams produced by an investment, and subsequently values each cash flow separately. At each reversion, the rent steps up as the DCF approach *explicitly* allows for rental growth on top of market rent by applying a rental growth rate known as the Amount of £1:

Example 1

Shop A was let 2 years ago on a 20-year FRI lease with 5-yearly rent reviews. The passing rent is £50,000 per annum and the market rent is currently £55,000 per annum. Using a rental growth rate of 1.5 per cent, calculate the rental cash flow over the life of the lease (Figure 18.7.1).

Cash flow 1 (Present day to Review 1)

Passing rent	£50,000 per annum

Cash Flow 2 (Review 1 to Review 2)

Market rent	£55,000 per annum
Amount of £1 @ 1.5% for 3 years	1.0457
Inflated market rent	*£57,512*

Cash flow 3 (Review 2 to Review 3)

Market rent	£55,000 per annum
Amount of £1 @ 1.5% for 8 years	1.1265
Inflated market rent	*£61,957*

Cash flow 4 (Review 3 to Lease expiry)

Market rent	£55,000 per annum
Amount of £1 @ 1.5% for 13 years	1.2136
Inflated market rent	*£66,745*

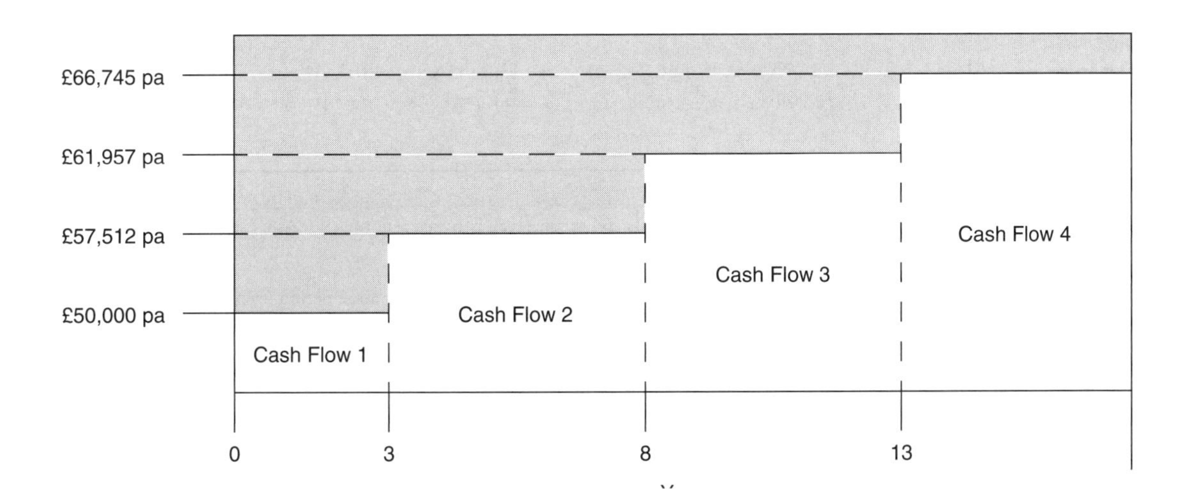

Figure 18.7.1 Cash flow analysis.

Due to the fact that we apply a rental growth rate to the market rent, we cannot use the ARY as a capitalisation rate, as it already implicitly accounts for rental growth and we would be double counting.

Therefore, we use the equated yield within a DCF. When calculating the equated yield, we use a 'safe' investment as a benchmark. This is the rate of return from UK government bonds (known as 'gilts') which provide fixed interest and eventual capital repayment at the end of a set period. The gilt rate is then adjusted to allow for the risk of property investment by the addition of a *risk premium* which has fluctuated between 4 and 6 per cent over recent years.

A discounted cash flow clearly lays out all the different income streams of an investment in a table. The income streams are separated into periods of time and the net cash flow for each period is grown, capitalised and then discounted back.

Example 2

Shop A was let 2 years ago on a 15-year FRI lease with 5-yearly rent reviews. The passing rent is £50,000 per annum and the market rent is currently £55,000 per annum. Assuming a rental growth rate of 1.5 per cent, an equated yield of 8.7 per cent and an ARY of 7.5 per cent, calculate the value of the freehold using a DCF (Table 18.7.1).

Table 18.7.1: Discounted cash flow analysis

A	B	C	D	E	F	G
Periods (years)	*Income (£)*	*Amount of £1 @ 1.5%*	*Cash flow (£)*	*YP @ 8.7%*	*PV of £1 @ 8.7%*	*Net present value (£)*
3–5	50,000	–	50,000	2.5449 *(3 years)*	–	127,245
6–10	55,000	1.0457 *(3 years)*	57,512	3.9201 *(5 years)*	0.7786 *(3 years)*	175,538
11–15	57,512	1.0773 *(5 years)*	61,957	3.9201 *(5 years)*	0.5131 *(8 years)*	124,610
Exit value	61,957	1.0773 *(5 years)*	66,745	13.3333 *(YP perp 7.5%)*	0.3381 *(13 years)*	300,868
						728,261

Notes:
Column A: each cash flow is split into reversionary time periods, i.e. the intervals between rent reviews. As we are already 2 years into the lease the first cash flow starts in Year 3.
Column B: rental income before rental growth is added.
Column C: we use the Amount of £1 to explicitly reflect rental growth for the length of the time period (except the first cash flow, as we start receiving this income immediately).
Column D: rental income once growth has been added.
Column E: we capitalise each cash flow using YP single rate at the equated yield for the length of the time period, i.e. in the first cash flow, the £50,000 pa is received for 3 years, so we use YP @ 8.7% for 3 years etc.
Column F: although we have accounted for rental growth, we still need to discount each cash flow back to its present day value, so we use present value of £1 over the number of years in the cash flow. The first income starts immediately, so we don't need to discount it back to the present day.
Column G: after multiplying across Columns D–F, each cash flow produces a *net present value*.

Exit value

A DCF should only go on for as long as is necessary. However, the period beyond the DCF cannot be ignored as the investment will still exist and perform, so to account for this we capitalise the final rental cash flow into perpetuity. We use *YP in perp* at an appropriate *ARY* to calculate the capital value of the asset at this final point.

In Example 2, we capitalise the income with YP in perp at the ARY in Column E, but are still not due to receive this sum for 13 years, so we need to discount using PV of £1 @ 8.7% (equated yield) for 13 years.

Net present value

We total up each of the separate cash flows, including the exit value, which gives us our final net present value.

Further reading

Blackledge, M. (2009) *Introducing Property Valuation*, Routledge, London, pp. 240–259.

18.8 Valuing vacant property

Key terms: void; vacant rates; rent free; market rent; all-risks yield; vacant costs

Historically, valuers have built risk of vacant property through the yield applied to the income. This has, however, become increasingly difficult to evaluate in current market conditions as the risk of a property becoming vacant has increased.

According to Jones Lang LaSalle (2013) at the end of 2012 total supply of immediately available industrial floor space across Great Britain stood at 327.4 million sq ft, while at the beginning of 2013 vacant offices stood at 24.8 million sq ft.

When a property is generating no income or is owner occupied how can we apply the investment method of valuation to find the market value?

We need to establish the market rent for the property and then consider how long it may remain vacant. We also need to make an assumption on how much rent free to offer in order to achieve market rent. Using comparable analysis of the market, the valuer needs to establish market rent and also typical void/marketing periods and associated rent free periods.

Case study 1

Figure 18.8.1 Unit 1 Morton Palms, Darlington.

Morton Palms is located approximately 1.6 km (1 mile) east of Darlington town centre on the A66, which links the town with both the A1(M) to the west and the A19 to the east.

Constructed in 2006, the estate comprises 12 detached two-storey office buildings providing modern open plan accommodation. Unit 1 extends to a net internal area of 418.45 sq m (4,508 sq ft) and has been vacant for approximately 6 months. It is being actively marketed by a local agent.

The market rent is £71,000 per annum and the agent marketing the premises has advised that the property will let within 24 months, although a 12-month rent free period will need to be offered to achieve the market rent.

ARYs for similar vacant investments are in the order of 10.5 per cent.

Valuation

The market rent will be achieved in perpetuity, but the landlord will not receive this for 3 years. Therefore, the income has been deferred over the two year marketing void and the 12-month rent free period. A yield of 10.5 per cent has been used, as the risk has been built into the cash flow by deferring the income.

Market rent	£71,000 per annum
YP in perp @ 10.5%	9.5238
PV of £1 @ 10.5% for 3 years	0.7412
Market value	*£501,167*

Costs associated with vacant property

Once a property becomes vacant and in the case of tenanted property the liability for costs falls with the landlord. These costs may include:

Maintenance/repairs

This will depend on the lease terms that have been agreed with the vacating tenant. In most cases the lease will be subject to dilapidations, and the tenant must hand back the property in a tenantable state of repair. Therefore, the property may not need repairs in order to re-let. If, however, the tenant has

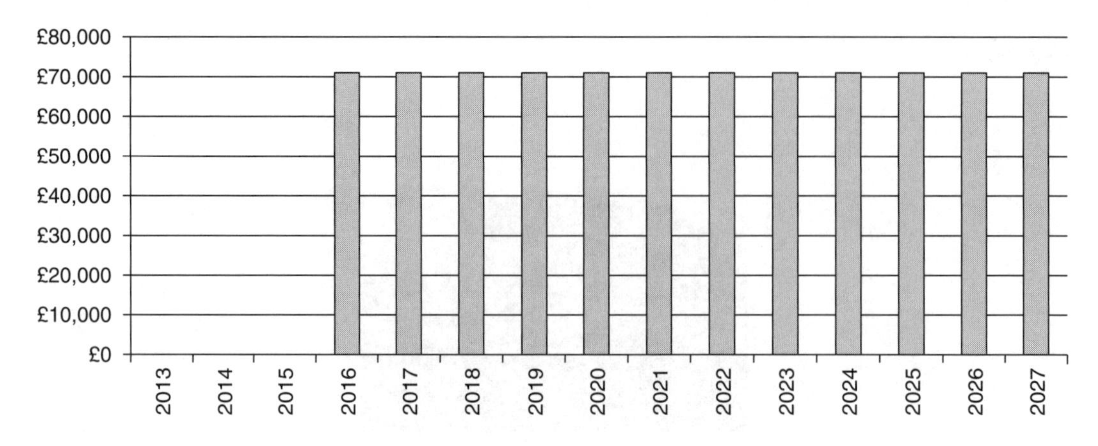

Figure 18.8.2 Cash flow graph.

vacated due to administration/liquidation then the landlord may need to carry out necessary repairs in order to bring the property into a tenantable state of repair.

Service charge

Within multi-let buildings such as office blocks or shopping centres, the landlord will cover the cost of the service charge for any void units. Service charge costs predominantly cover the upkeep, insurance and repair of all common areas.

Security

This depends on the property, but the landlord will want to maintain the value and marketability of the property and may therefore require security.

Services

Any essential services will be transferred over to the landlord/investor. The amount payable may be minimal but it is still a liability.

Refurbishment costs

Although the tenant may have carried out some dilapidation works, the property may need refurbishment if the landlord is hoping to re-let quickly and at a good rent.

Letting costs

The property will need to be marketed in order to re-let, and therefore expert advice is required from a surveyor at a cost. Agency fees are around 10 per cent of market rent, although some properties are subject to joint agency agreements and in this instance the fees are around 15 per cent of market rent.

Case study 1: Unit 1 Morton Palms, Darlington

There are a number of costs associated with the case study:

Vacant rates

The landlord will be liable to pay 2 years empty business rates.
Rateable Value – £52,000
UBR – 47.1 pence (standard business premises)
Rates payable = £52,000 × 0.471 = £24,492 per annum (£2,041 per month)
21 months (void period, allowing for 3 months rates free) × £2,041 = £42,861
The landlord will pay £42,861 in vacant rates over the void period.

Security/service charge

The site is subject to a service charge for the cleaning of common areas, landscaping and security on site. In practice, a valuer may be given the cost information by the client to incorporate into their valuation. For the purposes of this valuation, we have assumed 7.5 per cent of the market value will account for service charge expenditure.

£71,000 × 7.5% = £5,325 per annum

£5,325 × 2 years (void period) = £10,650

The landlord will pay £10,650 in vacant service charge over the void period.

Utilities

The unit will be subject to water, gas and electricity rates, and the costs of these services will be transferred to the landlord. In practice, a valuer may be provided with this information, but in this example we have assumed 1 per cent of market rent.

£71,000 × 1% = £710 per annum

£710 × 2 years = £1,420

The landlord will pay £1,420 in vacant services over the void period.

Agency fees

The property is already being marketed by a local agent and the landlord will have agreed to pay a fee once the property is let. The fee for a single agent is 10 per cent of the market rent agreed.

£71,000 × 10% = £7,100

The landlord will therefore accrue the following vacant costs over the 2 years:

Vacant rates	£42,861
Service charge	£10,650
Utilities	£1,420
Agents fees	£7,100
Total	*£62,031*

This is a cost to any potential landlord/investor, and should therefore be deducted from the market value of the property:

Value of Unit 1	£501,167
Less vacant costs	£62,031
Market value	*£439,136*
	Say £439,000

Vacant costs will vary depending on the property to be valued. Any additional works (i.e. refurbishment) which may be required should be deducted from the market value.

Further reading

Jones Lang LaSalle (2013) *UK Industrial Property Trends Today*, Issue 4.

18.9 Valuation and sustainability

Key terms: energy efficiency; corporate social responsibility; green buildings

> A sustainable building should be one which meets peoples' needs – as a home, or as a workplace – in ways which enhance its positive impacts and minimise its negative impacts, environmentally and socially, both locally and globally over time.
>
> (UK Green Building Council, 2009)

Due to the increased development of sustainable 'green buildings' in the commercial property market over the last decade, there is now increasing pressure on the valuation profession to incorporate the additional value of sustainability into valuation appraisals.

Over the past few years, there has been a rapid shift in market sentiment with regard to sustainable methods of constructing buildings and energy efficiency. Legislation such as The Energy Performance of Buildings (Certificates and Inspections) Regulation 2007 and the revised Building Regulations (Part L, Conservation of Fuel and Power) mean that developers are being forced to build or redevelop green buildings. Company directors also have new duties under Section 172 of the Companies Act 2006 to have regard 'to the impact of the company's operations on the community and the environment'.

Occupiers have their own corporate social responsibility (CSR) policies and will wish to demonstrate that such policies are being applied in relation to the management of their property. There is also the added financial incentive of cost savings in relation to energy efficient buildings which benefits both landlord and tenants

A great deal of research has already gone into the creation of green leases which should complement the day-to-day operation of sustainable buildings. A green lease is a lease that has additional provisions for both the landlord and tenant to undertake specific responsibilities with regard to the sustainable operation of a property. Such leases are already used in Australia where the Australian Government has published a series of Green Lease Schedules that put the onus on tenants to operate efficiently within their premises and maximise the environmental benefits of their building.

Some would argue that it may be possible to encourage the sustainable operation of buildings by voluntary agreements between landlord and tenant, but many believe that the best way to get both parties to act responsibly with regard to energy efficiency and sustainability of buildings is to confer obligations on them via their lease. Otherwise, both parties are not necessarily incentivised to take positive action.

On 2 March 2007, practitioners from the international valuation industry met to discuss the crossover between valuation and sustainability at the Vancouver Valuation Summit. Acknowledgement of the fact that sustainability is now globally impacting on the process of valuation involved the signing of the Vancouver Accord – a commitment by global professional bodies to begin the process to embed sustainability into valuation and appraisals.

Following on from the Vancouver Accord, the Green Building Council of Australia commissioned a report entitled Valuing Green in 2008 which examined the valuation of green leases in Australia. This is one of the first documents to conduct detailed research into how the green agenda affects the value of buildings, and the report identified the following factors that can potentially impact on value:

1. *Obsolescence*
   Sustainable green buildings potentially have a lower risk of obsolescence and depreciation due to the fact that they are essentially 'future-proofed' against growing legislation and regulation with regard to sustainability. By reacting to future demand and the changing green agenda of many potential tenants, green buildings are more likely to retain their value.

2 *Penalties*
There is the potential future threat of increased insurance costs for non-green buildings, in addition to possible legal implications for poor energy performing buildings.

3 *Incentives*
To encourage the efficient management and occupation of sustainable buildings, it is likely that tax and financial incentives will be introduced for sustainable buildings.

4 *Net rent*
An increase in net rent is likely due to savings in outgoings.

5 *Lease terms*
There is the potential to secure longer lease terms and retain tenants in a sustainable building, along with the ability to attract good tenant covenants from large companies with a similar green agenda.

6 *Rental growth*
Research suggests that green buildings have a higher growth potential which could lead to lower ARYs and risk premiums.

Although the direct comparison method is the most widely used, the Green Building Council Australia highlighted the difficulty in finding comparably 'green' buildings to make this method viable.

The Council's recommended method for valuing green buildings was using DCF. Although the method is still problematic with regard to comparable evidence, it is not so heavily reliant on comparables, and the timing of cash flows can be accurately modelled. This allows the valuer to factor in future shifts in the market for green buildings and incorporate the additional incentives/penalties that may be applicable.

This is an emerging market and a relatively new area for valuers, so a lack of knowledge and experience is likely to affect valuation standards. Knowledge comes with experience, and valuers therefore need to be exposed to market transactions which are very limited at present, particularly in the UK. Valuers will also need a degree of additional training subject to guidance from their relevant professional body, but in the meantime we must ensure that we keep up to date with developments in the market.

The RICS have published a guidance note on sustainability and commercial property valuation which provides best practice when assessing a building's sustainability characteristics and reflecting these characteristics in market value, fair value, market rent or investment value.

Further reading

RICS (2014) *Global Guidance: Sustainability and Commercial Property Valuation*, RICS Books, London.

18.10 The Valuer Registration Scheme

Key terms: VRS; Registered Valuer; RICS; Professional Standards; Red Book

From 2011 all UK practitioners undertaking valuations within RICS Valuation – Professional Standards 'the Red Book' are required to join the RICS Registered Valuers Scheme (also known as the Valuer Registration Scheme, or VRS).

The scheme was introduced to ensure quality valuations as well as raise the credibility of valuers and provide clear and consistent application of the valuation standards internationally.

Valuation services are used throughout the business world and include investment and corporate finance transactions, assessment of company accounts and stock exchange listings. It is important that any organisations relying on the expertise of a valuer are confident that the valuation is of the highest quality.

The registration scheme ensures consistency throughout the world and that all registered valuers:

- adhere to the RICS Valuation – Professional Standards;
- are committed to openness and transparency;
- are experts in their field;
- deliver credible high quality reports.

Any member of the professional body that has achieved valuation competency to level 3 and is actively involved with the valuation of property under the RICS Valuation – Professional Standards can apply to the RICS. Applications can be through their firm in which they work, individually or as a sole practitioner.

Those professional members who qualified from 1 January 2012 who have not achieved level 3 must undertake a 'Top up' assessment if they wish to become registered.

The candidate is required to submit further evidence of experience to the RICS which will be assessed by an appointed RICS professional registered valuer. The candidate must provide:

- evidence of work experience supervised by a registered valuer;
- case study using work-based evidence;
- continuing professional development (CPD) record.

Further reading

RICS (2010) *Regulation, Rules for the Registration of Schemes*, RICS, London.
RICS (2010) *Regulation, Appendix A – RICS Valuer Registration*, RICS, London.
RICS (2011) *Why use a RICS registered valuer*, RICS, London.

18.11 The comparative method

Key terms: comparable evidence; transactions; open market; lease terms; rent reviews; specification; location; quantum

The comparative (or comparison) method is one of the five core methods of valuation and is the most widely used by the surveying profession. When understating a comparative valuation you are directly comparing the subject property with similar properties that have recently transacted – these are known as comparables.

In a buoyant market the method is relatively straightforward as there are regular and plentiful transactions. In a poor market (such as that which we are experiencing today) it becomes much harder and there is a greater emphasis on a valuer's experience and skill to analyse the comparable evidence effectively.

One of the greatest difficulties when valuing commercial property using the comparative method is that it compares like with like, but no two properties are identical. Even properties within the same developments can differ slightly.

Sourcing comparative information

The collection of comparable evidence begins on inspection. It is important to understand the area surrounding the property you are valuing. Are there any adjoining units available to the market? Have there been any new lettings, refurbishments or developments that may aid in the search for recent transactions? A good tip on any inspection is to take note of agent's boards and people to contact who may help you in your search. Other sources of information may include your employer's database and records, national property websites and databases, local agents and auction results.

Analysing rental comparables

Once a thorough search has been undertaken the comparables must be analysed. The comparables should be very similar to the property being valued and comprise recent transactions that have occurred in the open market and on arm's length terms, i.e. between two parties who are not materially related. These transactions also need to be verifiable and consistent with current market activity. Provided that this criteria is met, the method can provide an accurate indication of value.

When analysing each comparable we need to consider a number of different aspects:

Location

When considering rental comparables we need to prioritise those within the same location. If we are considering an office suite within the central business district (CBD) of a city centre, then we need to consider similar office properties within the same city centre location. The disparity is too great between rental values within a CBD and rental values on the edge of a city centre.

Transaction date

The best evidence is formed through recent transactions. This is because the most recent rents agreed will provide a good indication of current market conditions for the specific type of property that you are valuing.

Arm's length

Comparable transactions need to be unbiased and reflect the current market. Therefore you should only analyse transactions that have taken place at 'arm's length'. For example, if a retail unit is let to an adjoining occupier for expansion, the landlord and tenant have an existing relationship so the new lease may be agreed at a lower rent. In this instance, the transaction has not been completed at arm's length terms.

Specification and layout

The internal layout of each comparable and how this compares to the subject property should also be considered. In terms of offices, tenants require modern space that accommodates their technological needs. Many older office suites are compartmentalised, lacking in modern fixtures (such as air conditioning) and have unsightly perimeter trunking; while newer offices are open plan and have raised access floors and suspended ceilings. In terms of industrial properties, factors such as high eaves height, good access and yard space command higher rental values. Older buildings including listed buildings may have restrictions that prevent them being adapted to suit modern requirements and, therefore, this must also be considered.

Size

The larger a property, the less rent (per square metre/foot) a tenant will be expected to pay. Industrial units provide a good example of the quantum that a tenant may be offered, i.e. an established industrial estate in the north-east of England commands in the order of £75.35 per sq m (£7.00 per sq ft) for small units of approximately 464.50 sq m (5,000 sq ft) while the larger units of over 4,645 sq m (50,000 sq ft) may only achieve £43.06 per sq m (£4.00 per sq ft).

Offices also benefit from quantum, and size should also be taken into consideration, while retail property, larger supermarkets and department stores are valued on a per square metre/per square foot basis and therefore benefit from quantum allowance, but smaller shops are let on a Zone A rate basis.

Lease terms

When compiling comparable data an important factor is the lease terms which have been agreed on any letting. The length of the lease, rent review pattern and repairing obligations can all impact on the rent achieved for that particular property, i.e. a tenant agreeing to take a 5-year lease will often pay more rent than a tenant taking a 15-year lease, as the landlord is agreeing to a shorter lease which will impact on the value of the investment. Frequent rent reviews (every 3 years) are detrimental to a tenant but beneficial for the landlord, so a landlord may offer a lower rent to a tenant that is prepared to agree to more rent reviews in the lease. Conversely, a rent may be higher if there is longer between rent review periods or no rent review at all. Restrictive covenants in a lease could also impact on value, i.e. due to increased competition for a retail unit, a tenant may agree to a tight user clause.

In a perfect world a valuer could obtain all of this information, but unfortunately this is rarely the case in practice. Therefore, valuers regularly have to make a number of assumptions that can impact on valuation accuracy.

Further reading

Blackledge, M. (2009) 'Introducing Property Valuation', Routledge, London.

18.12 Valuation accuracy

Key terms: valuation report; RICS; Mallinson Report; Red Book; Carsberg Report

> A valuation is a professional opinion at a single point in time.
>
> (RICS, 2011)

The accuracy of valuation is a topic area that has been discussed and analysed at length. When reporting a valuation, any information provided must not be misleading but simply present the valuer's professional matter of opinion. It is a valuer's duty to highlight any factors that could impact on certainty, but the extent of the commentary will depend on the purpose of valuation and report format, and also what information the valuer has agreed to present to their client.

Following the 1990 property market crash there was an increase in litigation against valuers due to the fact that many had over-valued property in a transitional market. The Royal Institution of Chartered Surveyors (RICS) commissioned the Mallinson Report in 1994 which made a total of 43 recommendations on how regulation of valuation could be improved, and was the catalyst for the modern version of the RICS Red Book.

Mallinson's recommendations included a focus on the improvement of communications between clients and valuers, the need for consistent reporting, the express statement of operational parameters for valuers, an improvement of skills and mathematical models and improved access to data. Recommendation 34 was perhaps one of the most important – 'common professional standards and methods should be developed for measuring and expressing valuation uncertainty'. The RICS subsequently published the *RICS Appraisal and Valuation Manual* (4th edition) Sept 1995 which combined all previous sets of standards in one improved volume.

Another important review of valuation accuracy took place in the Carsberg Report which was published in 2002 in response to studies revealing client influence on investment valuations. The major recommendation was that the RICS should establish an acceptable method by which uncertainty could be expressed but not confuse users of the valuation. This would mean agreeing with appropriate representative bodies the circumstances and format in which the valuer would convey uncertainty in a uniform and useful manner.

The result of this is that the current version of the Red Book, the RICS Valuation – Professional Standards 2014, includes Valuation Practice Guidance – Applications (VPGA) 9 on Valuations in markets susceptible to change: certainty and uncertainty.

> *The status of the valuer* – any valuers undertaking valuations in accordance with the Red Book should be suitably qualified and have relevant expertise in their field. They should also be acting independently of any influence and without any conflict of interest.
>
> *Inherent uncertainty* – a valuer should be open about inherent uncertainty caused by factors such as unusual or unique property character, the quantification of hope value, i.e. development potential, a special purchaser or certain assumptions that they have been asked to make.
>
> *Restrictions on enquiries* – if a valuer is unable to obtain certain information, then this could compromise the accuracy of their valuation. The Red Book requires all sources of information to be stated in a report, so attention must be drawn to any limitations.
>
> *Market instability* – valuers need to accept that there may be unforeseen financial circumstances which can be impacted by macro-economic, legal, political or natural events. Working in such circumstances can be testing for the valuer, so judgements should be expressed clearly.

In order to ensure their opinion of value is as accurate as it can be, and to minimise their liability, valuers must remain compliant with not only the RICS Code of Conduct Rules for Members and Firms but also the RICS Valuation – Professional Standards 2014. It is also recommended that valuers refer to any additional guidance such as the RICS Valuation Information Papers and ensure that they are covered by appropriate levels of professional indemnity insurance.

Further reading

RICS (2011) 'Report on PII for Valuations in the UK' available at: https://consultations.rics.org/gf2.ti/f/278818/6396741.1/PDF/-/16377_RICS_PII_Report%204.pdf (accessed 26/11/2013).

RICS (2014) *Valuation Professional Standards*, RICS, London.

18.13 Depreciated replacement cost

Key terms: depreciation; obsolescence; market value; comparable evidence

Depreciated replacement cost (DRC) is used to value properties where there is no active market, e.g. schools, libraries, museums, power stations etc. There is a lack of demand for the property in its existing use, other than from the party using it.

The method assumes that the owner cannot buy another similar product on the market, so it is necessary to purchase a site and rebuild the property in order to replace the asset. It is based on the concept that a buyer would not pay more to buy the subject property than the cost of constructing an equivalent new one. This cost then becomes the measure of the value of the property.

The main purpose for valuing such specialist properties is because they are assets, and their value still needs to be recorded in the balance sheets of company accounts. If a specialist property was surplus to company requirements, i.e. unused and waiting to be sold off, then this method would be not be appropriate and a valuation method reflecting the alternative non-specialist use would be used.

DRC is often referred to as a method of the last resort, and it should only be used if other methods such as comparison, investment or profits do not give a reliable value.

The methodology can be summarised as follows:

Cost of constructing the building
(Including fees)
Less
Allowance for age and depreciation
Equals
Net replacement costs
Plus
The cost of land
Equals
Depreciated replacement cost.

The valuer is costing the erection of a modern equivalent building, from a greenfield site to a completed building which provides the same level of service (as at the date of the valuation).

When the subject property is a new building, costing the equivalent building will be relatively simple. However, if the subject property is an older building, then its replacement will be an alternative building offering the same facility but not the same physical qualities. The valuer will have to make decisions about the appropriate size and specification of the equivalent building (which must give comparable performance and usefulness to the owner). Both size and specification may differ in comparison to the subject property, due to current building techniques and layout design being more efficient than at the original construction date. The following elements will make up the total cost of construction.

Building costs

These costs include site works, installation of services, infrastructure, construction of premises, finishes, fitting out and a contingency (3%). Costing may be obtained from actual costs of the subject building with adjustments for inflation; standard costs from the RICS Building Cost Information Service; or a specialist quantity surveyor.

Professional fees

These fees are usually assumed at 12–15 per cent of the total build costs (including statutory fees for planning and building regulations).

Interim finance

Allowance is made for financing during the construction period taking into account the likely method of payment. We account for interest charges by multiplying the total build costs by the formula $(1 + i)^{n-1}$.

VAT

Irrecoverable VAT may be allowed as a cost item.

Allowance for age and depreciation

In theory, a purchaser would pay less for an imperfect older property than for a modern equivalent. Therefore, in order to value an older specialist property using cost-based valuation, it is necessary to depreciate the construction costs of the equivalent modern asset. The valuer is now reflecting the differences between the modern equivalent and the actual subject property.

The basis of value for the cost of land is market value, but for the current use. There may be problems in achieving a figure from the market as there is unlikely to be direct comparable evidence for the actual user because the user is specialist. This is therefore often the most difficult aspect of a DRC valuation as you need market evidence to support land values for a specialist property type that does not have a market! The cost of land may also be high compared with the cost of construction, e.g. public sector property in city centres or schools with large sites in residential areas may have very high land values due to alternative use.

The valuer is estimating the cost of a suitable site for a modern equivalent to their subject property. This may be the actual site on which the current property stands, or it may be a different site because the assumption is that the current operator would buy the least expensive option for operational purposes. With regard to location, it may be that the current location is essential for operational purposes or that the current location is now considered too expensive, e.g. a fire station in a city centre could be relocated to an out-of-town location close to local road networks.

Example 1

You have been asked to value the freehold interest in a specialist storage facility which is government owned and used to store national treasures. The building was purpose built 15 years ago and it has been agreed that the method of valuation will be the DRC approach.

The property has an area of 8,200 m² GIA (gross internal area) and an estimated remaining life of 60 years. The site of 2.1 ha is well located on a site close to the M40 with neighbouring business park and residential land uses. Only 1 ha is built on – the rest is open space and is not expected to be developed by the owner. Comparable evidence shows land in this location selling for £1,200,000 ha for business use and twice this for residential use. The build costs are estimated at £1,300 per m² for a modern equivalent building and the build period would be 2 years. You have been informed that finance costs are 6 per cent per annum and VAT is recoverable.

Total building costs	£10,660,000
Professional fees *(15% of TBC)*	£1,599,000
Contingency *(3% of TBC + fees)*	£367,770
Finance costs at 6% for 2 years *(0.1236 × (TBC + fees))*	£1,515,212
Total cost of construction	*£14,141,982*
Less	
Allowance for age and obsolescence (straight line approach)	
£14,141,982 × (60 ÷ 75) = £11,313,586	
Net replacement cost	£11,313,586
Plus	
The cost of land	

Only 1 ha would be included in the DRC valuation as that is all that has been built on. The remaining 1.1 ha would be valued on the basis of MV.

1 ha @ £1,200,000 based on comparable evidence **£1,200,000**

Depreciated replacement cost

Net replacement cost + Cost of land = DRC

£11,313,586 + £1,200,000 = £12,513,586

(subject to the prospect and viability of the continuance of the occupation and use)

Further reading

Shapiro, E., Mackmin, D. and Sams, G. (2013) *Modern Methods of Valuation*, Routledge, Abingdon, pp. 344–348.

18.14 Valuing leasehold interests

Key terms: profit rent; ARY; equated yield; premium

There may be more than one legal interest in a property, and it is the legal interest that produces a rental income and has a value. Freehold interests are *perpetual*, but leasehold interests have a set term and often have little or no value, because the rent being paid is market rent. However, it is possible for a leasehold interest to generate a profit rent.

A profit rent occurs when the rent paid to the landlord is less than the market rent that could be achieved were the property to be sublet, and the difference between the two accrues to the tenant. A leasehold interest that generates a profit rent has a value and the interest can be sold to realise a capital amount, often referred to as a *premium*.

As a lease has a fixed term, we therefore know that income will *not* continue into perpetuity and will only be receivable for a defined period of time. However, if a tenant occupies land or property under a long lease (commonly in excess of 99 years) with a low ground rent and permission to sub-let, a valuer will assume that they have the security of a *virtual freehold* and use the freehold valuation approaches. If a long lease has an *unexpired* term of 50 years or more and the subtenant is paying market rent, we can value the income into perpetuity:

Example 1

A large shopping centre is let on a 99-year head lease that has 68 years unexpired. The head rent is fixed at £1,000 per annum and there is provision for subletting. Market rent is currently £367,000 per annum. Comparable freehold retail investments let at market rent are achieving yields of 6 per cent at sale. Calculate the value of the leasehold interest:

Profit rent	£366,000 per annum
Term of profit rent	68 years
Yield adjustment	6% (+2% for additional risk) = 8%
Profit rent	£366,000 per annum
YP in perp @ 8%*	12.5
Premium	*£4,575,000*

* Even though we are utilising the same methodology, this is a *leasehold* investment rather than a *free-hold* and therefore we need to adjust the yield up by 1–2 per cent to reflect this.

If a long lease has an unexpired term of *less* than 50 years, then we apply YP single rate for the number of years that profit rent will be receivable:

Example 2

A large shopping centre is let on a 99-year head lease that has 24 years unexpired. The head rent is fixed at £1,000 per annum and there is provision for subletting. Market rent is currently £367,000.

Comparable freehold retail investments let at market rent are achieving yields of 6 per cent at sale. Calculate the value of the leasehold interest.

Profit rent	£366,000 per annum
Term of profit rent	24 years
Yield adjustment	6% (+2% for additional risk) = 8%
Profit rent	£366,000 per annum
YP single rate @ 8% for 24 years	10.5288
Premium	*£3,853,541*

Short-term leasehold interests rarely have a profit rent, and if they do, it tends to be much lower than long–medium term leases. Short-term profits rents are also receivable for a much shorter period of time (a short-term leasehold interest is subject to rent reviews which eradicate profit rent). This type of interest is far riskier than a freehold investment, so traditionally it has been valued using the *dual rate approach*.

However, given the characteristics of leasehold interests, the most appropriate and effective way of valuing them, particularly if the profit rent is likely to vary during the term of the lease, is to use a DCF approach:

Example 3

A city centre office building owned by Granite Properties is let on a 99-year head lease to Grey Mare Asset Management. The lease has 20 years unexpired at a fixed head rent of £10,000 per annum. It was sublet 5 years ago to Chipchase Accountants (with Granite Properties' consent) on a modern 25-year lease with 5-yearly rent reviews, and the passing rent was recently reviewed to £90,000 per annum FRI. The freehold ARY for similar property is 7.5 per cent and the freehold target rate is currently 10 per cent.

Stage 1 (Find evidence of ARYs from recent market transactions)
7.5% – provided in the question, but derived from analysis of recent market transactions

Stage 2 (Calculate the freehold investor's target rate of return)
10% – provided in the question, but determined with reference to a secure low risk rate (10–15 year gilt redemption yield plus property risk premium)

Stage 3 (Calculate the rate of rental growth expected by investors)
Equated yield (10%) – ARY (7.5%) + 0.5% = implied rental growth (3%)

Stage 4 (Use this rate of rental growth to calculate the cash flow generated by the leasehold interest for the duration of the profit rent)

Table 18.14.1 assumes the same growth rate persists over the duration of the unexpired term.

Table 18.14.1 Rental cash flow

Period	*Initial rent*	*Amount of £1 @ 3%*	*Rent (£)*
Years 1 to 5	£90,000	–	£90,000
Years 6 to 10	£90,000	1.1593	£104,337
Years 11 to 15	£104,337	1.1593	£120,958
Years 16 to 20	£120,958	1.1593	£140,227

Stage 5 (Look at the risks to the leasehold income and decide on an appropriate leasehold target rate to reflect those risks)

This is influenced by a complex range of factors, including current market conditions, the terms of the leases and the covenant strength of the tenants. In this worked example, we can assume a 'mark-up' of 1 per cent to the freehold target rate. If, however, in the valuer's judgement the leasehold interest is judged to be of greater risk, the degree of mark-up may be 2 per cent or more.

Stage 6 (Discount the profit rent cash flow at the leasehold target rate to calculate the present value of the leasehold cash flow)

The market value of the leasehold interest, assuming an investor's leasehold target rate of 11 per cent, is £747,500 (Table 18.14.2). This is the price that Grey Mare Asset Management should seek to sell their leasehold interest for.

Table 18.14.2 Present value of cash flow

Period	*Income*	*Less head rent*	*Profit rent*	*YP @ 11%*	*PV of £1 @ 11%*	*Net present value*
1 to 5	90,000	10,000	80,000	3.6959	–	295,672
6 to 10	104,337	10,000	94,337	3.6959	0.5934	206,895
11 to 15	120,958	10,000	110,958	3.6959	0.3522	144,434
16 to 20	140,227	10,000	130,227	3.6959	0.2090	100,593
Total						£747,593

Further reading

Blackledge, M. (2009) *Introducing Property Valuation*, Routledge, Abingdon, pp. 240–259.

18.15 Asset valuations

Key terms: existing use value; market value; depreciated replacement cost; valuation date

The RICS Valuation – Professional Standards (2014) covers the valuation for financial statements under UK Valuation Standard 1 (UKVS 1).

Asset valuations are carried out for companies who own and occupy properties and include them as fixed tangible assets in financial statements in accordance with UK General Accepted Accounting Principles (UK GAAP).

Assets held by a company may include owner occupied property, investment property, vacant property, development land or vacant land.

Once a valuer is instructed to undertake an asset valuation, the basis of value for each asset should be agreed before any further work is undertaken. The basis of value in accordance with RICS Valuation – Professional Standards UKVS 1 include:

- existing use value (EUV) defined under UKVS1.3;
- market value (MV) defined under VS3.2;
- depreciated replacement cost (DRC).

Existing use value

EUV is applied to property other than specialist property that is owner occupied and solely used for the purpose of the business.

The definition is very similar to market value; however, it assumes that the willing buyer is granted vacant possession and any potential alternative use is disregarded – i.e. the asset would be purchased in the market by a company continuing to use the property for the same purpose.

Market value

MV is applied to property that is surplus to requirements or held as an investment by the company. The investment will be generating an income and can be sold in the market at market value.

Depreciated replacement cost

DRC basis of value is applied to property that is classed as specialist. This basis of value will be discussed at some length by the valuer and their client before being implemented.

Specialist properties are those that are rarely sold in the market and do not generate any profit, therefore there are no market transactions with which to compare. Examples of these include hospitals, schools, libraries, museums, power stations or churches.
(DRC is covered in Concept 18.13).

Frequency of valuations

Although annual revaluations are not a requirement, Financial Reporting Standard 15 (FRS 15) provides guidance on revaluation and recommends the following good practice:

- Full valuations are undertaken at no less than every 5 years with interval valuation in every three.
- If there is a material change in the property then revaluation should occur in years one, two and four.
- A rolling programme should be put in place for portfolio properties in order that each property is covered over a 5-year cycle. Any material changes to any of the properties within the portfolio will be covered by interim valuations.

Valuation date

The date of valuation is agreed between the valuer and company before the valuation is undertaken. The date is usually dictated by the company's financial year. A valuer may find they are working to a date in advance of the preparation of the valuation and therefore this must be referred to in any published reference.

Case study 1: Industrial Estate, Eaglescliffe

The property is an owner-occupied industrial property providing a GIA of 23,225 sq m (250,000 sq ft) (Figure 18.15.1).

A substantial part of the factory, approximately 4,645 sq m (50,000 sq ft), is unused by the occupier and is surplus to requirement. The unused section can be accessed separately and has been partitioned by the occupier in order to let it as a separate entity. The valuation date is 31 March 2014.

Existing use value for the occupied part of the factory

Market rent
18,580 sq m × £21.53 = £400,000 per annum
200,000 sq ft × £2.00 per sq ft = £400,000 per annum

Market rent	£400,000 per annum
YP in perp @ 11%	9.0909
Existing use value	*£3,636,364*
	Say £3,640,000

Void periods have been ignored, as a similar company would take occupation of the property with immediate effect from the date of valuation on 31 March 2014.

Figure 18.15.1 Industrial Estate, Eaglescliffe.

Provide a market value for the surplus accommodation

The surplus accommodation benefits from a separate access to that of the main factory and has been partitioned from the remainder of the used floor space. It can be let to another occupier without impinging on the company's business and occupation. It can therefore be valued on the basis of market value.

Market rent
4,645 sq m × £21.53 per sq m = £100,000 per annum
50,000 sq ft × £2.00 per sq ft = £100,000 per annum

Market rent	£100,000 per annum
YP in perp @ 11%	9.0909
PV for 1 years @ 11%	0.9009
Market value	*£818,999*
	Say £820,000

On this basis the valuer must consider void periods and rent free offered in order to let the property at the market rent of £21.53 per sq m (£2.00 per sq ft). The market rent is deferred over this period.

Costs

In accordance with the RICS Valuation – Professional Standards a valuer must not include acquisition or disposal costs in the valuation although if requested by the client these are stated as a separate figure.

Further reading

RICS Valuation – Professional Standards (2014), UK Valuation Standard 1 Valuations for financial statements, RICS, London.
RICS Valuation – Professional Standards (2014), UK appendix 1 Accounting Concepts and terms used in FRS15 and SSAP 19, RICS, London.

18.16 Valuing trading properties

Key terms: goodwill; fair maintainable operating profit; profit and loss

A trading property is a specialist type of property which has certain defining characteristics:

- *The property is designed/adapted for a specific trading use*, e.g. a cinema can really only be used as a cinema – another type of business could not simply occupy it and operate successfully without making any alterations to the building design.
- *The property enjoys an element of monopoly*, e.g. any property that requires a licence, such as a pub, has a monopoly over other types of property in that area that do not hold a licence.
- *Ownership of the property passes with the sale of the business* – some freehold interests will be sold *as a going concern* which means that the property comes with the benefit of a trading business, e.g. a publican purchasing a new pub will gain the benefit of the business as well as ownership of the property itself.
- *There are other assets in addition to land and buildings* – a trading property will have other valuable assets such as fixtures/fittings and goodwill which are fundamental to the success of the trader and are discussed in more detail below.

- *Value is based directly on trading potential for the specific use* – instead of using other comparable properties or rental income as our basis of value, we look at the potential for profit when valuing trading property.

The difference between an operational trading property and other types of commercial property is that the asset can often include more than land and buildings. A trading property is typically made up of the following elements:

1. *Legal interests in land*
   This can include either freehold or leasehold title.
2. *Trade fixtures and fittings*
   In addition to the actual 'bricks and mortar', a trading property will also have a number of permanent fixtures and fittings that will be fundamental to trading. For example, a bar and beer pumps would be essential to a property trading as a pub; beds and bedroom furniture would be fundamental to the successful trading of a hotel etc. Such fixtures and fittings carry an additional value that has to be accounted for in a profits method valuation.
3. *Transferable goodwill*
   When valuing a trading property, it can be difficult for valuers to distinguish between the trading potential of a property, and the goodwill that is associated with a current owner. For example, a pub or restaurant with an excellent reputation will have some goodwill value, but when the current owner sells the property and moves on, the goodwill could go with them. Transferable goodwill is inherent in the property itself and brings economic trading benefit e.g. due to location or a monopoly position.
4. *The benefit of any transferable licences*
   The valuer should assume in their valuation that the proposed purchaser has the ability to obtain or renew any existing licences, consents, certificates or permits.

The majority of operators will decide whether or not to take a specialist trading property based on the profits that they can earn from running their business – if they cannot make an adequate profit, they will not take the premises. Therefore, the profits method is the most appropriate method for valuing the majority of specialist trading property, and can be used to find both market rent and market value. In order to find market value using the profits method, the valuer has to determine the potential profitability that could be achieved by a competent operator of the business. This is known as the fair maintainable operating profit (FMOP). This figure is then capitalised using a YP multiplier (in practice, surveyors will refer to multipliers rather than yield percentages when discussing profits valuations).

The basic methodology is outlined below:

Turnover

Income from all the potential trading strands within the operational entity

Less

Purchases

Costs that support business turnover i.e. food, drink etc.

Equals

Gross profits

Less

Reasonable working expenses

Running costs for the operational entity e.g. wages, fuel, repairs and maintenance etc.

Equals

Fair maintainable operating profit

Multiplied by

YP multiplier

Equals

Market value

Even when using the profits method, the comparison method forms the basis of all judgements regarding the individual components, e.g. adjustments to the accounts, choice of multiplier etc. Once market value has been established using the profits method, it may be checked by comparable sales evidence using direct capital comparison.

To carry out a profits valuation, the valuer will need to draw up a hypothetical profit and loss account for the subject trading property based on a reasonably competent operator running the business.

This mean they will require a full set of accounts for at least three years of trading so that they can determine the reasonable levels of turnover, purchase costs and fixed costs for the business in order to arrive at the FMOP. Where there are no accounts available, the valuer would need to forecast their own based on any business plans for the subject operational entity and their market knowledge of similar businesses. The valuer must have confidence in the reliability of the accounts provided by the client and seek verification or auditing if in doubt of their accuracy.

The accounts provided by the current operator reflect only the existing business under current management at the current time. If the income and expenditure figures are above or below the norm for that type of business, the valuer will need to make adjustments to the actual figures. This is a critical stage in a profits valuation and requires a great deal of expertise on the part of the valuer.

Although a YP multiplier is used to capitalise the profit, it is not determined as an ARY per cent but as a multiplier that is analysed from comparable sales transactions. However, the multiplier still reflects the valuer's opinion of the associated risks, so the more profitable the business, the higher the YP multiplier.

Further reading

Baum, A., Nunnington, N. and Mackmin, D. (2007) *The Income Approach to Property Valuation*, 5th edn, Estates Gazette Books, London, ch. 12, p. 231.

Blackledge, M. (2009) *Introducing Property Valuation*, Routledge, Abingdon, ch. 14, p. 289.

Day, H. and Kelton, R. (2007) 'The valuation of licensed property', *Journal of Property Investment & Finance*, 25(3): 321–327.

Index

Pages in *italic* denote a table/figure.